# KINDLING 4

# ADVANCES IN BEHAVIORAL BIOLOGY



## *Recent Volumes in this Series*

Volume 23 THE AGING BRAIN AND SENILE DEMENTIA
Edited by Kalidas Nandy and Ira Sherwin

Volume 24 CHOLINERGIC MECHANISMS AND PSYCHOPHARMACOLOGY
Edited by Donald J. Jenden

Volume 25 CHOLINERGIC MECHANISMS: Phylogenetic Aspects, Central and Peripheral Synapses, and Clinical Significance
Edited by Giancarlo Pepeu and Herbert Ladinsky

Volume 26 CONDITIONING: Representation of Involved Neural Functions
Edited by Charles D. Woody

Volume 27 THE BASAL GANGLIA: Structure and Function
Edited by John S. McKenzie, Robert E. Kemm, and Lynette N. Wilcock

Volume 28 BRAIN PLASTICITY, LEARNING, AND MEMORY
Edited by B. E. Will, P. Schmitt, and J. C. Dalrymple-Alford

Volume 29 ALZHEIMER'S AND PARKINSON'S DISEASES: Strategies for Research and Development
Edited by Abraham Fisher, Israel Hanin, and Chaim Lachman

Volume 30 DYNAMICS OF CHOLINERGIC FUNCTION
Edited by Israel Hanin

Volume 31 TOBACCO SMOKING AND NICOTINE: A Neurobiological Approach
Edited by William R. Martin, Glen R. Van Loon, Edgar T. Iwamoto, and Layten Davis

Volume 32 THE BASAL GANGLIA II: Structure and Function—Current Concepts
Edited by Malcolm B. Carpenter and A. Jayaraman

Volume 33 LECITHIN: Technological, Biological, and Therapeutic Aspects
Edited by Israel Hanin and G. Brian Ansell

Volume 34 ALTERATIONS IN THE NEURONAL CYTOSKELETON IN ALZHEIMER DISEASE
Edited by George P. Perry

Volume 35 MECHANISMS OF CEREBRAL HYPOXIA AND STROKE
Edited by George Somjen

Volume 36 NOVEL APPROACHES TO THE TREATMENT OF ALZHEIMER'S DISEASE
Edited by Edwin M. Meyer, James W. Simpkins, and Jyunji Yamamoto

Volume 37 KINDLING 4
Edited by Juhn A. Wada

---

A Continuation Order Plan is available for this series. A continuation order will bring delivery of each new volume immediately upon publication. Volumes are billed only upon actual shipment. For further information please contact the publisher.

# KINDLING 4

Edited by

**Juhn A. Wada**
Neurosciences and Neurology
University of British Columbia
Vancouver, British Columbia, Canada

**PLENUM PRESS • NEW YORK AND LONDON**

**Library of Congress Cataloging-in-Publication Data**

International Kindling Symposium (4th : 1989 : Vancouver, B.C.)
Kindling 4 / edited by Juhn A. Wada.
p. cm. -- (Advances in behavioral biology ; v. 37)
"Proceedings of the Fourth International Kindling Symposium, held June 15-17, 1989, in Vancouver, British Columbia, Canada"--T.p. verso.
Includes bibliographical references and index.
ISBN-13: 978-1-4684-5798-8 e-ISBN-13: 978-1-4684-5796-4
DOI: 10.1007/978-1-4684-5796-4
1. Kindling (Nuerology)--Congresses. I. Wada, Juhn A. II. Title. III. Title: Kindling four. IV. Series.
RC372.5.I524 1989
616.8'5307--dc20 90-7589
CIP

Proceedings of the Fourth International Kindling Symposium,
held June 15–17, 1989, in Vancouver, British Columbia, Canada


Softcover reprint of the hardcover 1st edition 1990

A Division of Plenum Publishing Corporation
233 Spring Street, New York, N.Y. 10013

Dedicated to the original kindler,
and our friend
the late Graham Goddard

# PREFACE

This is the proceedings of the fourth in a series of international gatherings on kindling held on the University of British Columbia campus, June 15-17, 1989. Since the last gathering in 1985, kindling continues to attract an ever-increasing number of investigators as reflected by the phenomenal increase in the number of kindling-related research reports. No other acute or chronic model has been exploited so extensively using electrophysiological, pharmacological, biochemical and behavioural approaches involving a variety of brain sites and animal species. The continuing search, during the past few years, for the mechanism underlying the enduring change induced by kindling is beginning to shed some light on aspects of its molecular basis and to suggest the future direction of research.

The late Graham Goddard, the original kindler, would have been delighted with this development. We were all shocked and saddened by his tragic death, but his spirit is very much alive among all of us who gathered together to share new information and collectively reassess the present state of knowledge at this symposium. I must say that we miss him very, very much. We know he is happy in his heaven knowing that he accomplished more in a short lifetime than most and that his love for mankind, and life's challenges, remain as a legacy, and goal, for us all.

By the nature of this gathering, the proceedings represent a collection of data, overviews, historical perspectives, speculation, and prejudices of some of the most active researchers from many different disciplines in the field. Although tantalizing clues to the potentially fruitful future avenues of investigation were revealed, we must admit that we are still a long way from grasping a clear and unequivocal view of the substrate of kindling-induced enduring change.

Not all the intimate discussions can be replicated, but, as in the previous proceedings in this series, the gist of the formal discussion has been included to share our concern and interest with a wider audience beyond the immediate symposium participants. Although a bibliography on kindling between 1985 and 1989 was initially compiled, it has not been included, since computerization of literature references and their retrieval has become almost universal and the exclusion of a rather extensive bibliography prevents unnecessary escalation of the price of this volume.

We hope that this volume will bring the landscape of creative kindling research during the past five years to the attention of clinical and basic neuroscientists who have a fundamental interest in the mechanism of learning, memory, and epilepsy, and thereby stimulate further innovative research on the workings of the human brain mechanism through this robust model, the precious gift left with us by the late Graham Goddard.

Juhn A. Wada

ACKNOWLEDGEMENTS

The editor wishes to express his appreciation to the Medical Research Council of Canada for unfailing support enabling him to organize this symposium; to the University of British Columbia Health Sciences Center and to Ciba-Geigy, Canada; Abbott, Canada; and Nihon-Kohden, America, for their generous support.

Finally, but not least, it is a great pleasure to acknowledge the superb secretarial assistance of Mrs. Mary Woodland and the meticulous editorial assistance of Ms. Sue Calthrop.

CONTENTS

# KINDLING AS A MODEL OF LONG-LASTING PLASTICITY :

# NEW PERSPECTIVES IN THE SEARCH FOR MOLECULAR MECHANISMS

**Gildas LE GAL LA SALLE**

Centre National de la Recherche Scientifique, Laboratoire de Physiologie Nerveuse, Département de Neuropharmacologie Moléculaire, 91198, Gif-sur-Yvette, FRANCE

Twenty years after its discovery, kindling continues to give rise to an increasing interest as attested by this fourth symposium and the impressive number of papers dealing with this subject in the literature. But, as kindling studies advance, the objectives tend to differ. Although kindling was initially recognized as a suitable model for studying plasticity, it served primarily in epilepsy research. A considerable amount of works were devoted to the analysis of antiepileptic drugs and the study of the neurobiological basis of epileptogenesis. Now, the overall tendency is to reconcile both aspects in again considering kindling as a model of plasticity. A paper on this subject is presented in the book Kindling 3 (27).

At the present time, new data, in particular those concerned with kindling-induced structural changes, and new tools such as those available in molecular biology, allow us to define the scope of this model and to outline new perspectives for the search of mechanisms.

In this paper I will focus on the reflection in three directions : first, on the unique feature of the kindling phenomenon when compared with other models of plasticity; then, on the recent data showing that morphological alterations might subserve endurance of kindling; and finally, I will discuss some perspectives of a new molecular framework for further research in long-term plasticity.

Models of neuronal plasticity are numerous and differ in many respects. A basic feature which characterizes them is the time during which plastic changes are retained. How long "memory" is can vary greatly from seconds to days, weeks or even longer according to the model. Among long-lasting models, one of the most appreciated to date is the so-called long-term potentiation phenomenon (LTP), initially discovered by Bliss and Lomo in 1973 (4). LTP has received considerable attention because it is generally considered to be the best model available to give insights into the mechanisms of learning and memory (33).

The characteristics and the possible relevant mechanisms of both the LTP and kindling have been compared in recent reviews (2, 6). Both models share many interesting features in common, but, as noted by Cain (6), the most striking difference between the two models is "the fact that LTP decays relatively rapidly, within a few hours to a few weeks, whereas kindling is permanent". In fact, kindling can last for months or even for years depending on the life span of the species being considered (44).

*Kindling 4*
Edited by J. A. Wada
Plenum Press, New York, 1990

Thus, kindling stands apart from the other models of plasticity because of its particularly long time course of retention, and its primary characteristic of long-lasting robustness and endurance.

Therefore, due to the stability of the plastic changes occurring in kindling, this phenomenon seems to represent a particularly suitable model for studying the process of engram formation and could be considered a caricatural situation of what happens in "memory".

When viewed as a model of long-term plasticity, however, a possible drawback of kindling stems from its very close relationship with epilepsy. The question remains to whether or not kindling-like processes are dissociable from the epileptic phenomenon ? In spite of the number of studies on kindling this point is still a matter of debate and merits further discussion.

**KINDLING AS A MODEL OF FUNCTIONAL PLASTICITY**

The primary value of a model is its conformity with what is to be studied, and secondly it has to be amenable to experimental manipulations. The kindling model satisfies these two requirements. The development and the long-term effects of kindling are measured according to well-defined and reproducible epileptic criteria. However, outside the epileptic field, the kindling procedure can also induce, or more generally modify, some specific behavioral manifestations which are not necessarily correlated with epilepsy. A number of studies in the literature present such examples, e.g. those concerning conditioned emotional response (24), passive avoidance task (5) aggressive behavior and responsiveness to handling, and predatory behavior (1), etc. In a previous study we also showed that the morphine withdrawal syndrome was reduced following kindling: rats which had previously been kindled by daily amygdaloid stimulations failed to exhibit the typical autonomic and behavioral signs of this syndrome (22).

Although it is difficult to determine the particular areas concerned with epilepsy, one can speculate, from a theoretical point of view, that if selected brain structures or pathways can be gradually and then permanently modified by kindling, the specific behaviors that are controlled by these anatomical sites would also be impaired. Therefore, the kindling procedure should also induce progressive and long-lasting non-epileptic behavioral changes. To demonstrate that such modifications are not related to epilepsy is difficult and generally not attempted. However, in the case of sleep, recent experiments confirming some previous results (40) show that the disturbance of sleep organization following amygdaloid kindling is transitory and independent of the newly acquired epileptic susceptibility due the kindling process itself (17).

As already noted, one of the hallmarks of kindling is the permanence of the phenomenon. Once achieved, kindling remains for months or years. However, although the kindling procedure in most studies is usually pursued until the stage of full secondarily generalized seizures, completion of this stage is not necessary for permanence (18, 20). It has been demonstrated that "endurance of the kindling effect is present early in the kindling process, and is not dependent upon the occurrence of generalized seizures" (18). With respect to this result and to the behavioral modifications in kindled animals, we also showed in our study on the morphine withdrawal syndrome a similar reduction of the symptoms in both fully and partially kindled rats (22).

Furthermore, the kindling procedure is known to lower the after-discharge threshold. Tress and Herberg (42) have demonstrated that the permanent changes in seizure threshold are a function of electrical stimulation rather than of epileptic seizure itself. Thus, early, lasting modifications can be induced in the absence of epileptic discharges.

On the other hand, electrical stimulation of some structures does not induce after-discharges in any way (14). As a consequence, kindling (with epileptogenic development) is not possible to obtain. Nevertheless, some long-term modifications can be induced either by repetition of subconvulsive stimulations or by stimulation of " non-kindlable " structures where after-discharges are normally not induced. The fact that some areas do not respond to the kindling procedure by expression of paroxysmal discharges or do not develop convulsive seizures by any means signifies that the repetition of stimulations is not durably modifying the neuronal activity from the cell population of the stimulated area. Such an assumption has been tested by Fernandez-Guardiola and colleagues by stimulating the dorsal raphe nucleus (11) and by Stevens and Livermore by "kindling" the ventral tegmental area in the cat (38).

To proceed in this direction should permit the development of new models of plasticity using the kindling procedure. Such strategies - allowing induction of long-lasting behavioral, electrographical, biochemical or molecular changes in specific subsets of neurons- are not limited and in some specific cases might offer some practical interest in clinical therapy. To give a single example, we can hypothesize the progressive and then permanent control of pain by repetitive electrical stimulations of structures supposed to play a key role in these functions.

## KINDLING AS A MODEL OF STRUCTURAL PLASTICITY

From the time of its discovery, due in most part to its endurance and robustness, a tangible substrate accountable for the establishment of the kindled state has been sought (25). Thus, the presence of morphological changes representing possible engrams have been studied at the cellular and subcellular levels. At the same time, morphological studies performed on the LTP have demonstrated that this long-term phenomenon was associated with growth of synaptic contacts (33). Structural changes in synaptic architecture have also been demonstrated in kindled animals. Although these results sometimes diverge according to the experimental material and in some instances need more consistent replications , some morphological alterations have been well documented. Changes in several aspects of synaptic morphology have been reported (13). This paper will concentrate on the discovery of sprouting in kindling.

Kindling was recently reported to induce nerve terminal sprouting of the hippocampal mossy fibers by establishing aberrant connections upon the granular cells layer of fascia dentata (37).

In addition to this result we also noted that amygdaloid, but not entorhinal cortical kindling, elicits further aberrant sprouting in the infrapyramidal cell layer of CA3 (34). The spatial distribution of the distal infrapyramidal terminal field of mossy fibers is known to be particularly prone to species variations and to different experimental or pathological situations. This could explain why our results have not been reproduced in other studies. Experiments at the electron microscopic level are currently being carried out for further description of this phenomenon.

Interestingly, positive correlations were found between intra and infrapyramidal mossy fiber terminal fields and learning abilities (23). Furthermore, a recent experiment has reported stronger LTP in inbred mouse strains having larger mossy fiber projections, suggesting that the extent of this anatomical system may be correlated with synaptic efficacy (16). That infrapyramidal synaptic alterations are observed in amygdaloid but not in entorhinal cortical kindling, at least signifies that the result is not due to the seizure activity itself. In addition, it is highly conceivable that such plastic morphological changes may extend to other structures. Useful markers for plasticity are still lacking, and today the hippocampal formation is the only structure where such alterations can be identified by indirect methods.

We have shown that mossy fiber sprouting is associated with a significant increase in the density of high affinity kainic acid binding sites in the aberrantly innervated region (34). Similar recurrent connections in the supragranular layer of the dentate gyrus were also elicited by kainic acid-induced CA4 lesions and have been correlated with abnormal excitatory responses (41).

Therefore, according to the structural synaptic reorganization and the correlated increase of kainic acid binding sites, an increased efficacy of synaptic transmission in the sprouted area would be expected. Such a dysfunction may facilitate or even promote seizure activity and may be part of the process of engram formation underlying the endurance of kindling. New perspectives for research will stem from these results.

## PERSPECTIVES IN THE SEARCH FOR MOLECULAR MECHANISMS

As shown in the preceding paragraph there is now abundant evidence for neuronal structural changes in some brain areas in response to the LTP or kindling. Furthermore, there seems to be general agreement that such alterations may form a basis for enduring plastic modifications in such areas as learning and memory (31). The long-lived persistence of kindling effects offers a very appreciable advantage since it permits the study of plasticity for a long time after its development, thus discarding possible interference due to short-term or transitory modifications, such as consequences of the seizures themselves, which are not necessarily relevant to long-term effects.

At this point, at least three features of this phenomenon will be described and briefly reported upon. First, it is important to recall that a very critical feature of kindling development concerns the intervals between stimulations. Limbic kindling develops efficiently only if interstimulus intervals are long enough; otherwise kindling is impeded (14,30,32). Hourly or shorter interstimulus intervals, as well as massed trial paradigms, disrupt kindling development although they do produce occasional and temporary convulsive phenomena. Therefore, we can outline the functional importance of a narrow time window during which the mechanisms occurring are likely to represent a sine qua non condition for the establishment of kindling. Permanence of kindling operates during this time window. This assumption is further substantiated by the second following comments.

A number of experiments performed on different models of long-term plasticity have stressed the requirement of protein synthesis. This is a very basic feature of these phenomena since formation of new proteins is not a prerequisite in short-term models such as short-term potentiation and post-tetanic potentiation (15). In these short-term models covalent modifications of pre-existing proteins are supposed to be the basis of plasticity. For kindling this topic is well documented. It has been repeatedly demonstrated that translational and transcriptional inhibitors retard or block this phenomenon (7,19,28). Interestingly, it was also reported that the action of protein synthesis inhibitors occurs mainly, if not exclusively, in the time window discussed above. Exposure to inhibitors delayed by as little as one hour after stimulation impairs kindling development.

As our knowledge of cellular biology progresses there is increasing evidence to assume that the structural changes involved in long-lived neuronal kindling plasticity are likely to share some similarities with the growth process and differentiation occurring during normal development. In this respect, growth of the cell can be considered to be a natural plastic change. Today, as we penetrate even further into the innermost mechanisms of the cell, frontiers between biological disciplines become more and more tenuous and neurobiological research has much to gain by taking advantage of all the fields of cell biology. In recent years mechanisms underlying cellular growth and differentiation have become much better understood. This rapid outgrowth of knowledge is mainly due to the explosive development of new techniques and biological tools. Mechanisms supporting kindling plasticity should and have to benefit from this technological advance.

Thus, with the above remarks in mind, by taking advantage of what is currently known about the molecular mechanisms of cellular growth processes and differentiation, I would now like to discuss the hypothesis suggesting that the mechanisms which could account for the stable anatomical alterations in kindling will undoubtedly involve genomic modifications (15). Most of that which is not relevant to protein synthesis will be deliberately omitted from this discussion.

Kindling stimulations inducing synaptic reorganization are likely to trigger a cascading series of events, occurring in the days or weeks following stimulations and including phenomena such as differentiation and neurite extension. Thus, a basic question remains as to how the incoming signal represented by kindling stimulus may convert short-term events into long-term responses such as nerve terminal sprouting. The cascade of events is likely so complex, well organized and spread over time that we are obliged to admit modifications influencing pre-existing genetic programs involved in differentiation in such a way that they lead to synaptic remodeling. A general picture of the molecular basis of plasticity is beginning to emerge. The schema is still rudimentary but several elements are currently under study.

Studies are now in progress by several groups to further the understanding of the early mechanisms which trigger the cascade of transcriptional events. Recent interest in brain oncoprotein expression has focused primarily on *c-fos*. *C-fos* is a cellular oncogene member of the so-called "early immediate gene family" (8, 43). Owing to its DNA binding property, it is supposed to control the expression of a number of other genes involved in growth process and differentiation. The regulation of pre-existing programmed gene expression by these early genes may play a basic role in the formation of a set of proteins participating in the long-term engramation of neuronal plasticity.

Recently, paroxysmal activity has been shown to induce the expression of *c-fos* in restricted subsets of neurons (9,10,21,26,35),particularly those of the hippocampal granule cells giving rise to the sprouted mossy fibers. This has been well documented for kindled seizures by Dragunow and Robertson (9,10). We have also confirmed this result while analysing a number of other experimental models of epilepsy in order to see whether limbic *c-fos* expression is a general feature of epilepsy or whether it is only some forms of epilepsies that are related to *c-fos* expression. In spite of minor changes depending on the models, the most striking difference concerns genetic models such as the audiogenic DBA/2 mice. Following auditory stimulation this strain of mice exhibits a characteristic seizure response. *C-fos* immunoreactivity was absent from limbic structures but clearly positive in the subcortical auditory nuclei, i.e. the cochlear nuclei, superior olivary complex, inferior colliculi and medial geniculate bodies (Le Gal La Salle and Naquet, in preparation ). Similar experiments are currently being performed on other genetic models of epilepsy.

Further analysis of early transcription factors such as *c-fos* should improve our understanding of the conversion of short-term signals into long-term responses. The role of other nuclear growth-related genes or oncogenes will certainly constitute a promising approach in the near future. Some of them are already known (*fos*-related genes, *jun*, *myb*, *ets1*, *ets2*, *ski*, *sno*, *erg*, etc.) and others will undoubtedly be discovered as a result of the available powerful new molecular biological techniques such as differential hybridation.

Genes which are induced very rapidly but transiently following kindling stimulation may control other genes participating in the regulation of molecules playing a necessary step in the mechanisms leading to long-lived plastic modifications. The participation of a great number of such molecules is suspected. However, I will limit this chapter to those substances that are currently under investigation, i.e. growth associated proteins, adhesion molecules and trophic factors and their related receptors.

A recent hypothesis supported by a number of promising findings is that expression of growth-associated proteins (GAP) is a prerequisite for axonal growth and synaptic remodeling. The GAP constitute a family of neuron-specific developmentally regulated phosphoproteins which are synthesized at the highest levels during axonal outgrowth or regeneration following axotomy (3). They are enriched in the growth cones of nerve endings and may play a prominent role in the formation and reorganization of neuronal circuitry. Increased phosphorylation of GAP-43 has been demonstrated in the LTP. Furthermore, in the adult brain GAP subsides into subsets of cortical and hippocampal neurons that are supposed to be "specialized for synaptic remodeling" (29).

In addition, there is now evidence that particular molecules are required to generate the specificity of neuronal circuitry in the brain. Among them an important family is represented by the adhesion molecules. Usually substrate-adhesion molecules (such as laminin, fibronectine, collagens,..) and cell adhesion molecules (N-cadherin, N-CAM,...) are considered in this group. Both categories are supposed to play a role in neuronal recognition during axonal growth (36). These molecules might also interact in the plastic remodeling process such as those observed in kindled animals. For example, modifications of the molecular form of neuronal cell adhesion molecules (such as the degree of sialylation) could regulate sprouting. Transduction, expression (or re-expression) and regulation of these molecules into sprouted areas in kindled animals undoubtedly constitute an interesting approach for further research.

There is now considerable support for the view that newly forming or transforming synapses require a complex environment and that, subsequently, target cells may release specific neurotrophic molecules of critical importance. It has been hypothesized that the amount of trophic material could be regulated by the activity of the nerve cell itself. In the context of molecular research, it is interesting to stress here the suggestion that, outside the classical process of protein synthesis in the soma of the cell, limited synthesis of trophic proteins at the post-synaptic site could be synthesized locally on polyribosomes present in the immediate vicinity of the spine (39). A number of substances promoting survival and neurite outgrowth have been identified in the brain : nerve growth factor (NGF), fibroblast growth factor, neuroleukin, insulin, insulin-like growth factor, etc... Other unidentified neurotrophic substances have also been found in the conditioned medium of sprouted fibers following axotomy. It has already been demonstrated that NGF may be important for the establishment of long-lasting neuronal changes induced by kindling since intraventricular injection of antiserum to NGF ( which is not an anticonvulsant by itself ) prevents the normal development of kindling (12).

In connection with the last experiment reported here, it is interesting to stress the necessity for new strategies in pharmacological studies. Substances devoid of anticonvulsant properties could however prevent kindling by acting at specific steps all along the cascade of modifications of which we have tried to give some examples. The experiment with the antiserum against NGF is of particular interest from this point of view (12). Other similar agents modifying the development of kindling have to be found. The transduction signals inducing nuclear growth-related genes or oncogenes would also constitute an interesting field of research. Substances acting at the level of the early immediate genes could have clinical pharmacological properties.

In conclusion, kindling appears to be a valuable model of long-lasting plasticity whose morphological synaptic changes might provide a stabilized substrate. Whether kindling is necessarily related to epilepsy remains open to debate. In any case, molecular research using this model would be very promising. An interesting approach concerns the study of early inducible genes. In fact, these immediate nuclear genes are at the beginning of the cascade of molecular events that may contribute to the enduring alterations. A number of regulated genes now needs to be discovered and studied. Further study in this direction would lead to a comprehensive schema of how early regulatory genes initiate and control later effector ones involved in long-lived modifications responsible for enhanced excitability. Knowledge of this sequential gene activation which is within the critical time frame of the long-term modifications would promote new perspectives for the understanding of mechanisms underlying long-term plasticity.

## REFERENCES


1 ADAMEC,R. (1976). Behavioral and epileptic determinants of predatory attack behavior in the cat. In : Kindling, J.A. Wada (Ed).Raven Press, New York, pp 135-154.

2 BAUDRY,M. (1986). Long-term potentiation and kindling: similar biochemical mechanisms? In : Advances in Neurology, 44, pp 401-410. A.V. Delgado-Escueta, A.A.Ward. Jr.; D.M. Woodbury and R.J. Porter (eds) Raven Press. New York.

3 BENOWITZ,L.I. and ROUTTENBERG,A. (1987). A membrane phosphoprotein associated with neural development, axonal regeneration, phospholipid metabolism, and synaptic plasticity. **TINS**, 10, 527-532.

4 BLISS, T.V.P. and LOMO,T. (1973). Long-lasting potentiation of synaptic transmission in the dentate area of the anaesthetized rabbit following stimulation of the perforant path. **J. Physiol.** (London), 232, 331-356.
5 BOAST,C.A. and McINTYRE,D. (1977). Bilateral kindled amygdala foci and inhibitory avoidance behavior in rats: a functional lesion effect. **Physiol. Behav.**,18, 25-28.
6 CAIN, D.P.(1989). Long-term potentiation and kindling: how similar are the mechanisms? **TINS**, 12, 6-10.
7 CAIN,D.P.; CORCORAN,M. and STAINES,W. (1980). Effects of protein synthesis inhibition on kindling in the mouse. **Exp. Neurol.** 68, 409-416.
8 CURRAN,T. and MORGAN,J.I. (1987). Memories of fos. **Bio Essays**, 7, 255-258.
9 DRAGUNOW, M. and ROBERTSON,H.A. (1987). Kindling stimulation induces c-fos protein(s) in granule cells of the rat dentate gyrus. **Nature**. 329, 441-442.
10 DRAGUNOW, M. and ROBERTSON,H.A. and ROBERTSON,G.S. (1988). Amygdala kindling and c-fos protein(s). **Exp. Neurol.**, 102, 261-263.
11 FERNANDEZ-GUARDIOLA,A.; JURADO,J.L. and CALVO,J.M. (1981). Repetitive low-intensity electrical stimulation of cat's nonlimbic brain structures: dorsal raphe nucleus kindling. in : Kindling 2. J.A.Wada (Ed.) Raven Press, New York, pp 123-135.
12 FUNABASHI,T.; SASAKI,H. and KIMURA,F. (1988). Intraventricular injection of antiserum to nerve growth factor delays the development of amygdaloid kindling. **Brain Res.** 458, 132-136.
13 GEINISMAN,Y.; MORREL,F. and de TOLEDO-MORREL,L. (1988). Remodelling of synaptic architecture during hippocampal 'kindling'. **Proc. Natl. Acad. Sci.** USA., 85, 3260-3264.
14 GODDARD,G.V.; McINTYRE,D.C. and LEECH,C.K. (1969). A permanent change in brain function resulting from daily electrical stimulation. **Exp. Neurol.** 25, 295-330.
15 GOELET,P.; CASTELLUCCI,V.F.; SCHACHER,S. and KANDEL,E.R. (1986). The long and the short of long-term memory - a molecular framework. **Nature**, 322, 419-422.
16 HEIMRICH,B.; CLAUS,H.; SCHWEGLER,H. and HAAS,H.L. (1989). Hippocampal mossy fiber distribution and long-term potentiation in two inbred mouse strains. **Brain Res.** 490, 404-406.
17 HIYOSHI,T.; NORI,N. and WADA,J.A. (1989). Feline amygdaloid kindling and sleep. **Electoencephal. Clin. Neurophysiol**, in press.
18 HOMAN,R.W. and GOODMAN,J.H. (1988). Endurance of the kindling effect is independent of the degree of generalization. **Brain Res.**,447, 404-406.
19 JONEC,V. and WASTERLAIN,C.G. (1979). Effect of inhibitors of protein synthesis on the development of kindled seizures. **Exp. Neurol.** 66, 524-532.
20 LE GAL LA SALLE,G. (1982). Amygdaloid organization related to the kindling effect. In: Kyoto Symposia. P.A. Buser; W.A. Cobb and T. Okuma (Eds) Elsevier Biochemical Press. pp 239-248.
21 LE GAL LA SALLE,G. (1988). Long-lasting and sequential increase of c-fos oncoprotein expression in kainic acid-induced status epilepticus. **Neurosci. Letts.**,88, 127-130.
22 LE GAL LA SALLE,G. and LAGOWSKA,J. (1980). Amygdaloid kindling procedure reduces severity of morphine withdrawal syndrome in rats. **Brain Res.**, 184, 239-242.
23 LIPP,H.P.; SCHWEGLER,H.; HEIMRICH,B. and DRISCOLL,P. (1988). Infrapyramidal mossy fiber and two-way avoidance learning: developmental modification of hippocampal circuitry and adult behavior of rats and mice. **J. Neurosci.** 8, 1905-1921.
24 McINTYRE,D.C. and MOLINO,A. (1972). Amygdala lesions and CER learning: long-term effect of kindling. **Physiol. Behav.**,8, 1055-1058.

25 MESSENHEIMER,J.A.; HARRIS,E.W. and STEWARD,O. (1979). Sprouting fibers gain access to circuitry transsynaptically altered by kindling. **Exp. Neurol.**, 64, 469-481.
26 MORGAN,J.I.; COHEN,D.R.; HEMPSTEAD,J.L. and CURRAN ,T. (1987). Mapping patterns of c-fos expression in the central nervous system after seizure. **Science,** 237, 192-197.
27 MORRELL,F. and TOLEDO-MORRELL,L. (1986). Kindling as a model of neuronal plasticity. In: Kindling 3. J.A.Wada (Ed) Raven Press, New York, pp 17-35.
28 MORRELL,F.; TSURU,N.; HOEPPNER,T.J.; MORGAM, D. and HARRISON,W.H. (1975).Secondary epileptogenesis in frog forebrain : effect of inhibition of protein synthesis. **Canad. J. Neurol. Sci.** 2, 407-416.
29 NEVEU,R.L.; FINCH,E.A.; BIRD,E.D. and BENOWITZ,L.I. (1988). Growth-associated protein GAP-43 is expressed selectively in associative regions of the adult human brain. **Proc. Natl. Acad. Sci.** 85 , 3638-3642.
30 PETERSON,S.L.; ALBERTSON,T.E. and STARK,L.G. (1981). Intertrial intervals and kindled seizures. **Exp. Neurol.** 71, 144-153.
31 PETIT,T.L. and MARKUS,E.J. (1987). The cellular basis of learning and memory:the anatomical sequel to neuronal use. In : Neuroplasticity, learning and memory. Alan R. Liss,Inc. pp 87-124.
32 RACINE,R.J.; BURNHAM,W.M.; GARTNER,J.G. and LEVITAN,D. (1973). Rates of motor seizure development in rats subjected to electrical brain stimulation: strain and inter-stimulation interval effects. **Electroencephal. Clin. Neurophysiol.** 35, 553-556.
33 RACINE, R.J. and KAIRISS,E.W. (1987). Long-term potentiation phenomena: The search for the mechanisms underlying memory storage processes.In : Neuroplasticity, Learning and Memory. Alan R. Liss Inc. pp 173-197.
34 REPRESA,A.; LE GAL LA SALLE,G. and BEN ARI,Y. (1989). Hippocampal plasticity in the kindling model of epilepsy in rats. **Neurosci. Letts,** 99, 345-350.
35 SAFFEN,D.W.; COLE,A.J.; WORLEY,P.F.; CHRISTY,B.A.; RYDER,K. and BARABAN,J.M.(1988). Convulsant-induced increase in transcription factor messenger RNAs in rat brain. **Proc. Natl. Acad. Sci.** 85, 7795-7799.
36 SUNSHINE,J.; BALAK,K.; RUTISHAUSER,U and JACOBSON,M. (1987). Changes in neural cell adhesion molecule (NCAM) structure during vertebrate neural development. **Proc. Natl. Acad. Sci.** 84, 5986-5990.
37 SUTULA,T.; XIAO-XIAN,H., CAVAZOS,J. and SCOTT,G. (1988). Synaptic reorganization in the hippocampus induced by abnormal functional activity. **Science,** 239, 1147-1150.
38 STEVENS,J.R. and LIVERMORE,A.JR. (1978). Kindling of the mesolimbic dopamine system: animal model of psychosis. **Neurology,** 28, 36-46.
39 STEWARD,O. and FALK,P.M. (1986). Protein-synthetic machinery at postsynaptic sites during synaptogenesis: A quantitative study of the association between polyribosomes and developing synapses. **J. Neurosc.** 6, 412-423.
40 TANAKA,T and NAQUET,R. (1975). Kindling effect and sleep organization in cats. Electroencephal. **Clin. Neurophysiol.**,39, 449-454.
41 TAUCK,D.L. and NADLER,J.V. (1985). Evidence of functional mossy fiber sprouting in hippocampal formation of kainic acid-treated rats. **J. Neurosc.** 5, 1016-1022.
42 TRESS,K.H. and HERBERG,L.J. (1972). Permanent reduction in seizure threshold resulting from repeated electrical stimulation. **Exp. Neurol.** 37, 347-359.
43 VERMA,I.M. and GRAHAM,W.R. (1987). The fos oncogene. In : Advances in cancer research. 49, pp 29-52.
44 WADA.J.A. (1978). Kindling as a model of epilepsy. **Electroencephalogr.Clin. Neurophysiol.**, 34. 309-316.

# SPONTANEOUS ELECTROGRAPHIC SEIZURES IN THE HIPPOCAMPAL SLICE: AN *IN VITRO* MODEL FOR THE STUDY OF THE TRANSITION FROM INTERICTAL BURSTING TO ICTAL ACTIVITY

Darrell V. Lewis and Wilkie A. Wilson

Duke University and Veterans Administration Medical Centers
Durham, North Carolina 27710 USA

Recently, several methods of provoking robust seizure discharges in the hippocampal slice preparation have been reported from our laboratory and from others. These seizure discharges are of long duration, complex morphology and have a stereotyped temporal evolution. Therefore, they are clearly different from the interictal bursts previously studied in the slice preparation. They resemble the seizures recorded from the intact brains of animals and humans, as for example, when recording from hippocampal depth electrodes in patients being considered for epilepsy surgery. Because of these similarities, we refer to these events as electrographic seizures or EGSs. Our hope is that the mechanisms of the EGSs in the slice are similar to the mechanisms of actual ictal events in the intact brain. If this is the case, the simpler slice preparation can be used to help us understand the transition from interictal activity to seizures as it occurs in the whole brain.

The various methods used to elicit EGSs in brain slices include exposure to low magnesium medium (1,9,17,19) high potassium medium (7,18) potassium channel blockers (2,5) low calcium medium (8) and, in slices from immature brain, exposure to common convulsants such as penicillin (16). In addition, tetanic stimulation of slices will elicit brief electrically induced seizures or afterdischarges (13). Any of these models could provide opportunities to study the interictal to ictal transition in the slice preparation and some new information has already been published in this regard utilizing high potassium medium (18).

However, we have been searching for an in vitro model that would exhibit spontaneous seizures as well as interictal bursts in the absence of both convulsants and unphysiological ionic manipulations. Such a model of ictal onset might more closely resemble the events occurring in the intact brain when seizures begin. In previous studies we have found that kindling-like electrical stimulation of hippocampal slices produces a long lasting hyperexcitable state. One manifestation of the hyperexcitability is spontaneous interictal bursting, termed stimulustrain induced bursting or STIB, that persists for the life of the slice (14). In addition to the development of spontaneous bursting, the afterdischarges following each train in these slices become more

*Kindling 4*
Edited by J. A. Wada
Plenum Press, New York, 1990

prolonged and complex with repeated trains (15). These epileptogenic changes are more easily induced in slices from younger animals than from mature animals.

Therefore, to develop a model of spontaneous EGSs in the slice, we have used slices from young animals and used repeated, kindling-like stimulus trains to induce a long lasting, hyperexcitable state. In some slices, we have observed the evolution of spontaneous EGSs triggered by interictal bursts. This method of inducing spontaneous seizures in the slice has marked similarities to the kindling model of epileptogenesis, where only electrical stimulation is used to induce a long lasting epileptic state.

The experimental methods were as follows. Male Sprague-Dawley rats, 10 to 30 days old, were anesthetized with chloroform and sacrificed by decapitation. The hippocampi were dissected free from the rest of the brain and hippocampal slices (625 microns thick) were prepared using a McIlwain tissue chopper. After one hour in a holding chamber, slices were placed in a submersion chamber of 2.5 ml volume and perfused at 6 ml/min with artificial cerebrospinal fluid (ACSF) containing in mM; NaCl 120, KCl 3.3, CaCl2 1.2, MgSO4 0.9, NaHCO3 25, NaH2PO4 1.23, and dextrose 10 at pH 7.4 and 33.5$^{o}$C. The concentration of magnesium of 0.9 mM was used because this is the reported value of magnesium in rat CSF (Stasheff, et al., 1989). The ACSF was constantly bubbled with a gas mixture of 95% O2 and 5% $CO_2$. The 3rd to the 6th slice from the temporal end of each hippocampus was used whenever possible. Only one slice was used from each animal. Extracellular recordings were made with 2 - 5 megohm glass microelectrodes containing 2 M NaCl. Stimulation was performed using a monopolar tungsten stimulating electrode with rectangular 100 to 800 uA, constant current, monophasic pulses of 0.1 msec duration. The stimulation electrode was placed in the stratum radiatum of CA3b in order to stimulate the Schaffer collaterals. This is the same stimulation location used in our previous studies of stimulus train induced bursting (15).

Extracellular potentials were recorded by electrodes placed in stratum pyramidale of both CA3b and CA1b. An input-output series was done to determine the stimulus strength giving the maximum amplitude orthodromic population spike in CA3b, and twice this stimulus intensity was used for the stimulus trains. Slices were not used if the maximum CA3 orthodromic population spike was less than 0.5 mV or if more than one orthodromic population spike occurred in the response to a single stimulus. Trains of stimuli, 60Hz, 2 sec duration, were given every 10 minutes to elicit electrical afterdischarges or EGSs.

Each EGS had a early rapidly firing or tonic phase, followed by a final, slowly firing or clonic phase (Figure 1). As described by Stasheff et al. (15), the morphology of these EGSs evolved gradually with successive stimulus trains. The tonic phase became more robust, and the number of bursts per EGS increased dramatically. After about 2 to 10 trains, the morphology of the EGSs stabilized, and each one was quite similar to the preceding one (Figure 1).

At this point, the EGSs also exhibited all or none behavior (see 15). When the stimulus duration was reduced to the threshold for eliciting an EGS, the slice responded with a fully developed EGS in spite of the reduced stimulus. The threshold for eliciting the full EGS was sharp, with stimuli just below threshold producing no EGS. The threshold was also consistent over time, and waiting for several hours without stimulation did not change the to the stimulation train.

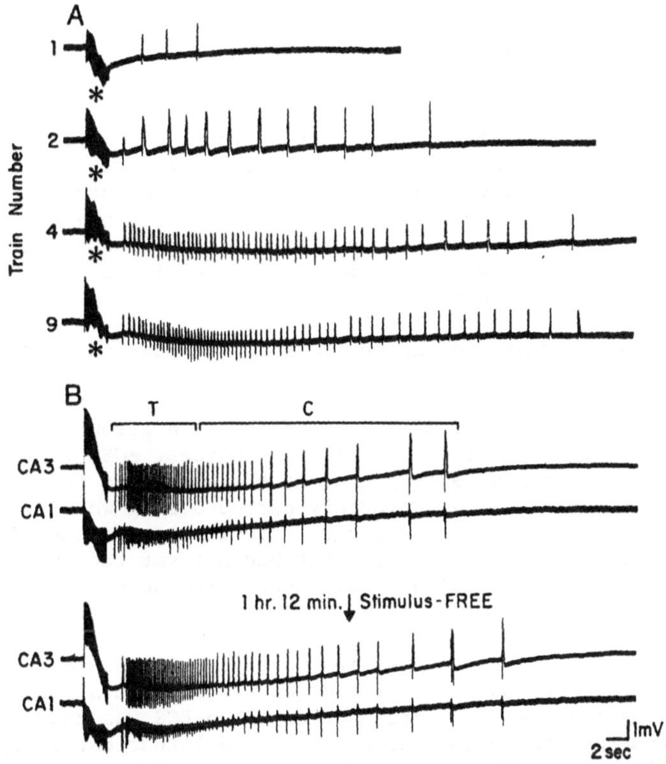


Figure 1. Induction of EGSs. A) Field recordings from area CA3 of the hippocampal slice show the progressive enhancement of afterdischarges that follows successive stimulus trains (bars). After two to ten trains, these develop into a stereotypical pattern characterized by a constant duration and two distinct phases and are then designated as electographic seizures (EGSs). B) Recordings from another slice (upper traces) show that seizures were concurrent in CA3 and CA1 and illustrate more clearly the tonic (T) phase of rapidly firing population spikes and the clonic-like (C) phase of less frequent population bursts. (Lower traces) After more than an hour (72 min) without additional stimuli, a similar seizure was elicited (Reproduced with permission from Stasheff et al., 1989).

In these slices from young animals, only 4 to 6 trains were needed to reach the fully developed, all or none EGSs. In addition, after 2 or 3 trains, the slices from animals older than 15 days developed stimulus train induced bursting or STIB (14). The spontaneous bursting continued for hours after delivery of trains had ceased. The interictal bursts were consistently higher in amplitude in CA3 compared to CA1 and appeared to begin a few milliseconds earlier in CA3 than in CA1 (Figure 2). Interestingly, slices from animals aged from 10 to 15 days did not develop spontaneous interictal bursting.

When the electrically induced afterdischarges had become stereotyped and all or none, the stimulus trains were stopped and the slices were left unstimulated for 1 to 2 hours. In 4 of 20 slices observed in this fashion, we noted the onset of spontaneous EGSs (Figure 2). These EGSs would occur repetitively, every 10 to 20 minutes and go on for 30 min to 2 hr. Prior to each EGS, there was a gradual increase in the frequency and complexity of the interictal bursts. These spontaneous EGSs were very similar to the EGSs or electrical afterdischarges that followed the stimulus trains. They had an initial rapidly firing, or tonic, phase and a final clonic phase. After the spontaneous EGSs, there was a postictal silence without spontaneous bursting lasting 1 to 2 min.

After seeing that spontaneous EGSs could occur in these slices, we made deliberate attempts to trigger EGSs in five other slices that did not have spontaneous EGSs after train stimulation. To trigger EGSs, single stimuli were delivered at 1 Hz at the same stimulus intensity that was used for the trains. In all five slices, EGSs were generated when interictal bursts were triggered by these stimuli (Figure 3). After each EGS, there was a refractory period during which interictal bursts could not be elicited by single stimuli. Attempts to elicit EGSs with stimulation at 1 Hz in slices that had not yet received tetanic stimulation was unsuccessful and only repetitive normal field potentials were found.

The relationship of epileptiform activity in CA3 to that in CA1 during the interictal to ictal transition was studied also. The onset of spontaneous interictal bursting prior to an EGS was characterized by high amplitude bursts in CA3 which led low amplitude bursts in CA1 by several milliseconds (Figure 2). However, the bursts in CA1 became larger and more complex prior to the EGSs, and during the EGSs, the bursts in area CA1 often led those in CA3.

A similar evolution was seen before the onset of the EGSs triggered by driving the interictal bursts. After driving the bursts at 1 Hz for a minute or so, the burst morphology typically became more complex in both CA1 and CA3 (Figure 3). Simultaneously, a negative baseline DC shift began in both areas as well. Just prior to seizure onset, the bursts developed afterdischarges. At this point the triggering stimuli were stopped and the EGS continued. During the initial tonic phase of the EGS, the bursts in CA1 usually led the bursts in CA3 (Figure 3C), as was observed in the slices with spontaneous EGSs. We have seen these same changes in the relative timing of intereictal bursts during the onset of seizure activity in hippocampal slices maintained in low magnesium ACSF (unpublished data). Perhaps CA1 has an important role in the transition into and maintenance of EGSs, as has been suggested by others (7,18) using high potassium induced epileptiform activity.

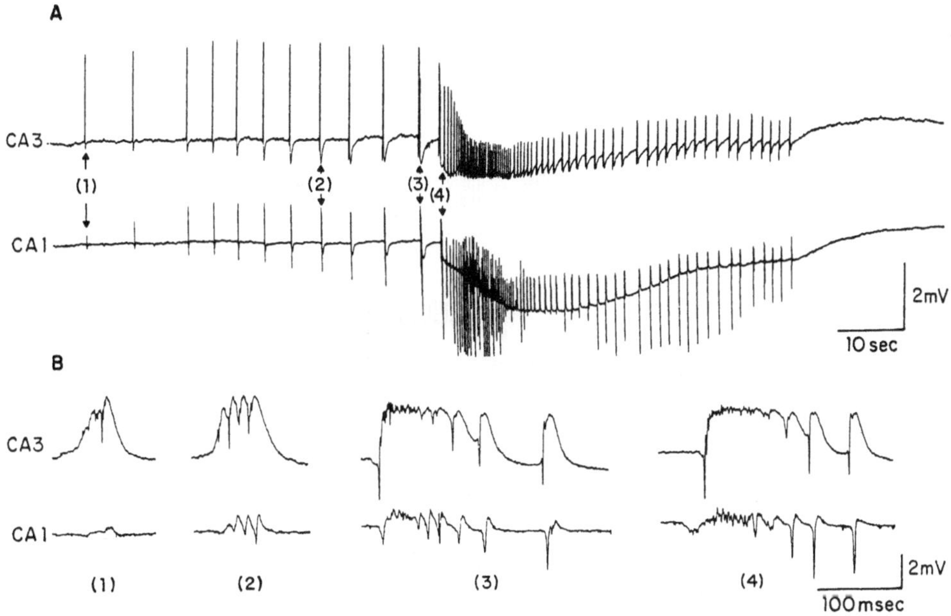


Figure 2. Spontaneous seizure in the hippocampal slice. A) The two records labeled CA3 and CA1 were recorded simultaneously from the strata pyramidale during the evolution of a spontaneous seizure. The interictal bursts were initially very low in amplitude in area CA1, but increased in size during the transition into ictal activity. The negative DC shift during the seizures and the division into tonic and clonic phases can be seen clearly. B) These expanded traces of the interictal bursts correspond to the numbered bursts in (A; 1 - 4). Burst (1) is the first burst in the series preceding the seizure. Note that it begins in CA3 many milliseconds before being reflected in CA1 as a very low amplitude negative wave. Burst (2) shows less latency and higher amplitude in CA1. By the time of burst (3) was sampled, just prior to seizure onset, the bursts in both CA1 and CA3 have developed afterdischarges and there is little difference in latency. Burst (4) is the first burst of the seizure. Note that now the burst begins in CA1 shortly before CA3.

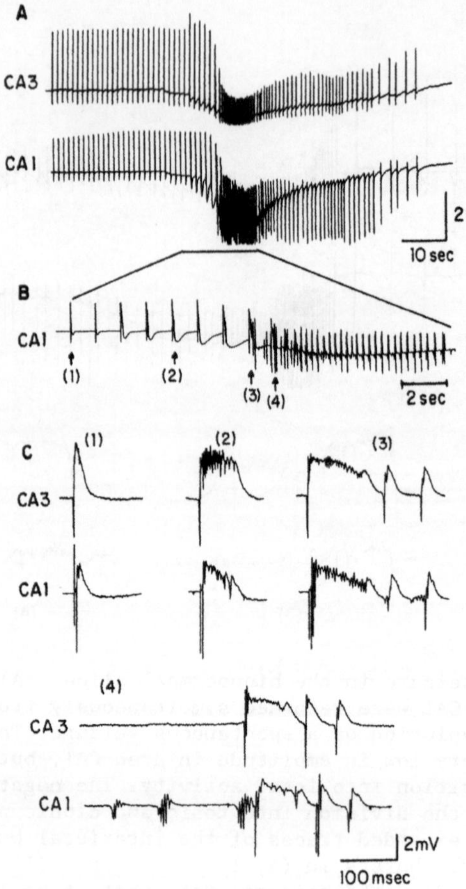


Figure 3. Triggered Seizure in the Hippocampal Slice. A) Simultaneous CA1 and CA3 records of a seizure occurring while interictal bursts are being triggered at 1 Hz. The triggering stimulus (0.2 mA) was discontinued at the onset of the seizure. B) This is an expanded record of the preictal and early, tonic ictal activity in CA1 shown for the purpose of demonstrating the exact points at which the bursts in part (C) were sampled. C) Expanded view of the bursts (1 - 4) indicated in part (B). Bursts 1,2, and 3 were triggered by stimuli and occur nearly simultaneously in areas CA1 and CA3. Note the preictal build up of the afterdischarge. Burst 4 occurs when the seizure has become self-sustaining and the stimulus has been discontinued. Note in the trace of burst 4, the leading, small bursts in area CA1 preceding the burst in area CA3.

The ability to study the interictal to ictal transition in the slice preparation, in the absence of convulsants and in physiological medium, would represent a significant advance in experimental models of epilepsy. Early studies of the transition from interictal to ictal activity were limited almost exclusively to in vivo preparations (11). Using penicillin to induce hyperexcitability, Matsumoto and Ajmone Marsan (10) in cat cerebral cortex and Dichter and Spencer (3) in cat hippocampus described how neuronal behavior during the interictal bursts changed just before the onset of a seizure. Intracellular recordings revealed that the burst afterhyperpolarization disappeared and, in its place, there appeared a series of brief afterdepolarizations corresponding to the extracellularly recorded burst afterdischarges. The build up of the spike afterdischarge seen in these earlier in vivo studies appears to be the same phenomenon we are seeing in these slices prior to an EGS. As in these slices, the seizures in the animals were either spontaneous or elicited by triggering interictal bursts at slightly higher than spontaneous firing rates. The mechanisms of these transitional changes described two decades ago are still not understood.

Attempts to clarify this type of interictal to ictal transition by studying it in smaller neural aggregates have been few. Dichter et al. (4) showed that both interictal and ictal activity could be generated by areas of hippocampus disconnected by knife cuts from the rest of the brain. Hoffer et al. (6) transplanted fragments of hippocampus to the anterior chamber of the eye and recorded both interictal and ictal discharges from them. After intense electrical stimulation, the implants became chronically hyperexcitable and ictal events could be elicited by simply triggering a few interictal bursts. The behavior of these implants was quite similar to the activity we see in slices exposed to stimulus trains.

It is important to note that our experiments were done with slices from relatively young animals, which seem to be more prone to generate EGSs. Swann and Brady (16) have shown that penicillin will produce EGSs in slices from immature rats, whereas in slices from mature rats, penicillin produces only interictal bursts. Schwartzkroin (12) has demonstrated that the propensity for seizures in the immature hippocampus might be partly explained by the earlier development of excitatory synaptic transmission compared to inhibition using slices from immature rabbits. We are currently studying the ontogenesis of these spontaneous and triggered EGSs in detail.

Another important variable is the magnesium concentration of the ACSF. These experiments were done in physiological magnesium concentration of 0.9 mM, which was used in previous studies from this laboratory on stimulation induced EGSs (see 15) because this level of magnesium may be closer to that found in vivo. We do not yet know if spontaneous or triggered EGSs would occur at higher levels of magnesium.

These preliminary observations demonstrate a new and important aspect of the long term hyperexcitability induced in hippocampal slices by repeated tetanic stimulation. We have previously shown that tetanic stimulation will induce interictal bursting and EGSs. These changes are similar to the progressive enhancement of excitability in the kindling model of epilepsy. Now, based on the present data, we believe that tetanic stimulation can also induce a state wherein spontaneous seizures can be generated by hippocampal slices. To our knowledge, this is the first demonstration of spontaneous EGSs occurring in hippocampal slices bathed in physiological medium and not exposed to epileptogenic agents.

Using this model, we will attempt to clarify the mechanisms of the interictal to ictal transition and the role of anticonvulsants in regulating these processes. The circuitry needed for the transition to seizure is apparently present in these slices. This suggests that feedback of activity from other brain regions may not be necessary for hippocampal seizure onset, maintenance, or termination. In addition, drastic alterations of the ionic environment and the presence of convulsants do not appear to be necessary to sustain seizure activity in the hippocampal slice, just as they are not necessary to sustain seizures in the intact brain.

References


1. Anderson, W.W., D.V. Lewis, H.S. Swartzwelder, and W.A. Wilson (1986) Magnesium-free medium activates seizure-like events in the rat hippocampal slice. Brain Res. 398:215-219.
2. Chesnut, T.J., and J.W. Swann (1988) Epileptiform activity induced by 4-aminopyridine in immature hippocampus. Epil. Res. 2:187-193.
3. Dichter, M., and W.A. Spencer (1969) Penicillin-induced interictal discharges from the cat hippocampus. II. Mechanisms underlying origin and restriction. J. Neurophysiol. 32:663-687.
4. Dichter, M., C. Herman, and M. Seltzer (1973) Penicillin epilepsy in isolated islands of hippocampus. Electroenceph. clin. Neurophysiol. 34:631-638.
5. Galvan, M., P. Grafe, and G. Ten Bruggencate (1982) Convulsant actions of 4-aminopyridine on the guinea pig olfactory cortex slice. Brain Res. 241:75-86.
6. Hoffer, B.J., A. Seiger, D. Taylor, L. Olson and R. Freedman.(1977). Seizures and related epileptiform activity in hippocampus transplanted to the anterior chamber of the eye. Exptl. Neurol. 54:233-250.
7. Jensen, M.S., and Y. Yaari (1988) The relationship between interictal and ictal paroxysms in an in vitro model of focal hippocampal epilepsy. Ann. Neurol. 24:591-598.
8. Konnerth, A., U. Heinemann, and Y. Yaari (1986) Nonsynaptic epileptogenesis in the mammalian hippocampus in vitro I. Development of seizure-like activity in low extracellular calcium. J. Neurophysiol. 56:409-423.
9. Lewis, D.V., L.S. Jones, and H.S. Swartzwelder (1989) The effects of baclofen and pertussis toxin on epileptiform activity induced in the hippocampal slice by magnesium depletion. In press. Epil. Res.
10. Matsumoto, H. and C. Ajmone Marsan (1964) Cortical cellular phenomena in experimental epilepsy:ictal manifestations. Exp. Neurol. 9:305-326.
11. Ralston, B.L. (1958) The mechanism of transition of interictal spiking foci into ictal seizure discharges. Electroenceph. Clin. Neurophysiol. 10:217-232
12. Schwartzkroin, P.A. (1982). Development of rabbit hippocampus: Physiology. Devel. Brain Res. 2:469-486.
13. Somjen, G.G., P.G. Aitken, J.L. Giacchino, and J.O. McNamara (1985) Sustained potential shifts and paroxysmal discharges in hippocampal formation. J. Neurophysiol. 53:1079-1097.
14. Stasheff, S.F., A.C. Bragdon, and W.A. Wilson (1985) Induction of epileptiform activity in hippocampal slices by trains of electrical stimuli. Brain Res. 344:296-302.
15. Stasheff, S.F., W.W. Anderson, S. Clark, and W.A. Wilson (1989) NMDA antagonists differentiate epileptogenesis from seizure expression in an in vitro seizure model. Science 245:648-651.

16. Swann, J.W., and R.J. Brady (1984) Penicillin-induced epileptogenesis in immature rat CA3 hippocampal pyramidal cells. Dev. Brain Res. 12:243-254.
17. Swartzwelder, H.S., D.V. Lewis, W.W. Anderson, and W.A. Wilson (1987) Seizure-like activity in brain slices: suppression by interictal activity. Brain Res. 410:362-366.
18. Traynelis, S.F., and R. Dingledine (1988) Potassium-induced spontaneous electrographic seizures in the rat hippocampal slice. J. Neurophysiol. 59:259-276.
19. Walther, H., J.D.C. Lambert, R.S.G. Jones, U. Heinemann and B. Hamon. (1986). Epileptiform activity in combined slices of the hippocampus, subiculum and entorhinal cortex during perfusion with low magnesium medium. Neurosci. Lett. 69:156-161.
20. Wilson, W.A., H.S. Swartzwelder, W.W. Anderson, and D.V. Lewis (1988) Seizure activity in vitro: a dual focus model. Epil. Res. 2:289-293.

# IS THE PYRIFORM CORTEX IMPORTANT FOR LIMBIC KINDLING?

Dan C. McIntyre and Mary Ellen Kelly

Department of Psychology, Carleton University
Ottawa, Ontario, K1S 5B6, Canada

Interest in the contribution of the pyriform cortex to complex partial seizures is not new. In the 1890s Hughlings Jackson and colleagues (9, 10) described a lesion limited to the human uncus, the homologue of the rodent pyriform cortex (2), which they believed initiated 'uncinate fits'. The development of elaborate behavioral symptoms during the uncinate seizure was presumed to be a result of seizure spread beyond this area, perhaps to the frontal cortex via the uncinate fasciculus (25). Occasionally these spontaneous uncinate seizures developed secondarily into full generalized convulsions, an outcome frequently observed after electrical stimulation of the uncus (21). Thus, it seems that provocation of the uncus is able to directly trigger, or gain access to mechanisms necessary to trigger, secondarily generalized convulsions.

Our current interest in the pyriform cortex concerns its role in the development and maintenance of limbic kindling. This interest is based primarily on four observations. First, the interictal discharge, which many consider to be the hallmark of epilepsy, was reported by Racine et al. (27) to originate in the pyriform cortex, independent of which structure served as the kindled focus. In addition, when interictal discharges occurred elsewhere they generally followed the pyriform response. Second, Piredda and Gale (23) described an area in the prepyriform cortex in which injections of small amounts of convulsants like bicuculline, kainic acid or carbachol initiated seizure discharges that progressed finally into a generalized convulsion. This part of the pyriform cortex was apparently far more sensitive to such pharmacological manipulations than other forebrain sites. Third, when comparing various structures in their ability to recruit and propagate seizure activity, namely, their kindling rates, stimulation of the pyriform cortex provided the most rapid genesis of convulsive activity, except for its primary afferent the high threshold olfactory bulb (e.g., 7, 15). And fourth, one hour after provocation of partial status epilepticus (SE) by continuous stimulation of a previously kindled focus in the amygdala, considerable uptake of radioactive 2-deoxyglucose was observed in the lateral half of the amygdala and entire pyriform cortex (McIntyre et al., in preparation). Two weeks following the spontaneous offset of SE, massive loss of cells was noted in the basolateral amygdala, entorhinal cortex and entire pyriform cortex (14). Possibly the offset of SE is a result of the progressive neural dysfunction created by the dying neurons of the pyriform lobe.

*Kindling 4*
Edited by J. A. Wada
Plenum Press, New York, 1990

With these points in mind, in an effort to examine the excitability of the pyriform cortex, we developed a new in vitro slice preparation of the area, using coronal sections that preserved communication between the amygdala and pyriform cortex (18, 19). Intracellular recordings from pyriform cortex neurons in tissue taken from nonkindled control rats showed strong burst responses to stimulation of the adjacent amygdala. These evoked burst events were significantly increased in duration in tissue taken from rats which had experienced amygdala kindling up to 8 weeks previously. In addition, many of the slices from kindled rats showed spontaneous discharges, while this was rarely observed in the control tissue. Thus, for many weeks after amygdala kindling, the in vitro excitability of the pyriform cortex remained enhanced.

One of the more robust effects in the kindling literature is the suppressive role played by norepinephrine (NE) in kindling development (4, 11). After stage-5 seizures have been fully developed, however, NE has little effect on their expression (31). Compared to all other cortices, the pyriform contains the highest concentration of NE. In our amygdala-pyriform slice preparation, superfusion of the chamber with 1-4 uM of NE blocked the burst response in control tissue, while at least twice the concentration was needed in the kindled tissue to affect the event (19). Thus, we observed a loss of NE effectiveness in the pyriform cortex as a consequence of amygdala kindling. Presumably this loss would have the permissive effect of releasing the pyriform lobe from NE-mediated inhibition.

Interest in N-methyl-D-aspartate (NMDA) receptor involvement in the kindling process is clearly evident in the present symposium. Several recent studies have shown that antagonism of the glutamate-sensitive NMDA receptors with the NMDA antagonists 2-amino phosphonovaleric acid (APV) or MK-801 substantially retarded kindling development (e.g., 3, 6, 29), but may have a lesser effect on mature kindled seizures (e.g., 22). In addition, much of the pathology in the pyriform cortex which results from SE seems to be a consequence of excessive glutamate innervation of NMDA receptors (e.g., 30) and can be prevented by NMDA antagonists.

The NMDA receptors are normally gated in a voltage-dependent fashion by the cation magnesium. With sufficient depolarization, as with the tetanic kindling stimulus, the gate is removed and the channel becomes fully operative. In the in vitro preparation, the gate may be removed simply by eliminating magnesium from the perfusate. This has the effect in the hippocampus of providing enhanced excitability in control tissue, which is further exacerbated by previous kindling of the tissue donor rat (e.g., 30).

We (24, 15) have examined the excitability of the amygdala-pyriform cortex in vitro under conditions of normal versus zero magnesium perfusion. In this series of experiments, we changed our strain of rat from the very seizure-prone Wistar of Charles Rivers, Canada (Quebec City), used by McIntyre and Wong (18, 19), to the far less seizure-prone Long-Evans Hooded. In addition, we now place each amygdala-pyriform slice in a preheated recording chamber without repositioning or manipulation, rather than adjusting each slice in the chamber to the desired orientation, and subsequently raising the temperature over time to 32°C. Under these new conditions, and as was mentioned earlier, in normal solution evidence of spontaneous pyriform discharges in control tissue was rarely observed. However, after a few minutes of perfusion with the zero

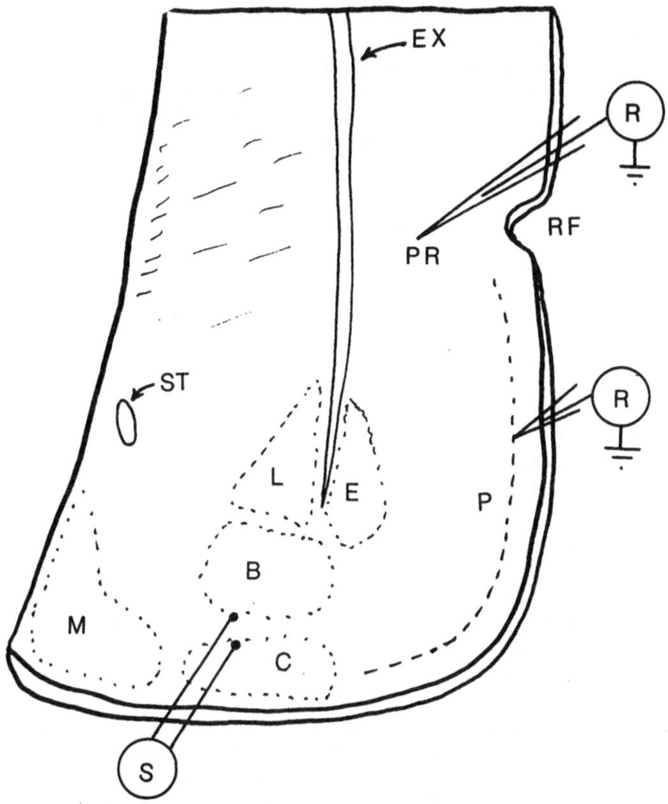


Fig. 1. Diagrammatic representation of the amygdala-pyriform slice preparation. S represents a typical stimulation site in the amygdala, while responses were recorded either intracellularly or extracellularly (R) from the layer-2 of the pyriform cortex (P) and/or perirhinal cortex (PR). B, basal nucleus of the amygdala; C, cortical nucleus; E, endopyriform nucleus; EX, external capsule; L, lateral nucleus; M, medial nucleus; RF, rhinal fissure; ST, stria terminalis.

magnesium solution, substantial burst responses with afterdischarges were provoked in the pyriform cortex by stimulation of the adjacent amygdala. Eventually these discharges appeared spontaneously. Because of our interest in the pyriform cortex, the spontaneous events were of particular interest, and we attempted to determine their origin. By comparing the temporal onset of the discharges in control tissue using simultaneous or paired extracellular recordings in different sites, we determined their origin was not in the pyriform cortex, or basolateral amygdala (5), but rather in the perirhinal cortex medial to the rhinal fissure (see Figure 1). Intracellular recordings from these cells indicated large amplitude, protracted depolarizations, and a characteristic response profile in the extracellular record (see Figure 2). The very different but characteristic response profiles of the perirhinal area and pyriform cortex during paired extracellular recordings may be seen in Figure 3. In normal magnesium solution, we often recorded small (0.2-0.5 mV), infrequent discharges in the perirhinal cortex of control slices, but never in the neighboring pyriform cortex. On the other hand, in rats previously kindled from the amygdala, nearly all slices in normal magnesium manifested spontaneous events initiated in the pyriform cortex, while the perirhinal area was either unresponsive or merely followed the pyriform event. Clearly, amygdala kindling increased the excitability of the pyriform cortex and altered the relationship between the pyriform and perirhinal cortices.

In zero magnesium, the spontaneous pyriform discharge in kindled slices continued to precede the perirhinal area, as well as the basolateral amygdala, which was the in vivo kindling site! The fact that the pyriform response continued to lead the perirhinal response was not surprising since long exposure to zero magnesium similarily altered the relationship between these two areas in control tissue. That is, with sufficiently long exposure to zero magnesium, the pyriform discharge in all control slices eventually assumed and maintained the lead compared to the perirhinal cortex (Plant and McIntyre, in preparation). Therefore, as suggested by Hoffman and Haberly (8), protracted exposure to zero magnesium seems to 'kindle' the tissue.

The specificity of the altered pyriform-perirhinal relationship to kindling is indicated by our observation (Plant and McIntyre, in progress) that only the pyriform lobe ipsilateral to the kindled amygdala exhibited the change, while the contralateral lobe from the same kindled rat behaved like control tissue. This is consistent with observations of kindling transfer between the two amygdalae in vivo (e.g., 13), where only a brief local after-discharge is triggered by stimulation of the contralateral amygdala after full kindling of the ipsilateral amygdala. Following several daily repetitions, however, stimulation of the second amygdala elicited long afterdischarges, which eventually triggered a stage-5 convulsion. Such secondary site kindling results in an approximate 50% positive savings in delivered stimuli compared to the original primary kindling site.

When examining transfer of kindling between homotopic sites in the dorsal hippocampi, we observed near-complete epileptogenesis in the contralateral site following kindling of the initial site (e.g., 12). This is likely a result of facilitative innervation through the massive hippocampal commissure, since its bisection prior to primary site kindling prevented any evidence of contralateral epileptogenesis (17). The effect of the strong excitatory innervation between the two dorsal hippocampi can be observed also in our SE preparation, where SE induction from a unilateral site in the posterior-ventral hippocampus resulted in bilateral pathology in the hippocampus and the amygdala-pyriform cortex. The pathology in the hippocampus and amygdala-pyriform cortex was lateralized when the hippocampal commissure was sectioned (16). These data suggest that, unlike the amygdala, the kindling of one hippocampus dramatically alters the excitability of both hemispheres, likely through the hippocampal commissure. Since the altered excitability produced by hippocampal kindling seems to be bilateral, we (McIntyre, Plant and Kelly, in progress) questioned (a) whether kindling of the dorsal hippocampus would alter the temporal relationship between the spontaneous discharges in the pyriform and perirhinal areas in vitro similar to amygdala kindling, (b) would that alteration be unilateral or bilateral, and (c) would that alteration occur if generalized seizures had not yet been provoked. Four rats were kindled in the dorsal hippocampus until they exhibited 6 stage-5 seizures. One to 3 weeks later their pyriform lobes were sliced and the

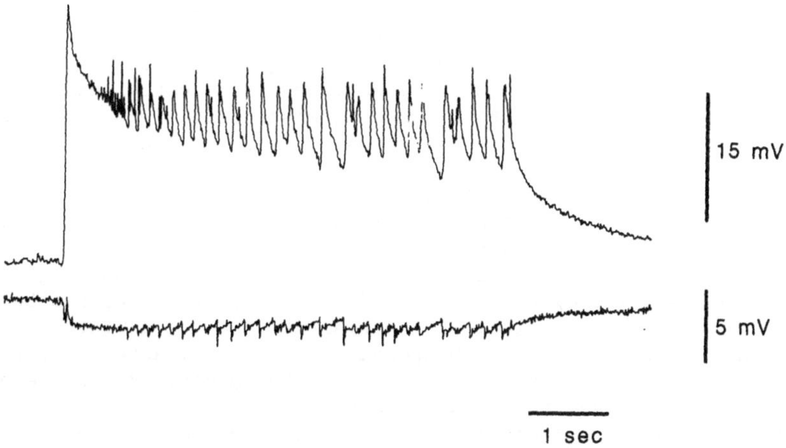


Fig. 2. Simultaneous intracellular (upper trace) and extracellular (lower trace) recordings of a spontaneous epileptic discharge from the perirhinal cortex of control tissue during zero magnesium perfusion.

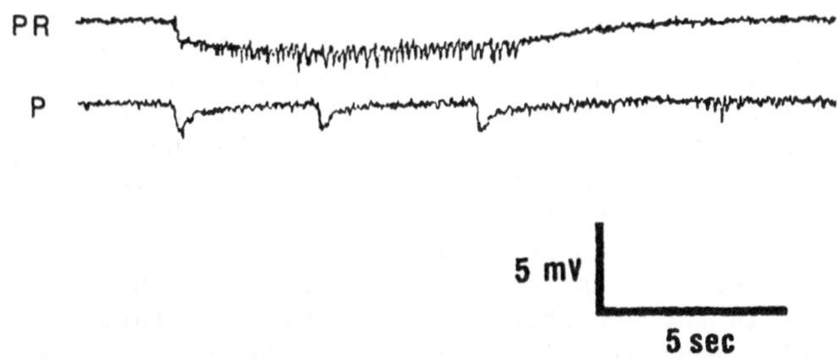


Fig. 3. Simultaneous extracellular recordings of a spontaneous discharge in the perirhinal (PR) and pyriform (P) cortex in control tissue during zero magnesium perfusion. Note the different but characteristic response profiles of each area.

pyriform-perirhinal relationship assessed. In all 10 slices (2-3 per rat) of the amygdala-pyriform area ipsilateral to the kindled focus, the pyriform cortex exhibited small spontaneous discharges in normal magnesium, as did 3 of 5 slices (with 2 silent) from the contralateral amygdala-pyriform area. On the other hand, in 2 additional rats, which experienced 30 or more afterdischarges without generalization, all 6 slices (3 from each hemisphere) in normal magnesium showed small spontaneous discharges in the perirhinal area, which preceded the pyriform discharge when the latter was present. One hour of exposure to zero magnesium dramatically increased the amplitude and frequency of these perirhinal and pyriform events, but did not affect their various relationships. Clearly the development of stage-5 seizures from the dorsal hippocampus induced an alteration in pyriform cortex excitability similar to amygdala kindling but in both hemispheres. This alteration was observed only if secondary generalization had previously occurred.

If alteration of pyriform lobe excitability is central to the production of generalized seizures from the kindled hippocampus, and if amygdala kindling produces an alteration in pyriform lobe excitability which is similar to hippocampal kindling, then one might expect good kindling transfer between the two structures. Such transfer has been reported by Burnham (1). Also, it is apparent that the preferred seizure output pathway from the hippocampus is ventral-anterior through the entorhinal (26) and pyriform cortices. The amygdala discharge also utilizes anterior projections via the amygdalofugal path (28). Perhaps then both hippocampal and amygdala kindled seizures make use of the same common final pathways. If so, primary site discharges from the hippocampus, after reaching the temporal lobe, should alter the amygdalofugal and pyriform pathways, and result in near-complete transfer to subsequent secondary site kindling of the amygdala. Furthermore, primary site kindling of the amygdala should facilitate secondary site kindling of the hippocampus by approximately the number of stimulations necessary for the amygdala discharge to modify the pyriform lobe and produce generalized seizures, i.e, the primary site amygdala kindling rate. Burnham (1) has provided data relative to these questions. Primary site kindling of the dorsal or ventral hippocampus produced near-complete transfer to secondary site kindling of the amygdala (see Table 1), while primary site kindling of the amygdala facilitated hippocampal kindling by nearly exactly the number of stimulations required to kindle the primary amygdala site. These data suggest that hippocampal kindling initially proceeds by facilitating the production of afterdischarges in local circuits, which over several repetitions slowly recruit and modify sufficient extrahippocampal circuits, perhaps in the entorhinal and amygdala-pyriform areas, to trigger secondarily generalized convulsions. Subsequent stimulation of the amygdala requires only one or two applications to access these altered circuits and trigger a stage-5 response.

If the pyriform lobe plays a critical role in the development or expression of seizures from kindled sites in the amygdala or hippocampus, then its destruction should prevent the development of secondarily generalized discharges, or interfere with previously kindled responses. In an attempt to directly test the involvement of the pyriform lobe in kindling, Racine et al. (28) produced bilateral electrolytic lesions of the amygdala-pyriform cortex prior to septal stimulation. Such lesions did not prevent kindling from the septum, but did delay appearance of the stage-5 responses. Unfortunately, the lesions were far from complete. To test the hypothesis satisfactorily lesions need to involve the <u>entire</u> pyriform lobe in <u>both</u> hemispheres, since local seizure production can readily find commissural access to the contralateral hemisphere, and trigger the generalized response.

TABLE 1

Primary and Secondary Site (Transfer) Kindling Between Hippocampus and Amygdala

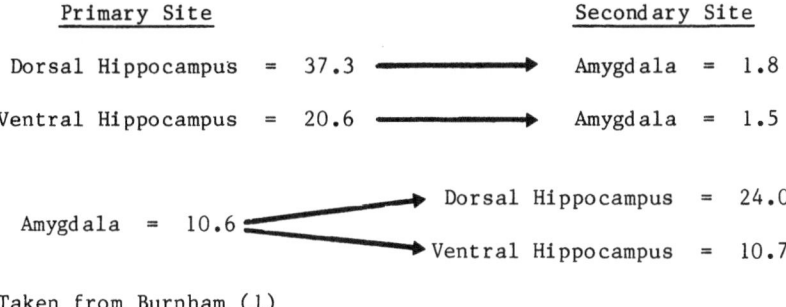

| Primary Site | | Secondary Site |
|---|---|---|
| Dorsal Hippocampus = 37.3 | → | Amygdala = 1.8 |
| Ventral Hippocampus = 20.6 | → | Amygdala = 1.5 |
| Amygdala = 10.6 | → | Dorsal Hippocampus = 24.0 |
| Amygdala = 10.6 | → | Ventral Hippocampus = 10.7 |

Taken from Burnham (1)

The likelihood of completely destroying the pyriform cortex bilaterally by electrolytic, aspiration or radiofrequency lesions is very low. On the other hand, the offset of SE is nearly always associated with complete loss of the pyriform area, extending from the olfactory bulbs anteriorly to the entorhinal cortex posteriorly. Unfortunately, with our amygdala-kindling model of SE (e.g., 14) this damage is only unilateral, and convulsions can be provoked again from a kindled focus in the dorsal hippocampus (likely via commissural access to the intact hemisphere) (McIntyre and Kelly, in progress). Data from a previous SE experiment, however, can be utilized to address this question. In split-brain rats, with discharge restricted to the ipsilateral hemisphere, SE was provoked by stimulation of a kindled focus in the posterior-ventral hippocampus (16). In 2 rats developing partial SE (#41R and 70L), temporal lobe pathology was restricted to the ipsilateral pyriform cortex with no evidence of damage to hippocampus, entorhinal area or substantia nigra. During 2 weeks of testing after SE, both rats exhibited hippocampal afterdischarges which were indistinguishable in duration and complexity to those before SE, yet no convulsive response was evident; this was also true for two rats (#85R and #94L) with pyriform plus entorhinal damage. This interference with convulsive activity was not the product of additional damage to the hippocampus, since one rat (#61R), with minimal pyriform but substantial CA1 and CA3 pathology, exhibited unaltered generalized convulsions to hippocampal stimulation after SE. It seems, therefore, that hippocampal discharges may need to access the pyriform lobe before a generalized seizure can be provoked.

Presently, we (Kelly and McIntyre, in progress) are testing this hypothesis in two SE experiments by producing bilateral pyriform cortex pathology using systemic kainic acid, or stimulation-induced SE from kindled foci in both amygdalae, in rats with previously kindled sites in the dorsal hippocampus and/or olfactory bulb. Preliminary results from the kainic acid SE experiment completely support our pyriform lobe hypothesis. By two weeks following kainic acid SE, rats with complete loss of several midline thalamic nuclei, including the paraventricular, paratenial, dorsal medial and reuniens nuclei, as well as the endopyriform nucleus and medial anterior olfactory nuclei, but no damage to the pyriform cortex, readily reacquired generalized seizures.

Although their focal seizures were normal, only those rats with additional bilateral loss of the pyriform cortex failed to reacquire generalized seizures. Further results from these experiments will answer more definitively the question of pyriform cortex involvement in limbic kindled seizures.

An interesting and important result materializing from this kainic acid SE experiment was that most of the kindled rats, compared to non-kindled control rats, exhibited little or no pyriform cortex damage. It seemed that kindling provided a protection for the pyriform cortex which was not available to the non-kindled animal. The source of this protection should be a significant disclosure in the continuing search for kindling mechanisms and consequences.

REFERENCES

1. Burnham, W. M., 1975, Primary and transfer seizure development in the kindled rat, Can. Neurol. Sci., 2:417.
2. Brodal, A., 1969, "Neurological Anatomy," Oxford University Press, London.
3. Cain, D. P., Desborough, K. A., and McKitrick, D. J., 1988, Retardation of amygdala kindling by antagonism of NMD-Aspartate and Muscarinic cholinergic receptors: evidence for the summation of excitatory mechanisms in kindling, Exp. Neurol., 100:179.
4. Corcoran, M. E., 1981, Catecholamines and kindling, in: "Kindling 2," J. A. Wada, ed., Raven Press, New York.
5. Gean, P-W., Shinnick-Gallagher, P., 1988, Epileptiform activity induced by magnesium-free solution in slices of rat amygdala: antagonism by N-methyl-D-aspartate receptor antagonists, Neuropharmac., 27:556.
6. Gilbert, M. E., 1988, The NMDA-receptor antagonist, MK-801, suppresses limbic kindling and kindled seizures, Brain Res., 463:90.
7. Goddard, G. V., McIntyre, D. C. and Leech, C. K., 1969, A permanent change in brain function resulting from daily electrical stimulation, Exp. Neurol., 25:295.
8. Hoffman, W. H., and Haberly, L. B., 1989, Bursting induces persistent all-or-none EPSPs by an NMDA-dependent process in pyriform cortex, J. Neurosci., 9:206.
9. Jackson, J. H., and Colman, W. S., 1898, Case of epilepsy with tasting movements and "dreamy state": very small patch of softening in the left uncinate gyrus, Brain, 21:580.
10. Jackson, J. H., and Stewart, P., 1899, Epileptic attacks with a warning of crude sensation of smell and with intellectual aura (dreamy state) in a patient who had symptoms pointing to gross organic disease of the right temporo-sphenoidal lobe, Brain, 22:334.
11. McIntyre, D. C., 1981, Catecholamine involvement in amygdala kindling of the rat, in: "Kindling 2," J. A. Wada, ed., Raven Press, New York.
12. McIntyre, D. C., and Edson, N. E., 1987, Facilitation of secondary site kindling in the dorsal hippocampus following forebrain bisection, Exp. Neurol., 96:569.
13. McIntyre, D. C. and Goddard, G. V., 1973, Transfer, interference and spontaneous recovery of convulsions kindled from the rat amygdala, Electroencephalogr. Clin. Neurophysiol., 35:533.

14. McIntyre, D. C., Nathanson, D., and Edson, N., 1982, A new model of partial status epilepticus based on kindling, Brain Res. 250:53.
15. McIntyre, D. C., and Plant, J. R., 1989, Pyriform cortex involvement in kindling, Neurosci. Biobehav. Rev., (in press).
16. McIntyre, D. C., Stokes, K. A., and Edson, N., 1986, Status epilepticus following stimulation of a kindled hippocampal focus in intact and commissurotomized rats, Exp. Neurol., 94:554.
17. McIntyre, D. C., and Stuckey, G. N., 1985, Dorsal hippocampal kindling and transfer in split-brain rats, Exp. Neurol., 87:86.
18. McIntyre, D. C., and Wong, R. K. S., 1985, Modification of local neuronal interactions by amygdala kindling in vitro, Exp. Neurol., 88:529.
19. McIntyre, D. C., and Wong, R. K. S., 1986, Cellular and synaptic properties of amygdala-kindled pyriform cortex in vitro, J. Neurophysiol., 55:1295.
20. Mody, I., Stanton, P. K., and Heinemann, U., 1988, Activation of N-methyl-D-aspartate receptors parallels changes in cellular and synaptic properties of dentate gyrus granule cells after kindling, J. Neurophysiol., 59:1033.
21. Penfield, W., and Kristiansen, K., 1951, "Epileptic Seizure Patterns," Charles C. Thomas, Springfield.
22. Peterson, S. L., and Boehnke, L., 1989, Anticonvulsant effects of MK-801 and gylcine on hippocampal afterdischarge, Exp. Neurol., 104:113.
23. Piredda, S., and Gale, K., 1985, A crucial epileptogenic site in the deep prepiriform cortex, Nature, 317:623.
24. Plant, J. R., and McIntyre, D. C., 1988, Zero Mg++ induced seizure discharge in the perirhinal-pyriform slice preparation, Soc. Neurosci. Abstr., 14:573.
25. Quesney, L. F., and Gloor, P., 1985, Localization of epileptic foci, in: "Long-term Monitoring in Epilepsy," J. Gotman, J. R. Ives, and P. Gloor, eds., Electroencephalogr. Clin. Neurophysiol., suppl. 37:165.
26. Ribak, C. E., and Khan, S. U., 1987, The effects of knife cuts of hippocampal pathways on epileptic activity in the seizure-sensitive gerbil, Brain Res., 418:146.
27. Racine, R. J., Mosher, M., and Kairiss, E. W., 1988a, The role of the pyriform cortex in the generation of interictal spikes in the kindled preparation, Brain Res., 454:251.
28. Racine, R. J., Paxinos, G., Mosher, M., and Kairiss, E. W., 1988b, The effects of various lesions and knife-cuts on septal and amygdala kindling in the rat, Brain Res., 454:264.
29. Sato, K., Morimoto, K., and Okamoto M., 1988, Anticonvulsant action of a non-competitive antagonist of NMDA receptors (MK-801) in the kindling model of epilepsy, Brain Res., 12:20.
30. Sloviter, R. S., and Dempster, D. W., 1985, 'Epileptic' brain damage is replicated qualitatively in the rat hippocampus by central injection of glutamate or aspartate but not by GABA or acetylcholine, Brain Res. Bull., 15:39.
31. Westerberg, V. S., Lewis, J., and Corcoran, M. E., 1984, Depletion of noradrenaline fails to affect kindled seizures, Exp. Neurol., 84:237.

## Discussion of Dr. McIntyre's Presentation

DR. CAIN: Those are very interesting data, Dan. You focused on the forebrain, and I am intrigued by the story you tell about the pyriform, but I have a two part comment. First, would you care to comment on any deeper structures in midbrain or brainstem. Do you think that there might be changes taking place there in the transfer phenomenon in particular. I think there are a bunch of data suggesting that there might be, and I actually reviewed those data in my last talk in Kindling Three. And secondly, with respect to Mary Allan's study with status epilepticus, I wonder if there are changes further down, if they might not be going on as a result of the status epilepticus treatment, and if that might not be a possible confound to what you might find subsequently in that status brain.

DR. MCINTYRE: Well, let me answer the second question first. The status that we are doing is of two kinds. We're doing kainate status which will knock out the pyriform quite nicely, but it also, very often, the literature says, interferes with the nigra. And so, in fact, downstream you can have modifications that could interfere with your seizure production as well, and you're quite right, it could confound any interpretation of the pyriform. The other kind of status that we're doing, is kindling status. We're doing it by having 2 amygdala foci, both hemispheres, we're provoking status in the one hemisphere, we're then allowing a few weeks off; we're then kindling the other hemisphere, and going to produce status from that. We have never seen any pathology downstream with that particular model. And so I hope that if there is an alteration of a negative nature downstream, that it's only in the kainate model and it's not in the other model, and therefore we won't have a particular confound related to that. Now whether the seizures themselves are making subtle changes that would be noticeable biochemically or in some other way with receptor numbers, or something to that effect, is very very possible, and we aren't going to see it until we see it. The first question as to whether or not these downstream structures are important - clearly they are important. There are so much data to suggest that they're important. I just have not yet worked down there, and so my interest continues to be in the forebrain. The animal's mobility is all down there, and so the final common pathways have to involve brainstem mechanisms. Mac Burnham's data so beautifully show the clonic disposition, is a cord disposition, and in fact you can hack that cord all up with semi or hemisections, and the animal still has a clonic response going on. So, indeed, all of those brain stem structures are going to finally elaborate these seizures. Exactly how they're doing it, don't get me to do that today.

DR. ADAMEC: When you were mentioning the zero magnesium studies, and you were talking about manipulating the NMDA receptor, I don't think you mentioned whether those changes in excitability were blockable by NMDA receptor blockers?

DR. MCINTYRE: Yes. I am embarrassed to say, having written this chapter in one week, I also did not include those data in the chapter, but yes, we blocked that with APV.

DR. ADAMEC: I'm wondering how important you think the NMDA receptor changes are in the pyriform, because studies that have done systemic administration of various NMDA blockers suggest that NMDA receptors are contributory but not critical for kindling.

DR. MCINTYRE: Right, but then there are also studies, like Peter has done, and Mary as well, and others, where if you get the blockade a little closer to the action, that in fact there is a powerful, fairly powerful effect on interfering with the kindling disposition of the animals. So, systemic administration does not excite me quite as much as perhaps sticking it directly into the focus, for example, which has a very strong effect.

DR. ADAMEC: So you'd be willing to predict that if you could put enough of let's say, MK 801 or something into the pyriform you could block kindling entirely.

DR. MCINTYRE: I would think you might. These systems are obviously, everything we've ever seen suggests very clearly, that this is a balanced system, with many transmitters operating. We've acetylcholine operating, we've NMDA operating, we've got all of our little antagonists like noradrenaline and so on and so forth. And GABA which we rarely get around to talking about. But, these stories are all interwoven and clearly they're all going to be contributing in one sense or another. I don't think there has been a single system yet, however, that has shown to be absolutely pivotal. Seizures will be generated, I think, if there's living tissue. It has the ability to be seizure prone.

DR. OKAMOTO: I have a great deal of interest in your study. I made kindling in cats, and the data showed that epileptic transfer to the contralateral amygdala was protected by lesioning of the fornix but not by section of the commissure pathway, including the hippocampus commissure. Do you have some comments on this?

DR. MCINTYRE: Isn't that interesting. Well, that just blows me away. I mean we could hang it on the usual, and that is of course, that is an entirely different species, and bla, bla, bla, the cat is such a unique animal, and so on and so forth. But that is surprising that you got complete protection, or the elimination of complete protection if you sectioned the fornix. So there is no transfer from one amygdala to the other, if you sectioned the ipsilateral fornix. Well, what it would suggest is that somehow the amygdala discharge - to create the transfer in the cat - is somehow going perhaps not back into the hippocampus, coming through the fornical system into hippothalamic structures, but what it does from there commissurally to affect that other amygdala I haven't got the vaguest notion.

DR. WADA: We will discuss the matter further after Dr. Okamoto's presentation.

# METABOLIC MAPS OF KINDLING: A SEQUENTIAL DOUBLE-LABEL DEOXYGLUCOSE STUDY

Kenji Ono[1], Hiroshi Baba[2], Shuhei Nishimura[3] and Juhn A. Wada[3]

Departments of Physiology[1] and Neurosurgery[2], Nagasaki University School of Medicine, Nagasaki, 852 Japan; and Divisions of Neurosciences and Neurology[3], Faculty of Medicine, The University of British Columbia, Vancouver Canada V6T 2A1

## INTRODUCTION

Ever since the first description of the kindling phenomenon(6), it has been widely believed that kindling can induce disseminated and long-lasting functional changes in the brain. This is evidenced, for example, by the evolution of new seizure manifestations implicating different functional categories along with the development of kindling, and by the transfer phenomenon at various brain sites directly or indirectly innervated from the focus. Such disseminated functional alterations must relate not only to the seizure phenomenon but also to the secondary brain dysfunction manifested in some epileptics. However, we do not yet have a comprehensive knowledge of the anatomical system of kindling and induced alterations.

The 2-deoxyglucose(2DG) method(16) has been widely used to measure the regional metabolic rate of glucose and to map regional functional activity in the brain. Potentially, autoradiographic determination of the distribution of radioactively labeled 2DG may provide a comprehensive map of the neural structures activated by a particular brain function. With this method, the average uptake of different brain regions in an experimental group is usually compared to the average uptake of the control group. However, the method often requires a large group of animals to yield statistically significant differences due to the considerable variations of uptake in such studies. It may be difficult, on occasion, to detect a small but significant change even if the measures are frequently repeated. In addition, the unavoidable inaccuracy of sampling the anatomical regions in animals may conceal a potential intra-structural organization of glucose uptake. Therefore, a sequential double-label autoradiographic 2DG method has been proposed(2,4,7,9,11,15) which allows the comparison of regional metabolic activity of the same brain on two separate occasions, thus eliminating most of the non-specific influences including inter-subject variability due to inherent variance of uptake in the same region and the uncontrollable differences of the experimental conditions. This advantage provides a tool for investigating the spatial distribution of the altered metabolic rate related to a particular brain function. It is possible to select tracers that differ sufficiently in their half-life or in the energy of the emitted radiation, so that the

*Kindling 4*
Edited by J. A. Wada
Plenum Press, New York, 1990

distribution of each tracer alone can be visualized. So far reported, most of quantitative double-label 2DG studies employed a combination of $^{18}F$ and $^{14}C$ (7,11) or $^{18}F$ and $^{3}H$ labeled tracers(15), because the half-life of $^{18}F$ is very short(110 min) so that a pure $^{14}C$ or $^{3}H$ image can be easily obtained after sufficient decay of $^{18}F$ activity. This may be the simplest way for isotopic separation, but an $^{18}F$ labeled tracer is available only in a limited number of institutions having a medical cyclotron. In other studies using a combination of $^{14}C$ and $^{3}H$ labeled tracers, the X-ray film image was simply assumed to be a pure $^{14}C$ image, although the brain slice was exposed to the X-ray film with an additional sheet interposed to block $^{3}H$ radiation(4,9). Sometimes such a potential cross-contamination of the tracers was not so seriously considered because of its qualitative nature.

The purpose of the present study was to obtain a comprehensive map of neurons specifically relating to establishment of premotor cortical kindling. We selected a pair of commercially available tracers, $^{14}C$ and $^{3}H$ labeled 2DG, and estimated the uncontaminated tracer images through a digital image procedure after having precisely accounted for cross-contamination. In a previous report(7), we devised a split brain preparation where the side of the task-irrelevant brain in both uptake periods served as a reference to estimate test-retest variability due to non-specific factors. In the present study, the non-specific influences were evalutated based on the statistics obtained from distribution of the uptake differences in the whole brain during the two uptake periods.

## METHOD

### Experiment

Male Wistar rats weighing approximately 300 g underwent an electrode implantation under pentobarbital anesthesia. A bipolar stimulation electrode made from twisted stainless steel wires was stereotaxically implanted in the right premotor cortex(8). One week following surgery, some of the rats were acutely kindled with an electrical pulse train of 10 Hz at an intensity strong enough to induce self-sustained seizure discharges every 20 min for 4 hours(a total of 12 trials). These animals were assigned to a "partially-kindled" group. Other rats which had not been stimulated prior to the labeling experiment were assigned to a "naive" group. Two weeks later, two rats from each group underwent the sequential double-label 2DG experiment. The acute kindling procedure was also carried out on the rest of animals, and the induced seizure duration was measured to compare the long-term effects of the acute kindling.

The double-label procedure consisted of three phases of control labeling by $^{14}C$-2DG, acute kindling, and test labeling by $^{3}H$-2DG, respectively. At the beginning of the experiment, 50 uCi of $^{14}C$-2DG (specific activity : 300 mCi/mmol) were injected intraperitoneally to label the glucose metabolic rate in the control brain(Phase 1). Immediately after injection, 6 Hz electrical stimulation, weak enough not to induce any seizure response, was continued for an hour. After sufficient clearance of the free tracer from the precursor pool(the half-life in the precursor pool has been estimated as 2.39 min by Sokoloff et al., 1977), the rat underwent the acute kindling procedure, consisting of 12 self-sustained seizures induced every 20 min by low frequency premotor cortical stimulation(Phase 2). The animal was then allowed to have an hour's rest. Seven hours after the injection of the first tracer, 1 mCi of $^{3}H$-2DG(10 Ci/mmol) was injected intraperitoneally and the metabolic rate in the kindled brain was labeled(Phase 3). Exactly the same stimulation as that of Phase 1 was repeated during the second uptake period to depict a potentially enhanced neurotransmission. Immediately after the end of Phase 3, the animal was sacrificed and the brain was processed for the subsequent autoradiography. Care was taken to reduce self-absorption by defattening the brain tissue(5).

Autoradiography and Image Processing

Each brain slice was exposed to a standard radiographic film(X-ray film, Fuji RX) having a thin protective coating, and subsequently to uncoated $^{3}H$ sensitive film(Ultrofilm, LKB) together with polymer standards. Thus, we obtained two autoradiograms for each brain slice, i.e., an X-ray film image and an Ultrofilm image. These were then digitized through a CCD TV camera and converted to optical density (OD) images, being exactly aligned for the subsequent digital image processing. Calibration curves for relating tissue concentration to film OD were obtained from simultaneously exposed polymer standards. On a double logarithmic domain, each tissue concentration vs film OD relation was well approximated by a straight line with a high correlation coefficient above 0.99(Fig 1). In addition, calibration lines relating to the same kind of tracers had virtually the same slope. Therefore, the inter-film OD ratio, defined as the ratio of X-ray film OD vs Ultrofilm OD made from the same radioactive source, was constant at least in a range of actual tissue concentration. In the present study, the ratios were 0.66 for $^{14}C$ and 0.11 for $^{3}H$. Now, the OD image of the same brain slice on X-ray film and Ultrofilm was plotted as X and U, respectively, and each film image was expressed as a sum of cross-contamination from

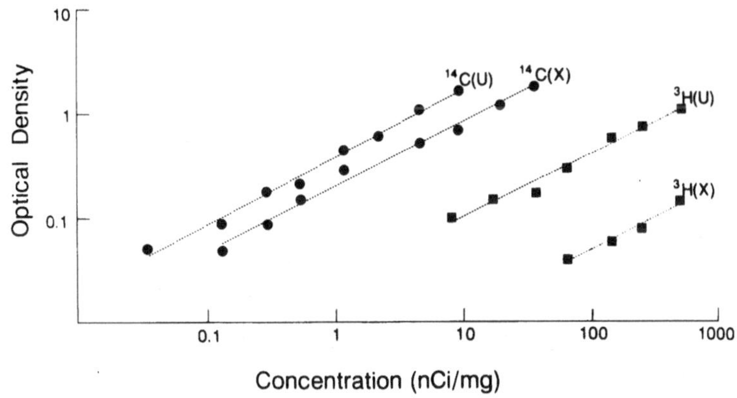


Figure 1. Calibration curves for relating tissue $^{14}C$ and $^{3}H$ (nCi/mg) to optical density (OD) on either an X-ray film (X) or a LKB Ultrofilm (U) obtained from simultaneously exposed polymer standards. Each relation was well approximated by a regression line in the double logarithmic domain (r>0.99). These lines have virtually the same slope coefficient so that the inter-film OD ratio generated by the same radioactive source could be assumed constant, i.e., 0.66 for $^{14}C$ and 0.11 for $^{3}H$, respectively. It was also estimated that, under an effective dose ratio of $^{14}C$:$^{3}H$=1:40, X-ray film autoradiogram might be cross-contaminated as much as 14% from $^{3}H$ in the same brain slice, indicating that X-ray film image did not merely reflect $^{14}C$ label.

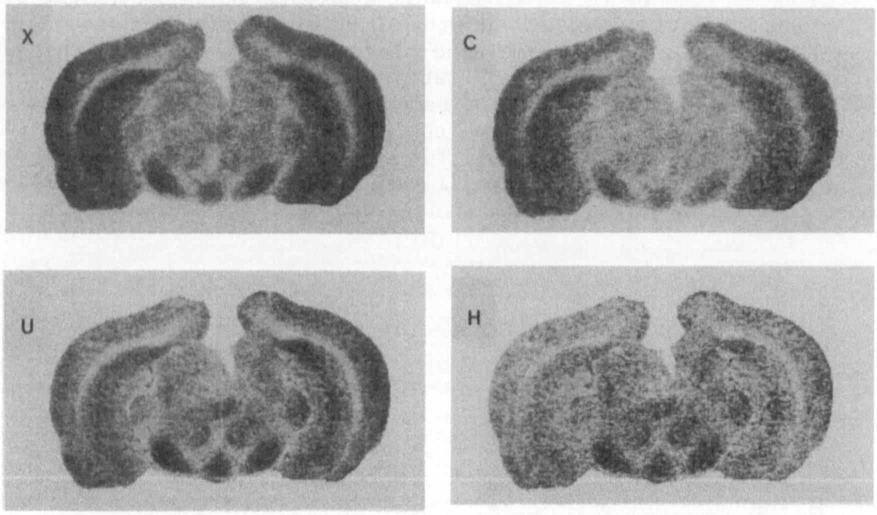


Figure 2. An example of isotopic separation. X and U show OD images on an X-ray film and a LKB Ultrofilm, respectively. Based on inter-film OD ratio, non-contaminated, the first tracer ($^{14}C$-2DG) image C and the second tracer ($^{3}H$-2DG) image H were separately estimated. Note that all images were normalized relative to the mean and standard deviation within each image.

$^{14}C$ and $^{3}H$ as follows:

$$U = U(c) + U(h) \qquad \text{( Eq. 1 )}$$
$$X = X(c) + X(h)$$

where **(c)** and **(h)** represent a portion of the film OD contributed from $^{14}C$ and $^{3}H$ in the same brain slice, respectively. Further, the inter-film OD ratios might be:

$$X(c)/U(c) = a \qquad \text{( Eq. 2 )}$$
$$X(h)/U(h) = b$$

Each uncontaminated tracer image could then be estimated through the following equations:

$$U(h) = (aU-X)/(a-b) \qquad \text{( Eq. 3 )}$$
$$U(c) = U - U(h)$$

Fig 2 shows an example of X-ray film image **X** and Ultrofilm image **U** on the left side, and each tracer image thus estimated on the right side(C and H). The pure tracer OD images were then converted to an activity image of glucose uptake so that a given density of the images represents the same glucose uptake irrespective of the kind of tracers, based on (1) Ultrofilm sensitivity against $^{14}C$ and $^{3}H$, (2) injected dose of tracers and (3) loss of phosphorylated products of the first tracer during Phase 2 and Phase 3, estimated with a reported half-life of approximately 8 hours(16).

Subtraction of the first tracer image from the second tracer image was then accomplished to extract differences in glucose uptake prior to and following kindling(Fig 3) where positive and negative values represent increased and decreased uptake, respectively. The difference image might be expected to contain test-retest variability or random metabolic variability. The differences were distributed as shown by the histogram in the lower column of Fig 3. The distribution was very well approximated by the sum of two normal distributions , one being around zero and the other around 0.2. Accordingly, it was hypothesized that the population distributed around zero represented an unchanged metabolism with random fluctuations and served as a reference to estimate test-retest variability. Z-transformation of the difference image was thus accomplished against the mean and standard deviation of the unchanged population, yielding density-coded maps of significant differences of neural activity specifically related to kindling without a priori assumptions about the anatomical system.

## RESULTS AND DISCUSSIONS

As shown in Fig 4, the induced seizure response was consistently enhanced, along with repetition even at a relatively short inter-trial interval of 20 min as revealed by prolonged duration and augmented convulsive behavior, although stimulus intensity was gradually raised to overcome the threshold increment concurrently induced. As previously noted with premotor cortical kindling (14), limbic involvement was sometimes observed in later succession. More importantly, animals which had once experienced the same stimulation manifested more intense seizure response than naive rats, confirming that the acute kindling procedure used in the present study could induce a long-lasting enhancement of seizure susceptibility at least to some extent.

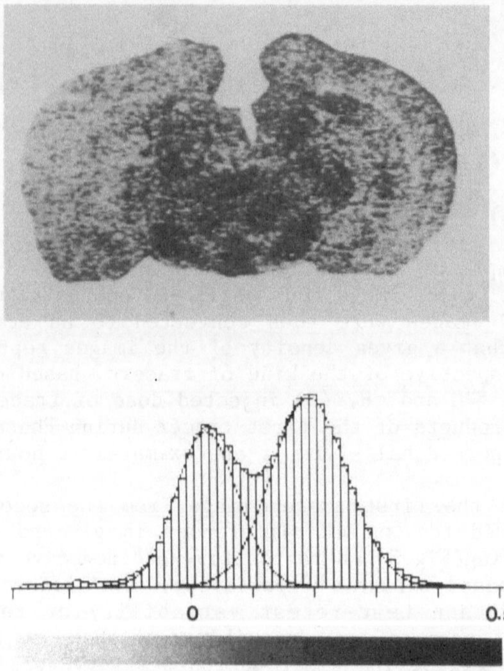


Figure 3. Difference image and histogram of the differences. Non-contaminated tracer images were corrected in terms of film efficacy, injected dose and loss of phosphorylated products so that a given density of each image was representative of the same uptake rate, irrespective of the kind of tracers. The first tracer image C was then subtracted from the second tracer image H. The distribution of the differences was well approximated by two normal distributions, where one around zero represents a portion which did not change and the other having a mean of 0.20 represents a portion of increased uptake after kindling.

The metabolic rate prior to and following acute kindling was compared after having precisely accounted for the contribution of nonspecific influences by means of a sequential double-label 2DG method by commercially available tracers, $^{14}C$ and $^{3}H$ labeled 2DG. Isotopic separation was attained through a newly proposed digital image procedure based on the different efficacy of each tracer against two different films. This method made it possible to use easily available tracers instead of $^{18}F$ labeled tracer, and may also be applicable to other double-label approaches, e.g., a combination of 2DG and iodoantipyrine(IAP) for simultaneous measurement of local glucose uptake and blood flow, for example.

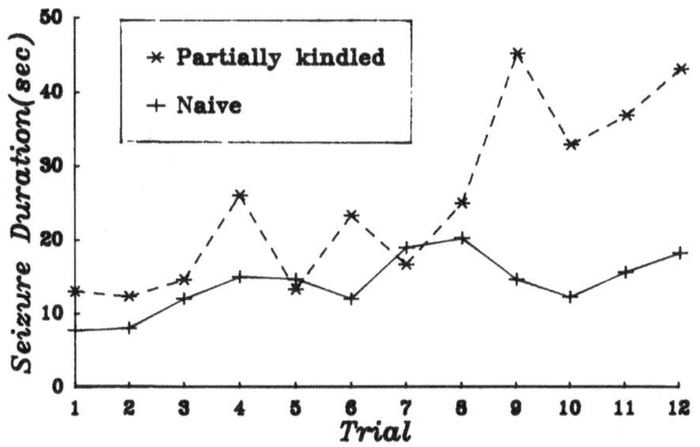


Figure 4. Seizure development during acute kindling. Duration of self-sustained seizure discharges induced by low frequency cortical stimulation was measured in two groups of rats, one that had been stimulated with the same stimulation paradigm 14 days prior to the test session (partially-kindled group), and another that had never been stimulated (naive group). The partially kindled rats showed obviously longer seizure duration and faster development than that of naive rats.

Fig 5 shows Z-transformed difference images at three different coronal levels(i.e., the striatum, thalamus and midbrain) from a rat which had been partially kindled two weeks prior to the labeling. The arrow indicates the side of cortical stimulation. The differential images were transformed relative to the reference population around zero of OD differences in the whole brain(right lower column). The number of each quantified brain region(pixel) which belongs to the population of increased metabolic rate was estimated through approximated normal distributions(broken lines) and it was suggested that more than 60% of brain regions might be involved in the kindling effect showing an increase of metabolic rate. Significant increases were found in disseminated but well organized structures of the brain, including the perifocal cortex and its contralateral site, the dorsal striatum, thalamus, entopeduncular nucleus, substantia nigra, interpeduncular nucleus, red nucleus, periventricular gray substance, superior colliculus, and cingulate and

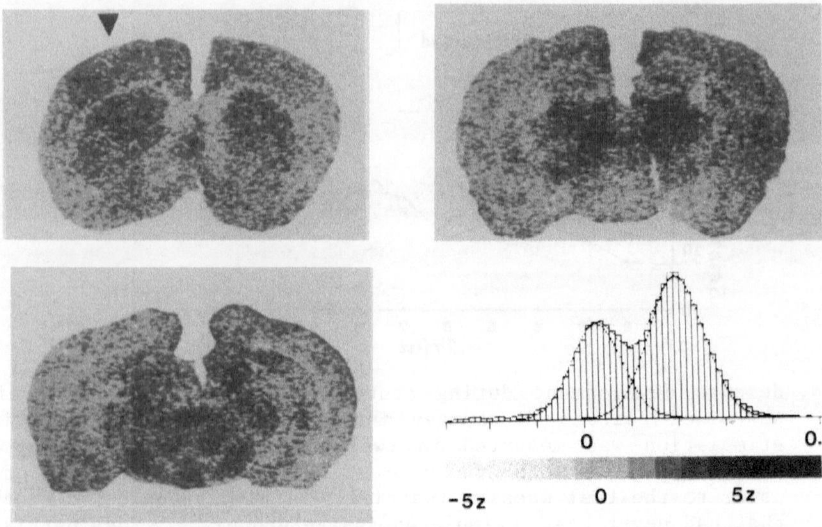


Figure 5. Z-transformed difference image in a partially kindled rat. The degree of metabolic change was density-coded relative to a statistical distance from the unchanged population around zero. Arrow indicates the side of kindling stimulation. Increased uptake was found in more than 60% of the brain region investigated, including the cingulate and hippocampal cortices in addition to the perifocal and its contralateral cortices, the striatum, the thalamus and the midbrain.

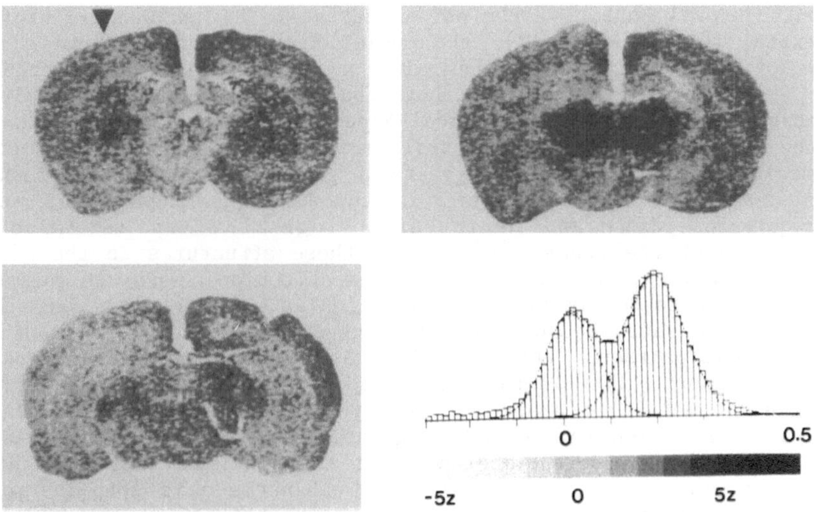


Figure 6. Z-transformed difference image in a naive rat. The general feature was essentially the same as that of a partially kindled rat (Fig. 5). However, hippocampal uptake virtually did not increase, probably relating to the degree of kindling effect. Metabolic enhancement in the midbrain (the substantia nigra and the red nucleus) was also moderate.

temporal cortices, in addition to discrete regions of the dorsal hippocampus. It was further noted that the metabolic alterations were generally bilateral except for the contralateral dominance of the temporal and hippocampal cortices. Results obtained from a naive rat were essentially the same as those shown in Fig 6. However, metabolic enhancement in some structures was less intensive while other structures showed more prominent change. Such metabolic enhancement measured one hour after the last seizure is not a direct influence of the convulsive seizure itself since it has been reported that postictally-induced metabolic change is generally depressed and dissipates rapidly(1,10), Rather, the enhancement is more likely related to the kindling effect.

Although it may be difficult, at present, to directly relate these differences to the degree of kindling, the results suggest that widespread and persistent metabolic enhancement could be the result of kindling or repeatedly induced seizures. In addition to the structures in which glucose uptake increased during the ictal phase, i.e., the stimulated and its contralateral cortices, the dorsal striatum, the substantia nigra and the ventral thalamus(3,12,13), the medial frontal cortices and discrete regions of the dorsal hippocampus also showed a significant increase of uptake, probably correlating to behavioural limbic involvement evolved in the later phase of premotor cortical kindling(14). Furthermore, the fact that the red nucleus and the ventrolateral nucleus of the thalamus were also metabolically activated suggests a potential participation of the cerebro-cerebellar neural circuit in the kindling phenomenon. Further electrophysiological and neuropharmacological study is needed for clarification of the functional role of these structures in the kindling phenomenon. Besides, the observed enhancement could include an increase of metabolic response to electrical stimulation as well as an increase in the basal metabolic rate, because the kindling site was stimulated during the uptake periods, although the intensity was weak enough not to evoke any convulsive seizure. It will thus be needed to distinguish whether kindling induces basal metabolic change or solely enhances metabolic response to the focal stimulation reflecting augmented neurotransmission. This will be attained by an uptake label without stimulation. Another important issue might be whether such a metabolic enhancement is long-lasting and parallel with the concurrently induced kindling effect. We are not able to test this hypothesis at present, because, when the inter-label period is very long relative to the half-life of phosphorylated products, the sequential double-label method may no longer be applicable, and an alternate experimental strategy should be devised. However, our method may be applicable beyond the example of the present study with minor modification of the experimental design.

## ACKNOWLEDGMENTS

This work was supported by grants from the Ministry of Education, Science and Culture, National Center of Neurology and Psychiatry (NCNP) of the Ministry of Health and Welfare in Japan( K.O.) and the Medical Research Council of Canada(J.A.W.).

## REFERENCES

(1)Ackermann, R.F., Chugani, H.T., Handforth, A., Moshe, S.L., Caldecott-Hazard, S., and Engel,J.,Jr., 1986, Autoradiographic studies of cerebral metabolism and blood flow in rat amygdala kindling, in:Kindling 3, J.A. Wada, ed., Raven Press, New York.

(2)Altenau, L.L. and Agranoff, B.W., 1978, A sequential double-label 2-deoxyglucose method for measuring regional cerebral metabolism, Brain Res., 153:375-381.
(3)Collins, R.C., Kennedy, C., Sokoloff, L., and Plum, F., 1976, Metabolic anatomy of focal motor seizure, Arch Neurol., 33:536-542.
(4)Friedman, H.R., Bruce, C.J., and Goldman-Rakic, P.S., 1984, Double-label 2-DG technique yields double dissociation of functional states with an image differencing method, Soc. Neurosci. Abstr., 1002.
(5)Geary II, W.A., Toga, A.W., and Wooten, G.F., 1985, Quantitative film autoradiography for tritium: Methodological Considerations, Brain Res., 337:99-108.
(6)Goddard,G.V., McIntyre, D.C., and Leech, C.K., 1969, A permanent change in brain function resulting from daily electrical stimulation, Exp. Neurol., 25:295-330.
(7)John, E.R., Tang, Y., Brill, A.B., Young, R., and Ono, K., 1986, Double-label metabolic maps of memory, Science, 233:1167-1175.
(8)Krieg, W.J.S., 1946, Connections of the cerebral cortex. I.The albino rat. A. Topography of the cortical areas, J. Comp. Neurol., 84:221-275.
(9)Livingstine, M.S., and Hubel, D.H., 1981, Effect of sleep and arousal on the processing of visual information in the cat, Nature, 291:554-561.
(10)Namba, H., Iwasa, H., Kubota, M., Yamaura, A., Sato, T., Hagihara, Y., and Makino, H., 1989, Local cerebral glucose utilization in the post-ictal phase of amygdaloid kindled rats, Brain Res., 486:221-227.
(11)Olds, J.A., Frey, K.A., Erhenkaufer, R.L., and Agranoff, B.W., 1985, A sequential double-label autoradiographic method that quantifies altered rates of regional glucose metabolism, Brain Res., 361:217-224.
(12)Ono, K., Mori, K., Baba, H., Seki, K., and Wada, J.A., 1986, A new chronic model of partial onset generalized seizure induced by low frequency cortical stimulation: Its relationship to the kindling phenomenon, in: Kindling 3, J.A. Wada, ed., Raven Press, New York.
(13)Ono, K., Mori, K., Baba, H., and Wada, J.A., 1987, A role of the striatum in premotor cortical seizure development, Brain Res., 435:84-90.
(14)Ono, K., Baba, H., Mori, K., and Wada, J.A., 1988, Premotor cortical kindling interferes with subsequent hippocampal kindling, Brain Res., 475:182-186.
(15)Redies, C., Diksic, M., Evans, A.C., Gjedde, A., and Yamamoto, Y.L., 1987, Double-label autoradiographic deoxyglucose method for sequential measurement of regional cerebral glucose utilization, Neuroscience, 22: 601-619.
(16)Sokoloff, L., Reivich, M., Kennedy, C., Des Rosiers, M.H., Patlak, K.D., Pettigrew, K.D., Sakurada, O., and Shinohara, M., 1977, The [$^{14}C$] deoxyglucose method for the measurement of local cerebral glucose utilization: Theory, procedure, and normal values in the conscious and anesthetized rat, J. Neurochem., 28:897-916.

Discussion of Dr. Ono's Presentation

DR. WADA: One of the slides showed some enhancement in the midline thalamic area. Would you like to comment on that, on its significance, or is it an artifact?

DR. ONO: I'm afraid I don't know. It's just an artifact.

DR. WADA: It just happens that this slide shows, as you noted, enhancement of the red nucleus and the ventral thalamus, and you mentioned cerebellum. Have you had the chance to look at the cerebellum itself?

DR. ONO: Unfortunately we have not yet investigated the cerebellar uptake of glucose.

DR. ACKERMAN: This is beautiful work. Would you please repeat how long after the kindling was the second injection made, and would you repeat why you don't think that the increases which you saw, are due to the effect of the stimulation which preceded. Because I did not understand that. My first question is, you said it but I don't remember, how long after the kindling sessions were over, did you do the second injection?

DR. ONO: One hour.

DR. ACKERMAN: And I think you said that you don't think that the increases you saw with the second injection in cooperation are due to the effect of the kindling on the seizures per se, is that correct? Yes. And if it is correct, why is that so?

DR. ONO: I'm afraid I don't understand.

DR. ACKERMANN: Let me try and answer my own question, not necessarily with the right answer, but with an answer. This morning when I met Dan McIntyre we immediately resumed our running bull session, and in the bull session he asked me if I thought - we were discussing a study that we might do - he asked me if I thought you had to do quantification in order to get meaningful results, and my answer is no, which might strike you as funny because I'm going to give you a quantification talk, but I think that the answer will appear I hope. If not, you can ask me a question.

# SEIZURE MECHANISMS IN THE PIRIFORM CORTEX

Jeffrey S. Stripling and Doris K. Patneau*

Department of Psychology
University of Arkansas
Fayetteville, Arkansas 72701

## INTRODUCTION

We are currently involved in a series of studies whose purpose is to compare and contrast the effects of kindling and long-term potentiation (LTP) in the same cortical system, the olfactory cortex. Our findings regarding LTP are presented in detail elsewhere (20,21,22,23). This paper will present some initial observations regarding seizure mechanisms in the olfactory cortex and discuss their implications.

### Anatomical and Functional Organization of the Olfactory Cortex

We chose the olfactory cortex for study because of the numerous advantages it offers. It is an ideal cortical structure in which to study seizure activity: it is very seizure-prone (10), kindles rapidly (1), is activated by kindling at many limbic sites (8), and may play a facilitating role in all limbic kindling (15). In addition, its anatomical and functional organization make it particularly suitable for electrophysiological analysis using evoked potentials, enabling us to monitor functional changes over long periods of time in chronically implanted animals.

The primary olfactory cortex in mammals consists of those areas receiving monosynaptic input from the olfactory bulb (OB) via the lateral olfactory tract (LOT) (5). The largest of these areas is the piriform cortex (PC), a 3-layered paleocortical structure that occupies a substantial portion of the ventral surface of the forebrain in rodents. Layer II of the PC is densely packed with the somata of pyramidal cells. The apical dendrites of these cells receive excitatory synapses in layer Ia from OB mitral cells, whose axons form the LOT. This constitutes the principal input to the PC. The axons of PC pyramidal cells form a massive association fiber system that interconnects various parts of the PC, making excitatory synapses in layer Ib and layer III on other pyramidal cells and on inhibitory interneurons. These axons also provide feedback to the OB, where they activate granule cells in the deep part of the granule cell layer (9,11), which in turn inhibit mitral cells via dendrodendritic synapses in the external plexiform layer (11). In addition, the axons of PC

*Current address: Unit of Neurophysiology and Biophysics, Laboratory of Developmental Neurobiology, National Institutes of Health, Bethesda, Maryland 20892.

*Kindling 4*
Edited by J. A. Wada
Plenum Press, New York, 1990

pyramidal cells provide output from the PC to other parts of the olfactory cortex, including the contralateral PC, as well as the endopiriform nucleus, thalamus, and neocortex (5,12). Although the results presented here will be limited to the PC, other parts of the olfactory cortex such as the anterior olfactory nucleus have a similar anatomical organization and relationship with the OB (9).

These anatomical considerations emphasize the network properties of the olfactory forebrain. Information transmitted to the olfactory cortex by the LOT is elaborated upon by the association fiber system and is represented in the cortex in a parallel distributed fashion. Furthermore, negative feedback loops within the cortex and from the cortex to the OB control the timing and synchronization of electrophysiological activity throughout the system (3,5).

The network attributes of olfactory cortex can best be monitored by evoked potential techniques that reflect the response of a population of cells, rather than any single cell, to activation of afferent pathways. Single-pulse electrical stimulation of the LOT or the granule cell layer of the OB activates mitral cell axons and produces a monosynaptic EPSP in PC pyramidal cells. Those cells that fire in response to this input generate activity in the PC association fiber system, producing a second, disynaptic EPSP in PC pyramidal cells and activating inhibitory interneurons that then produce IPSPs in the pyramidal cells (6,26). These events are reflected in the potential evoked in the PC by this stimulation (Figure 1). An initial surface-negative wave (period 1) consists of two components, $A_1$ and $B_1$, which represent the mono- and di-synaptic EPSPs triggered by LOT activation (6). Period 1 is followed by period 2, which is temporally associated with the IPSPs produced in PC pyramidal cells following their activation (2,6), and appears to reflect inhibitory processes in the PC (17). All components of the evoked potential are generated by current sources and sinks in the membrane of PC pyramidal cells, and reverse polarity between layer I and layer III (17), making it easy to discriminate between potentials generated within the PC and those volume-conducted from other sites.

The research that follows examines the characteristics of seizures triggered by electrical stimulation of the OB, PC, or LOT, and uses evoked potentials to determine the circuit elements activated by stimulation of these sites at various current intensities. This information is used to assess which circuit elements must be activated to trigger a seizure in the olfactory forebrain.

## METHODS

All experiments utilized freely-moving male Long-Evans rats (Blue

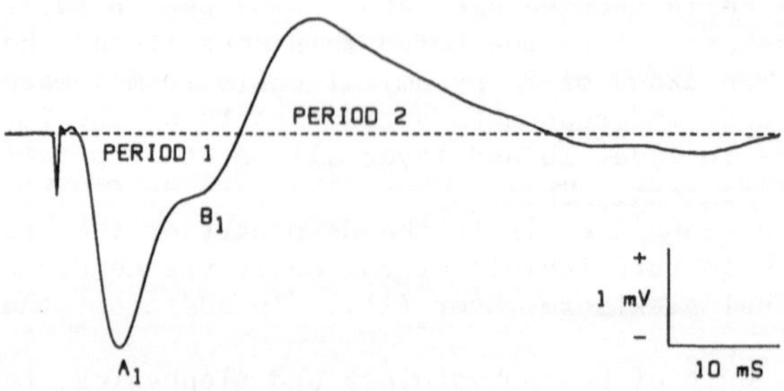


Figure 1. A potential evoked in layer I of the PC by 400 $\mu$A stimulation of the granule cell layer of the OB.

Spruce Farms) with chronically implanted electrodes. Stainless-steel wires with a diameter of 75-125 μm were used for stimulation and recording, and were implanted under electrophysiological guidance. Electrode position was measured accurately at the end of each experiment using the Prussian blue method. A detailed description of the surgical, recording, and histological techniques used will be presented elsewhere (23). All stimulation consisted of 0.2 ms constant-current monophasic cathodal square waves. Evoked potentials were elicited by stimulation at 0.2 Hz and averaged ($n$=12) for analysis. The amplitude of the $A_1$ component in PC potentials was estimated from the peak amplitude of period 1. Seizures were triggered by a single train of 100 pulses at 100 Hz. Further details will be provided with the results for each experiment.

The animals in these experiments had previously been used in studies of LTP (20,21,22, in preparation). Some had received control stimulation at a low frequency (1 Hz) that produced no change in the PC evoked potential, while others had received potentiating stimulation in the form of 10-pulse 100-Hz trains repeated at 5 or 10 s intervals at a current intensity well below seizure threshold. We have found that this pattern of stimulation, when applied to the granule cell layer of the OB, leaves the EPSP components of the PC evoked potential unaltered but produces a selective LTP of period 2 (22,23). This phenomenon will not be discussed further here, but because much of the research to be presented below was conducted in animals that had previously been exposed to LTP stimulation, it is important to note that this had no effect on the monosynaptic EPSP produced by LOT fibers in PC pyramidal cells, as measured by the $A_1$ component of the evoked potential (22,23). In this regard the results contrast dramatically with the effects of LTP in the hippocampal formation and elsewhere (13).

## WHAT CIRCUIT ELEMENTS MUST BE ACTIVATED BY ELECTRICAL STIMULATION TO PRODUCE A SEIZURE?

Both the OB and the PC readily generate an electrographic seizure in response to high-frequency electrical stimulation. Examination of the relationship between the site of stimulation, the potentials evoked by the stimulation, and the afterdischarge (AD) threshold can provide insight into the circuit elements responsible for generating a seizure.

### Stimulation of the Olfactory Bulb

The OB has a high AD threshold relative to other limbic structures, but kindles very rapidly if suprathreshold stimulation is used (1). As Cain first observed (1), the AD threshold of the OB varies with electrode placement: it is lowest in the granule cell layer and rises with increasing distance from that layer. Our own data (Figure 2) confirm this finding and indicate that AD threshold rises in a gradual fashion as the site of stimulation shifts from the core of the granule cell layer to its outer boundary. Thus there appears to be a circuit element in the granule cell layer that must be stimulated to produce a seizure.

One possibility is that stimulation in the granule cell layer produces an AD by activating circuit elements intrinsic to the OB. However, the OB does not readily sustain an AD if it has been isolated from the olfactory cortex, either surgically (1) or by cryogenic blockade (4). This emphasizes the importance of functional connections between the OB and other structures in producing seizure activity. The most obvious of these is the LOT. However, the LOT can be effectively activated by stimulation outside the granule cell layer, as indicated by the amplitude of the $A_1$ component of the potential evoked in the PC by OB stimulation (see Table 1 and

Table 1. Characteristics of potentials evoked in layer I of the PC by stimulation in the granule (GR), mitral (M), or external plexiform (EP) layer of the OB. Values for $A_1$ are median (range).

| | Stimulation Site in OB | | |
|---|---|---|---|
| | GR ($\underline{n}$=97) | M/EP ($\underline{n}$=15) | $\underline{p}$ |
| $A_1$ amplitude (mV) at 400 $\mu$A | 3.05 (0.73 - 6.45) | 2.70 (1.76 - 4.21) | >0.17[a] |
| Proportion of animals exhibiting X wave at 800 $\mu$A | 0.62 (60 of 97) | 0.07 (1 of 15)[c] | <0.001[b] |

[a]Mann-Whitney U test (2-tailed)
[b]Chi-square test
[c]The one animal exhibiting an X wave was stimulated in the mitral cell layer.

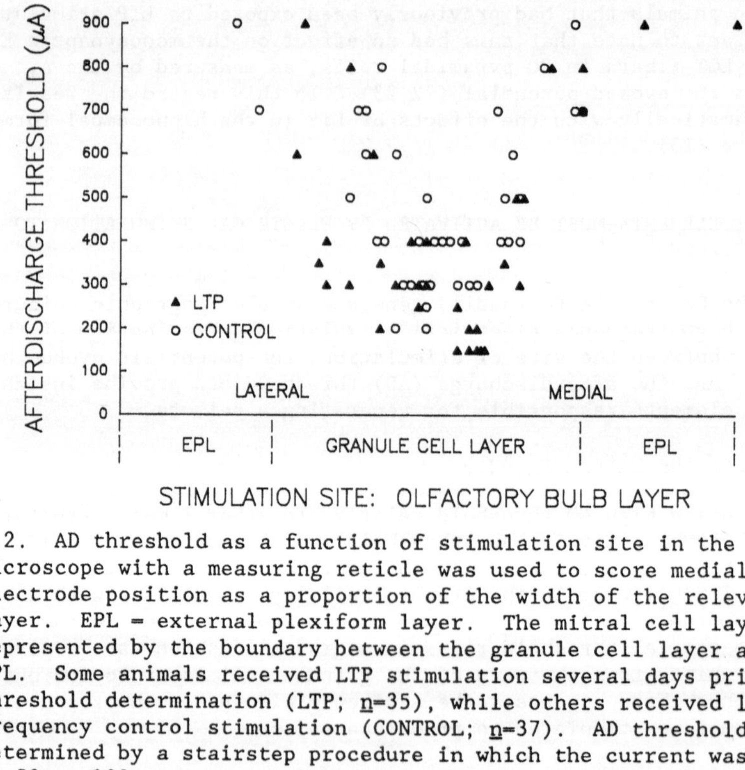


Figure 2. AD threshold as a function of stimulation site in the OB. A microscope with a measuring reticle was used to score medial-lateral electrode position as a proportion of the width of the relevant OB layer. EPL = external plexiform layer. The mitral cell layer is represented by the boundary between the granule cell layer and the EPL. Some animals received LTP stimulation several days prior to AD threshold determination (LTP; $\underline{n}$=35), while others received low frequency control stimulation (CONTROL; $\underline{n}$=37). AD threshold was determined by a stairstep procedure in which the current was increased in 50 or 100 $\mu$A steps at 1-4 min intervals. Animals not exhibiting an AD by 800 $\mu$A received an AD threshold score of 900 $\mu$A. There was a significant correlation, pooled across the two groups, between AD threshold and electrode distance from the core of the granule cell layer, $\underline{r}(68) = 0.74$, $\underline{p}<0.001$. An analysis of covariance, with electrode position as the covariate, indicated that LTP animals had a significantly lower AD threshold than CONTROL animals, $\underline{F}(1,69) = 11.69$, $\underline{p}<0.01$.

Figure 3); furthermore, the amplitude of this component (measured in layer I) is only weakly correlated with AD threshold, $r(63) = -0.21$, $p<0.10$ (pooled across LTP conditions).*

A stronger correlation, $r(63) = -0.42$, $p<0.001$ (point biserial correlation coefficient pooled across LTP conditions), exists between AD threshold and the presence of a late surface-negative wave (here referred to as X) that is evoked in the PC by stimulation of the granule cell layer of the OB, but not by stimulation outside that layer (see Table 1 and Figure 3). This potential appears to correspond to that reported by Racine et al. (13), and is enhanced by kindling (13,18). A number of considerations have led us to suggest (21, in preparation) that this component reflects activity at association fiber synapses on PC pyramidal cells due to antidromic activation of the association fiber system by stimulation of axon terminals in the granule cell layer of the OB.

These observations are far from conclusive, but taken as a whole they point to the possibility that the critical element that must be activated by OB stimulation to produce a seizure is the PC association fiber system. With this in mind we now turn to the more direct approach of stimulating specific fiber systems in the PC.

Stimulation of the LOT and the PC Association Fiber System

On first consideration the most plausible way to trigger a seizure in the PC might seem to be stimulation of its primary afferent pathway, the LOT. In fact, several studies have shown that seizures can be triggered by

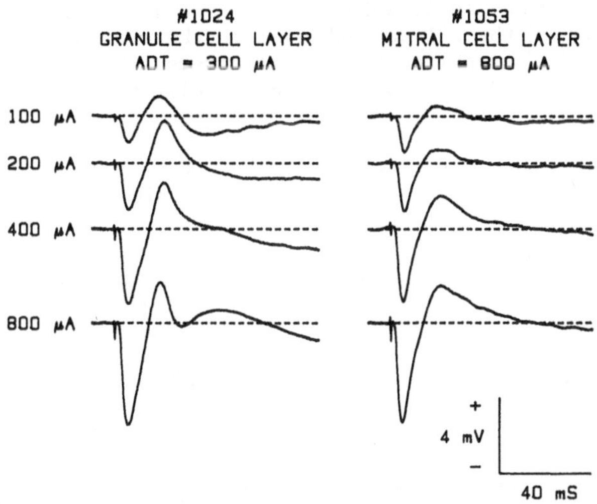


Figure 3. Potentials evoked in layer I of the posterior PC by stimulation in the granule (#1024) or mitral (#1053) cell layer. Both animals have comparable period 1 amplitude but the animal stimulated in the granule cell layer has a lower AD threshold and exhibits a late negative wave at high current intensities (here referred to as the X wave).

*The numbers of animals represented in Table 1, Figure 2, and this correlation differ because not all animals were measured for AD threshold or had a recording site in layer I of the PC. Each analysis included all animals with complete data for the relevant variables.

repeated high-intensity stimulation through electrodes placed in the LOT (3,13). Because the LOT forms a discrete bundle that runs over the ventral surface of the olfactory cortex, it is possible to stimulate it selectively. However, in order to achieve focal stimulation of the LOT, it is necessary not only to place the stimulating electrode near the center of the LOT, but also to keep the current intensity low enough to avoid current spread to the adjacent olfactory cortex. In the case of the PC, current spread beyond the LOT would activate the association fiber system running in layer Ib (immediately adjacent to LOT terminals) and, with higher current intensities, fibers running in layer III as well.

For these reasons we have carefully examined what happens when one tries to trigger an AD by LOT or PC stimulation. The experiments that follow used rats with recording electrodes in the OB and posterior PC, and a stimulating electrode placed in the LOT at the level of the anterior PC. The stimulating electrode had three leads: a 75 μm wire placed in the LOT by electrophysiological guidance, a second 75 μm wire in layer Ib or layer III of the anterior PC, and a 125 μm wire, bare for 3 mm, whose tip was 3 mm dorsal to the LOT and was used as a stimulation reference. This configuration permitted selective activation of the LOT or PC association fibers (see Figure 4). These animals had previously been used in one of two studies of LTP in the PC (21, in preparation). AD threshold was determined by a stairstep procedure using a train of 100 pulses at 100 Hz and current intensities of 40, 60, 80, 100, 120, 140, 160, 200, 300, 400, and 600 μA. Stimulation began at the lowest current intensity and was repeated at the next highest intensity every 3 min until an AD occurred. Animals not exhibiting an AD by 600 μA were assigned an AD threshold score of 800 μA.

Stimulation of the LOT. AD threshold for stimulation via the LOT electrode was determined in 12 animals. As Figure 5 indicates, near-maximal activation of the LOT was achieved in these animals by current intensities of 40-80 μA applied to this electrode. Thus if LOT activation is able to trigger an AD, it should do so at relatively low current inten-

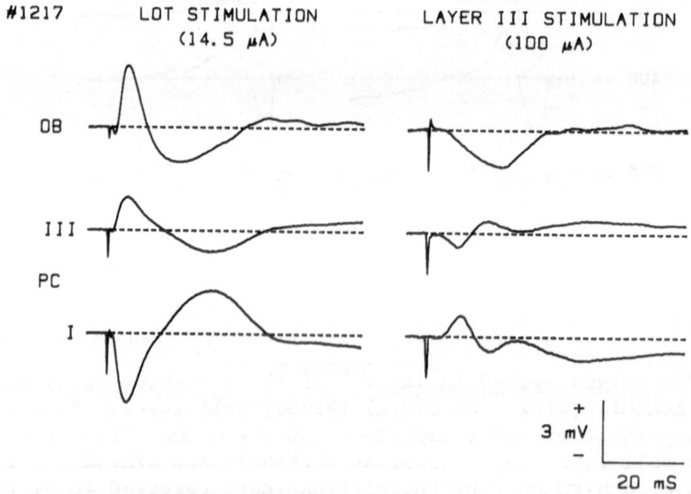


Figure 4. An illustration of potentials evoked in the OB and layer I and III of the posterior PC by stimulation of the LOT or layer III in the anterior PC. The LOT current was chosen to produce half the maximal amplitude of $A_1$. Note that even much stronger stimulation of layer III failed to activate the LOT, as indicated by the absence of $A_1$. The large stimulation artifacts are due to the monopolar stimulation configuration and have been clipped at 2 mV.

Table 2. AD threshold and characteristics of seizures triggered by stimulation of the LOT or adjacent association fibers in layer I or III of the anterior PC. ADs were recorded in the posterior PC and the granule cell layer of the OB. Values are median (range).

| | Stimulation Site | | | | |
|---|---|---|---|---|---|
| | n | LOT | n | Layer I/III | p |
| AD Threshold (μA) | 12 | 200<br>(60-800) | 30 | 120<br>(40-400) | * |
| AD Duration (s) | 8[a] | 2.5<br>(1-14) | 30 | 9.5<br>(1-15) | ** |
| AD Amplitude (mV)[b] | | | | | |
| OB | 7 | 8.0<br>(4.0-10.0) | 24 | 8.3<br>(3.5-11.5) | - |
| PC, layer I | 7 | 2.0<br>(1.0-5.0) | 24 | 5.0<br>(1.0-9.0) | ** |
| PC, layer III | 7 | 1.5<br>(0.7-5.5) | 24 | 6.0<br>(1.0-11.5) | *** |

Mann-Whitney U test (2-tailed): * = $p<0.05$ ** = $p<.01$ *** = $p<.001$

[a]Four of the 12 LOT animals did not have an AD.

[b]Animals that did not have recording electrodes in the granule cell layer of the OB were excluded from this analysis.

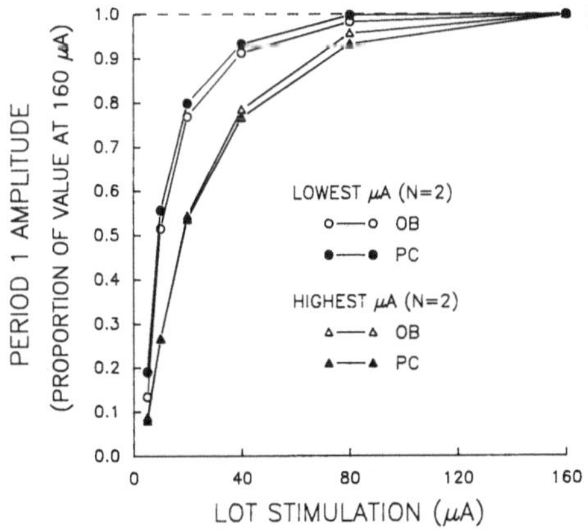


Figure 5. Input-output curves for the amplitude of the initial component of potentials evoked in the OB and posterior PC by LOT stimulation of 5, 10, 20, 40, 80, and 160 μA. Values shown are proportions of the amplitude produced by 160 μA. To illustrate the range of effectiveness in LOT activation among the 12 animals for which AD threshold was determined via stimulation of the LOT electrode, separate curves are plotted for the two animals requiring the lowest and the highest current intensity to evoke a potential one-half the amplitude of that evoked by 160 μA. The potentials produced by 160 μA were substantial, with median values of 6.49 mV in the granule cell layer of the OB and 6.72 mV in layer I of the PC. In comparison, 400 μA stimulation of the granule cell layer of the OB, which is sufficient to trigger an AD in the majority of animals (Figure 2), produces a potential with a median amplitude of 3.05 mV in layer I of the PC (Table 1).

sities. As Table 2 shows, this was not the case. Only 2 of the 12 animals had an AD at 80 μA or less, and 4 of the 12 had no AD at current intensities up to 600 μA. (These 4 animals were subsequently tested for AD threshold using stimulation of layer I or III and all exhibited an AD, with AD thresholds ranging from 80 to 300 μA [median = 180 μA].) The ADs that were produced by stimulation through the LOT electrode were brief (only one lasted longer than 3 s) and small in amplitude in the PC, but not the OB (see Table 2). An example of such an AD is given in Figure 6. Thus, in the vast majority of these animals, maximal activation of the LOT was not sufficient to trigger an AD. To verify that the stairstep procedure used to determine AD threshold had not somehow hindered the expression of seizure activity, 5 of the 8 animals that had an AD were given a single stimulation train the following day using the AD threshold current determined by the stairstep procedure. This triggered a seizure in only one animal, which had a 3 s AD. Thus if anything the data presented above are overly optimistic about the ability of LOT stimulation to trigger an AD.

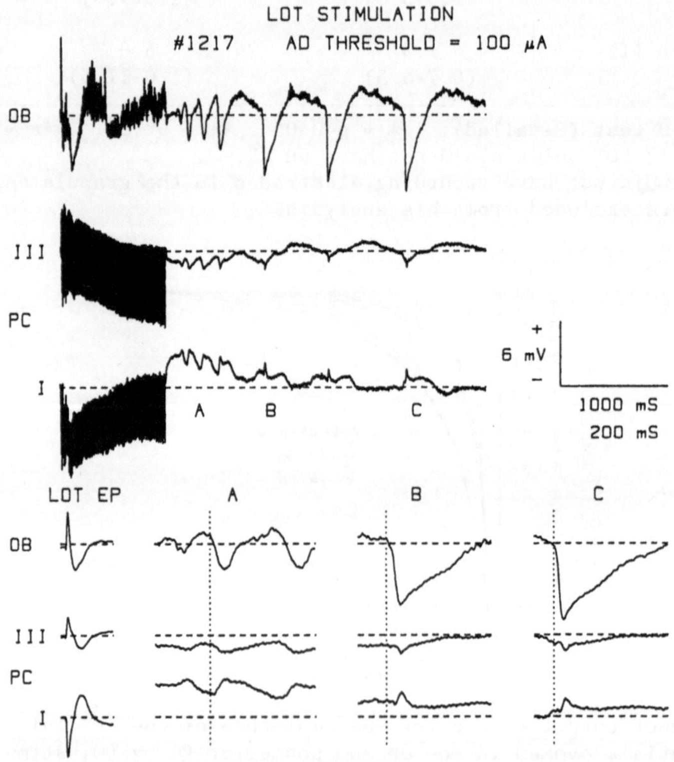


Figure 6. An illustration of an AD triggered by a 100-pulse 100-Hz train delivered through the LOT electrode. The AD was 2.5 s long, was triggered at 100 μA, and was recorded in the granule cell layer of the OB and in layers I and III of the posterior PC. The regions indicated by the letters A-C are plotted on an expanded time scale below, with vertical lines drawn at the approximate onset of a spike in the OB. Potentials evoked by stimulation of the LOT are plotted on the same time scale for comparison (see also Figure 4); the current intensity was set to produce half the maximal amplitude of the LOT evoked potential.

Stimulation of PC association fibers. The AD threshold for stimulation via the electrode in layer I or III of the anterior PC was determined in 30 animals. In comparison to LOT stimulation, stimulation of PC association fibers through these electrodes was far more effective in producing seizures (see Table 2). All 30 animals had an AD. The ADs were triggered at significantly lower current intensities, were significantly longer, and were of significantly higher amplitude in the PC than those triggered by stimulation through the LOT electrode. A representative AD is shown in Figure 7.

The wide range in AD duration and amplitude within each group in Table 2 can be explained in large part by current spread from the site of stimulation. Only one animal receiving LOT stimulation had an AD longer than 3 s. This animal had the highest AD threshold (400 μA) of the 8 that had an AD, and the largest AD amplitude in the PC. It is likely that in this case the stimulation was intense enough to activate substantial extra-LOT elements in the PC and thus trigger an AD in a manner similar to layer I/III stimulation. Conversely, the animal with the briefest AD (and lowest AD amplitude) triggered by layer I/III stimulation had its electrode in layer I only 75 μm from the LOT; the stimulation in this animal may have had effects similar to high-intensity stimulation through an LOT electrode.

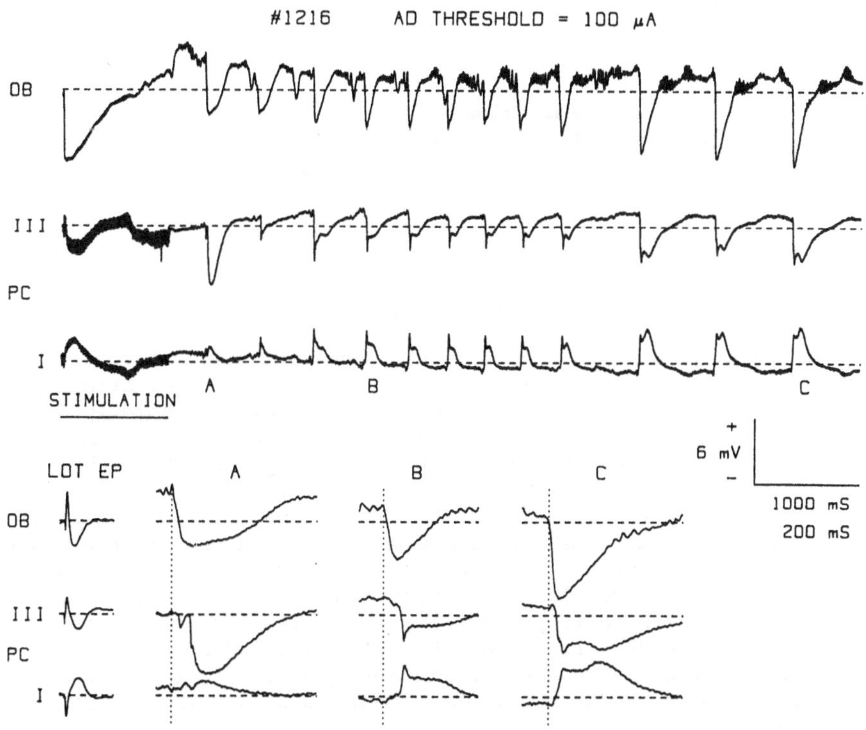


Figure 7. An illustration of an AD triggered by stimulation of layer III in the anterior PC. The AD was 6 s long and was triggered at 100 μA. It is plotted with the same format and time scale used in Figure 6 to facilitate comparison.

Figure 8 shows the AD threshold in animals receiving layer I/III stimulation in relationship to the ability of the stimulating electrode to activate the LOT. No significant relationship exists. The majority of animals exhibited an AD at a current intensity too weak to activate the LOT at all, much less maximally. These data, in combination with the data from animals receiving LOT stimulation, indicate that activation of the LOT is neither necessary nor sufficient to trigger an AD in the OB and PC. Rather, the critical element appears to be activation of the PC association fiber system.

## SEIZURE PATTERNS IN THE OLFACTORY CORTEX

Having examined the circumstances necessary to trigger a seizure in the olfactory cortex, we now briefly turn to a description of the seizure

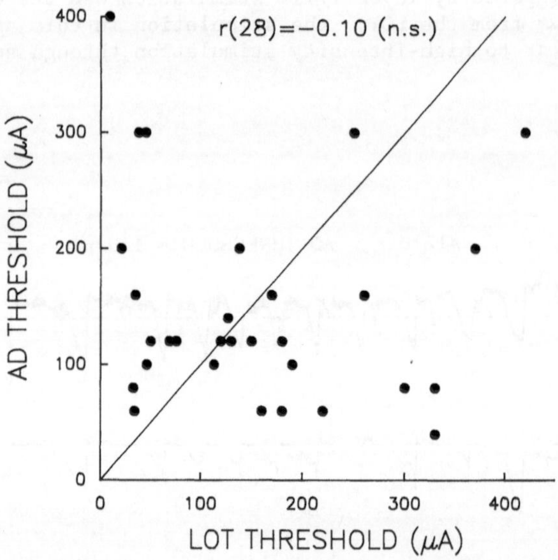


Figure 8. The AD threshold for 30 animals receiving stimulation of layer I or III in the anterior PC is unrelated to the ability of stimulation through that electrode to activate the LOT. LOT threshold is defined as the minimum current intensity required to produce an $A_1$ amplitude of 1 mV in layer I of the posterior PC. Data points above the diagonal line represent animals that had an AD threshold higher than the current required to activate the LOT (n=14), while data points below the line represent animals that had an AD at a current that did not activate the LOT (n=15). (The one remaining data point lies on the line.) The electrode placements varied throughout layers I and III. Because LOT threshold is primarily a predictor of distance of the stimulating electrode from the LOT, those electrodes with low LOT thresholds are close to the LOT and consequently are most likely to activate it when the AD threshold current is used. The 9 leftmost data points represent animals with electrode placements in layer I. These observations are consistent with data from previous studies (19,24) in which AD threshold was determined by stimulation of the posterior PC, where the LOT is no longer a discrete bundle; there we also found no significant relationship between the depth of the stimulating electrode within the PC and AD threshold, $r(68) = -0.17$, $p>0.15$.

pattern created by this stimulation. Figures 6 and 7 illustrate features that were common to ADs produced in these experiments by either LOT or PC stimulation. In all these seizures, which were focal ADs triggered for the first time in each animal, the individual spikes in the field potential were predominantly monophasic and relatively simple in form. Although they varied considerably in amplitude and duration, the major component of each spike was always the polarity of period 2 in the potential evoked at the same site by LOT stimulation. Thus the spikes do not themselves reflect activation of the PC via the LOT (which is represented by the $A_1$ component of period 1).

The spikes were clearly a network phenomenon that never occurred in isolation at one recording site, although during the brief seizures triggered through the LOT electrode the OB spikes were much larger than those in the PC. The most typical pattern was for each spike to occur slightly earlier in the OB than in the PC (e.g., see Figure 7), but in some cases the PC spike occurred first (not shown). This variability in direction of AD propagation occurred not only across animals but also for two different spikes within the same AD. Racine et al. (14) reported a similar variability for interictal spikes in the PC.

## DISCUSSION

The results presented here point clearly to the association fiber system, rather than the LOT, as the critical element that must be activated to produce a seizure in the OB and PC. Stimulation confined to the LOT was ineffective in triggering an AD, and even supramaximal stimulation through the LOT electrode, which would also activate association fibers near the LOT, triggered only a very brief AD that was poorly expressed in the PC. In contrast, direct stimulation of association fibers in layer I or III was consistently effective in triggering a robust AD, regardless of whether it activated the LOT or not. If activation of the LOT alone has any ability to trigger seizure activity, such activity is most likely to occur in the OB rather than the PC, based upon the relative AD amplitudes at these sites (Table 2). Although indirect, the evidence regarding seizures triggered by OB stimulation also points to a critical role for the cortical association fiber system rather than the LOT.

Interpretation of what happens once a seizure is triggered is more problematic. The field potentials during an AD have the polarity of period 2 rather than period 1 in the evoked potential. This observation, among others, led Freeman (3) to propose that the AD represents synchronous inhibitory processes, rather than excitation. In fact, his model requires the failure of transmission in the LOT before a seizure can occur. Whether the AD in the PC represents inhibitory or excitatory events cannot be determined with certainty from the field potential alone; its polarity could result either from a hyperpolarizing conductance in layer I or a depolarizing conductance in layer III, such as might be provided by input from deep pyramidal cells and the endopiriform nucleus (see 7,25). Furthermore, membrane conductances that are widely distributed in the cell membrane may not be clearly evident in the field potential. Interpretation of the AD recorded in the OB also poses problems. Anatomical considerations indicate that the negative wave in the AD at this site is most probably due to an EPSP in granule cells in the granule cell layer (11,16), which in turn is most likely to originate from activity in cortical association fibers. These considerations suggest that the AD represents activation of inhibitory elements in the OB by the cortical association fiber system. However, mitral cells do not contribute substantially to the field potential (16), and consequently their activity level during seizure activity cannot be reliably inferred from the AD. As these comments indicate, additional research is needed to clarify the sequence of events that underlies the AD at each site.

The olfactory cortex provides us with an excellent model for the study of cortical function in a relatively uncomplicated system. The patterns of activity it exhibits are governed by the interplay of influences transmitted over the LOT and association fibers, and cannot be clearly understood without considering the entire olfactory cortex and OB as a unified network. Seizure activity can be regarded as one manifestation of the powerful influence of these interconnections. The OB and PC are perhaps the most seizure-prone and rapidly kindling structures in the brain. We feel that this is due in part to the elaborate interconnections provided by the cortical association fiber system. This system is undoubtedly involved in information processing as well, and evidence indicates that it is the critical element that must be activated to produce the LTP that we have seen in the OB and PC (21). This form of LTP causes a selective increase in the amplitude and duration of period 2 at each site (22,23). As noted above, AD spikes are also the polarity of period 2, although much larger and more prolonged. Whether seizure activity and LTP are separate manifestations of the same mechanisms in these structures is a question we cannot answer at present, but the association fiber system seems to be the key to understanding both.

ACKNOWLEDGEMENTS

This research was supported by NSF Grant BNS 85-19700 and by a grant from the Marie Wilson Howells Fund.

REFERENCES

1. Cain, D. P. Seizure development following repeated electrical stimulation of central olfactory structures. Ann. N. Y. Acad. Sci. 290: 200-216, 1977.
2. Freeman, W. J. Relations between unit activity and evoked potentials in prepyriform cortex in cats. J. Neurophysiol. 31: 337-348, 1968.
3. Freeman, W. J. Petit mal seizure spikes in olfactory bulb and cortex caused by runaway inhibition after exhaustion of excitation. Brain Res. Rev. 11: 259-284, 1986.
4. Gray, C. M., Freeman, W. J., and Skinner, J. E. Induction and maintenance of epileptiform activity in the rabbit olfactory bulb depends on centrifugal input. Exp. Brain Res. 68: 210-212, 1987.
5. Haberly, L. B. Neuronal circuitry in olfactory cortex: Anatomy and functional implications. Chem. Senses 10: 219-238, 1985.
6. Haberly, L. B., and Bower, J. M. Analysis of association fiber system in piriform cortex with intracellular recording and staining techniques. J. Neurophysiol. 51: 90-112, 1984.
7. Hoffman, W. H., and Haberly, L. B. Bursting induces persistent all-or-none EPSPs by an NMDA-dependent process in piriform cortex. J. Neurosci. 9: 206-215, 1989.
8. Kairiss, E. W., Racine, R. J., and Smith, G. K. The development of the interictal spike during kindling in the rat. Brain Res. 322: 101-110, 1984.
9. Luskin, M. B., and Price, J. L. The topographic organization of associational fibers of the olfactory system in the rat, including centrifugal fibers to the olfactory bulb. J. Comp. Neurol. 216: 264-291, 1983.
10. McIntyre, D. C., and Wong, R. K. S. Cellular and synaptic properties of amygdala-kindled pyriform cortex in vitro. J. Neurophysiol. 55: 1295-1307, 1986.

11. Mori, K. Membrane and synaptic properties of identified neurons in the olfactory bulb. Prog. Neurobiol. 29: 275-320, 1987.
12. Price, J. L. Beyond the primary olfactory cortex: Olfactory-related areas in the neocortex, thalamus, and hypothalamus. Chem. Senses 10: 239-258, 1985.
13. Racine, R. J., Milgram, N. W., and Hafner, S. Long-term potentiation phenomena in the rat limbic forebrain. Brain Res. 260: 217-231, 1983.
14. Racine, R. J., Mosher, M., and Kairiss, E. W. The role of the pyriform cortex in the generation of interictal spikes in the kindled preparation. Brain Res. 454: 251-263, 1988.
15. Racine, R. J., Paxinos, G., Mosher, J. M., and Kairiss, E. W. The effects of various lesions and knife-cuts on septal and amygdala kindling in the rat. Brain Res. 454: 264-274, 1988.
16. Rall, W., and Shepherd, G. M. Theoretical reconstruction of field potentials and dendrodendritic synaptic interactions in olfactory bulb. J. Neurophysiol. 31: 884-915, 1968.
17. Rodriguez, R., and Haberly, L. B. Analysis of synaptic events in the opossum piriform cortex with improved current source-density techniques. J. Neurophysiol. 61: 702-718, 1989.
18. Russell, R. D., and Stripling, J. S. Effect of olfactory bulb kindling on evoked potentials in the pyriform cortex. Brain Res. 361: 61-69, 1985.
19. Stripling, J. S., Gramlich, C. A., and Cunningham, M. G. Effect of cocaine and lidocaine on the development of kindled seizures. Pharmacol. Biochem. Behav. 32: 463-468, 1989.
20. Stripling, J. S., and Patneau, D. K. Selective long-term potentiation in the pyriform cortex: Role of the lateral olfactory tract. Soc. Neurosci. Abstr. 12: 508, 1986.
21. Stripling, J. S., and Patneau, D. K. Role of association fibers in selective long-term potentiation in the pyriform cortex. Soc. Neurosci. Abstr. 14: 1189, 1988.
22. Stripling, J. S., Patneau, D. K., and Gramlich, C. A. Selective long-term potentiation in the pyriform cortex. Brain Res. 441: 281-291, 1988.
23. Stripling, J. S., Patneau, D. K., and Gramlich, C. A. Selective long-term potentiation in the piriform cortex: Characterization and anatomical distribution. In preparation.
24. Stripling, J. S., and Russell, R. D. Twenty-four-hour post-seizure inhibition during limbic kindling requires seizure generalization. Neurosci. Let. 99: 208-213, 1989.
25. Tseng, G.-F., and Haberly, L. B. Paired shocks induce a very high amplitude facilitation of late EPSPs in deep neurons in piriform cortex. Soc. Neurosci. Abstr. 13: 156, 1987.
26. Tseng, G.-F., and Haberly, L. B. Characterization of synaptically mediated fast and slow inhibitory processes in piriform cortex in an in vitro slice preparation. J. Neurophysiol. 59: 1352-1376, 1988.

## Discussion of Dr. Stripling's Presentation

DR. RACINE: Just one comment. I think that the data are very nice. I really like the story you're building. I just have one suggestion, though. In your talk you made a number of comparisons between the responses evoked by single pulses and the paroxysmal event recorded during the epileptiform AD. It might be informative to make the comparison with train-evoked responses rather than pulse evoked responses.

DR. STRIPLING: Thank you. If you recall the animal had an after-discharge triggered by LOT stimulation - there was a brief 2.5 second after-discharge. On the slide, this is the train that did it, 100 microamps right here, and this is the single discharge that I said you can see in the train. Here's the LOT stimulation. You'll notice, first of all, that you see this at sub-threshold stimulation currents; 40 doesn't do it, 60 does, 80 does, 100 you do it and you get a seizure, brief but there. So there is something going on throughout the olfactory tract stimulation at relatively lower currents. Here, for example is this spike, in the OB, there it is in the PC. But in terms of talking about what's going on during the trains, that is what you're talking about, right? Let's expand these trains and take a closer look and see what's happening, if I may. This is the first 660 milliseconds of the same 4 trains in the olfactory bulb. You can see the spikes here, and what's going on is you see period 1 in the olfactory bulb which represents an EPSP between the mitral cells and the granule cells in the external plexiform layer. Here's the first stimulation, the second, the third, after about 5 or 6 stimulations it starts to alternate. Every other response is totally inhibited. And the reason for that is fairly clear. The antidromic activation of the mitral cells triggers recurrent inhibition in them that then blocks further antidromic activation, and so you only get response every other time. I might add that the sampling right here is too low to show the stimulation artifact very clearly. But that's what's going on. Now, what you're saying though, let's look at the lateral olfactory tract evoked potentials in the pyriform cortex, and that's on the next slide. Here's the same animal, LOT stimulation, there's no alternation whatsoever. For every stimulus pulse we get an evoked response. And this is period 1, and since it is only 10 milliseconds apart the period 2 is only visible by the envelope of activity, but you get an evoked response each time. What you'll notice is the A1 component is dying out; it does it the first time, it did it at 40 microamps; it dies out but you don't get a seizure. It dies out again at 60, at 80 and at 100, it's almost completely fatigued. Long trains really wipe it out. In fact, Walter Freeman in his model proposes that you have to have fatigue at the LOT synapse, between the LOT and the pyriform cortex pyramidal cells in order to get a seizure. He proposes that it turns the association fiber system loose. And by the way, if you look at the circuitry of the olfactory cortex, they're relatively few inhibitory interneurons but there's massive convergence of the association fiber system upon them. And so they get activated fairly strongly. I don't know whether Freeman is right about that, I have thoughts both ways, but as far as what happens to the potential, it simply dies out, the LOT activation. Does that address your question?

DR. MCINTYRE: Jeff, what are the transmitters of the association system?

DR. STRIPLING: The leading candidate is glutamate, and we can block our long-term potentiation with ketamine. We haven't looked at the effect of ketamine on seizures. But that certainly fits with everything we know about long-term potentiation and glutamate.

# DOUBLE-LABEL QUANTITATIVE AUTORADIOGRAPHIC STUDIES OF ANAEROBIC GLUCOSE METABOLISM IN LIMBIC SEIZURES

Robert F. Ackermann[1] and James L. Lear[2]

[1]Division of Nuclear Medicine and Biophysics
Department of Radiological Sciences; Department of Neurology
Laboratory of Biomedical and Environmental Sciences
and Brain Research Institute
UCLA School of Medicine
Los Angeles, CA 90024

[2]Division of Nuclear Medicine, Department of Radiology
University of Colorado Center for Health Sciences
4200 E. 9th Ave., Denver, CO 80262

## RATIONALE

Sokoloff's (1) radiolabeled 2-deoxyglucose (2DG) method for calculating the local cerebral metabolic rate for glucose (LCMR) has been employed to determine those structures most involved in a variety of experimental and clinical paradigms. It has proven particularly valuable in identifying the structures most involved in a number of human and animal seizure states (2).

With seizures induced by kainic acid (KA), increased values for the "lumped constant" (LC), a conversion factor that appears in the demominator of Sokoloff's operational equation (1), can account for apparent increases in 2DG utilization as great as 25%; however, 2DG utilization has been reported to increase by as much as 500% in the hippocampus of KA-treated rats (3), much too large an increase to be accounted for by so small an increase in the LC. Here we suggest that these large increases in 2DG utilization result from the induction of "aerobic glycolysis" in seizure-involved structures (4).

In its normal, "unstimulated" state, the brain derives almost all its energy from mitochondrial oxidation of glucose via Krebs' tricarboxylic acid cycle (5). Consequently, lactate, the end-product of glycolysis (the Emden-Meyerhoff pathway), is produced by the normal "resting" brain at a low rate relative to other tissues, such as muscle (5). However, brain lactate production increases sharply whenever glycolyis is induced, such as during ischemia (6), or during seizures (4). Importantly, recent reports have suggested that such extremely intense "stimuli" are not required; glycolysis can be induced by acute sensory stimulation, under conditions of unimpeded mitochondrial oxidation (7, 8).

## METHODOLOGY

For these experiments we used the fluorinated form of 2DG, [F-18]-fluorodeoxyglucose (FDG), in order to obtain two sets of autoradiograms from each

*Kindling 4*
Edited by J. A. Wada
Plenum Press, New York, 1990

brain, one from FDG and a second from [C-14]-6-glucose (GLC). Our method and kinetic model described below are based on the following notions concerning the effect of stimulation on glucose metabolism.

In a manner similar to 2DG (9), FDG radiolabel accumulates as FDG-6-phosphate, before the point of divergence between oxidative and glycolytic glucose metabolism. Thus, FDG-derived LCMR necessarily represents total glucose utilization, oxidative plus glycolytic (Fig. 1). On the other hand, because GLC is metabolized like normal non-radioactive glucose, its mode of label accumulation depends on a cell's glucose metabolic state. Under normal conditions, when glucose metabolism is predominantly oxidative, GLC-derived label accumulation occurs primarily in Krebs-cycle-related metabolites, particularly the large glutamate pool (10). Thus, to the extent that glucose metabolism is predominantly oxidative, GLC-derived label is effectively trapped, so long as the experiment remains brief (5-10 min) (9).

However, under any condition that stimulates glycolysis, a substantial fraction of GLC-derived label is diverted to lactate, which is then transported out of the brain (Fig. 1) at approximately half the rate for glucose transport (12, 13). Therefore, after subtraction of a factor representing the labeled lactate that remains in the brain at the end of the experiment, the GLC-derived LCMR represents the rate of oxidative glucose metabolism. Since the FDG-derived glucose metabolic rate represents oxidative and glycolytic glucose metabolism combined, the difference between the FDG-derived LCMR and the GLC-derived LCMR (corrected for retained lactate) represents the rate of glycolysis.

We have tested these notions about glucose metabolism with a double-label autoradiographic method that employs sequentially administered FDG and GLC. Two sets of autoradiograms are produced from each brain: one representing predominantly F-18 accumulation; and the other representing C accumulation.

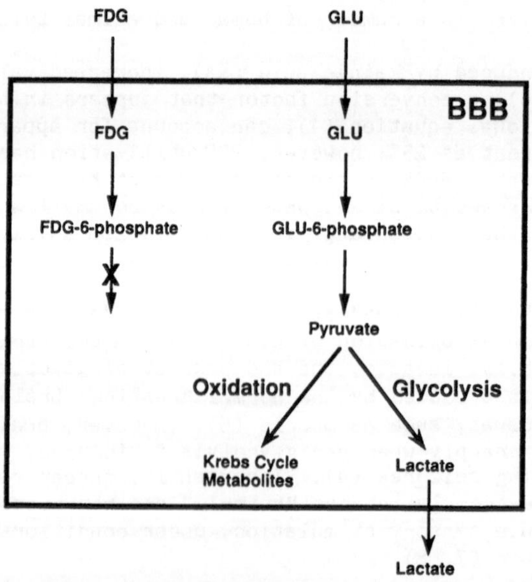


Figure 1. Schematic depicting differential accumulation of FDG and GLC depending on whether glucose metabolism is predominantly oxidative or predominantly glycolytic. BBB = blood-brain barrier. (From ref. [11]; used with permission)

Each autoradiogram is digitized with a solid-state image scanner. Combining the autoradiographic data with blood plasma tracer time-activity functions and plasma glucose levels, a computer then applies a kinetic model we have developed (see Appendix) to produce two images of LCMR from each brain section: one derived from the FDG data, and the second from the GLC data (Figs. 2-6).

Our kinetic model is derived from two previously published methods for LCMR determination, from either 2DG (1) or GLC (14), with modifications for simultaneous determination of GLC-based and FDG-based LCMR (9). Our model presumes that both methods are accurate within the bounds detailed by their authors, and that glucose metabolism remains stable over the 36 min duration of the experiments.

## APPLICATION AND RESULTS

We have applied our autoradiographic method and kinetic model to normal "resting" rat brains and to brains stimulated in various ways.

In normal rats, the glycolytic component of brain glucose metabolism is small compared to the oxidative component. Consequently, the FDG- and GLC-derived estimates of baseline LCMR are virtually identical (7, 10) (Fig. 2). However, any condition that stimulates glycolysis will, according to the scheme in Fig. 1, cause the FDG-derived LCMR to increase sharply, out of proportion to the corresponding GLC-derived increase in LCMR (Figs. 3-5).

In one set of experiments, we stimulated the brain by exposing the rats to 16 Hz light flashes during the FDG and GLC infusions, and found that FDG- and GLC-derived estimates of LCMR corresponded in all structures except the optic tectum, which receives direct retinal projections. There, FDG accumulation was markedly greater than GLC accumulation, the data translating to a significant glycolytic component when applied to our kinetic model (Fig. 3).

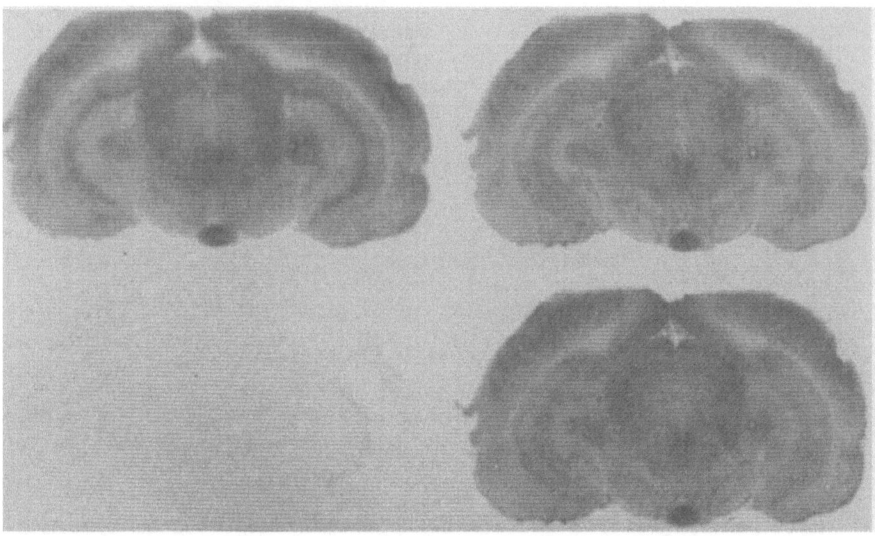

Figure 2. Multiple images from the same brain section of a control animal. Upper left panel: FDG-derived LCMR; Upper right panel: GLC-derived LCMR; Lower left panel: computed glycolytic LCMR component; Lower right panel: computed oxidative LCMR component. Lower left image is virtually invisible, indicating that FDG- and GLC-based LCMR rates are nearly identical. (From ref. [11]; used with permission).

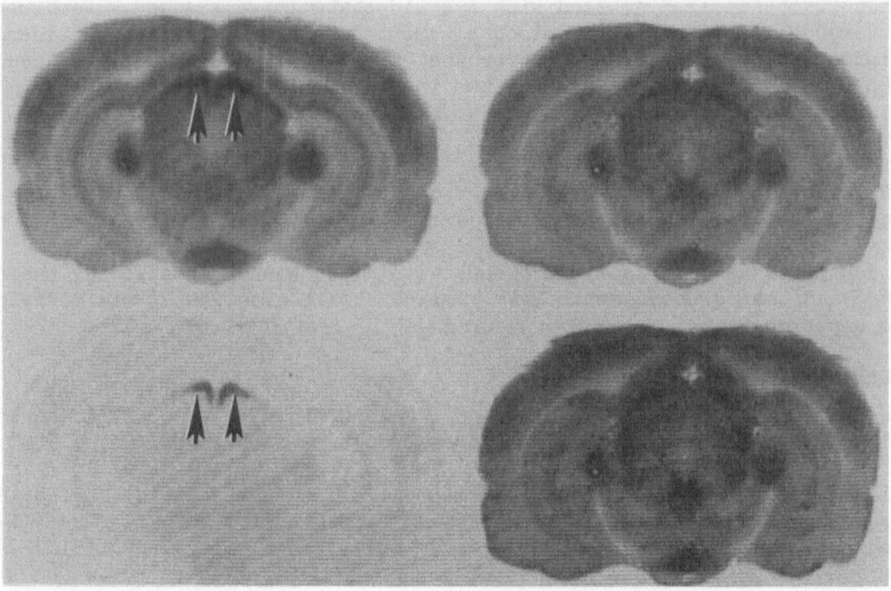

Figure 3. Multiple images from brain section of rat exposed to 16 Hz light flashes with both eyes open; presentation scheme the same as Fig. 2. FDG-derived LCMR (upper left) is markedly greater than GLC-derived LCMR (upper right) in the optic tectum (arrows), resulting in computed tectal glycolysis (arrows, lower left). Computed tectal oxidation (lower right) remains relatively unaffected. (From ref. [11]; used with permission).

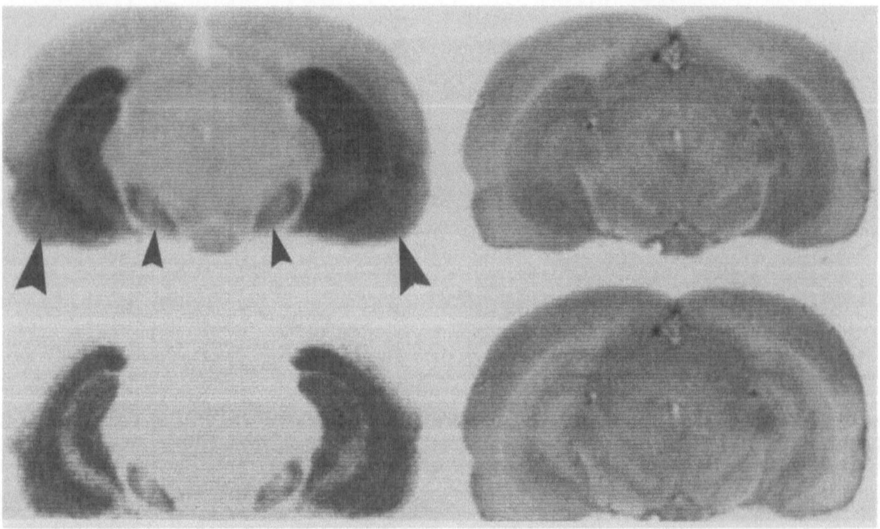

Figure 4. Multiple images from section of an animal subjected to systemic KA (7 mg/kgm, i.p.) 1 hr before FDG infusion; presentation scheme same as Fig. 2. The KA activated hippocampus, entorhinal cortex (large arrows) and substantia nigra (small arrows) (upper left panel). These structures have stimulation-induced glycolytic LCMR component; structures with intrinsically high LCMR (e.g., mamillary nuclei) do not (lower left). (From ref. [11]; used with permission).

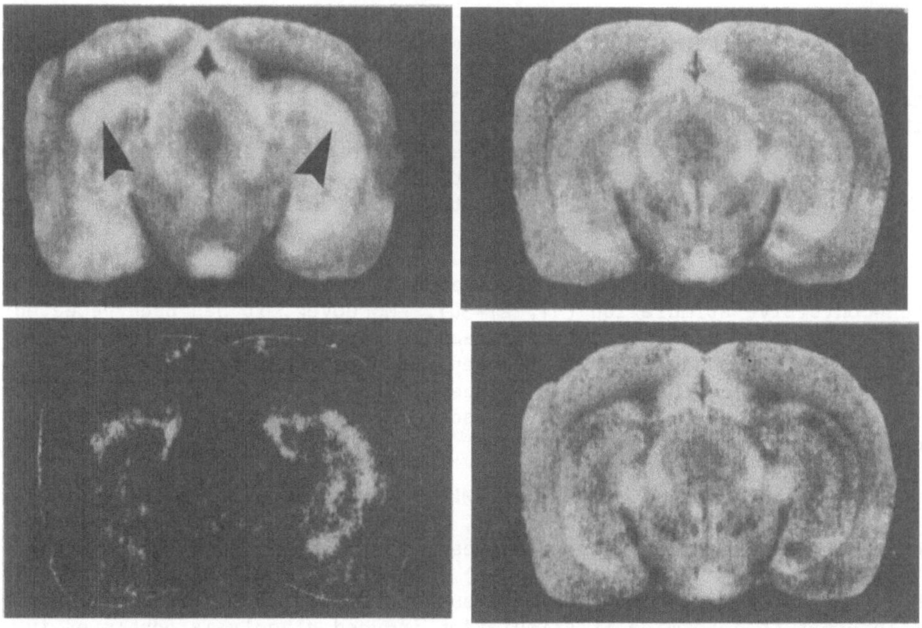

Figure 5. Multiple images from the same brain section of an animal subjected to two entorhinal kindled seizures; presentation scheme same as Fig. 2. In this brain, there was enhanced FDG accumulation in both the ipsilateral and contralateral hippocampi (arrows, upper left panel), resulting in a glycolytic component to hippocampal glucose metabolism (lower left panel).

In a second set of experiments, we stimulated the brain with systemic KA about one hr before the FDG/GLC infusions, and found that FDG- and GLC-derived estimates of LCMR corresponded in all structures except the hippocampus and several other limbic structures, which are activated by the KA. In the affected limbic structures, FDG accumulation was markedly greater than GLC accumulation, the data once again translating to a significant glycolytic component when applied to our kinetic model (Fig. 4).

We have recently begun to apply this double-label methodology to kindled seizures. Rats with entorhinal electrodes are first kindled to Stage 5 criterion (15). They are then given two additional stimulations during the double-label experiment, one immediately after the FDG infusion, the second immediately after the GLC infusion. Preliminary results are qualitatively similar to those for the KA experiments described above. Glycolysis appears to be induced in the hippocampus (Fig. 5), in this case presumably by propagation of seizure activity from the entorhinal cortex via the perforant path.

## IMPLICATIONS

Other workers have noted discrepancies between 2DG and GLC accumulation in stimulated brain similar to those described here (7, 9). Taken together, their and our results suggest that the most probable explanation for the markedly increased stimulation-induced 2DG accumulation is that glycolysis is disinhibited in stimulated structures, and this occurs despite the presence of adequate tissue oxygen (4, 16) ("aerobic glycolysis" [17-19]). Glycolysis consumes approximately 15 times more glucose for a given rate of ATP production than does oxidation (5).

At the cellular level, "stimulation" is movement of $Na^+$, $K^+$, and other ions across the cell membrane. The activity of $Na^+$-$K^+$ ATPase has been shown in muscle cells and in tumor cells to be linked to aerobic glycolysis via the ATP breakdown products, ADP and $P_i$ (18, 19). Brain 2DG accumulation is sensitive to oubain, indicating that it also is linked to $Na^+$-$K^+$ ATPase activity (20, 21). Excited neurons release $K^+$ into the extracellular space (22, 23), and extracellular $K^+$ is known to accelerate both glycolysis and oxidation in brain tissue (24, 25), and to induce increased 2DG accumulation (21, 26).

It has been established that brain lactate levels rise under both convulsive and non-convulsive stimulation (17, 27-30). Moreover, it has been repeatedly suggested that elevated intracellular lactate levels promote cell damage and death (31, 32). For example, it has been shown recently that N-methyl-D-aspartate (NMDA) receptor blockade reduces ischemic cell damage (33) and also reduces lactate production induced by ischemia, electroconvulsive seizures, or local NMDA administration (34).

It has also been demonstrated that NMDA receptor activation is an important factor in the kindling process (35, 36), and while there is no direct evidence that kindled seizures produce cell loss, it has been reported that they do induce axonal sprouting (37, 38). Sprouting implies antecedent cell death (39), and it may promote seizure susceptibility (40). Thus, it would be of interest to determine if stimulation-induced glycolysis continues in seizure-mediating structures post-ictally, and if so, for how long.

ACKNOWLEDGEMENTS

This work was supported by Department of Energy contract DE-AC03-76-SF00012; National Institutes of Health grants NS-15654, NS 26657, and MH-037916; and by a Scholar's Grant from the Radiological Society of North America to Dr. Lear. The authors gratefully acknowledge the efforts of Mr. Wrother Meredith and Mrs. Charna Nissenson, whose technical expertise made collection of these data possible. The authors also gratefully acknowledge the support and encouragement of Dr. Michael E. Phelps; and the technical support of Dr. Jorge Barrio and the Chemistry Section, and Dr. N. Satymurthy and the Cyclotron Section, of the UCLA Division of Nuclear Medicine and Biophysics, who provided the FDG for these experiments.

APPENDIX

FDG-based LCMR was measured using the 2DG method of Sokoloff et al. (1), with small but significant modifications to account for differences in FDG vs. 2DG kinetics. Values used for $k_1$ and $k_2$ of FDG were obtained by multiplying published 2DG values (1) by approximately 1.4, to account for the more rapid transport of FDG compared to 2DG (9). Similarly, the values of $k_3$ and LC for FDG were set to approximately 1.3 times the published values for 2DG (9).

THUS,

$$dC_e/dt = (k_1 \cdot C_p) - [(k_2 \cdot C_e) + (k_3 \cdot C_e)] \qquad (1)$$

$$dC_m/dt = (k_3 \cdot C_e) - (k_4 \cdot C_m) \qquad (2)$$

$$C_t = C_e + C_m \qquad (3)$$

WHERE:

$C_p$ = concentration of FDG or GLC in blood plasma
$C_e$ = concentration of FDG or GLC in brain
$C_m$ = concentration of FDG or GLC metabolites in brain
$C_t$ = total concentration of FDG- or GLC-derived label in brain
$k_1$ = rate constant of blood-to-brain transport of FDG or GLC
$k_2$ = rate constant of brain-to-blood transport of FDG or GLC
$k_3$ = rate constant of phosphorylation of FDG or GLC
$k_4$ = rate constant of loss of FDG or GLC metabolites

HENCE,

$$\text{LCMR(FDG)} = \frac{C_t - k_1 \cdot e^{-(k_2+k_3)T} \int_0^T C_p\, e^{-(k_2+k_3)t}\, dt}{LC \left[ \int_0^T (C_p/P)\, dt - e^{-(k_2+k_3)T} \int_0^T (C_p/P) \cdot e^{(k_2+k_3)t}\, dt \right]} \qquad (4)$$

WHERE,

$k_1$(FDG) (gray matter) = 0.3
$k_2$(FDG) (gray matter) = 0.4
$(k_2 + k_3)$(FDG)(gray matter) = 0.5
LC (FDG) = 0.6

GLC-based LCMR was determined by the method of Hawkins et al. (14). Equations 1-3 were solved numerically, with values for $k_1$, $k_2$, and $k_4$ fixed (see below), and the value for $k_3$ varied from 0.0 to 0.3. This generated a table of values for $C_m$ and $C_e$ as a function of the values for $k_3$. These values for $C_m$ and $C_e$ were then entered in equation (5) below to generate a look-up table of LCMR vs. $C_t$.

THUS,

$$\text{LCMR(GLC)} = C_m / \int_0^T (C_e/P) \qquad (5)$$

WHERE,

$k_1$(GLC) (gray matter) = 0.2
$k_2$(GLC) (gray matter) = 0.3
$k_4$(GLC) (gray matter) = 0.1 $k_3$

Each pair of FDG- and GLC-based LCMR values, derived from equations (4) and (5), was then applied to the following model which we developed, to obtain estimates of both the oxidative and glycolytic components of glucose metabolism.

FOR ANY GIVEN METABOLIC RATE, IF:

$GM_t$ = total glucose metabolism (oxidative + glycolytic)
$GM_o$ = oxidative glucose metabolism

$GM_g$ = glycolytic glucose metabolism
$F_{rl}$ = retained lactate fraction
$C_l$ = correction factor for retained $^{14}C$-lactate
LCMR(FDG) = LCMR estimate derived from FDG label accumulation
LCMR(GLC) = LCMR estimate derived from GLC label accumulation

THEN:

$$GM_t = LCMR(FDG) \quad (6)$$

$$GM_o = LCMR(GLC) - C_l \quad (7)$$

$$C_l = F_{rl} \cdot GM_g \quad (8)$$

$$\begin{aligned} GM_g &= GM_t - GM_o \\ &= LCMR(FDG) - [LCMR(GLC) - C_l)] \\ &= LCMR(FDG) - [LCMR(GLC) - (F_{rl} \cdot GM_g)] \\ &= LCMR(FDG) - LCMR(GLC) + (F_{rl} \cdot GM_g) \\ &= [LCMR(FDG) - LCMR(GLC)] / (1 - F_{rl}) \end{aligned} \quad (9)$$

$$GM_o = LCMR(FDG) - GM_g \quad (10)$$

A lookup table of $F_{rl}$ vs. $GM_g$ was created by modifying the GLC model's rate constant for label loss ($k_4$) to include loss due to lactate efflux, which was assumed to occur with a rate constant equal to 0.5 times the rate constant of glucose transport ($k_1$) (12, 13). Equations (9) and (10) were solved simultaneously with a computer to generate lookup tables relating F-18 and C-14 tissue tracer concentrations to values for the glycolytic and oxidative components of total cerebral glucose metabolism. The computer also was used to produce digitized images of FDG and GLC tissue concentration. These tracer-concentration images were then converted by the computer to images of total, oxidative, and glycolytic glucose metabolic rates (Figs. 2-5).

## REFERENCES


1. L. Sokoloff, M. Reivich, C. Kennedy, M. H. Des Rosiers, C. S. Patlak, K. D. Pettigrew, O. Sakurada, and M. Shinohara, The [$^{14}$C]deoxyglucose method for the measurement of local cerebral glucose utilization: Theory, procedure, and normal values in the conscious and anesthetized albino rat, J. Neurochem. 28:897 (1977).
2. R. F. Ackermann, J. Engel Jr., and M. E. Phelps, Identification of seizure-mediating brain structures with the deoxyglucose method: Studies of human epilepsy with positron emission tomography, and animal seizure models with contact autoradiography, in: "Advances in Neurology, vol. 44," A. V. Delgado-Escueta, A A. Ward Jr., D. M. Woodbury, and R. J. Porter, eds., Raven Press, New York (1986).
3. M. C. Evans and B. S. Meldrum, Regional brain glucose metabolism in chemically-induced seizures in the rat, Brain Res. 297:235 (1984).
4. F. Plum and T. E. Duffy, The couple between cerebral metabolism and blood flow during seizures, in: "Brain Work, Alfred Benzon Symposium VIII," D. H. Ingvar and N.A. Lassen, eds., Munksgaard, Copenhagen (1975).

5. M. Erecińska and I. A. Silver, ATP and brain function, J. Cereb. Blood Flow Metab. 9:2 (1989).
6. S. Rehncrona, H. N. Hauge, and B. K. Siesjö, Enhancement of iron-catalyzed free radical formation by acidosis in brain homogenates: Difference in effect by lactic acid and $CO_2$, J. Cereb. Blood Flow Metab. 9:65 (1989).
7. R. C. Collins, D. W. McCandless, and I. L. Wagman, Cerebral glucose utilization: Comparison of [$^{14}$C]deoxyglucose and [6-$^{14}$C]glucose quantitative autoradiography, J. Neurochem. 49:1564 (1987).
8. P. T. Fox, M. E. Raichle, M A. Mintun, and C. Dence, Nonoxidative glucose consumption during focal physiologic neural activity, Science 241:462 (1988).
9. J. L. Lear and R. F. Ackermann, Comparison of cerebral glucose metabolic rates measured with fluorodeoxyglucose labeled in the 1, 2, 3-4, and 6 positions using double label quantitative digital autoradiography, J. Cereb. Blood Flow Metab. 8:575 (1988).
10. C. J. Van den Berg and R. Bruntink, Glucose oxidation in the brain during seizures: Experiments with labeled glucose and deoxyglucose, in: "Glutamine, Glutamate, and GABA in the Central Nervous System," L. Hertz, E. Kvamme, E. G. McGeer, and A. Schousboe, eds., Alan R. Liss, New York (1983).
11. R. F. Ackermann and J. L. Lear, Glycolysis-induced discordance between glucose metabolic rates measured with radiolabeled fluorodeoxyglucose and glucose, J. Cereb. Blood Flow Metab. 9:774 (1989).
12. J. E. Cremer, V. J. Cunningham, W. M. Pardridge, L. D. Braun, and W. H. Oldendorf, Kinetics of blood-brain barrier transport of pyruvate, lactate and glucose in suckling, weanling and adult rats, J. Neurochem. 33:439 (1979).
13. W. H. Oldendorf, L. Braun, and E. Cornford, pH dependence of blood-brain barrier permeability to lactate and nicotine, Stroke 10:577 (1979).
14. R. A. Hawkins, A. M. Mans, D. W. Davis, J. R. Viña, and L. S. Hibbard, Cerebral glucose use measured with [$^{14}$C]glucose labeled in the 1,2, or 6 position, Am. J. Physiol. 248 (Cell Physiol. 17): C170 (1985).
15. R. J. Racine, Modification of seizure activity by electrical stimulation: II. Motor seizure, Electroenceph. clin. neurophysiol. 32:281 (1973).
16. E. Pinard, A. S. Rigaud, D. Riche, R. Naquet, and J. Seylaz, Continuous determination of the cerebrovascular changes induced by bicuculline and kainic acid in unanaesthetized spontaneously breathing rats, Neuroscience 23:943 (1987).
17. M. Ueki, F. Linn, and K.-A. Hossmann, Functional activation of cerebral blood flow and metabolism before and after global ischemia of rat brain, J. Cereb. Blood Flow Metab. 8:486 (1988).
18. R. J. Paul, Functional compartmentalization of oxidative and glycolytic metabolism in vascular smooth muscle, Am. J. Physiol. 244 (Cell Physiol. 13):C399 (1983).
19. E. Racker, Why do tumor cells have a high aerobic glycolysis?, J. Cell. Physiol. 89:697 (1976).
20. M. Mata, D. J. Fink, H. Gainer, C. B. Smith, L. Davidsen, H. Savaki, W. J. Schwartz, and L. Sokoloff, Activity-dependent energy metabolism in rat posterior pituitary primarily reflects sodium pump activity, J. Neurochem. 34:213 (1980).
21. N. Brookes and P. J. Yarowsky, Determinants of deoxyglucose uptake in cultured astrocytes: The role of the sodium pump, J. Neurochem. 44:473 (1985).
22. R. K. Orkand, J. G. Nicholls, and S. W. Kuffler, Effect of nerve impulses on the membrane potential of glial cells in the central nervous system of amphibia, J. Neurophysiol. 29:788 (1966).

23. R. D. Keynes and J. M. Ritchie, The movements of labelled ions in mammalian non-myelinated nerve fibres, J. Physiol. 179:333 (1965).
24. H. McIlwain, Phosphates of brain during in vivo metabolism: Effects of oxygen, glucose, glutamate, glutamine, and calcium and potassium salts, Biochem. J. 52:289 (1952).
25. R. Casteels and F. Wuytack, Aerobic and anaerobic metabolism in smooth muscle cells of taenia coli in relation to active ion transport, J. Physiol. 250:203 (1975).
26. M. Shinohara, B. Dollinger, G. Brown, S. Rapoport, and L. Sokoloff, Cerebral glucose utilization: Local changes during and after recovery from spreading cortical depression, Science 203:188 (1979).
27. W. G. Kuhr and J. Korf, Extracellular lactic acid as an indicator of brain metabolism, J. Cereb. Blood Flow Metab. 8:130 (1988).
28. L. J. King, O. H. Lowry, J. V. Passonneau, and V. Venson, Effects of convulsants on energy reserves in the cerebral cortex, J. Neurochem. 14:599 (1967).
29. W. P. Pulsinelli and R. P. Kraig, Photic stimulation causes enhanced glycolysis in the superior colliculus, Soc. Neurosci Abstr. 14:48 (1988).
30. D. Richter and R. M. C. Dawson, Brain metabolism in emotional excitement and in sleep, Am. J. Physiol. 154:73 (1948).
31. M. Nedergaard, S A. Goldman, and W A. Pulsinelli, Lactic-acid-induced intracellular acidification in primary cultures of mammalian brain, J. Cereb. Blood Flow Metab. 9 (Suppl. 1):S384 (1989).
32. J. H. Swan, M. C. Evans, and B. S. Meldrum, Long-term development of selective neuronal loss and the mechanism of protection by 2-amino-7-phosphonoheptanoate in a rat model of incomplete forebrain ischaemia, J. Cereb. Blood Flow Metab. 8:64 (1988).
33. E. Ozyurt, D. I. Graham, G. N. Woodruff, and J. McCulloch, Protective effect of the glutamate antagonist, MK-801 in focal cerebral ischemia in the cat, J. Cereb. Blood Flow Metab. 8:138 (1988).
34. W. G. Kuhr and J. Korf, N-methyl-D-aspartate receptor involvement in lactate production following ischemia or convulsions in rats, Eur. J. Pharmacol. 155:145 (1988).
35. N. Mori and J. A. Wada, Bidirectional transfer between kindling induced by excitatory amino acids and electrical stimulation, Brain Res. 425:45 (1987).
36. D. P. Cain, K. A. Desborouch, and D. J. McKitrick, Retardation of amygdala kindling by antagonism of NMD-Aspartate and muscarinic cholinergic receptors: Evidence for the summation of excitatory mechanisms in kindling, Exp. Neurol. 100:179 (1988).
37. T. Sutula, H. Xiao-Xian, J. Cavazos, and G. Scott, Synaptic reorganization in the hippocampus induced by abnormal functional activity, Science 239:1147 (1988).
38. A. Represa, G. Le Gal La Salle, and Y. Ben-Ari, Hippocampal plasticity in the kindling model of epilepsy in rats, Neurosci. Lett. 99:345 (1989).
39. T. L. Babb, W. R. Kupfer, and J. K. Pretorius, Synaptic reorganization of mossy fibers into inner molecular layer in human epileptic fascia dentata, Neurosci. Abstr. 14:881 (1988).
40. S. Feldblum and R. F. Ackermann, Increased susceptibility to hippocampal and amygdala kindling following intrahippocampal kainic acid, Exp. Neurol. 97:255 (1987).

Discussion of Dr. Ackermann's Presentation

DR. MCINTYRE: Bob, in your careful methodology you did not tell us how long these kainate animals were in the throes of seizure when you actually did those experiments.

DR. ACKERMANN: OK. Yes. I should have said that. Thank you for asking. First, did you ask me the dose - you should have asked me the dose. We used a moderate dose, we used 7 milligrams per kilogram IP which in the strains we used either don't produce, well, the animal's response is idiosyncratic. The animal's response ranges from virtually no effect to having occasional clonic seizures. In other words, the effect at that dose is neither negligible nor very severe in most animals, it's a moderate dose. Typically we start the experiments about 45 minutes to an hour after the animal has been injected with kainate. We don't want damage to have been done by the kainate, but the results are essentially identical within a wide time, say between 1/2 hour to 2 hours. The results we get on these experiments are essentially the same.

DR. MCINTYRE: I presume then that you don't have any electrographic data on these guys.

DR. ACKERMANN: No.

DR. MCINTYRE: Because you know those kainate seizures are kind of funny and how they behave and what structures are active and what time they turn on and off over the same period of time. It's quite a bizarre seizure pattern.

DR. ACKERMANN: The idea of using kainate was not to produce seizures. It was to produce selective limbic activation and I don't believe that seizures are required to get this phenomenon, that's why we then tried it with the visual stimulation. Because Collins et al. have already shown in separate group visual stimulation studies that glucose and deoxyglucose paralleled each other in uptake up to about 8 Hz. Beyond 8 Hz the glucose response flattened out but the deoxyglucose continued to increase. That already strongly suggested that there is a metabolic alteration in stimulated structures. So the idea of doing the visual stimulation was to replicate Collins et al. in single animals, and also to show that the phenomenon we're talking about is not dependent on producing seizures. But the explanation that I'm offering for the phenomenon is true, it holds even more so for seizures, if you assume that in seizures you're going to have more ion pumping per unit at time of seizure, than you would simply by stimulating the brain with a relatively mild stimulus.

DR. MOSHE: During the seizure do you think that there is now an increase in metabolism, or a decrease. If there's a change in the glycolytic pathway you produce less ATP, so the speculation that during a seizure you have tremendous increases in metabolic demands, may not be valid. If you switch to the glycolytic pathway you will produce much less ATP.

DR. ACKERMANN: That's right. So the implication, if this is right, is that, the apparent energy demands during seizures are much less than you would think if you just did a normal quantitative calculation based on DG accumulation, because of the differences in the production of ATP by the two mechanisms.

DR. MOSHE: You understand now that you're going to throw off all the lawyers in malpractice cases when they try to tell us that we have to treat very aggressively because the metabolic needs are increasing tremendously. You're showing that they are not increasing tremendously.

DR. ACKERMANN: But the other side of the coin is that we're producing a lot of lactate. So lawyers can now say, well there's not too much problem with the glucose, but this lactate is terrible stuff and so we have to sue on that basis.

DR. MOSHE: Well, actually on an age-dependent basis lactate should be able to be metabolized from the immature brain much better than the mature brain.

DR. ACKERMANN: That's correct.

DR. MOSHE: And that may be important again from the lawyers' point of view because they attribute damage to seizures which is not necessarily the case. That is, in the immature brain, babies or baby animals should be capable to deal with lactate much better than adults. That means that they should not have brain damage.

DR. ACKERMANN: That's correct.

DR. MOSHE: I think it's a tremendous contribution in understanding the pathophysiology of seizures. Up to now we used to think that in neonates if you have a seizure it's the worst possible thing that can happen to you. I think it is tremendous to show you that this is not the case.

DR. ACKERMANN: Of course, if I may just add to what you said, you know that the transport of lactate in neonates is much greater than it is in adults. So there is intracellular damage done by lactate by pH chains. Intuitively one would think that that effect would be much less in babies. On the other hand, I believe Wasterlain showed that babies can be depleted of glucose through seizure activity to a much greater extent than adults. Is that right? Does anyone know? So the best answer is that neither you nor I should become lawyers.

DR. WEISS: I have a colleague, Dennis Phiney, perhaps you know he does a lot of deoxyglucose related to brain injury, and we've debated. Now, in your technique you've restrained these animals in the process of doing the infusions. Yes? Yes. We found that those animals are very extremely stressed, so we measured levels. I've never seen them so high. So what you're really studying, I think, is that caution has to be exercised. Here is a controlled stressed animal and another animal that's stressed and kindled - so that overlay has to be considered, particularly related to kainic acid. Supulsky has shown that the effects in the hippocampus are very stress-related. I think that caution ought to be exercised.

DR. ACKERMANN: I agree with everything you've said, but have to add that I don't believe that the results of these experiments would be much different if you did them free moving. But I don't know. And the other thing is that results are essentially the same with the optic stimulation as with kainic acid, and in different structures. So my belief is that if these results are wrong, they're not wrong for that reason. Probably not.

DR. POST: Can lactate itself cause seizures?

DR. ACKERMANN: I don't know.

DR. POST: I guess I was wondering about the implications of your data in the induction of this pathway for lactate-induced panic attacks.

DR. ACKERMANN: Yes, someone else mentioned that to me. But I don't know the relationship. I think I'm stretching the data pretty far as it is.

# ALTERATIONS OF SYNAPTIC ULTRASTRUCTURE INDUCED BY HIPPOCAMPAL KINDLING

Yuri Geinisman*, Frank Morrell† and Leyla deToledo-Morrell†,§

*Department of Cell Biology and Anatomy, Northwestern University Medical School, Chicago, IL 60611 and Departments of †Neurological Sciences and §Psychology, Rush Medical College, Chicago, IL 60612

## INTRODUCTION

Kindling, first discovered by the late Graham Goddard[1], is widely regarded as a dramatic, reliable and robust form of neural plasticity[2-6]. One of the most remarkable features of kindling is that it induces a virtually permanent change in brain function. Synaptic responsiveness of the circuit stimulated during kindling undergoes an augmentation which persists, without further reinforcement, for many months[5-9]. This exceptionally enduring enhancement of synaptic efficacy caused by kindling and the dependence of the process on protein synthesis[10,11] and axonal transport[12,13] imply an underlying structural modification of the synapse itself.

Structural synaptic substrates of kindling may include increases in the number and/or size of excitatory synaptic junctions. In this respect, the findings of a few morphological studies published so far, though inconclusive, are nevertheless encouraging. The most consistent results were obtained with regard to the synaptic size. Kindling was reported to induce an increase in dimensions of presynaptic axon terminals[2,14,15] and postsynaptic dendritic spines[15,16] in the terminal field of stimulated axons. However, the density of synaptic contacts per unit area was found to be constant in kindled animals[2,14,16]. Only one study of an *in vitro* kindling model demonstrated an increase in the number of synapses involving dendritic shafts and sessile ("stubby") dendritic spines[17]. Consonant with the latter data are the observations concerning sprouting of mossy fiber axons and terminals in the inner molecular layer of the dentate gyrus after perforant path kindling or synchronous activation[18].

Unfortunately, the reported results of quantitative analyses of synapses during kindling are difficult to interpret. Since most of this work was done at the time when modern stereological and morphometric techniques were not available, the employed methodology involved a number of serious problems. One of them is related to the use of random, non-serial sections for synapse identification. An unequivocal identification of synaptic contacts, which is an absolute requirement for their subsequent quantitative analyses[19,20], can only be performed in consecutive serial sections: small and tangentially cut synaptic profiles produced by sectioning of over 20% of all synapses are not recognized in random sections[21]. Another problem is that synaptic samples were obtained without implementation of the sampling procedure of the disector technique[19,20,22] and were, therefore, biased by several factors. These include the size, shape and orientation of synapses, section thickness, the phenomenon of lost caps and Holmes' effect[19,20].

*Kindling 4*
Edited by J. A. Wada
Plenum Press, New York, 1990

Furthermore, the parameters used to characterize the synaptic number (quantity of profiles per unit sampling area) and size (area of profiles measured in random sections) were additionally biased by changes in the volume of brain tissue resulting from its processing for electron microscopy and, possibly, from experimental manipulations. The magnitude of these biases is unknown, and it might be different for the kindled and control groups compared.

Due to recent advances in the field of stereology and morphometry, it has become possible to reexamine the issue of structural synaptic alterations induced by kindling with the aid of unbiased technical approaches. Using such approaches, we have been able to characterize some changes in the number and size of synapses associated with kindling. The purpose of the present paper is to review these data and to discuss whether the changes we have found may represent structural synaptic substrates of kindling. The results concerning the estimation of synaptic numbers in the middle molecular layer of the dentate gyrus have been reported elsewhere[23].

## EXPERIMENTAL DESIGN

In designing the experiment described below, it was first necessary to select a system in which to study possible effects of kindling on synaptic morphology. Our choice was the projection from the entorhinal cortex to the hippocampal dentate gyrus, since it is an anatomically well defined monosynaptic system which is readily susceptible to kindling. An important advantage of this system for synapse morphometry is an exceptionally high density of synaptic contacts made by entorhinal axons in their terminal field in the dentate gyrus. Axons, which originate in the entorhinal cortex and travel in the perforant path, terminate in the outer and middle molecular layer of the dentate gyrus[24-27] where they form over 90% of the total synaptic population[28].

Young adult rats were stereotaxically implanted with bipolar electrodes into the medial perforant path on the right side. After a 2-week recovery period, they were randomly assigned to 3 groups. One group (kindled rats) was stimulated twice a day with 1 msec pulses at 60 Hz for 2 sec at a current level that initially produced an afterdischarge of 10 sec or less. The stimulation was terminated when a kindling criterion of 5 generalized seizures was reached. A second group of animals (coulombic controls) was stimulated with parameters (120 pulses at 2 Hz) that do not evoke kindling[6]. Each rat of this group was matched with a respective kindled rat according to the current level and the total amount of current delivered. The third group (unstimulated controls) consisted of animals that received the same handling as the other two groups, but were not stimulated. Three rats, one from each group, were coded and sacrificed 4 weeks after an animal belonging to the first group reached criterion. This allowed us to assess the long-lasting consequences of kindling, rather than the immediate effects of seizures.

The brains from each triplet were processed simultaneously for electron microscopy, since it is a necessary condition for processing-induced dimensional changes to be the same in animals under comparison[29]. The protocol of tissue preparation was described by us in detail earlier[23,30]. Briefly, rats were perfused intracardially with paraformaldehyde-glutaraldehyde fixatives. The right hippocampal formation was dissected free and cut perpendicular to its long axis into blocks of about 1 mm in thickness. Two blocks, with their rostral faces 1.5 and 3.5 mm caudal to the septal pole of the hippocampal formation, were osmicated, dehydrated, and embedded in Araldite. All blocks were assigned code numbers to be decoded after completion of morphological work. The rostral face of each block was trimmed down so as to include the whole width of the molecular and granule cell layer (GCL) in the central (in the mediolateral direction) segment of the hidden blade of the dentate gyrus. From each block, two complete series of ultrathin sections were prepared and stained with uranyl acetate and lead citrate.

Synapses were examined in the middle (MML) and inner (IML) molecular layer for the following reason. The MML is predominantly innervated by axons originating in the medial entorhinal cortex and reaching the dentate gyrus by way of the medial perforant path[24-27] which was stimulated in our experiment. The IML, on the other hand, does not receive perforant path axons, and its innervation is mainly provided by the commissural, associational and septal fibers (see ref. 27 for a review). Thus, the MML was a directly stimulated structure, while the IML was not. The two are immediately adjacent synaptic fields likely to be equally susceptible to the effects of generalized seizures (and possible transient hypoxia) which are end-products of kindling that could also alter synaptic morphology.

The same neuropil area of the MML or IML was photographed in all sections of each series. Electron micrographs were taken from the middle of the MML (in relation to the GCL and hippocampal fissure) and from the IML zone located 20-50 μm dorsal to the granule cell layer at a final magnification of ×20,000. Additionally, micrographs of the same GCL area were obtained from the first, several intermediate and last sections of a given series at a final magnification of ×2,500. A magnification standard (grating replica) was photographed and printed with each series of micrographs.

## KINDLING-INDUCED CHANGES IN THE NUMBER OF SYNAPSES PER NEURON

The final parameter to be estimated in the process of synapse quantitation was *the number of synapses per neuron*. It is advantageous to use this parameter rather than conventional indicators of synaptic density such as the number of synapses per unit area or volume, since the latter are influenced by dimensional changes in the reference space. Estimates of the number of synapses per neuron are unaffected by changes in tissue volume induced by its processing or experimental conditions when both synapses and neuronal cell bodies are sampled in the same reference space. However, in stratified structures, such as the dentate gyrus, synapses to be analyzed and cell bodies of postsynaptic neurons are placed in different layers and have to be sampled in different reference spaces. In such cases, a dimensional correction is made: the synapse to neuron ratio is multiplied by the estimator of the ratio of two reference volumes (i.e., by the ratio between the widths of synaptic and neuronal layers)[20]. This correction takes into account possible effects of volumetric alterations (including those which may result from changes in the number of neuronal cell bodies and dendrites) on estimates of the synaptic or neuronal numerical density.

The disector technique[22] as applied for synapse quantitation[20] was employed to obtain estimates of the number of synapses per neuron unbiased by size, shape or orientation and independent of the section thickness or the phenomenon of lost caps[19,20]. Each disector for synapse sampling consisted of two sections, a reference section and a look-up section immediately above it. All synaptic profiles contained in an electron micrograph of the reference section were identified with the aid of micrographs of adjacent serial sections by the presence of synaptic vesicles in a presynaptic axon terminal and a postsynaptic density (PSD) in a postsynaptic element. Using the sampling frame and unbiased two-dimensional sampling rule of Gundersen[31], only those synapses were sampled in which the PSD was seen within the sampling frame superimposed on the reference section micrograph, but was not observed in the look-up section micrograph. To obtain representative synaptic samples, synapses were counted in 12 disectors randomly selected from each section series. The same procedure was followed to sample granule cells in the disector consisting of the first and last sections of a series. These sections were used, in turn, as reference and look-up ones, the granule cell nucleus being employed as a counting unit.

The unbiased estimate of the number of synapses per neuron "n/N" was obtained by the formula[20]: $n/N = [(\Sigma q^{-} \cdot \Sigma A \cdot k)/(\Sigma Q^{-} \cdot \Sigma a)] \cdot (w/W)$. In this formula, "$q^{-}$" and "$Q^{-}$" are numbers of synapses and neurons sampled in areas "a" and "A", respectively; "k" represents the number of sections in a series minus one; "w" designates the width of the middle or inner third of the molecular layer and "W" indicates the width of the GCL (measured in 1 µm-thick sections stained with methylene blue); the summation "$\Sigma$" is over all disectors examined in a series. A final n/N value per animal was assessed by averaging n/N estimates obtained from 4 section series. Dendritic spine (axospinous) and dendritic shaft (axodendritic) synapses were differentially analyzed, since perforant path axons terminate mainly on spines and occasionally on shafts of dentate granule cells[32]. The data presented below for each individual animal were derived from series of 25-73 (mean = 42) ultrathin sections in which a total number of 448-742 (mean = 585) axospinous synapses, 24-63 (mean = 42) axodendritic synapses and 53-96 (mean = 68) neuronal cell bodies were sampled.

Analyses of synapses involving dendritic shafts in the MML and IML revealed no significant differences in their number per neuron between kindled and control animals (data are not shown). Axospinous synapses, however, were significantly ($p < 0.02$) diminished in number following kindling in both subdivisions of the molecular layer. In the MML, a decrease of 17.6 and 19.4% was observed in kindled rats relative to unstimulated or coulombic controls, respectively. In the IML, this change reached a magnitude of 22.9 and 24.1%.

In order to determine whether this loss of synapses selectively involved only a certain synaptic type, all axospinous synaptic contacts were further subdivided into two types according to the appearance of their PSD. Axospinous synapses exhibiting a discontinuous or perforated PSD profile in at least one serial section were attributed to the first type, and those without PSD discontinuities to the second type (Fig. 1). These will be referred to as perforated and nonperforated synapses, respectively. Since the axospinous synaptic population of both the MML and the IML is composed of the two synaptic types, their differential analysis was performed in each of these zones of the molecular layer.

Examination of the MML showed that the number of perforated synapses per neuron did not change significantly in kindled animals, although a trend towards an increase was observed (Table 1). However, nonperforated synapses of the MML were significantly reduced in numbers in kindled animals relative to either unstimulated or coulombic controls (Table 1). In the IML, the number of *both* perforated and nonperforated synapses per neuron was significantly diminished following kindling (Table 2).

Additionally, the ratio of perforated to nonperforated synapses was estimated. This parameter was calculated from numbers of synapses per neuron (although it could also be directly assessed from raw synaptic

→

Fig. 1. Electron micrographs of consecutive serial sections (A-C) demonstrating two distinct morphological types of axospinous synapses in the middle molecular layer of the rat dentate gyrus. Perforated synapses (open arrows) exhibit a discontinuous PSD profile in at least one serial section. Nonperforated synapses (solid arrows) do not show PSD discontinuities in any consecutive section. Due to their increased electron density, PSD profiles are especially noticeable in electron micrographs. Calibration bar - 0.5 µm.

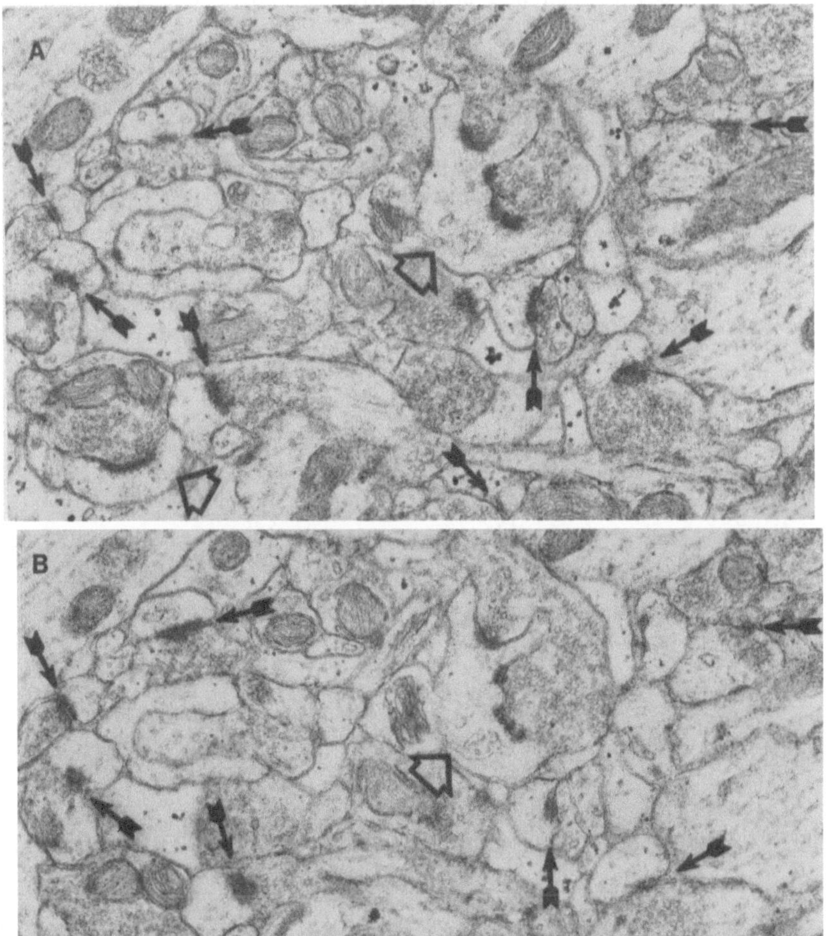
A
B
C

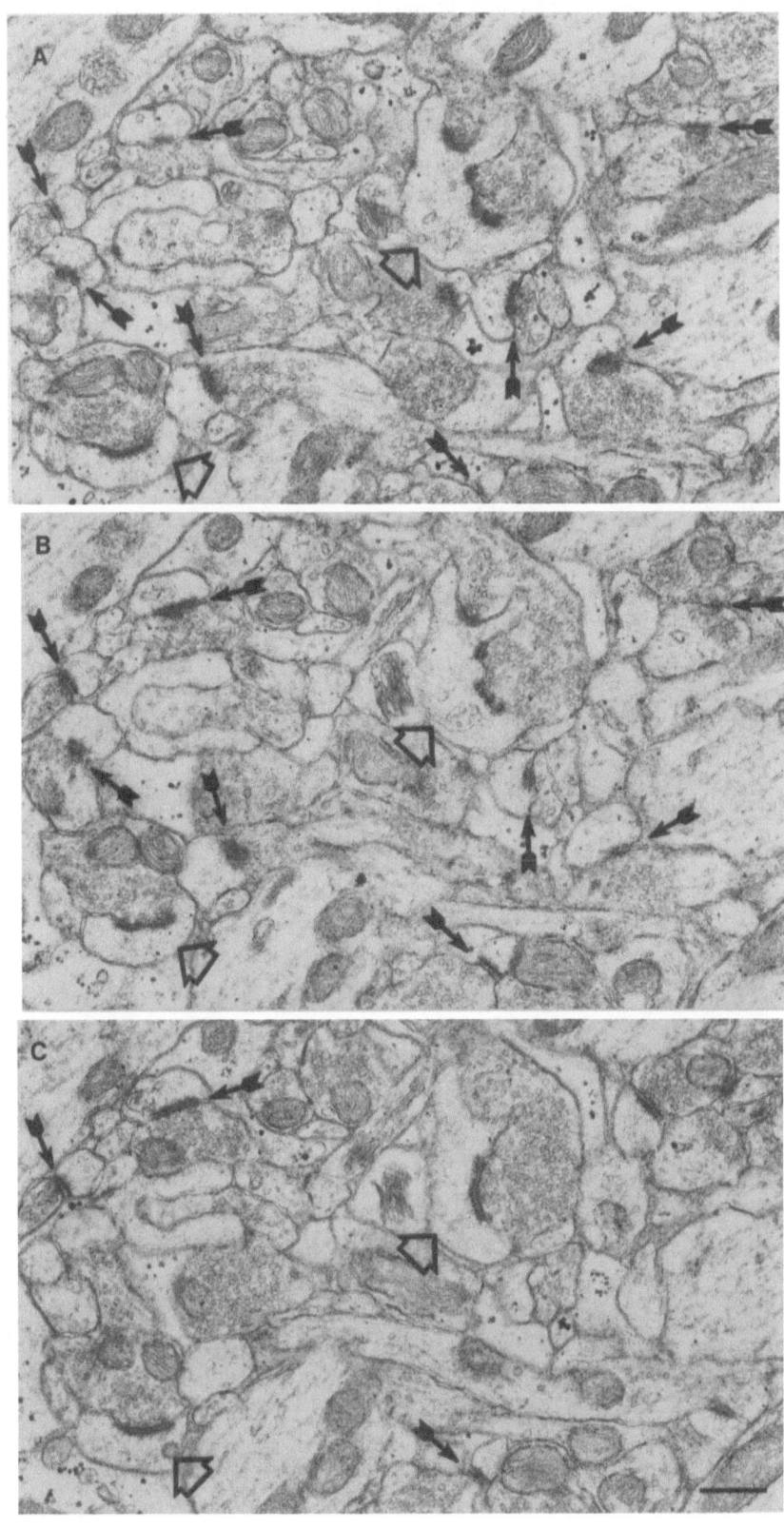

Table 1. Number of perforated and nonperforated axospinous synapses per neuron and the ratio of the two synaptic types in the middle molecular layer of the dentate gyrus of control and kindled rats.

Designations: $\Delta$ - difference between group means; $^{*}p < 0.05$ and $^{**}p < 0.005$, two-tailed Mann-Whitney $U$ test.

| Rat | A. Unstimulated control | B. Coulombic control | | C. Kindling | | |
|---|---|---|---|---|---|---|
| | Value | Value | ΔB-A | Value | ΔC-A | ΔC-B |
| | NUMBER OF PERFORATED SYNAPSES PER NEURON | | | | | |
| 1 | 174 | 250 | | 227 | | |
| 2 | 220 | 250 | | 203 | | |
| 3 | 187 | 177 | | 243 | | |
| 4 | 150 | 187 | | 231 | | |
| 5 | 198 | 182 | | 212 | | |
| 6 | 264 | 251 | | 184 | | |
| 7 | 170 | 149 | | 256 | | |
| Group mean | 195 | 207 | +6.2% | 222 | +13.8% | +7.2% |
| | NUMBER OF NONPERFORATED SYNAPSES PER NEURON | | | | | |
| 1 | 1792 | 2160 | | 1305 | | |
| 2 | 1812 | 2040 | | 1465 | | |
| 3 | 1667 | 1822 | | 1778 | | |
| 4 | 2163 | 1701 | | 1633 | | |
| 5 | 1889 | 1937 | | 1806 | | |
| 6 | 2277 | 2279 | | 1140 | | |
| 7 | 1920 | 1822 | | 1576 | | |
| Group mean | 1931 | 1966 | +1.8% | 1529 | -20.8%* | -22.2%* |
| | RATIO OF PERFORATED TO NONPERFORATED SYNAPSES | | | | | |
| 1 | 0.097 | 0.116 | | 0.174 | | |
| 2 | 0.121 | 0.123 | | 0.139 | | |
| 3 | 0.112 | 0.097 | | 0.137 | | |
| 4 | 0.069 | 0.110 | | 0.142 | | |
| 5 | 0.105 | 0.094 | | 0.117 | | |
| 6 | 0.116 | 0.110 | | 0.161 | | |
| 7 | 0.088 | 0.082 | | 0.162 | | |
| Group mean | 0.101 | 0.105 | +4.0% | 0.147 | +45.5%** | +40.0** |

counts in the same disectors) and used as a measure of the relative quantities of the two synaptic types in the neuropil. Following kindling, the ratio of perforated to nonperforated synapses was found to be markedly increased only in the MML where numbers of the two synaptic types shifted in opposite directions (Table 1). In marked contrast to that, no change in this ratio was observed in the IML of kindled rats, since perforated and nonperforated synapses of this subdivision of the molecular layer were decreased in numbers to an approximately equal degree (Table 2).

Thus, the quantitation of synaptic junctions provided an unexpected result. It seemed reasonable to assume that an increase in the number of axospinous synapses, which are generally considered to be excitatory in their action, could be induced by kindling in the terminal synaptic

Table 2. Numbers of perforated and nonperforated axospinous synapses per neuron and the ratio of the two synaptic types in the inner molecular layer of the dentate gyrus of control and kindled rats.

Designations are the same as in Table 1.

| Rat | A. Unstimulated control | B. Coulombic control | | C. Kindling | | |
|---|---|---|---|---|---|---|
| | Value | Value | ΔB-A | Value | ΔC-A | ΔC-B |
| | | NUMBER OF PERFORATED SYNAPSES PER NEURON | | | | |
| 1 | 154 | 217 | | 101 | | |
| 2 | 128 | 109 | | 80 | | |
| 3 | 133 | 160 | | 152 | | |
| 4 | 182 | 192 | | 102 | | |
| 5 | 151 | 82 | | 117 | | |
| 6 | 123 | 197 | | 99 | | |
| 7 | 154 | 157 | | 137 | | |
| Group mean | 146 | 159 | +8.9% | 113 | -22.6%* | -28.9%* |
| | | NUMBER OF NONPERFORATED SYNAPSES PER NEURON | | | | |
| 1 | 1984 | 2838 | | 1385 | | |
| 2 | 2015 | 1798 | | 1105 | | |
| 3 | 1945 | 1964 | | 1382 | | |
| 4 | 2254 | 2209 | | 1915 | | |
| 5 | 1749 | 1827 | | 2179 | | |
| 6 | 2687 | 2358 | | 1474 | | |
| 7 | 1849 | 1634 | | 1710 | | |
| Group mean | 2069 | 2090 | +1.0% | 1593 | -23.0%* | -23.8%* |
| | | RATIO OF PERFORATED TO NONPERFORATED SYNAPSES | | | | |
| 1 | 0.077 | 0.076 | | 0.073 | | |
| 2 | 0.064 | 0.061 | | 0.072 | | |
| 3 | 0.068 | 0.081 | | 0.111 | | |
| 4 | 0.081 | 0.087 | | 0.053 | | |
| 5 | 0.086 | 0.045 | | 0.053 | | |
| 6 | 0.046 | 0.084 | | 0.067 | | |
| 7 | 0.083 | 0.096 | | 0.080 | | |
| Group mean | 0.072 | 0.076 | +5.6% | 0.073 | +1.4% | -3.9% |

field of stimulated axons (MML). Contrary to this expectation, a loss of nonperforated axospinous synapses was found to occur in the MML of kindled rats (Table 1). If some of these synaptic contacts have an inhibitory action, their loss would be consistent with the idea that an augmentation of synaptic responsiveness during kindling could result from a diminution of inhibitory synaptic action[33-35]. At present, however, there is no direct evidence for or against an inhibitory role of nonperforated synapses. Most importantly, electrophysiological studies have not revealed a loss of inhibition in the dentate gyrus with perforant path kindling[36,37].

Another possibility is that the decrease in the number of nonperforated synapses in the MML represents a consequence of generalized seizures which are characteristic of fully developed kindling. Such a notion may be supported by our data obtained during quantitative analyses of synaptic junctions in the IML. Although this zone of the molecular layer was not directly stimulated in our experiment, it was susceptible, as was the MML, to the effects of generalized seizures. If the latter are indeed followed by a loss of nonperforated axospinous synapses, one would predict that this change should occur in both subdivisions of the molecular layer. In accordance with this prediction, nonperforated axospinous synapses of the IML were found to diminish in numbers following kindling (Table 2), and the magnitude of this loss was practically equal to that observed in the MML (Table 1). In the IML of kindled rats, however, a decrease in numbers involved not only nonperforated synapses, but also perforated ones (Table 2), a finding consistent with non-selective effects of generalized seizures. Interestingly enough, perforated synapses of the MML did not exhibit any loss with kindling and, in fact, showed a trend towards an increase. It is conceivable that a decrement in their numbers due to generalized seizures was compensated for by the new formation of those particular synaptic contacts.

Whether or not this is the case, a pronounced increase in the ratio of perforated to nonperforated synapses in the MML of kindled animals (Table 1) appears to represent a structural rearrangement which may contribute to the enduring augmentation of synaptic efficacy following kindling. It is particularly important in this respect that such a change was restricted to the terminal synaptic field of stimulated axons (MML) and did not involve an adjacent synaptic field (IML) innervated by nonstimulated fibers (Table 2). In the MML, the vast majority of nonperforated and virtually all perforated axospinous synapses are formed by stimulated axons of the medial perforant path[28]. Although it is generally accepted that the major consequence of perforant path activation is excitatory to dentate granule cells[38,39], it does not necessarily follow that all excitatory synaptic junctions are equally effective in their action. Perforated synapses are believed to be more efficacious than nonperforated ones[40,41]. It is possible, therefore, that the observed shift in the preponderance of perforated over nonperforated synapses may contribute to the sustained enhancement of excitatory synaptic "gain" which characterizes kindling.

## KINDLING-INDUCED CHANGES IN DIMENSIONS OF THE POSTSYNAPTIC DENSITY

To elucidate possible kindling-induced changes in the size of axospinous synapses, we estimated *PSD dimensions*, since a highly significant positive correlation exists between the size of the PSD, on one hand, and that of the presynaptic axon terminal or the postsynaptic dendritic spine, on the other[42-44]. The PSD is a plate(s) of various shape extending along the postsynaptic membrane. In sections prepared for electron microscopy, PSD profiles can be readily identified (Fig. 1). PSDs were sampled using the procedure of the disector technique as described above. Therefore, samples of PSDs were taken independently of their size, shape or orientation and were unaffected by section thickness or lost caps[20]. Each of the sampled PSDs was traced in consecutive serial sections below the reference one, and the length of all its profiles was measured on micrographs with a digital tablet. The total length of each discontinuous PSD profile was expressed as the sum of lengths of its fragments.

These measurements were used to derive two independent parameters characterizing PSD dimensions. One of them was the PSD maximal profile length which provides a robust estimate of the PSD mean caliper width (integrated over all orientations) in the plane of the PSD plate[20]. Such a parameter, however, might be inappropriate for those axospinous synapses that exhibit a segmented PSD consisting of 2-5 discrete plates[45]. To circumvent this problem, the PSD area was also assessed, although the accuracy of this measure was limited by that of section thickness estimation. For each of the sampled PSDs, the area of its

Table 3. Maximal profile length ($\mu m \times 10^{-3}$) and area ($\mu m^2 \times 10^{-4}$) of the postsynaptic density in axospinous synapses from the middle molecular layer of the dentate gyrus of control and kindled rats.

Designations: $\Delta$ - difference between group means; $^*p < 0.005$, two-tailed Mann-Whitney *U* test.

| Rat | A. Unstimulated control | B. Coulombic control | | C. Kindling | | |
|---|---|---|---|---|---|---|
| | Value | Value | ΔB-A | Value | ΔC-A | ΔC-B |
| | MAXIMAL PROFILE LENGTH OF PERFORATED POSTSYNAPTIC DENSITIES | | | | | |
| 1 | 343 | 365 | | 379 | | |
| 2 | 377 | 324 | | 405 | | |
| 3 | 369 | 345 | | 400 | | |
| 4 | 414 | 407 | | 479 | | |
| 5 | 348 | 359 | | 364 | | |
| 6 | 375 | 426 | | 446 | | |
| 7 | 354 | 373 | | 437 | | |
| Group mean | 369 | 371 | +0.5% | 416 | +12.7%* | +12.1%* |
| | MAXIMAL PROFILE LENGTH OF NONPERFORATED POSTSYNAPTIC DENSITIES | | | | | |
| 1 | 162 | 169 | | 161 | | |
| 2 | 173 | 167 | | 176 | | |
| 3 | 179 | 169 | | 169 | | |
| 4 | 171 | 156 | | 179 | | |
| 5 | 154 | 154 | | 153 | | |
| 6 | 179 | 176 | | 166 | | |
| 7 | 179 | 160 | | 175 | | |
| Group mean | 171 | 164 | -4.1% | 168 | -1.8% | +2.4% |
| | AREA OF PERFORATED POSTSYNAPTIC DENSITIES | | | | | |
| 1 | 934 | 994 | | 1032 | | |
| 2 | 896 | 851 | | 1076 | | |
| 3 | 962 | 952 | | 1193 | | |
| 4 | 1143 | 1042 | | 1294 | | |
| 5 | 914 | 946 | | 1025 | | |
| 6 | 1006 | 1065 | | 1306 | | |
| 7 | 1029 | 1106 | | 1296 | | |
| Group mean | 983 | 994 | +1.1% | 1175 | +19.5%* | +18.2%* |
| | AREA OF NONPERFORATED POSTSYNAPTIC DENSITIES | | | | | |
| 1 | 250 | 260 | | 250 | | |
| 2 | 250 | 246 | | 250 | | |
| 3 | 266 | 258 | | 265 | | |
| 4 | 264 | 233 | | 268 | | |
| 5 | 231 | 228 | | 224 | | |
| 6 | 294 | 292 | | 276 | | |
| 7 | 268 | 250 | | 268 | | |
| Group mean | 260 | 252 | -3.1% | 257 | -1.2% | +2.0% |

plate was ascertained by multiplying the total length of all PSD profiles by the mean section thickness. The thickness of ultrathin sections was estimated using the small-fold technique[46] as described by us earlier[30]. It was not possible to evaluate the thickness of all sections examined, since a number of them did not contain minimal folds. However, the mean section thickness for the groups of animals under comparison was the same (equal to 0.08 μm). This value, therefore, was used to calculate the PSD area. A differential analysis of perforated and nonperforated PSDs was performed in axospinous synapses from the MML and IML. The values for each individual animal presented below were derived from measurements of 32-68 (mean = 50) perforated PSDs and 45-71 (mean = 60) nonperforated PSDs.

The results showed that both the maximal profile length and the area of PSDs were selectively increased following kindling *only in perforated synapses of the MML* (Table 3). Such a change was observed in each kindled rat examined relative to respective unstimulated and coulombic controls. In marked contrast, nonperforated synapses in the MML were not significantly changed with respect to the measures used (Table 3). In the IML, PSD dimensions were not altered in either of the two synaptic types (data are not shown).

The above findings demonstrate that kindling results in an expansion of the PSD along the postsynaptic membrane. This alteration is likely to involve such PSD components as neurotransmitter receptors and ion channels[47,48] which could enhance the potency of synaptic transmission. Additionally, there is a close relationship between the PSD area and the number of synaptic vesicles in a presynaptic bouton[44]. PSD size is also positively correlated with that of the axon terminal[42] and of the dendritic spine head[43,44]. It has been postulated that any increase in PSD dimensions may be followed by an enlargement of the entire synapse[48], and a larger synaptic junction would be expected to release more neurotransmitter[49]. Thus, it appears likely that an increase in the PSD size may represent a structural modification particularly appropriate for the sustained augmentation of synaptic efficacy typical of kindling.

A striking feature of this structural synaptic alteration is its high selectivity. In kindled animals, PSD dimensions were increased only in the stimulated synaptic field (MML), but were unchanged in a neighboring region (IML) which was not directly stimulated. Moreover, the PSD expansion along the postsynaptic membrane following kindling was restricted to perforated axospinous synapses (Table 3). Enlargement of the PSD in these presumably efficacious synaptic contacts may provide a mechanism for the enduring and selective amplification of impulse transmission within the polysynaptic circuit involved in the kindling process.

## SUMMARY AND CONCLUSIONS

The major findings of this study are that the ratio of perforated to nonperforated axospinous synapses and the dimensions of perforated PSDs are significantly increased following kindling in the terminal synaptic field of stimulated axons (MML). Such changes, however, were not found in an immediately adjacent synaptic field (IML) which was not directly stimulated, but was equally susceptible to the effects of generalized seizures. This indicates that the observed structural alterations are a manifestation of synaptic plasticity associated with kindling rather than being a consequence of nonspecific effects of generalized seizures.

The obtained results, taken together, strongly suggest that *perforated* axospinous synapses formed by stimulated axons may play a special role in the process of kindling. The observations of others showing that kindling cannot be elicited in the cerebellar cortex[33], which lacks perforated axospinous synapses[44], are consistent with this notion. There are two major hypotheses concerning the possible functional significance of perforated synaptic contacts. The first of these suggests that perforated PSDs may be associated with an augmented synaptic transmission due to the presence of additional edges around their perforations (assuming that the edges are the most active sites of transmitter release)[40] or to a closer apposition of pre- and postsynaptic membranes at perforations[41]. This idea seems to be supported by our data demonstrating that the PSD area is much larger in perforated than in nonperforated synaptic contacts (Table 3). The PSD delineates the most concentrated area of postsynaptic neurotransmitter receptors and ion channels[46,47], and large perforated PSDs may contain a relatively high amount of these components which would facilitate synaptic transmission. According to the second view, a perforated PSD may be a structural intermediate in the process of synaptogenesis during which a relatively large perforated synapse splits into smaller nonperforated ones[28,50,51]. Both hypotheses imply that perforated synaptic junctions represent a structural modification likely to result in a marked and sustained augmentation of synaptic responsiveness. Therefore, the increases in the relative proportion of perforated axospinous synapses and in the dimensions of perforated PSDs found by us in the terminal field of stimulated axons may explain the physiologically defined, durable enhancement of synaptic efficacy which characterizes kindling.

Kindling is related to another, less enduring model of neuronal plasticity, long-term potentiation (LTP). LTP can be elicited in the dentate gyrus by high frequency electrical stimulation of the perforant path[39,52]. LTP appears to be an early and invariant feature of the kindling process[53], and its prior induction in a pathway subsequently stimulated to evoke kindling results in more rapid kindling of that circuit[54]. One may consider kindling as an outgrowth of LTP, a manifestation of augmented synaptic efficacy that is not confined, as is LTP, to the first synaptic relay, but extends gradually and successively to other synaptic stations of the stimulated circuit[6]. In fact, Siekevitz[48] had predicted, on theoretical grounds, that both kindling and LTP would be associated with comparable modifications of synaptic morphology. It is of special interest, therefore, that morphological alterations similar to those we describe here have been reported in LTP[55,56] as well as in other circumstances associated with storage of information, i.e. visual discrimination training[57]. Thus, the marked shift towards the preponderance of perforated over nonperforated synapses and the significant expansion of perforated PSDs, herein demonstrated for the kindling model, may represent principal morphological substrates of learning and other forms of synaptic plasticity.

ACKNOWLEDGEMENT

We thank Susan Evers, William Goossens, Nicholas Kriho and Diane L. Scholz for their skillful technical assistance. This work was supported by NSF Grants BNS 86-07272 and 87-19107.

REFERENCES

1. G. V. Goddard, Development of epileptic seizures through brain stimulation of low intensity, *Nature* 214:1020 (1967).
2. G. V. Goddard and R. M. Douglas, Does the engram of kindling model the engram of normal long term memory?, *Can. J. Neurol. Sci.* 2:385 (1975).
3. R. Racine, L. Tuff, and J. Zaide, Kindling, unit discharge and neural plasticity, *Can. J. Neurol. Sci.* 2:395 (1975).
4. R. Racine, Kindling: the first decade, *Neurosurgery* 3:234 (1978).
5. G. V. Goddard, The kindling model of epilepsy, *Trends Neurosci.* 6:275 (1983).
6. F. Morrell and L. deToledo-Morrell, Kindling as a model of neuronal plasticity, in: "Kindling 3", J. A. Wada, ed., Raven, New York (1986).

7. G. V. Goddard, D. C. McIntyre, and C. K. Leech, A permanent change in brain function resulting from daily electrical stimulation, *Exp. Neurol.* 25:295 (1969).
8. J. A. Wada and M. Sato, Generalized convulsive seizures induced by daily electrical stimulation of the amygdala in cats: correlative electrgraphic and behavioral seizures, *Neurology (Minneap.)* 24:565 (1974).
9. F. Morrell, Goddard's kindling phenomenon, in: "Chemical Modulation of Brain Function", H. C. Sabelli, ed., Raven, New York (1973).
10. F. Morrell, N. Tsuru, T. J. Hoeppner, D. Morgan, and W. H. Harrison, Secondary epileptogenesis in frog forebrain: effect of inhibition of protein synthesis, *Can. J. Neurol. Sci.* 2:407 (1975).
11. V. Jonec and C. G. Wasterlain, Effects of inhibitors of protein synthesis on the development of kindled seizures in rats, *Exp. Neurol.* 66:524 (1979).
12. F. Morrell, Biochemical alterations in secondary epileptogenic lesions, in: "Secondary Epileptogenesis", A. Mayersdorf and R. P. Schmidt, eds., Raven, New York (1982).
13. F. Morrell, Callosal mechanisms in epileptogenesis, in: "Epilepsy and the Corpus Callosum", A. Reeves, ed., Plenum, New York (1985).
14. R. Racine and J. Zaide, A further investigation into the mechanisms underlying the kindling phenomenon, in: "Limbic Mechanisms", K. L. Livingston and O. Hornykiewicz, eds., Plenum, New York (1978).
15. M. Langmeier and J. Mares, Changes in some ultrastructural parameters of cortical synapses in the initial phases of kindling, *Physiol. Bohemoslov.* 33:367 (1984).
16. A. J. Cronin, T. P. Sutula, and N. L. Desmond, Morphological changes in the hippocampal dentate gyrus accompany kindling of the entorhinal cortex, *Soc. Neurosci. Abstr.* 13:947 (1987).
17. N. Hawrylak, F.-L. Chang, D. Treacy, K. R. Isaaks, and W. T. Greenough, Synaptogenesis in kindling, *Soc. Neurosci. Abstr.* 14:881 (1988).
18. T. Sutula, H. Xiao-Xian, J. Cavazos and G. Scott, Synaptic reorganization in the hippocampus induced by abnormal functional activity, *Science* 239:1147 (1988).
19. H. J. G. Gundersen, Stereology of arbitrary particles, *J. Microsc.* 143:3 (1986).
20. H. Brændgaard and H. J. G. Gundersen, The impact of recent stereological advances on quantitative studies of the nervous system, *J. Neurosci. Meth.* 18:39 (1986).
21. C. A. Curcio and J. W. Hinds, Stability of synaptic density and spine volume in dentate gyrus of aged rats, *Neurobiol. Aging* 4:77 (1983).
22. D. C. Sterio, The unbiased estimation of number and sizes of arbitrary particles using the disector, *J. Microsc.* 134:127 (1984).
23. Y. Geinisman, F. Morrell, and L. deToledo-Morrell, Remodeling of synaptic architecture during hippocampal kindling, *Proc. Natl. Acad. Sci. U.S.A.* 85:3260 (1988).
24. A. Hjorth-Simonsen, Projection of the lateral part of the entorhinal area to the hippocampus and fascia dentata, *J. Comp. Neurol.* 146:219 (1972).
25. A. Hjorth-Simonsen and B. Jeune, Origin and termination of the hippocampal perforant path in the rat studied by silver impregnation, *J. Comp. Neurol.* 144:215 (1972).
26. O. Steward, Topographic organization of the projections from the entorhinal area to the hippocampal formation of the rat, *J. Comp. Neurol.* 167:285 (1976).
27. D. L. Rosene and G. W. Van Hoesen, The hippocampal formation of the primate brain, in: "Cerebral Cortex", Vol. 6, E. G. Jones and A. Peters, eds., Plenum, New York (1987).

28. M. Nieto-Sampedro, S. F. Hoff, and C. W. Cotman, Perforated postsynaptic densities: probable intermediates in synapse turnover, *Proc. Natl. Acad. Sci. U.S.A.* 79:5718 (1982).
29. Y. Y. Geinisman, V. N. Larina, and V. N. Mats, Changes of neurones dimensions as a possible morphological correlate of their increased functional activity, *Brain Res.* 26:247 (1971).
30. Y. Geinisman, L. deToledo-Morrell, and F. Morrell, Aged rats need a preserved complement of perforated axospinous synapses per hippocampal neuron to maintain good spatial memory, *Brain Res.* 398:266 (1986).
31. H. J. G. Gundersen, Notes on the estimation of the numerical density of arbitrary profiles: the edge effect, *J. Microsc.* 111:219 (1987).
32. D. A. Matthews, C. Cotman, and G. Lynch, An electron microscopic study of lesion-induced synaptogenesis in the dentate gyrus of the adult rat. I. Magnitude and time course of degeneration, *Brain Res.* 115:1 (1986).
33. J. O. McNamara, M. Byrne, R. Danshieff, and J. Fitz, The kindling model of epilepsy: a review, *Prog. Neurobiol.* 15:139 (1980).
34. C. E. Ribak, R. M. Bradburne, and A. B. Harris, A preferential loss of gabaergic symmetric synapses in epileptic foci: a quantitative ultrastructural analysis of monkey neocortex, *J. Neurosci.* 2:1725 (1982).
35. R. S. Sloviter, decreased hippocampal inhibition and a selective loss of interneurons in experimental epilepsy, *Science* 235:73 (1987).
36. L. P. Tuff, R. J. Racine, and R. Adamec, The effects of kindling on GABA-mediated inhibition in the dentate gyrus of the rat. I. Paired pulse depression, *Brain Res.* 277:79 (1983).
37. G. V. Goddard and E. Maru, Forces for and against the kindled state as revealed by EEG and field potential analysis in the hippocampal dentate area of perforant path kindled rats, in: "Kindling 3", J. A. Wada, ed., Raven, New York (1986).
38. P. Andersen, B. Holmquist, and P. E. Voorhoeve, Entorhinal activation of dentate granule cells, *Acta Physiol. Scand.* 66:448 (1966).
39. T. Lømo, Patterns of activation in a monosynaptic cortical pathway: the perforant path input to the dentate area of the hippocampal formation, *Exp. Brain Res.* 12:18 (1971).
40. A. Peters and I. R. Kaiserman-Abramof, The small pyramidal neuron of the rat cerebral cortex. The synapses upon dendritic spines, *Z. Zellfosch.* 100:487 (1969).
41. A. M. Sirevaag and W. T. Greenough, Differential rearing effects on rat visual cortex synapses. II. Synaptic morphometry, *Dev. Brain Res.* 19:215 (1985).
42. S. E. Dyson and D. G. Jones, Quantitation of terminal parameters and their interrelationships in maturing central synapses: a perspective for experimental studies, *Brain Res.* 183:43 (1980).
43. C. J. Wilson, P. M. Groves, S. T. Kitai, and J. C. Linder, Three-dimensional structure of dendritic spines in the rat neostriatum, *J. Neurosci.* 3:383 (1983).
44. K. M. Harris and J. K. Stevens, Dendritic spines of rat cerebellar Purkinje cells: serial electron microscopy with reference to their biophysical characteristics, *J. Neurosci.* 8:4455 (1988).
45. Y. Geinisman, F. Morrell, and L. deToledo-Morrell, Axospinous synapses with segmented postsynaptic densities: a morphologically distinct synaptic subtype contributing to the number of profiles of "perforated" synapses visualized in random sections, *Brain Res.* 423:179 (1987).
46. D. M. G. De Groot, Comparison of methods for the estimation of the thickness of ultrathin tissue sections, *J. Microsc.* 151:23 (1988).
47. C. W. Cotman and P. T. Kelly, Macromolecular architecture of CNS synapses, in: "The Cell Surface and Neuronal Function", C. W. Cotman, G. Poste, and G. L. Nocolson, eds., Elsevier, Amsterdam (1980).

48. P. Siekevitz, The postsynaptic density: a possible role in long-lasting effects in the central nervous system, *Proc. Natl. Acad. Sci. U.S.A.* 82:3494 (1985).
49. A. A. Herrera, A. D. Grinnell, and B. Wolowske, Ultrastructural correlates of naturally occurring differences in transmitter release efficacy in frog motor nerve terminals, *J. Neurocytol.* 14:193 (1985).
50. P. K. Carlin and P. Siekevitz, Plasticity in the central nervous system: do synapses divide?, *Proc. Natl. Acad. Sci. U.S.A.* 80:3517 (1983).
51. S. E. Dyson and D. G. Jones, Synaptic remodelling during development and maturation: junction differentiation and splitting as a mechanism of modifying connectivity, *Dev. Brain Res.* 13:125 (1984).
52. T. V. P. Bliss and A. R. Gardner-Medwin, Long-lasting potentiation of synaptic transmission in the dentate area of the unanaesthetized rabbit following stimulation of the perforant path, *J. Physiol (Lond.)* 232:357 (1973).
53. T. Sutula and O. Steward, Quantitative analysis of synaptic potentiation during kindling of the perforant path, *J. Neurophysiol.* 56:732 (1986).
54. T. Sutula and O. Steward, Facilitation of kindling by prior induction of long-term potentiation in the perforant path, *Brain Res.* 420:109 (1987).
55. N. L. Desmond and W. B. Levy, Changes in the numerical density of synaptic contacts with long-term potentiation in the hippocampal dentate gyrus, *J. Comp. Neurol.* 253:466 (1986).
56. N. L. Desmond and W. B. Levy, Changes in the postsynaptic density with long-term potentiation in the dentate gyrus, *J. Comp. Neurol.* 253:476 (1986).
57. G. Vrensen and J. Nunes Cardozo, Changes in size and shape of synaptic connections after visual training: an ultrastructural approach of synaptic plasticity, *Brain Res.* 218:79 (1981).

Discussion of Dr. Geinisman's Presentation

DR. CAIN: Have you thought of using a group that has not been kindled but has had a generalized seizure? This might help confirm some of your interpretations of your results.

DR. GEINISMAN: We have thought about a variety of experiments. However, it's a rather time-consuming task to conduct any of them. So our next step probably would be to study early stages of kindling in order to define whether the changes we observed are associated with long-term potentiation-induced modifications or not, because LTP was shown to evoke similar changes in synaptic morphology.

DR. CAIN: My pitch for generalized control is that it would involve a generalized seizure and you might be able to see different effects in the inner versus the middle layers there. It's just a thought.

DR. GEINISMAN: OK. We'll go along with your pitch but since our morphological studies are extremely time-consuming, it seems more efficient to examine earlier stages of kindling which are not complicated by generalized seizures.

DR. ADAMEC: You did mention LTP and I guess you have not done studies just inducing LTP with trains. You did not mention that. It would be very interesting, to see, though, particularly because LTP induced by trains is relatively short-lived but LTP produced by kindling is relatively permanent. It's possible that the kinds of structural changes you're observing here may be a correlate or perhaps an underlying mechanism of the difference between those two aspects of long-term potentiation. Are you planning to do LTP studies?

DR. GEINISMAN: Yes. With regard to your first question, we indeed have already started LTP studies which would allow us to ascertain LTP-induced changes in synaptic morphology with the aid of unbiased stereological and morphometric techniques. It is conceivable that the structural synaptic alterations we observe following kindling may be due to LTP which accompanies the earlier stages of the kindling process. In order to assess possible effects of kindling-associated LTP on synaptic morphology, we also plan to conduct experiments in which synapses will be examined morphologically during earlier stages of hippocampal kindling, and the results obtained will be compared with those reported here for the fully developed kindled state. As far as your second question is concerned, all perforated and most nonperforated axospinous synapses in the middle molecular layer of the dentate gyrus are formed by axons originating in the medial entorhinal cortex. These synapses are presumed to be glutamatergic, and at present there is no reliable marker to immunocytochemically label such synaptic contacts.

DR. APPLEGATE: Would you be willing to speculate about the relationship between your morphological findings and those increases in Timm's staining in the molecular layers following kindling reported by Dr. Sutula?

DR. GEINISMAN: Dr. Sutula and his colleagues have reported that perforant path kindling results in sprouting of mossy fiber axons and terminals in the supragranular zone of the inner molecular layer of the dentate gyrus. This synaptic reorganization, however, cannot be attributed, at least entirely, to the kindling process, since synchronous, low-frequency stimulation also was found by Dr. Sutula to be accompanied by such a change. Our study has shown that a significant decrease in the numbr of axospinous synapses per neuron occurs in the supragranular zone with kindling, but not with low-frequency activation of the perforant path. Assuming that axospinous synaptic contacts are, nevertheless, newly formed by mossy fibers in the supragranular zone of kindled animals, one would expect these synaptic junctions to exhibit ultrastructural features typical of mossy fiber synapses on dendritic spines in hippocampal field CA3. However, such typical mossy fiber synapses have not been observed by us in the dentate supragranular zone of kindled animals. In spite of these observations, the possibility cannot be ruled out that mossy fiber axons and terminals do sprout in the supragranular zone following kindling and establish there atypical axospinous synaptic contacts in response to the partial deafferentation.

DR. BURNHAM: I'm most interested in your material. I wonder about the functional implications of it. In this pathway I tend to think of the effect of kindling as being increased inhibition which lasts 2 or 3 weeks, and gradually dies away leaving no change. Wouldn't your data seem to predict increased excitation?

DR. GEINISMAN: Stimulation of the perforant path with parameters that induce kindling may result in both an increased excitatory synaptic drive via input to the granule cell dendrite and in an augmented inhibition of the granule cell due to an activation of intercalated inhibitory neurons. However, our observations are limited to axospinous synapses which are presumed to be excitatory in their action on the dentate granule cell. Since we find an increase in the relative proportion of perforated axospinous synapses formed by perforant path axons and an enlargement of the postsynaptic density in these synaptic contacts, our findings carry the functional implication of an augmented synaptic drive at that particular circuit element and at that particular stage of kindling. Such an interpretation is consistent with electrophysiological data of Maru and Goddard showing that the efficacy of excitatory synaptic transmission at the perforant path-granule cell synapse increases with perforant path kindling and remains augmented for at least one month after the last kindling trial. Nevertheless, our observations do not exclude the possibility that an enhancement of inhibition may occur or has occurred at other sites of the circuit and at other stages of kindling.

DR. BERMAN: If I read your slides correctly the nonperforated synapses far outnumber the perforated synapses and if you are losing 20-23% of those and only gaining 8% of the perforated, you would expect a much larger gain in the perforated synapses if it's a simple conversion.

DR. GEINISMAN: Our data demonstrate that numbers of both perforated and nonperforated axospinous synapses per neuron are significantly diminished in the inner molecular layer of the dentate gyrus of kindled rats. This change is likely to result from generalized seizures, since the inner molecular layer was not directly stimulated in our experiment. In the middle molecular layer, however, only the number of nonperforated synapses was decreased with kindling. Contrary to that, perforated synapses of this subdivision of the molecular layer showed a trend towards an increase in their number per neuron. A possible interpretation of these findings is that a loss of perforated synapses may also occur in the middle molecular layer of kindled rats as a consequence of generalized seizures. However, this decrement in the number of perforated synapses may be compensated for by the new formation of such synaptic contacts. If this is the case, then an actual kindling-induced increase in the number of perforated synapses in the middle molecular layer would exceed 30% which would be consistent with the hypothesis of a conversion of nonperforated synaptic contacts into perforated ones.

DR. MODY: We have had the electrophysiological proof, if you like, for some of the enhancement of the excitatory activity in this pathway because in this pathway where normally NMDA receptors do not participate in synaptic transmission, they suddenly become involved in kindling, which could be accounted for by this increase in the post-synaptic densities. I have a question, however. How much do you think the spine neck configuration changes or the spines change with such a 20% enhancement in the post-synaptic density areas?

DR. GEINISMAN: Thank you very much for your comment. It certainly gives credence to the argument that kindling produces a durable enhancement of granule cell responsiveness to the excitatory synaptic drive. I would also like to answer your question concerning possible effects of an increase in the area of the postsynaptic density (PSD) on the size and shape of dendritic spines. While there are no data regarding relationships between PSD dimensions and the spine configuration, the PSD area has been reported to correlate in a highly significant and positive manner with the size of the dendritic spine head. Therefore, an expansion of the PSD area is likely to be accompanied by an enlargement of the spine head. As far as the neck of spines is concerned it is a very difficult proposition to study. I'm not going to do that.

DR. ADAMEC: Are you planning to do LTP studies? One might expect not to see the structural changes you describe following LTP produced by non-convulsant trains, which is not permanent. But the structural changes you do see after kindling may be a correlate or a mechanism of the more permanent LTP observed following kindling. Is it possible to immunohistochemically label the perforated and non-perforated synapses? Is it possible that the favourable ratios of perforated to non-perforated synapses following kindling represent a rebalancing of excitatory and feedforward inhibitory inputs?

DR. GEINISMAN: With regard to your first question, we indeed have already started LTP studies which would allow us to ascertain LTP-induced changes in synaptic morphology with the aid of unbiased stereological and orphometric techniques. It is conceivable that the structural synaptic alterations we observed following kindling may be due to LTP which accompanies the earlier stages of the kindling process. In order to assess possible effects of kindling-associated LTP on synaptic morphology, we also plan to conduct experiments in which synapses will be examined morphologically during earlier stages of hippocampal kindling, and the results obtained will be compared with those reported here for the fully developed kindled state. As far as your second question is concerned, all perforated and most nonperforated axospinous synapses in the middle molecular layer of the dentate gyrus are formed by axons originating in the medial entorhinal cortex. These synapses are presumed to be glutamatergic, and at present there is no reliable marker to immunocytochemically label such synaptic contacts.

DR. WADA: A synaptic remodelling has been suspected from the outset, I believe, by the discoverer of this phenomenon, and I am sure, wherever he is now, he must be very pleased to know of Dr. Geinisman's work, which has opened a new chapter for the evidence of structural plasticity induced by kindling. Another study on hippocampal slice subjected to kindling-like treatment by Hawrylak and his colleagues reported an increase in sessile spine and shaft synapses. The question remains whether or not these changes are truly permanent.

# FOREBRAIN AND BRAINSTEM MECHANISMS GOVERNING KINDLED SEIZURE DEVELOPMENT: A HYPOTHESIS

James L. Burchfiel and Craig D. Applegate

Comprehensive Epilepsy Program, Department of Neurology
University of Rochester School of Medicine, Rochester NY 14642

## INTRODUCTION

What is the nature of the kindling process? "Kindling" was coined by Goddard and his colleagues [20] to describe an observed phenomenon - the progressively increasing seizure generalization seen with repeated, temporally spaced stimulations of certain brain structures. From this phenomenological point of view, kindling is well named, since it appears to be a self-perpetuating process which once started continues inexorably to completion.

But is kindling what it appears to be? Is the external appearance of an apparently continuous and unitary phenomenon mirrored in the internal neural events which drive the kindling process? Specifically, is kindling a continuous, cumulative process involving many small, subtle changes of neuronal organization which are distributed more or less evenly over time and space? Or is kindling a more discontinuous process involving discrete, independent transitions from one state of neuronal organization to another? Such a fundamental understanding of the nature of kindling is necessary if we are to rationally investigate its neural mechanisms. The answer will provide a framework which can guide us in deciding where to look for kindling mechanisms and what to look for.

We believe that kindling is a discontinuous process involving a stepwise progression of seizure generalization. This conclusion is based on our investigations using kindling antagonism, a modification of the standard kindling model. These investigations have unmasked an underlying architecture of kindled seizure development which provides significant insight into the phenomenon. From this insight, we have developed an hypothesis concerning the nature of the neural events which drive the kindling process.

## THE KINDLING ANTAGONISM MODEL

Kindling antagonism is a modification of kindling which involves the concurrent, alternate stimulation of two limbic system structures [9,16]. The consistent outcome of this alternating, two-site stimulation paradigm is a fundamental disparity in the seizure development induced from the two sites (Figure 1). One site undergoes a typical kindling progression culminating in a fully generalized motor convulsion. By contrast, kindled

*Kindling 4*
Edited by J. A. Wada
Plenum Press, New York, 1990

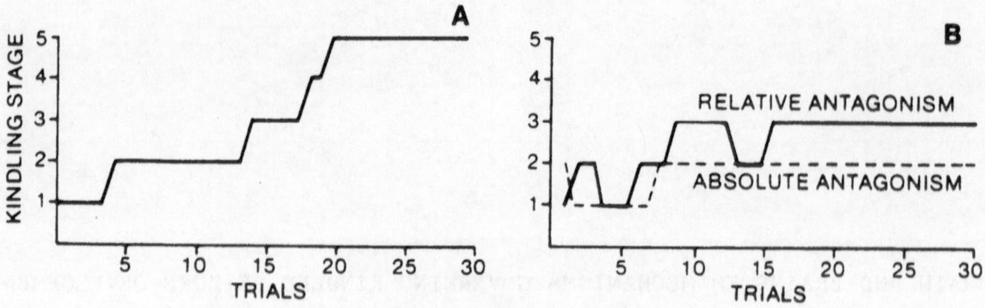


Figure 1. Kindling Antagonism. Stimulation of two limbic system sites in alternation on a trial-by-trial basis leads to a consistent disparity of kindled seizure development from the two sites. One site (dominant site, A) exhibits a typical progressive increase of seizure manifestations culminating in a fully generalized stage 5 seizure. By contrast, seizure development from the other site (suppressed site, B) arrests at an incomplete stage of generalization. Two distinct classes of kindling antagonism can be distinguished based on the kindling stage at which the process of seizure generalization is arrested: (1) Absolute antagonism which arrests at stage 1-2, and (2) Relative antagonism which arrests at stage 3. All animals exhibit one of these two classes of antagonism, and the arrest of the suppressed site is stable as long as the alternating two-site stimulation is continued.

seizure development from the other site is incomplete; seizure generalization proceeds to a certain stage, but then goes no further, despite continued stimulation.

This failure of kindled seizure development from one of the two sites is not due to lack of adequate stimulation. Stimulation of either site consistently elicits epileptiform afterdischarge (AD) activity, and the characteristics of these ADs are essentially identical. In the two-site, alternating stimulation paradigm, the site which fails to kindle can experience a number of ADs which is several times more than that required to induce fully generalized seizures if the site were stimulated alone.

Thus, it is the <u>effect</u> of the AD which fails, not the AD itself. This impotence of the AD to induce the necessary neural alterations to further its propagation represents a basic failure or antagonism of the kindling process. We have termed this phenomenon "kindling antagonism". In describing the condition of kindling antagonism, we refer to the site which exhibits complete kindling as the "dominant" site, and the site which fails to kindle as the "suppressed" or "antagonized" site.

Our investigations have consistently distinguished two classes of antagonism which we have termed "absolute" and "relative". These two classes are delineated by the kindling stage at which the process of seizure generalization becomes arrested (Figure 1). In absolute antagonism, seizure development is arrested at stage 1-2 behavior. Animals exhibit immobility during the AD with occasional episodes of facial twitching, but they exhibit no overt signs of motor convulsive behavior. In relative antagonism, the process halts at stage 3. The distinguishing feature of relative antagonism is that the animals exhibit no bilateral seizure manifestations. They show head nodding, clonic chewing and occasional contralateral forelimb extension or clonus, but the convulsive activity never affects both forelimbs and there is no generalized tonus.

These two classes of antagonism are quite distinct and consistent. Each animal subjected to the alternating, two-site stimulation paradigm will exhibit one of these two classes of behavior or no antagonism at all; there are no other distinguishable classes of antagonism. Absolutely antagonized animals exhibit only stage 1 or 2 behavior; they never exhibit any seizure manifestations of stage 3 or higher. Likewise, relatively antagonized animals exhibit predominantly stage 3 behavior, although they may also exhibit occasional trials of lower stage 1 or 2 behavior. Relatively antagonized animals also may exhibit rare, isolated stage 4 behavior for a few seconds during a seizure, but this is never the predominant manifestation, and it is never repeated consistently from trial to trial. We have never observed a sustained arrest of seizure development at stage 4. We have also never observed an animal to exhibit a prolonged, well-developed stage 5 seizure and then return to a state of stable relative antagonism. In our experience, once an animal exhibits consistent or well-developed bilateral convulsive motor manifestations of stage 4 or 5, it always progresses to full generalization.

Thus, the kindling antagonism phenomenon defines two critical transitions in the process of kindled seizure development. The first is the transition from stage 1 and 2 behavior to stage 3, and the second is from stage 3 to stage 4 and 5. That seizure development becomes arrested exclusively at these two transition points strongly suggests that they represent major steps in the kindling progression at which some discrete process of neural reorganization must take place before seizure development can proceed to the next phase of generalization.

## THE RELATIONSHIP OF THE KINDLING ANTAGONISM MODEL TO STANDARD KINDLING

The relevance of these findings derived from the kindling antagonism model for understanding the process of kindled seizure development depends on whether the kindling antagonism model operates within the same mechanistic frame as standard, single-site kindling. Theoretically, one could envision two basic hypotheses to explain the failure of kindling from the suppressed site during the alternating, two-site antagonism paradigm. One is that kindling antagonism introduces a persistent, paradigm-specific neural reorganization which differs significantly from that occurring during standard, single-site kindling. According to this theory, the antagonism paradigm calls into play a unique and persisting inhibitory mechanism which suppresses kindled seizure development from one of the two stimulated sites.

The alternative hypothesis is that kindling antagonism involves the same mechanisms as single-site kindling. According to this view, the process of neural reorganization which occurs during standard kindling also occurs in the antagonism paradigm, but it is completely effective for only one of the two sites (the dominant site). For the suppressed site, this normal process of kindled seizure development halts and remains incomplete.

It needs to be emphasized that these two basic hypotheses lead to very different conclusions concerning the significance of the antagonism model for understanding kindled seizure development. The first hypothesis supposes a persistent neural reorganization which is unique to the antagonism paradigm. Such a paradigm-specific reorganization might have considerable intrinsic interest, but it would have little relevance to the processes underlying normal kindling. By contrast, the second hypothesis envisions that the suppressed site fails to develop completely generalized seizures because the normal kindling process fails to advance beyond a certain stage. In this view, kindling antagonism represents an arrest of the normal kindling process. By arrest we mean that the neural reorganization underlying kindling fails to progress. The fundamental mechanisms of neural reorganization underlying kindling, however, are not permanently altered. Kindled seizure development from the suppressed site stops without the introduction of a new, permanent mechanism specific to the antagonism paradigm. If this interpretation of the suppression of kindled seizure development is true, then the kindling antagonism model has direct relevance to standard, single-site kindling. Elucidation of the mechanisms underlying suppression could cast considerable light on basic kindling mechanisms.

The weight of evidence gathered in our laboratory over the last several years strongly supports the hypothesis that the mechanisms underlying kindled seizure development in the antagonism paradigm are the same as those underlying standard, single-site kindling, and that kindling antagonism constitutes an interruption of the normal process of kindled seizure development. These data are discussed in detail in a recent review [8]. Briefly, there are two lines of evidence favoring the conclusion that the mechanisms of kindling antagonism do not differ significantly from those of standard kindling.

The first involves consideration of the characteristics of kindling in the antagonism paradigm. During the antagonism paradigm, the behavioral and electrographic characteristics exhibited from either the dominant or suppressed site are indistinguishable from those exhibited from a single site during standard kindling [8-10,16]. Thus, there are no unique seizure manifestations elicited during the kindling antagonism paradigm which distinguish it from standard, single-site kindling. If one were to observe an animal having a stage 3 kindled seizure, one could not predict on the basis of any qualitative or quantitative characteristic of the seizure whether the animal were being kindled in a single-site paradigm or in an alternating, two-site antagonism paradigm.

The second line of evidence suggesting that the kindling process is arrested, rather than fundamentally altered, during kindling antagonism concerns the potential of the suppressed site for continued seizure development. If there were a persistent, paradigm-specific reorganization which inhibited seizure development from the suppressed site, then one would predict that the potential for its continued kindling would be handicapped. In other words, the suppressed site should show some sign of residual inhibition in the form of retarded kindling following the antagonism paradigm. Our experimental data do not confirm this prediction [6;8; Burchfiel & Applegate, submitted]. We have investigated the state of the suppressed site by stimulating it alone on consecutive trials after a condition of antagonism has been established. Under these conditions which mimic standard, single-site kindling, the suppressed site exhibits normal kindled seizure development. Both qualitatively and quantitatively, the behavioral and electrographic characteristics of this post-antagonism kindling are the same as those of standard kindling.

In conclusion, the evidence strongly suggests that during the condition of kindling antagonism the suppressed site exists in a state of arrested, but otherwise normal, kindled seizure development.

## THE ARCHITECTURE OF THE KINDLING PROCESS

The evidence reviewed above strongly favors the concept that kindling antagonism operates within the same mechanistic framework as standard, single-site kindling. There is no evidence that the antagonism paradigm introduces a persistent neural reorganization which differs fundamentally from the process driving standard kindling. Rather, this paradigm appears to generate the typical kindling process for both sites, but, for some as yet unknown reason, this process is carried to completion for only one of them. The process at the other site arrests at an intermediate stage. The reality of this arrested seizure development has major implications for an understanding of the mechanisms underlying kindling.

The most striking implication is that the process of seizure generalization engendered by kindling occurs in a stepwise manner, rather than as a smooth continuous progression. That kindled seizure development can be arrested in the antagonism paradigm implies that the process of neural reorganization induced by kindling involves discrete transitions from one state of neural organization to another. These transitions constitute major steps in kindled seizure development. We conceptualize these steps as being controlled by "gates" which attenuate the ability of the AD to effect the necessary neural reorganization which drives the kindling process. For kindling to progress from one phase of seizure generalization to the next, a gate must be opened to allow the AD to alter the functional organization of a new element of neural circuitry. The gate opens for the dominant site, but remains closed for the suppressed site.

Our data suggest that there are two such critical gates or transitions which must be surmounted during kindled seizure development. We postulate that the first gate controls the transition between early manifestations of kindled seizure behavior (stages 1 and 2) and a middle phase of kindling in which stage 3 seizures predominate. The second gate then controls access from this middle phase to a late phase of kindling which culminates the process of seizure generalization (stages 4 and 5). These gates are defined by the characteristic manner in which the kindling process arrests during the antagonism paradigm.

At this point we need to emphasize that we are using the term gate in a metaphorical sense to describe the stepwise nature of the transitions underlying kindled seizure development. We feel that it is an apt metaphor, since these transitions involve a relatively abrupt change in the functional state of neural circuitry. The actual anatomical and physiological bases of the gates remain to be elucidated. Basically, the gates could operate either by preventing the AD from reaching critical circuitry or by influencing the state of potentiation within these neural circuits. At present, the data provide no clear insight into the fundamental mechanisms underlying the gates. The evidence strongly suggests, however, that norepinephrine (NE) exerts a major influence over the gate mechanisms. Our data are most clear for the second gate involving the transition from stage 3 to stages 4-5.

## The Gate to Stages 4-5: Relative Antagonism and Brainstem Norepinephrine

This gate appears to be dependent on a norepinephrine (NE) mechanism in the brainstem. This conclusion is based on the results of experiments involving the expression of relative antagonism under various conditions of brain NE depletion induced by 6-hydroxydopamine (6-OHDA). We have demonstrated that treatment of adult rats with 6-OHDA, which results in whole brain depletion of NE, virtually eliminates the expression of kindling antagonism [2,5]. Under the condition of whole-brain NE depletion, the alternating two-site stimulation paradigm fails to produce the characteristic arrest of kindled seizure development from one of the sites. Rather, both sites show identical patterns of typical kindling progression resulting in stable fully generalized seizures.

In contrast to the effect in adult animals, 6-OHDA treatment of neonatal rat pups produces a differential pattern of NE concentrations when assayed at maturity [3,5]. In the forebrain, there is significant depletion, while in the hindbrain (brainstem and cerebellum), there is a profound hypertrophy of NE innervation. This pattern of NE alteration did not affect the expression of kindling antagonism. Animals subjected to the two-site, alternating stimulation paradigm following neonatal 6-OHDA treatment showed typical patterns of relative antagonism.

In summary, whole brain depletion of NE abolishes kindling antagonism, while a NE pattern of forebrain depletion and hindbrain hypertrophy maintains antagonism. These findings strongly suggest that a hindbrain NE-dependent mechanism is responsible for the arrest of kindled seizure development experienced by the suppressed site during relative antagonism.

If this postulate is correct, then one would predict that animals treated as neonates with 6-OHDA would not develop kindling antagonism if their remaining hindbrain NE were depleted by subsequent 6-OHDA treatment as adults. We have recently confirmed this prediction [7;Applegate & Burchfiel, submitted]. We treated a group of neonatal rat pups with 6-OHDA according to the regimen which produces almost total elimination of forebrain NE accompanied by significant increases of hindbrain NE levels [23,39]. Then, at adulthood these animals received one of three treatments: (1) a low dose of 6-OHDA (75 ug administered intracisternally in 3 spaced injections over 7 days); (2) a high dose of 6-OHDA (200 ug administered intraventricularly in one dose via bilateral injections); or (3) a control injection of drug vehicle (some administered intracisternally according to the low dose procedure, and some administered intraventricularly according to the high dose procedure).

The results of this experiment are summarized in Figure 2. Biochemically, the control animals exhibited the expected pattern of NE forebrain depletion and hindbrain hypertrophy. Both of the subsequent adult treatments with 6-OHDA produced further reductions of forebrain NE. In the hindbrain, however, the two doses had different effects. The low dose depleted the cerebellum of NE, but did not significantly affect the levels of NE in the pons-medulla. The high dose significantly reduced NE levels throughout the hindbrain.

Behaviorally, only the high dose of 6-OHDA significantly disrupted the expression of kindling antagonism. Most of the animals in both the vehicle-treated group and the low dose 6-OHDA group developed relative antagonism (6 of 7 and 6 of 8, respectively), and these proportions were not significantly different from our previous results following neonatal 6-OHDA treatment (7 of 9; Ref 3). By contrast, only 3 of 9 animals treated with the high dose of 6-OHDA exhibited any signs of arrested seizure development from one of the two alternately stimulated sites.

These results lead to several conclusions regarding the mechanisms underlying kindling antagonism. First, the data strongly support the hypothesis that the manifestation of arrested seizure development is dependent on NE. Second, the results clearly indicate that the critical compartment of brain NE is in the pons-medulla (caudal to the level of the superior colliculus). This conclusion is evident from comparison of the biochemical and behavioral profiles in Figure 2. The high dose 6-OHDA group is the only one which shows significant disruption of antagonism, and the only biochemical feature which distinguishes this group from the others is the depletion of NE in the pons-medulla. All three groups have significant forebrain depletion, and both the high and low dose groups have significant cerebellar depletion; therefore, none of these NE alterations can account for the unique outcome of the high dose treatment.

Finally, the data indicate that the kindling transition for which brainstem NE is important is the one defined by relative antagonism, the

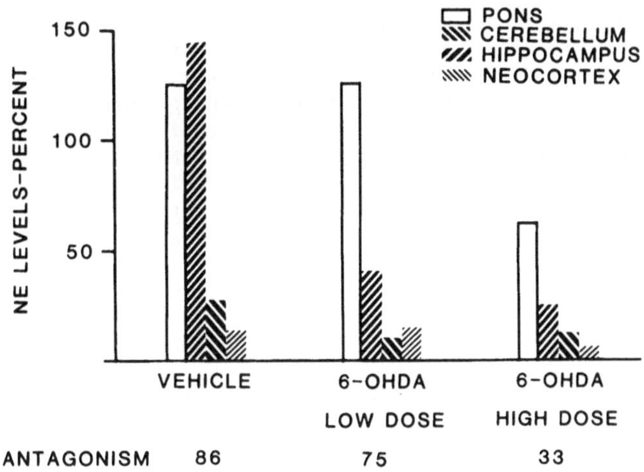


Figure 2. Results of combined neonatal and adult treatment with 6-OHDA. Animals received 6-OHDA as neonates and then had one of three different treatments as adults: (1) vehicle, (2) low dose 6-OHDA, or (3) high dose 6-OHDA (see text for experimental details). NE levels are expressed as percent of values measured in controls treated with vehicle both neonatally and as adults. Also listed is the percentage of animals in each of the adult treatment groups that exhibited relative antagonism. Note that only NE depletion of the pons-medulla (high dose 6-OHDA treatment) significantly affected the expression of relative antagonism.

transition from the phase of partial, predominantly stage 3 seizures to the phase of generalized seizures characterized by stage 4 and 5 behavior. All the animals in this recent study exhibited only relative antagonism following neonatal 6-OHDA treatment; therefore, it was exclusively this form of arrested seizure development which was disrupted by the subsequent depletion of brainstem NE.

## The Gate to Stage 3: Absolute Antagonism and Forebrain Norepinephrine

Absolute antagonism defines another gate in the kindling process which controls the transition from stage 1-2 seizure manifestations to stage 3 seizures. Evidence strongly suggests that this gate is also NE-dependent. The major evidence implicating NE in our antagonism model comes from depletion studies. Recently [Applegate & Burchfiel, submitted], we have performed robust whole brain NE depletions via bilateral intraventricular injections of 6-OHDA. This treatment completely abolished the expression of antagonism, both absolute and relative. Thus, absolute antagonism, like relative antagonism, appears to require normal levels of brain NE.

The weight of evidence favors a forebrain location for this NE-dependent kindling gate which fails to open during absolute antagonism. This contrasts with the brainstem location for the NE-dependent mechanism which governs the transition of kindled seizure development associated relative antagonism.

Evidence from the kindling antagonism model which supports a forebrain location for this early NE-dependent gate is presently somewhat equivocal. If this early gate, which remains closed for the suppressed site in absolute antagonism, were modulated by forebrain NE, then one would predict that selective depletion of this forebrain NE compartment should prevent the development of absolute antagonism. Unfortunately, our experimental results have so far neither clearly confirmed nor denied this prediction. In an early study, we reported that absolute antagonism did occur in a small number of animals following selective forebrain NE depletion induced by neonatal 6-OHDA treatment [3]. On the other hand, the results of our recent studies have been consistent with the predicted outcome [Applegate & Burchfiel, submitted]. Absolute antagonism was not observed in any of the animals who were treated as neonates with 6-OHDA and exhibited at maturity the typical pattern of forebrain NE depletion and hindbrain NE hypertrophy. This issue is currently under active investigation.

The major body of evidence supporting a forebrain NE compartment for the early kindling gate comes from the effects of NE depletion on the rate of seizure development in the standard kindling model. It is well established that selective depletion of brain NE significantly facilitates the rate of kindling [14,28]. It is also clear that forebrain NE plays a major role in this effect. Depletion of forebrain NE alone by lesions of ascending noradrenergic pathways accelerates the rate of kindling as significantly as does whole brain depletion [15,17]. A similar facilitation of kindling results from neonatal 6-OHDA treatment which depletes forebrain NE but produces an hypertrophy of noradrenergic innervation in the hindbrain [3,29]. Forebrain NE depletion also facilitates the rate of kindling displayed by the dominant site in the kindling antagonism paradigm [3].

The principal means by which forebrain NE affects the rate of kindling is by regulating the transition from stage 1 and 2 behavior to stage 3 seizures, the same behavioral transition which is blocked by absolute antagonism. Thus, the major effect of forebrain NE depletion is to reduce the number of trials during which the AD elicits only stage 1 and 2 seizure behaviors; the effect on the later stages of kindling is less significant [15;Applegate & Burchfiel, submitted]. A similar differential effect on the early stages of kindling is observed with an increase of forebrain NE induced by stimulation of the locus ceruleus prior to each kindling trial [21]. Such stimulation slows the rate of kindling, and this retardation is due exclusively to an increase in the number of trials spent in stage 1.

Thus, during the kindling process, forebrain NE affects predominantly the same transition in seizure development which is associated with the early gate expressed during absolute antagonism. This provides indirect, but strong, evidence for a NE-dependent mechanism regulating this gate.

## SUMMARY OF THE CRITICAL PHASES IN THE KINDLING PROCESS: AN HYPOTHESIS

The major insight that we have derived from the kindling antagonism model is that kindled seizure development occurs by a process of discrete, stepwise reorganization of neural mechanisms. The antagonism phenomenon defines two critical transitions of neural reorganization: one which advances kindling from stage 1-2 seizure manifestations to stage 3 seizures, and the other which advances the process from stage 3 to the culmination of seizure generalization at stages 4-5. In our conceptualization these transitions act like gates in the process; each must be opened in order for kindled seizure development to progress to the next phase of generalization. We believe that opening these gates is the basis of kindling, and that failure to open these gates is the basis of the kindling antagonism phenomenon.

The two gates defined by kindling antagonism divide the kindling process into three separate phases which are different from, but overlap, the traditional behavioral stages of kindling defined by Racine [33]: (1) an early phase during which the AD elicits non-convulsive motor behavior classified according to the traditional kindling behavior scale as stages 1 and 2; (2) a middle phase which involves partial convulsive motor behavior falling predominantly into stage 3; and (3) a late phase during which fully generalized, stage 4-5 convulsions are developed. Each phase involves an essential neural reorganization which advances the kindling progression to a more extensive degree of seizure generalization. We are still a long way from understanding the mechanisms which drive these reorganizations of neural function, but we believe that this conceptualization of kindling as a series of discrete, separate steps can provide a framework within which to search for mechanisms.

This hypothetical schema for the stepwise reorganization taking place in the kindling process is outlined in Figure 3, and in the following sections we will briefly summarize what we know or suspect about the three hypothesized kindling phases.

## Early Kindling Phase

In our conceptualization, this phase involves local events at the site of stimulation which determine AD characteristics, most notably threshold, and also probably a majority of the factors which determine duration. This AD phase appears to be independent of the kindling process per se. In other words, changes in the AD can occur independently of changes in kindled seizure development and vice versa.

This dissociation of AD characteristics and kindled seizure characteristics is most clearly evident for AD threshold. Racine [32] demonstrated many years ago that electrical stimulation alone could significantly lower AD threshold without affecting the subsequent characteristics of kindled seizure development. Likewise, a secondary, transfer site can exhibit significant signs of enhanced kindled seizure susceptibility without having any demonstrable change of AD threshold [11,32].

The mechanisms which govern AD duration also appear to be somewhat independent of those which drive kindled seizure manifestations. For example, we have demonstrated that enhancement of GABAergic inhibition in the amygdala by intracerebral injections of the GABA agonist, muscimol, or the GABA transaminase inhibitor, gamma vinyl GABA, completely block the motor expression of a kindled seizure without significantly affecting the local AD duration [1]. A similar dissociation was observed by Jimenez-Rivera, et al. [21] with locus ceruleus (LC) stimulation during kindling. They found that LC stimulation arrested kindled seizure development at stage 1, but despite this lack of behavioral progress, the local AD duration continued to lengthen.

The final distinguishing characteristic of this early phase is that, unlike the subsequent phases, it does not appear to involve any NE-dependent mechanisms. Depletion of brain NE does not significantly affect either AD threshold or duration [2,3,15,29]. AD characteristics are unaffected even after local NE depletion at the site of stimulation [26].

## Middle Kindling Phase

Entry into this phase is blocked by absolute antagonism. The data reviewed earlier in this paper suggest that this phase involves a breakdown of forebrain NE inhibitory influences which allows the local AD to propagate into other forebrain structures. The driving of these additional structures by the epileptiform discharge would then elicit progressively more advanced motor seizure behaviors. Specifically, there would be a transition from stage 1-2 seizure manifestations to stage 3.

In our conceptualization, this step involves the first substantial advancement of kindled seizure development. Our data suggest that prior to the development of stage 3 seizures, little, if any, actual progress of the kindling phenomenon has occurred. These data are derived from the state of the suppressed site during absolute antagonism. Even though the absolutely suppressed site has experienced many ADs during which stage 1-2 behavior is expressed, this does not significantly affect its susceptibility to further kindled seizure development. If the alternate, two-site stimulation paradigm is stopped, and the absolutely suppressed site is stimulated alone, its subsequent kindling development is not significantly different from that of a naive site.

It is the transition into this middle phase of the kindling process which appears to be chiefly responsible for differences in the rate of kindled seizure development. This issue has been addressed directly by Le Gal La Salle [24] who analyzed the stage-by-stage progression of kindled seizure development from a variety of sites in the amygdala complex. This analysis clearly demonstrated that the number of trials required to elicit the first generalized seizure varied in parallel with the number of trials

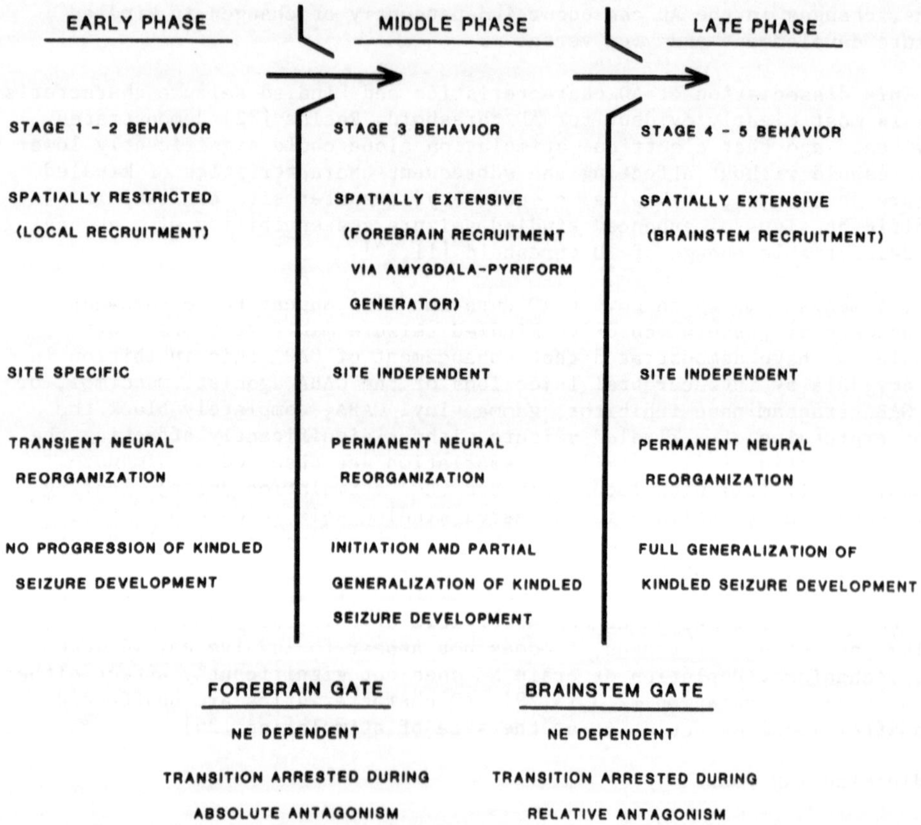


Figure 3. Hypothetical model of the stepwise progression of kindled seizure development. This model is based on data from the kindling antagonism paradigm which strongly suggests that the kindling process involves two discrete transitions or "gates", and these gates divide the process into three distinct phases of neural reorganization. See text for details.

spent in stages 1 and 2, but not stages 3 and 4. Animals with slow kindling rates spent proportionately more time in stages 1-2 than did animals with fast kindling rates; whereas, the number of trails spent in stages 3-4 was essentially constant regardless of the overall kindling rate.

We speculate that these findings of Le Gal La Salle may have general applicability to most, if not all, kindling sites. We suggest that, in general, the major distinguishing feature between slow kindling sites and fast kindling sites may be the number of trials spent in stages 1-2. The evidence supports this view. We have observed that the relatively protracted kindling from either the septum or the entorhinal cortex is due mainly to a greater number of stage 1-2 trials [16], and a similar situation has been reported for hippocampal kindling [40].

This concept that kindling rate depends significantly on the amount of time spent in the early stages of the process contributes an important element to our hypothesis. We theorize that the major factor which determines the kindling rate from a particular stimulation site is the facility with which the AD generated at that site can open the forebrain gate and effect the transition from stage 1-2 to stage 3, i.e. the transition from the early phase of kindling to the middle phase. Thus, the capacity of slow kindling sites to effect this transition would be weaker than that of faster kindling sites, and consequently, the slower sites would have to spend proportionately more trials in stage 1-2.

The corollary of this concept is that once the forebrain gate is surmounted, the subsequent process of kindled seizure development proceeds essentially the same for all kindling sites. Thus, the middle and late phases of the kindling process would be largely independent of the site of stimulation; once the AD from a given site has gained access to the circuitry which mediates the middle phase of kindling, then the rate of progression from stage 3 to stage 5 would depend little on the origin of the AD.

Another interesting question raised by this hypothesis is how the phenomenon of positive transfer fits into the schema? At present, we do not have any data bearing on this question. However, one can postulate various ways in which primary and secondary kindling sites might interact with the gates, and these postulates lead to specific, testable predictions regarding transfer characteristics. Such experiments are currently underway.

What are the critical forebrain structures in which the neural reorganization associated with the middle phase of kindling occur? A body of evidence suggests that the amygdala-pyriform region is likely to play a major role. This region fulfills many of the criteria one might predict for a structure intimately related to the middle phase. For example, kindling from the sites within the amygdala-pyriform region proceeds rapidly with a minimum of time spent in the stage 1-2 manifestations characteristic of the early phase of kindling [20,30,34]. The most impressive evidence implicating the amygdala-pyriform is that this region consistently leads other areas in generating electrophysiological signs of enhanced epileptogenicity during the kindling process. Racine and his co-workers [22,37,38] have searched extensively in the brain for the development of spontaneous interictal spikes during kindling and have found that the amygdala-pyriform area appears to be the generator of these epileptiform discharges, regardless of the actual site of stimulation. Comparatively, therefore, the neural organization of the amygdala-pyriform area appears to be more susceptible than other limbic areas to the development of chronic epileptogenesis during kindling. This conclusion has received support recently from McIntyre's work [27] on in vitro slices of the amygdala-pyriform region. He has demonstrated that slices from kindled animals show an increased propensity to generate epileptiform burst responses in comparison to slices from control animals. Such evidence of increased epileptogenicity is less clear in hippocampal slices from kindled animals.

In summary, we hypothesize that the middle phase of kindling involves the breaking down of a NE-dependent gate in the forebrain which allows the AD to activate epileptogenic-susceptible circuitry. The amygdala-pyriform region appears to be a major component of this susceptible circuitry, although it may not play an exclusive role (cf. 36,37). Within this susceptible circuitry the AD then promotes a process of neural reorganization which leads to reactive discharges and enhanced spontaneous epileptogenic activity. These processes of neural reorganization appear to be common for ADs, regardless of where they are originally triggered. Thus, once the NE-dependent gate is opened, an AD triggered from the hippocampus or septum or elsewhere gains access to a common epileptogenic-sensitive circuitry which initiates the process of kindled seizure development. A similar hypothesis which emphasizes the reactive properties of the pyriform lobe has been formulated by Racine and Burnham [35,36] and Racine and McIntyre [37].

Our hypothesis differs from these earlier, seminal conceptualizations in an important way. We view the activation of the epileptogenic-susceptible circuitry of the amygdala-pyriform region (and, possibly, other regions) as a discrete, stepwise event in the kindling process which actually initiates the mechanisms of permanent neural reorganization underlying the phenomenon. We hypothesize that prior to this event, kindling stimulation evokes site-specific, transient mechanisms associated primarily with development of AD characteristics. Our data from absolute antagonism indicate that no permanent advancement of kindled seizure development occurs in this early phase. Major advancement commences only when the process enters the middle phase with the predominant expression of stage 3 seizure behavior.

## Late Kindling Phase

This is the phase which is blocked by relative antagonism. Our data strongly indicate that the gate which controls access to this phase involves a NE-dependent mechanism in the brainstem. It should be emphasized, however, that our data do not exclude the additional, or perhaps exclusive, participation of more caudal, spinal cord mechanisms in the late phase of kindling (cf. 25).

We hypothesize that the major feature of this late phase is the activation of brainstem circuitry which mediates generalized convulsive motor behavior (stages 4 and 5). The cardinal characteristic of late phase seizures is that they involve bilateral motor activity. This contrasts which with middle phase seizures which are almost exclusively unilateral. Thus, the transition from the middle to the late phase involves a fundamental change in the seizure circuitry activated by the AD from circuitry mediating strictly unilateral motor manifestations to circuitry with bilateral motor projections. We view this transition, like the previous transition from the early to the middle phase, as a discrete, stepwise event in the kindling process. Also like the earlier transition, we believe that entry into the late phase occurs as a consequence of opening a NE-controlled gate. We hypothesize that the AD operates on this gate to break down an inhibitory NE-dependent mechanism which then allows the AD access to the brainstem circuitry mediating bilateral, generalized motor convulsions.

What structures are involved in the late phase of kindling? Our knowledge of brainstem kindling mechanisms is not as advanced as that of forebrain mechanisms; therefore, no leading candidate has emerged in the brainstem with a status comparable to that of the amygdala-pyriform in the forebrain.

Several structures, however, have been implicated. Wada and Sato [41-43] have implicated the midbrain reticular formation (MRF) based on three experimental findings in cats: First, generalized convulsions from unilateral amygdala stimulation readily develop in the absence of all forebrain commissures; second, lesions placed in the MRF ipsilateral to the stimulated amygdala significantly disrupt the development of generalized seizures in either forebrain bisected or intact animals; and third, the emergence of independent AD activity in the MRF is the most significant electrographic landmark for predicting the occurrence of generalization. These data strongly suggest that the MRF is involved in mediating the bilateral seizure manifestations of the late phase.

Browning and his co-workers [4] have more extensively probed brainstem seizure substrates, and they have confirmed the importance of the reticular core. Furthermore, they have identified different regions within the pontine RF which appear to mediate different aspects of clonic and tonic motor manifestations. These studies have been complemented by work from Burnham's laboratory [12] which has demonstrated the ability of the reticular core to produce self-sustained, generalized convulsions with high-intensity electrical stimulation, even in the absence of the forebrain.

Finally, Garant and Gale [19] have recently reported a region within the deep layers of the superior colliculus which appears to have a proconvulsive effect on generalized seizures. They speculate that this structure is the chief target of the GABAergic projections from the substantia nigra which are believed to play a major modulating role in the expression of generalized seizures [18,31].

In summary, there are considerable data suggesting the existence of brainstem mechanisms mediating the expression of generalized seizures. These data have been derived from a variety of seizure models, and their utility for guiding research into the brainstem mechanisms underlying the late phase of kindled seizure development is presently unclear. Whether a common brainstem substrate exists which mediates the expression of generalized seizures in kindling and other models remains to be determined. Recently, Burnham and Browning [13] have taken a significant step in this direction with the development of a theory which proposes such a common substrate within the reticular core of the brainstem and spinal cord.

## REFERENCES


1. Applegate, C.D. and J.L. Burchfiel. Microinjections of GABA agonists into the amygdala complex attenuates kindled seizure expression in the rat. Exp. Neurol. 102: 185-189, 1988.
2. Applegate, C.D., J.L. Burchfiel and R.J. Konkol. Kindling antagonism: Effects of norepinephrine depletion on kindled seizure suppression after concurrent, alternate stimulation in rats. Exp. Neurol. 94: 379-390, 1986.
3. Applegate, C.D., R.J. Konkol and J.L. Burchfiel. Kindling antagonism: A role for hindbrain norepinephrine in the development of site suppression following concurrent, alternate stimulation. Brain Res. 407: 212-222, 1987.
4. Browning, R.A. Effects of lesions on seizures in experimental animals. In: "Epilepsy and the Reticular Formation: The Role of the Reticular Core in Convulsive Seizures", edited by G.H. Fromm, C.L. Faingold, R.A. Browning and W.M. Burnham. New York: Alan R. Liss, 1987, pp. 137-162.
5. Burchfiel, J.L., C.D. Applegate and R.J. Konkol. Kindling antagonism: A role for norepinephrine in seizure suppression. In: "Kindling 3", edited by J.A. Wada. New York: Raven Press, 1986, pp. 213-229.

6. Burchfiel, J.L. and C.D. Applegate. Characteristics of the state of seizure suppression in the kindling antagonism model. Soc. Neurosci. 12: 70, 1986.
7. Burchfiel, J.L. and C.D. Applegate. Kindling antagonism: The effects of combined neonatal and adult 6-hydroxydopamine treatment. Epilepsia 29: 675, 1988.
8. Burchfiel, J.L. and C.D. Applegate. Stepwise progression of kindling: Perspectives from the kindling antagonism model. Neurosci. Bioehav. Rev. in press.
9. Burchfiel, J.L., K.A. Serpa and F.H. Duffy. Further studies of antagonism of seizure development between concurrently developing kindled limbic foci in the rat. Exp. Neurol. 75: 476-489, 1982.
10. Burchfiel, J.L., K.A. Serpa and F.H. Duffy. Kindling antagonism: Interactions of dorsal and ventral entorhinal cortex with septum during concurrent kindling. Brain Res. 236: 3-12, 1982.
11. Burnham, W.M. Primary and 'transfer' seizure development in the kindled rat. In: "Kindling", edited by J.A. Wada. New York: Raven Press, 1976, pp. 61-84.
12. Burnham, W.M. Electrical stimulation studies: Generalized convulsions triggered from the brain-stem. In: "Epilepsy and the Reticular Formation: The Role of the Reticular Core in Convulsive Seizures", edited by G.H. Fromm, C.L. Faingold, R.A. Browning and W.M. Burnham. New York: Alan R. Liss, 1987, pp. 25-38.
13. Burnham, W.M. and R.A. Browning. The reticular core and generalized convulsions: A unified hypothesis. In: "Epilepsy and the Reticular Formation: The Role of the Reticular Core in Convulsive Seizures", edited by G.H. Fromm, C.L. Faingold, R.A. Browning and W.M. Burnham. New York: Alan R. Liss, 1987, pp. 193-201.
14. Corcoran, M.E. Catecholamines and kindling. In: "Kindling 2", edited by J.A. Wada. New York: Raven Press, 1981, pp. 87-104.
15. Corcoran, M.E. and S.T. Mason. Role of forebrain catecholamines in amygdaloid kindling. Brain Res. 190: 473-484, 1980.
16. Duchowny, M.S. and J.L. Burchfiel. Facilitation and antagonism of kindled seizure development in the limbic system of the rat. Electroenceph. Clin. Neurophysiol. 51: 403-416, 1981.
17. Ehlers, C.L., D.K. Clifton and C.H. Sawyer. Facilitation of amygdala kindling in the rat by transecting ascending noradrenergic pathways. Brain Res. 189: 274-278, 1980.
18. Gale, K. Mechanisms of seizure control mediated by gamma-aminobutyric acid: Role of the substantia nigra. Fed. Proc. 44: 2414-2424, 1985.
19. Garant, D.S. and K. Gale. Substantia nigra-mediated anticonvulsant actions: Role of nigral output pathways. Exp. Neurol. 97: 143-159, 1987.
20. Goddard, G.V., D.C. McIntyre and C.K. Leech. A permanent change in brain function resulting from daily electrical stimulation. Exp. Neurol. 25: 294-330, 1969.
21. Jimenez-Rivera, C., A. Voltura and G.K. Weiss. Effect of locus ceruleus stimulation on the development of kindled seizures. Exp. Neurol. 95: 13-20, 1987.
22. Kairiss, E.W., R.J. Racine and G.K. Smith. the development of the interictal spike during kindling in the rat. Brain. Res. 322: 101-110, 1984.
23. Konkol, R.J., E.G. Bendeich and G.R. Breese. A biochemical and morphological study of the altered growth pattern of central catecholamine neurons following 6-hydroxydopamine. Brain Res. 140: 125-135, 1978.
24. Le Gal La Salle, G. Amygdaloid kindling in the rat: Regional differences and general properties. In: "Kindling 2", edited by J.A. Wada. New York: Raven Press, 1981, pp. 31-47.
25. Mason, S.T. and M.E. Corcoran. Seizure susceptibility after depletion of spinal or cerebellar noradrenaline with 6-hydroxydopamine. Brain Res. 166: 418-421, 1979.

26. McIntyre, D.C. Amygdala kindling in rats. Facilitation after local amygdala norepinephrine depletion with 6-hydroxydopamine. Exp. Neurol. 69: 395-407, 1979.
27. McIntyre, D.C. Kindling and the pyriform cortex. In: "Kindling 3", edited by J.A. Wada. New York: Raven Press, 1986, pp. 249-262.
28. McIntyre, D.C. and N. Edson. Facilitation of amygdala kindling after norepinephrine depletion with 6-hydroxydopamine in rats. Exp. Neurol. 74: 748-757, 1981.
29. McIntyre, D.C., M. Saari and B.A. Pappas. Potentiation of amygdala kindling in adult or infant rats by injection of 6-hydroxydopamine.. Exp. Neurol. 63: 527-544, 1979.
30. McNamara, J.O. M. Byrne, R. Dashieff and J. Fitz. The kindling model of epilepsy: A review. Prog. Neurobiol. 15: 139-159, 1980.
31. McNamara, J.O., M.T. Galloway, L.C. Rigsbee and C. Shin. Evidence implicating substantia nigra in regulation of kindled seizure threshold. J. Neurosci. 4: 2410-2417, 1984.
32. Racine, R.J. Modification of seizure activity by electrical stimulation: I. After-discharge threshold. Electroenceph. Clin. Neurophysiol. 32: 269-279, 1972.
33. Racine, R.J. Modification of seizure activity by electrical stimulation: II. Motor Seizure. Electroenceph. Clin. Neurophysiol. 32: 281-294, 1972.
34. Racine, R.J. Kindling: The first decade. Neurosurgery 3: 234-252, 1978.
35. Racine, R.J. and W.M. Burnham. The kindling model. In: "The Electrophysiology of Epilepsy", edited by P. Schwartzkroin and H. Wheal. London: Academic Press, 1984, pp. 153-171.
36. Racine, R.J., W.M. Burnham, M. Gilburt and E.W. Kairiss. Kindling mechanisms: I. Electrophysiological studies. In: "Kindling 3", edited by J. A. Wada. New York: Raven Press, 1986, pp. 263-282.
37. Racine, R.J. and D.C. McIntyre. Mechanisms of kindling: A current view. In: "The Limbic System: Functional Organization and Clinical Disorders", edited by B.K. Doane and K.E. Livingstone. New York: Raven Press, 1986, pp. 109-121.
38. Racine, R.J., M. Mosher and E.W. Kairiss. The role of the pyriform cortex in the generation of the interictal spikes in the kindled preparation. Brain Res. 454: 251-263, 1988.
39. Sachs, C.H., C. Pycock and G. Jonsson. Altered development of central noradrenaline neurons during ontogeny by 6-hydroxydopamine. Med. Biol. 52: 55-65, 1974.
40. Sato, M. and T. Nakashima. Kindling: Secondary epileptogenesis, sleep and catecholamines. Can. J. Neurol. Sci. 2: 439-446, 1975.
41. Wada, J.A. and M. Sato. Generalized convulsive seizure induced by daily electrical stimulation of the amygdala in cats: Correlative electrographic and behavioral features. Neurology 24: 565-574, 1974.
42. Wada, J.A. and M. Sato. The generalized convulsive seizure state induced by daily electrical stimulation of the amygdala in split brain cats. Epilepsia 16: 417-430, 1975.
43. Wada, J.A. and M. Sato. Effects of unilateral lesion in the midbrain reticular formation on kindled amygdaloid convulsion in cats. Epilepsia 16: 693-697, 1975.

## Discussion of Dr. Burchfiel's Presentation

DR. STRIPLING: I have a couple of questions. Is there any way to predict which site will become the dominant one?

DR. BURCHFIEL: No. I did not go into the details of this a lot for lack of time, but in fact, no. As you saw in the earlier slide the AD characteristics and the behavioural expression are exactly the same in the early stages, either initially or up to stage 3, for either the dominant or suppressed site. And it isn't until one of those two sites breaks and begins to actually exhibit 4/5 behaviour, that it becomes clear that now it is going to be the dominant site.

DR. STRIPLING: I wondered if you have looked at any other sites, in particular the pyriform and hippocampus, to get at the questions that Dan brought up.

DR. BURCHFIEL: When we initially started these studies we did look at a variety of sites. And we see this antagonism exhibited among a number of pairings: the septum, the entorhinal cortex, the hippocampus and the amygdala, each of them paired with one another, either in the same hemisphere or opposite hemispheres. They all exhibit this antagonistic behaviour. The interesting thing is that if you pair the amygdala with any of those other sites, the amygdala is always dominant. This is one of the cases in which there isn't an even distribution between the two sites. What we predict from that also is that there would be a greater expression of this absolute type of antagonism between the amygdala and another site for instance the hippocampus or the entorhinal cortex. So that the other site, the suppressed site would be exhibiting essentially stage 1/2 behaviour, which is something we want to explore in more detail.

DR. BURNHAM: Thank you, Jim, for a very nice paper. I think your phases in gates are a very useful concept. One comment on terms and two questions. With the terms, we have Racine's 5 point scale that we've had for some years. Now, you're going away from it to redefine stages 1 and 2. What we used to call automatisms you're calling 1 and 2. That's fine, but I'm beginning to be confused. When you talk about, say, McIntyre's or Corcoran's work and say stage 1 or 2, I'm seriously thinking is that Burchfiel's stage 1 or 2 or Racine's stage 1 or 2? Can we maybe settle on terms and decide that we're going to stick with Racine or abandon Racine? That's the first point. I also have two questions on the gates. The first is this: if you have a gate and you're dealing with an interactive situation, if you open it for one site, why do you close it for the other? Wouldn't you predict you'd open it for both? And second, if your gates are noradrenergic, when you do your 6-hydroxy depletions, whole brain, shouldn't you open both gates and have immediate kindling going all the way on stimulation onto stage 5? Or alternately, if you have a kindled animal, shouldn't something like amphetamine, which raises noradrenaline levels, close both gates and antagonise the seizures fully?

DR. BURCHFIEL: Let's start with the idea of the gate. Mac, will you repeat your second question?

DR. BURNHAM: If you open a gate to one site, shouldn't you open it to both?

DR. BURCHFIEL: It doesn't happen, and that's an interesting phenomenon. It appears that the process of going through that first gate seems to be site specific. I think it's an interesting phenomenon that the ability of one site to make connections with whatever that middle phase is, and go through that progression does not necessarily cause another site to be able to make the same connections. There does seem to be some site specificity. It's an interesting phenomenon, and I'm not sure exactly how that mechanistically can occur. But it does in fact occur. In response to your second question, if you were to eliminate all of these norepinephrine dependent gates, why don't you get immediate kindling? I think one of the things I want to emphasize is that really these gates are separating processes so there's something discrete that has to occur to move from this process to the next process. The processes themselves, like in this middle phase, may not involve and probably do not involve exclusively norepinephrine. There are probably other things that have to go on which do require time and will take up trials in order to cause them to move. There appears to be this transition that allows access to the mechanisms of this middle phase or middle process but those processes still have to go on. So, in fact, you wouldn't expect the thing to go to zero because we know very well that there are other neurotransmitters involved. There may be structural changes that need to occur in order for the kindling process to proceed. What I'm saying is that these norepinephrine dependent processes, or norepinephrine dependent gates, separate these processes, and allow access from one to the next. So there are both gates and processes, there are gates and phases. I think we have seen that one can, for instance, break down this gate. It was very nice to see Dan present some of his transfer data, because it can be interpreted through this framework. What you have done is you're stimulating in the amygdala itself which is bypassing this gate, breaking down this gate, so that the amygdala goes now to a generalized seizure but when you go over to stimulate the hippocampus it still has this gate opened up to get into the same processes that the amygdala started in. Therefore, it's going to take it the 20 stimulations that it normally takes to go from stage 1/2 to stage 3 before it gets into this process. Now, once it does that, though, this process has already been paired to completion. Once it gets to exhibiting stage 3 behaviour, from then on generalization goes very rapidly. That subsequent part has already been done, so that you get partial savings which is basically related to the number of trials spent in 1/2, but as soon as they begin to exhibit stage 3 behaviour they almost immediately go to stage 4/5. So you can effect these processes and gates independently.

DR. RACINE: Remind me - what happens if you impose a delay between the development of your antagonism and subsequent kindling of the suppressed site alone?

DR. BURCHFIEL: What happens is that there is the same progression, but it progresses at a little faster rate. In other words, if you go through the antagonism paradigm, stop the stimulation and then immediately start stimulating the suppressed site it looks from the data I showed that the animal starts at stage 3 and essentially goes on. If you allow them to rest for 2 or 3 weeks, they go at a slightly faster rate, so that there does appear to be some transient inhibitory phenomenon occurring that is expressed with the immediate transfer. I think it's something like the interference effect, and that does decay over a period of time, but the point is that there is no real residual or permanent residual of an inhibitory process.

DR. MOSHE: Your work, Jim, is terrific. Congratulations. You propose that the mechanisms underlying kindling antagonism are similar to the mechanisms responsible for regular kindling. What is the effect of forebrain NE depletion versus forebrain NE depletion and brainstem NE hypertrophy on single site kindling? According to your data you might expect an acceleration between early phase and middle phase and deceleration between middle to late phase in animals with brainstem hypertrophy. Have you looked at that? Did you do the lesions in the neonate as well as in the adult, and then single site kindling?

DR. BURCHFIEL: We've done that and we do get accelerated kindling in the case of single site.

DR. MOSHE: So how does this fit with opening and closing of the gates from the brainstem?

DR. BURCHFIEL: Well, with the neonate what you're doing is eliminating the forebrain norepinephrine, and that goes through the typical observation that one has with accelerated kindling where they spend fewer trials in the early stages. If we do that, that acceleration occurs with single site stimulation. It also occurs with the dominant site. The dominant site goes faster than it would in a non-depleted animal.

DR. MOSHE: Would you be expecting that you have a deceleration between the early phase and middle phase, and less stimulations between the middle phase and the late phase?

DR. BURCHFIEL: That's hard to assess. The problem is the number of trials an animal spends in stage 3 anyway under normal circumstances is not very few, so to be able to tell the difference between 3 or 4 trials spent there and 1 or 2 is very difficult on the population. Our impression is that when you do a whole brain depletion they do in fact accelerate more through all of these phases than if you leave the brainstem intact where they spend approximately a few more trials in 3 or 5. We can't prove that. There's too much overlap.

DR. MCINTYRE: Your final gate, Jim, you describe it as though it's controlling the output of that system, but it would seem like as if it almost would have to somehow be very interactive with the forebrain. I'm trying to see how you might explain something like a split brain animal who would have a generalized seizure on one side and be very calm on the other, if you've opened the gate it just lets the flood rush out.

DR. BURCHFIEL: Our data indicate that that gate is dependent upon the brainstem norepinephrine but it may be that that allows some reverberation to occur between the forebrain and the brain cell structures. I don't think that there's some sort of a key structure down in the brainstem that's controlling the whole thing. It may just be an access into a circuitry that involves both the brainstem and the forebrain. I don't mean to imply that there is a centre that controls 4/5 behaviour which has a door in front of it some place down there that opens up and allows you in. It's obviously more complicated than that. This is simply a schematic to sort of try to express the framework. The actual site for these circuits that are involved in the movement from one phase to another. I mean could involve both forebrain and brainstem structures, and probably do.

DR. ADAMEC: My first question is, is your norepinephrine gate alpha 2 based? The other question I have is, I am reminded when you talk of transition to stage 3 of Engel's and Woolson's 2DG data some time back showing the substantia nigra lighting up during the transition to stage 3, so my question is where does the substantia nigra fit into this model?

DR. BURCHFIEL: That is not my stage 3 - I haven't commented yet on whether we're going to throw Racine out or not. The reason I make that distinction is that for our purposes and our data, that appears to be a critical factor, as to how we define the stage 3 behaviour, and it is cleanest in the break between the effects of relative antagonism and the absolute antagonism, if we define the transition from stage 1 to stage 3 as the introduction of the real motor seizure convulsive manifestations. As long as the animal is still exhibiting, "species consistent" behaviour - grooming behaviours, automatisms - that seems in our framework is associated with really very little kindling actually occurring. That's why we make that break, and I don't know whether, if this framework is actually true, that is a consideration in definition that one would want to consider because it probably reflects a change of the underlying processes. All I would say is that if this concept is true, then the definition upon which it is based would be the one to use. So if this is substantiated then I think we ought to define stage 3 as the introduction of these clonic manifestations.

DR. FERNANDEZ-GUARDIOLA: I wonder what are your forebrain mechanisms? Where are they? I think it is too vague. Is only one neurotransmitter involved? What may be the specific anatomical counterpart of your proposed framework?

DR. BURCHFIEL: I can't give you any specific answers on the anatomical counterparts of this particular framework. I think that's run through all the answers to all these questions. The only thing I can comment on, where I think our data are clear, is in the processes upon which these transitions are dependent, the norepinephrine processes, the transition from stage 1/2 to stage 3 does appear to be dependent upon norepinephrine in the forebrain, whereas this later transition appears to be dependent upon norepinephrine in the brainstem. That I think holds up. Now, where the anatomical substrate is for these phases that are separated by these gates, that could be any place. As I was talking about with Dan, just because we define this gate as being dependent upon norepinephrine in the brainstem does not mean that the processes of this middle phase are all in the forebrain and the processes of the late event are all in the brainstem. It could be, in fact, and probably is, that this middle phase involves circuitry that projects from the forebrain to the brainstem, and probably back again, in some sort of a reverberatory way and it is the access of that after-discharge activity moving from the forebrain to the brainstem to begin to affect the brainstem motor mechanisms that is controlled by this gate. That access appears to be in the brainstem itself. But the processes themselves that are producing the various behaviours may involve both forebrain and brainstem circuits and may involve the substantia nigra at some stage.

DR. BERMAN: There are at least two sites for kindling, at least in our laboratory, that give nice after-discharges but the animals seem to get stuck behind the gate, and the two areas I'm thinking of are caudate nucleus and the dorsal hippocampus. You can get nice after-discharges and the animals will go, say, to stage 3, but only with great difficulty or not at all will they go further and I'm wondering whether that may relate to your gate?

DR. BURCHFIEL: My explanation for that would be that they do not ever make this transition and get into these structures. It would be an interesting thing, for people who are doing slice work, to look at it. For instance one would predict, if you stimulate those sites and produced a number of after-discharges it should not alter the characteristics of the circuitry in the pyriform if it is not exhibiting stage 3 behaviour. Only those sites that make that transition would alter the circuitry.

DR. WADA: Last comment, please.

DR. MCINTYRE: Let me just answer that question, Jim. I neglected in the talk I gave this morning to mention a couple of animals where we had kindled the dorsal hippocampus with over 30 stimulations and who showed no evidence of convulsive behaviour. When those animals were sectioned the pyriform cortex showed a disposition that was identical to the control animals and did not show the pattern of the animal who already had generalized convulsions.

DR. BURCHFIEL: Very good, very interesting.

# GENETIC CONTRIBUTIONS TO KINDLING: AN EXPERIMENTAL APPROACH

Craig D. Applegate,* James L. Burchfiel* and Paul E. Neumann**

*Comprehensive Epilepsy Program, Department of Neurology
University of Rochester School of Medicine and Dentistry
Rochester, New York, 14642 and **Department of Neurology
The Children's Hospital and Harvard Medical School, Boston
Massachusetts 02115

## INTRODUCTION

The development of kindled seizures has been shown to be influenced by genetic factors. The establishment of kindling-prone and kindling-resistant rat populations through selective breeding, along with the reported variations in kindling rates among inbred strains of mice strongly suggests the contribution of genetics to kindling [2,6,17,18]. The rapid kindling rates of genetically epilepsy prone rats [29] and El mice [13] in comparison with appropriate controls provide additional examples of the influence of genetic factors on kindled seizure development. However, to date, no systematic analysis of these factors has been reported for kindling-induced seizures.

In this paper we will describe the approach we have taken to analyzing the genetic contributions to kindled seizure development. The approach consists of three steps. The first step is choosing an experimental animal. Optimally, for an analysis of genetic factors in kindling, one should have two strains of animals which are genetically homogeneous and differ significantly in their susceptibility to kindled seizure development. We have chosen to use two inbred mouse strains, DBA/2J (D2) and C57BL/6J (B6) for our analysis. Establishing differences in seizure susceptibility between these inbred strains represents a critical component of the first step in the analysis. The second step in the analysis is to conduct a classical Mendelian genetic study between these strains to establish the dominance relationship between or among involved alleles and to generate hypotheses regarding the mode of inheritance of susceptibility to kindling stimulation. The third step in our approach involves defining the seizure phenotypes in a recombinant inbred strain series derived originally from crosses of D2 and B6 progenitor strains. This analysis can provide direct information on the number of genes involved in the difference between seizure susceptibility phenotypes and can be used for establishing linkage between genetic loci and the observed seizure phenotypes.

In the following section we will discuss each of these experimental steps in more detail. It should be emphasized that the genetic analysis of the kindling process is a new field of investigation. Therefore, this paper will focus primarily on a theoretical discussion of the experimental

*Kindling 4*
Edited by J. A. Wada
Plenum Press, New York, 1990

approach to this problem. Preliminary data gathered within our experimental framework will be presented in the context of this discussion.

## THE CHOICE OF AN EXPERIMENTAL ANIMAL

For practical reasons, inbred strains of mice represent a superior tool for examining questions surrounding genetic influences on seizure susceptibility or any complex trait of interest. There are a variety of readily available, genetically homogeneous mouse strains, and these strains are easily maintained. Inbred mice are the preferred species in mammalian genetics, and much is known about the mechanisms of genetic transmission in this species. In addition, extensive maps have been constructed of the mouse genome. These maps represent an invaluable resource for establishing genotype-phenotype associations (linkage). Knowledge regarding strain-dependent polymorphisms at identified marker loci can be used to formulate hypotheses about the genetic material involved in the expression of a complex trait such as seizure susceptibility.

Differences in the responsiveness to seizure-inducing agents among inbred mouse strains provide a striking example of the contribution of genetics to seizure susceptibility. Significant strain differences have been reported for seizures elicited by a variety of seizure-inducing agents [3,4,9,10,12,15,20,21,22,23,24,30,33,35,36], including kindling stimulation [2,17,18]. In a review of these studies, two strains consistently have been reported to exhibit clearly distinguishable phenotypes, the C57BL and DBA strains. In terms of either latency to seizure or severity of seizures elicited, DBA mice are significantly more sensitive than C57BL mice in virtually every seizure model in which they have been compared (Table 1).

Table 1. Selected comparisons of seizure susceptibility in DBA (D2) and C57BL (B6) strains.

| CONVULSANT AGENT | SUSCEPTIBILITY | REFERENCE |
|---|---|---|
| AUDIOGENIC STIMULATION | D2 > B6 | HENRY & BOWMAN, 1969 |
| | D2 > B6 | SEYFRIED, ET AL, 1980 |
| BICUCULLINE | D2 > B6 | PHILLIPS & DUDEK, 1984 |
| | D2 > B6 | FREUND, ET AL, 1987 |
| FLUROTHYL | D2 > B6 | MARLEY, ET AL, 1986 |
| KINDLING STIMULATION | D2 > B6 | LEECH, 1972 |
| | D2 > B6 | APPLEGATE, ET AL, 1988 |
| MERCAPTOPROPRIONATE | D2 > B6 | MARLEY, ET AL, 1986 |
| NICOTINE | D2 < B6 | TEPPER, ET AL, 1979 |
| MAXIMAL ELECTROSHOCK | D2 > B6 | TORCHIANA & STONE, 1959 |
| | D2 = B6 | HENRY & BOWMAN, 1969 |
| | D2 > B6 | APPLEGATE, ET AL, 1989 |
| PENTYLENETETRAZOL | D2 > B6 | APPLEGATE, ET AL, 1988 |
| | D2 = B6 | ENGSTROM & WOODBURY, 1988 |
| NMDA; KAINATE | D2 > B6 | ENGSTROM & WOODBURY, 1988 |
| STRYCHNINE | D2 > B6 | ENGSTROM & WOODBURY, 1988 |

The first step in this analysis then, will involve establishing the seizure phenotype of DBA/2J (D2) and C57BL/6J (B6) mouse strains to kindling stimulation. Observed differences in the rate of kindled seizure development and in the development of brain electrographic seizures will provide preliminary evidence for the influence of genetic factors on susceptibility to kindling stimulation.

## Preliminary Studies

In our initial study we have begun to define the seizure phenotype of D2 (N=17) and B6 (N=20) mice to kindling stimulation of the olfactory bulb (OB). The OB was chosen as a kindling site for several reasons. First, the OB is a prominent, easily accessible structure in the mouse. This feature facilitates reliable electrode placement and histological verification of the kindling site. Second, the OB exhibits a characteristic 6-10 Hz rhythm, which allows for an in vivo verification of the electrode placement and a simple means of establishing the functioning of the electrode assembly. Third, the OB kindles rapidly and reliably in the mouse. This feature minimizes the loss of animals due to failure or loss of the electrode assembly. Fourth, and perhaps most importantly, brain areas which receive afferents from the OB have been suggested to be critical components of forebrain circuitry involved in seizure generalization [28].

PROCEDURES. All mice, 8-12 weeks old (25-35g) at the time of surgery, were stereotaxically implanted with bipolar stimulating electrodes into the left or right OB under deep tribromoethanol anesthesia. The location of the OB is readily identified with reference to the anterior transverse sinus observable through the skull. The bulb occupies the space approximately 2.0 mm anterior to the sinus. Electrodes are introduced through a hole drilled 1.0 mm anterior to the sinus and 1.0 mm lateral from the midline. A stainless steel screw is implanted 1.0 mm posterior and 1.5mm lateral to bregma. The screw serves as mechanical support for the electrode assembly and as a cortical reference for recordings taken following each stimulation trial.

One week following surgery, thresholds for the elicitation of AD were determined. To establish threshold, current was set at 50 uA and increased at 50 uA increments until an AD is elicited. Stimulation intensity is set 50 uA above AD threshold for daily stimulation trials. Stimulation consisted of 1 s trains of 1 ms monophasic square wave delivered at 60 Hz. The progression of kindled seizures was scored using the standard seizure scale of Racine [27] which is consistent with behavioral description of kindled seizure development in mice originally reported by Leech [17]. On each trial the duration of AD was recorded along with the elicited motor seizure stage. Once seizure generalization occurred, the latency to bilateral forelimb clonus and the duration of motor seizure also were recorded. Animals were stimulated until 6 consecutive stage 5 seizures were elicited which represents our criterion for seizure stabilization. Data were analyzed by analysis of variance. Post-hoc pairwise comparisons were made using t-tests.

RESULTS. D2 and B6 mice exhibited clearly distinguishable phenotypes on virtually every measure of kindled seizure development and expression in this study. In comparison to B6 mice, D2 animals kindled significantly more rapidly (table 2; figure 1), exhibited significantly longer AD durations on both the first and last stimulation trials which elicited a stage 5 seizure, exhibited significantly longer latencies to motor seizure, and exhibited significantly longer durations of seizures. D2 and B6 mice did not differ in initial AD thresholds, initial AD duration or on qualitative assessments of the topography of the motor response at any stage in the kindling process. This preliminary strain comparison strongly suggests a genetic contribution to susceptibility to kindling stimulation. These data provide the basis for continuing to the second step in our analysis.

The differences in OB kindling profiles between D2 and B6 mice observed in this study are consistent with those reported by Leech [17] for amygdala kindling in these two strains. For amygdala kindling, D2 mice required significantly fewer AD's to elicit a generalized seizure than B6 mice; exhibited longer AD durations during the first generalized seizure than B6 mice; and, exhibited longer motor seizure durations and longer latencies to motor seizure than B6 mice. This pattern of strain variation in D2 and B6 mice is identical to that observed for OB kindling in these strains. In addition, we are currently conducting a pilot study on dorsal hippocampal kindling in D2 (N=5) and B6 (N=8) mice. Strain comparisons suggest that for hippocampal kindling, D2 mice kindle faster, exhibit longer AD and motor seizure durations and exhibit longer latencies to motor seizure than B6 mice. Together, results suggest that differences in the kindling profiles between these strains is independent of the site of kindled seizure initiation, and reinforce the conclusion that the seizure mechanisms in these strains exhibit inherent, genetically-based differences in sensitivity.

Table 2. Kindling characteristics of D2, B6 and $F_1$ mice. Values are means (SE).

| Group | N | Trials to first 5 | ADD (seconds) 1 | 5 | C | LMS (seconds) | DMS |
|---|---|---|---|---|---|---|---|
| D2 | 17 | 4.5 (0.5) | 17.7 (1.9) | 26.6 (1.8) | 27.5 (1.5) | 16.8 (1.7) | 13.5 (0.8) |
| B6 | 20 | 9.7* (0.6) | 14.7 (0.9) | 18.9* (0.7) | 23.0* (1.4) | 7.6* (0.3) | 11.2* (0.6) |
| $F_1$ | 19 | 4.9 (0.4) | 16.9 (1.0) | 24.8 (1.2) | 23.6* (0.8) | 14.5 (1.0) | 11.8* (0.5) |

ADD = afterdischarge duration for the first trial (1), the first stage 5 seizure (5) and the last stage 5 seizure (C). LMS = latency to bilateral forelimb clonus. DMS = duration of motor seizure (stages 4-5). * differs significantly from D2, $p < 0.05$).

Leech [17] reported that D2 mice required an mean of 2.5+1.6 AD's for kindled seizures to occur as compared with 6.9+3.1 AD's for B6 mice (cf, Table 2). The similarity in the rate of amygdala kindling reported by Leech with those of OB kindling reported here for either D2 or B6 animals are consistent with the similar and rapid kindling rates between these two structures in rats [1,7,14]. The remarkable similarity between our data and those of Leech both across laboratories and across decades illustrates the powerful reproducibility of experimental effects using genetically homogeneous populations. The hippocampus for either strain kindles significantly slower than the OB or amygdala. D2 animals require an average of 26 AD's (range 19-31) to kindle as compared with 35 AD's (range 31-38) for B6 animals. This finding is consistent with the differences in kindling rates among these structures in rats [25]. Thus the fundamental relationship of kindling rates among brain structures is maintained in these inbred mouse strains.

Finally, it should be emphasized that at the behavioral level, kindling in inbred mouse strains appears to represent an identical process to kindling in outbred rats. The progressive increase in stimulation elicited convulsive motor manifestations, as well as the motor seizures themselves are virtually identical in these two species. Furthermore, as discussed above, the relative rates of kindling among brain structures appears to be the same in inbred mice and their outbred rat cousins.

## CLASSICAL MENDELIAN ANALYSIS

The second step in this analysis is to define the seizure phenotypes in first ($F_1$) and second generation ($F_2$) hybrids as well as in backcross generations ($F_1$ x D2 and $F_1$ x B6). This analysis will establish that susceptibility to kindling stimulation is genetically transmissible and provide information on whether the trait of increased susceptibility is dominant, semi-dominant or recessive [11,19,31].

Classical analyses can provide information on the mode of inheritance. Predictable phenotypic ratios may be observed in $F_2$ and backcross segregating generations for traits with simple modes of inheritance, such as one or two gene traits. For example, if a single gene controls the expression of a trait and one allele is dominant, then all of the $F_1$ hybrids, which are heterozygous, will phenotypically resemble the parent with the dominant allele. The genetic ratios in $F_2$ hybrids will be 1:2:1, 25% having one parental genotype, 25% the other and 50% will be heterozygous. Phenotypically, $F_2$ hybrids will exhibit a ratio of 3:1 in favor of the dominant parent phenotype. Backcross generations will exhibit genotypic ratios of 1:1 and phenotypic ratios of either 1:1 or 1:0 depending on which parent is involved in the cross. Deviations from these ratios will suggest either the influence of more than 1 gene in the expression of a trait or the absence of complete dominance.

Two gene traits also may yield predictable ratios in $F_2$ and backcross populations. For example, if there is complete dominance at each locus and the loci are unlinked, classical Mendelian ratios of 9:3:3:1 and 1:1:1:1, may be observed in F2 hybrids and backcross progeny, respectively. These classic ratios are altered by more complex modes of gene interaction such as complementarity, epistasis, duplicate gene action, incomplete penetrance, variable expressivity, etc. However, the effects of these interactions produce characteristic phenotypic ratios. Thus the influence of more complex modes of inheritance can be identified within the context of a Mendelian analysis [11,19,31].

### Preliminary Studies

In this series of studies we have begun to define the seizure phenotypes of reciprocal crosses ($F_1$; N=19), backcrosses ($F_1$xD2; N=16, $F_1$xB6; N=25) and intercrosses ($F_2$; N=37) derived originally from the D2 and B6 parents to susceptibility to kindling stimulation of the OB. These animals represent those required for conducting a classical Mendelian analysis. All surgical and kindling procedures were the same as those described earlier.

Results. $F_1$ mice required 4.9+0.4 stimulations to develop stage 5 seizures indicating a clear dominance of the D2 phenotype for this measure in these hybrids (table 2; figure 1). Similarly, AD duration during the initial generalized seizure and the latency to motor seizure in $F_1$ mice again indicated a dominance of the D2 phenotype. In contrast, the duration of motor seizure in $F_1$ mice as well as the AD duration on the last criterion trial were consistent with the B6 phenotype. This tendency for $F_1$ hybrids to exhibit long local afterdischarges and relatively short motor seizures represents a dissociation of these two measures suggesting they may reflect different processes, which are influenced by different genetically based mechanisms.

Examinations of kindled seizure development in the $F_2$ and backcross segregating groups suggest a complex mode of inheritance for susceptibility to kindling (figure 1). These data are inconsistent with a simple single gene model. The almost complete absence of animals exhibiting a clear B6 phenotype in the $F_2$ hybrid group argues against a single gene mode of inheritance. The data are more consistent with the hypothesis that at least two genes are acting to influence this trait. However, the numbers of animals used in these studies is relatively small for a classical Mendelian analysis, and this inference can be made only tentatively.

The analysis of AD duration and motor seizure characteristics in these groups of animals is currently being completed. The ultimate utility of these measures for further characterizing the phenotypes in segregating generations as well as for providing insights into the genetic influences on the kindling process remain to be determined.

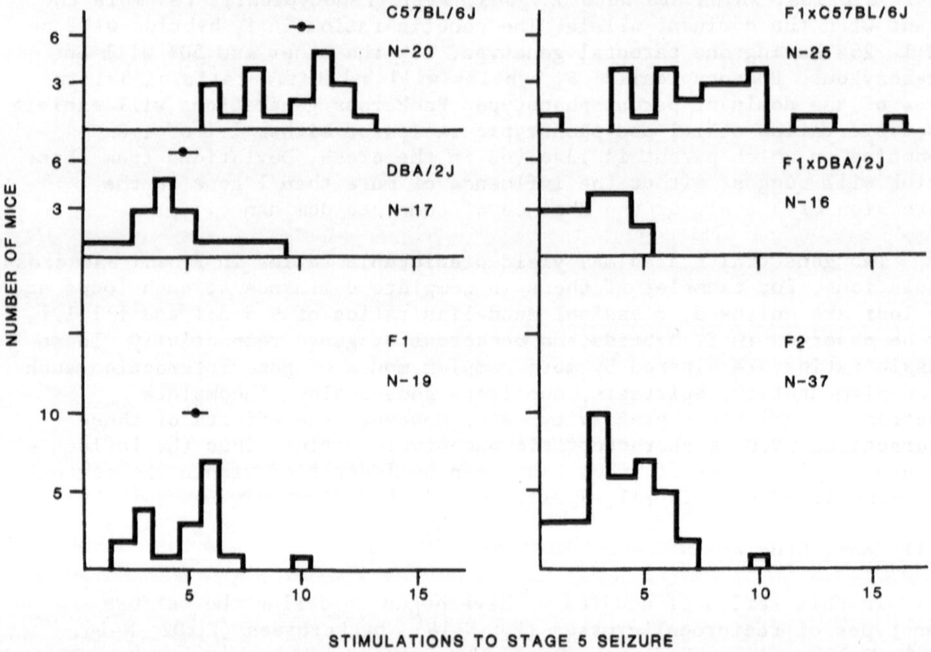


Figure 1. Phenotypic distributions of the number of stimulations necessary to elicit a stage 5 seizure in $F_1$, $F_2$ and backcross groups of mice derived from D2 and B6 inbred strains. Note the almost complete dominance of the D2 phenotype in $F_1$ generation hybrids for this measure of susceptibility. The distributions observed in the $F_2$ and backcross groups are inconsistent with a single gene hypothesis and suggest that at least 2 genes influence the trait of susceptibility to kindling stimulation.

## RECOMBINANT INBRED STRAIN ANALYSIS

Recombinant inbred (RI) strains represent a critically important component of our experimental strategy. To create an RI strain, $F_2$ hybrids derived from two highly inbred strains are randomly selected and bred. The progeny of this mating are then inbred (brother-sister matings) for 20 or more generations. Once inbred, the genetic variability among siblings approaches zero, and a new inbred strain is created. Typically , a number of RI strains are developed from different, randomly paired, $F_2$ generation individuals resulting in an RI strain set or series [5]. For each genetic locus at which the parental strains differ, each RI strain will have the exclusive contribution from one parent or the other. The pattern of recombination within the genome, however, will be different for each strain (Figure 2). Once inbred, these patterns become fixed and homogeneous within a given RI strain. Thus, RI strains represent stable, segregant populations. That is, an RI strain set represents a series of unique, random recombinations of the original parental genomes [5,16]. One of the largest RI strain series has been developed and maintained between the C57BL/6J and DBA/2J strains by B.A. Taylor at The Jackson Laboratory. Twenty-six, genetically distinct strains exist and represent the BXD RI strain series.

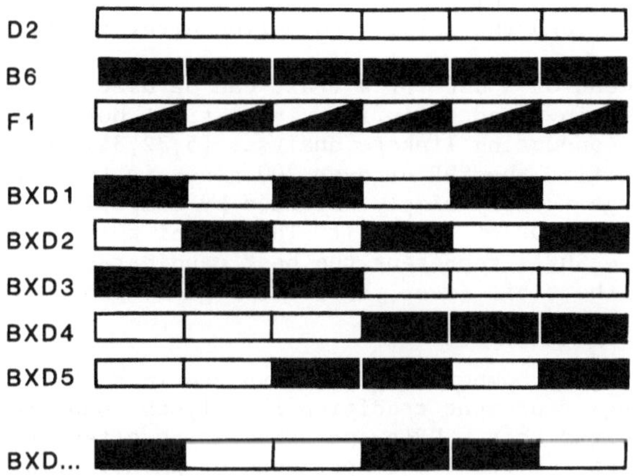


Figure 2. A simplified representation of the genome of mice from the BXD RI strain series. Randomly selected $F_2$ generation mice derived originally from D2 and B6 parent strains are bred, and then inbred, to produce a RI strain series. Each strain will receive approximately half of its genetic material from one or the other parent strain and, as a result of inbreeding, will be homozygous at any given genetic locus. The pattern of genetic rearrangement is unique for each strain. These characteristics make an RI strain series an valuable resource for conducting genetic analyses of complex traits such as seizure susceptibility. There are 26 strains in the BXD RI strain series.

A series of RI strains, each with a unique genetic makeup, is an invaluable resource for dissecting the underlying genetic basis for phenotypic differences observed in the parental strains. The basic tool used in this dissection is the strain distribution pattern, or SDP. The SDP is constructed by defining the phenotype of each strain for the trait of interest. For each strain this phenotype might be the same as that of one of the parent strains, or it might be a new, distinct phenotype, different from that of the parents. The total array of phenotypes observed across the RI strain series constitutes the SDP for a trait.

Knowledge regarding the SDP of a trait represents a powerful experimental tool for establishing the number of genes influencing that trait [5,8,16]. As stated earlier, for a given locus, each RI strain is equally likely to have inherited a gene from one or the other parent strain. From this, it would be predicted that if a single gene is in control of a trait, then all the RI strains would resemble phenotypically one or the other parent strain. If new phenotypes are present, it implies the existence of additional genes affecting the trait. Based on the number of new phenotypes which appear, it is possible to estimate the number of genes involved. For example, given a random reassortment and fixation of genes in the RI strains, if two genes were involved in the control of a trait, then we would expect the appearance of up to 4 phenotypes within the RI strain series, each with approximately equal representation. Two of the phenotypes would be identical to those of the parent strains and two would be different, reflecting the appearance of new genetic combinations. In general, if n genes control a trait, then the number of phenotypes in an RI series will equal $2^n$ [5,16]. From a practical standpoint, the limiting factors in analyses using RI strains are the number of strains available in an RI strain series, and the ability to define discrete phenotypic classes.

The development of a SDP for a trait can be used not only to estimate the number of genes involved in influencing a trait but can also serve as a powerful tool for conducting linkage analyses [5,32,34]. Genetic mapping studies have identified the SDP of over 200 genes in the BXD RI strains. Evidence for linkage is very simply obtained by comparing the SDP of a known marker with the SDP of the trait. Those loci which exhibit the best "match" between the SDP's represent the best candidates for contributing to the expression of the trait (i.e., have the highest probability of linkage).

Estimation of the number of genes involved in a trait and establishing linkage represent traditional analyses conducted using observed SDP's. Beyond this, SDP's can serve as an effective experimental framework for direct qualitative comparisons of phenotypically similar traits which may have different etiologies or different genetically based mechanisms. For example, clonic seizures can be elicited in mice using a variety of seizure-inducing agents. The motor expression as well as other characteristics of these seizures can be identical. The question is, are the seizures produced by one agent dependent on the same or different genetically based mechanisms as the seizures produced by another agent? Direct comparisons of the SDP's of seizure phenotypes produced by the two agents should yield a definitive answer to this question. It is highly unlikely that different genetic mechanisms of seizure susceptibility would result in exactly the same SDP. Given a random model of recombination, the _a priori_ probability of observing a specific SDP can be calculated as $1/2^k$, where k is the number of RI strains in the series [16]. Thus the probability of observing two identical SDP's by chance is extremely low, and identity or partial identity between SDP's would strongly suggest shared or partially shared mechanisms between the seizure inducing agents compared.

Defining seizure phenotypes among the BXD RI strains therefore, can serve as common general template on which to directly compare commonly used experimental seizure models. Within any seizure model, the generation of SDP's on multiple dependent measures can provide a tool for dissecting the process of seizure genesis and expression in that model. For example, within the kindling model, comparisons of the SDP's for measures of AD's to initial generalization, local AD duration, duration of motor seizure, etc. could provide insights into the processes involved in the neural reorganization produced using kindling protocols. These comparisons are potentially important as they can serve as the basis for hypotheses concerning which aspects of kindling exhibit the highest degree of genetic control and, therefore, may be more basic to seizure susceptibility in the kindling model. In addition, the comparison of phenotypic SDP's with the SDP's generated from biochemical, molecular biological or electrophysiological measures can be a potent tool for establishing the relationships between neural mechanisms and susceptibility [32].

In summary, RI strain analysis add a significant dimension of power to more traditional genetic studies. Once a phenotypic SDP has been defined, this analysis can provide information on the number of genes influencing a trait and can provide a basis for establishing linkage. Beyond this, RI strain analysis can (1) be used as a framework for directly comparing seizures elicited in a variety of seizure models, (2) serve as a tool for uncovering the underlying architecture of seizure processes and seizure susceptibility, and (3) serve as a means of more definitively relating measures of neural function with measures of seizure susceptibility.

One of the major caveats of this analysis is the ability to define phenotypes. The ability to distinguish and define discrete, homogeneous subgroups is critical to the development of an experimentally useful SDP. Our preliminary data suggest that we can clearly distinguish the seizure phenotypes of the D2 and B6 parental strains. However, establishing clear criteria for reliably distinguishing groups of animals displaying intermediate phenotypes is more difficult. Central to this problem is determining which of a variety of measures are the most basic to the seizure process and underlying susceptibility.

Preliminary Studies

To date, we have tested 10 of the 26 available BXD RI strains. The strains tested are BXD 1, 2, 5, 6, 12, 13, 14, 16, 25, and 27 (N=4-5/strain). Defining the seizure phenotypes in each of these strains constitutes a preliminary strain distribution pattern (SDP) for susceptibility to OB kindling. All experimental procedures are the same as those described previously.

Results. The seizure phenotypes for the 10 BXD RI strains are shown in figure 3. Both rapid kindling (D2-like) and slow kindling (B6-like) phenotypes are clearly present among the BXD RI strains tested. Strains BXD 5 and 13 required relatively few trials to kindle and exhibit a D2-like phenotype. Strains BXD 1, 2, 6 and 27 required a relatively greater number of stimulations to kindle and exhibit B6-like phenotype characteristics. A third, intermediate phenotypic class is suggested by BXD strains 12, 14, 16 and 25. The small N's and large variances in these latter groups however, make the precise phenotype classification of these strains difficult. Pairwise comparisons of these strains indicate that they all differ significantly from the D2 parent phenotype. None of these strains however, differ significantly from the B6 parent. The characteristic of a truly intermediate phenotypic class is that it is statistically dissociable from both parental phenotypes [8,16]. Strain BXD 14 approximates, but does not meet this criterion ($t=2.5$ vs D2, $p=.02$; $t=1.9$ vs B6, $p=.08$). However, data from strain BXD 14 suggest the possibility that a group of strains exhibiting an intermediate phenotype will emerge once all 26 BXD strains are tested in sufficient numbers.

Other evidence more strongly suggests the existence of more than two phenotypes among the RI strains. Post-hoc pairwise comparisons of strains BXD 5 and 13 with the D2 parent strain indicate that these strains kindle significantly faster than D2 mice. Similarly, strain BXD 2 is statistically dissociable from the B6 parent, kindling reliably more slowly on the average than B6 mice. These data would suggest the presence of at least three phenotypic classes among the BXD RI strains. A "super-D2" phenotype represented by strains 5 and 13; a B6-like phenotype represented by strains 1, 6 and 27; and a "super-B6" phenotype represented by strain 2. A true D2-

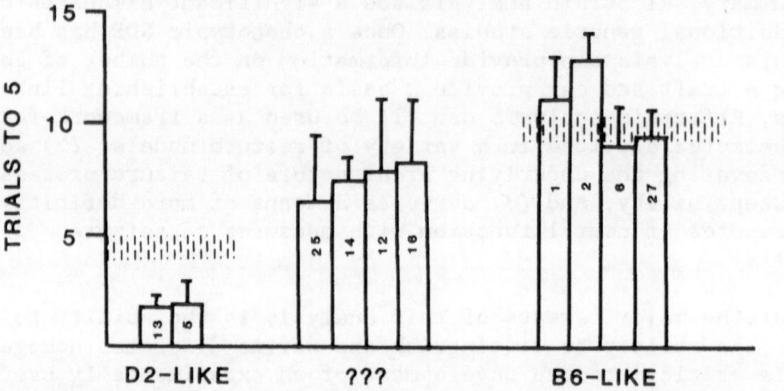


Figure 3. The mean (+SE) number of olfactory bulb stimulations to elicit a stage 5 seizure in the BXD RI strains tested to date. Shaded horizontal bars represent the mean (SE) kindling rate of the D2 (left) and B6 (right) parent strains. Strains have been grouped based on statistical criteria as either rapid kindling (D2-like) or slow kindling (B6-like). A third group that is presently unclassifiable (????) also is shown. The strain number is shown in the bars. N = 4-5 per group. See text for details.

like phenotypic class is anticipated to emerge once all strains have been tested. The presence of three or four phenotypic classes among the RI strains strongly suggests that two major genes are influencing susceptibility to kindled seizure development and support the interpretation of the Mendelian analysis described earlier.

These preliminary data suggest that the D2 and B6 phenotypes may represent intermediate phenotypes in the final SDP. The observations that some RI strains kindle more rapidly than the D2 parent and that others may kindle more slowly than the B6 progenitor strain further suggest that within the genome of either parent strain there is a mosaic of both pro- and anti-convulsant factors which contribute to the expression of the parental phenotypes.

Analyses of other aspects of kindled seizure development and expression in the BXD RI strains are currently underway. We are assessing the utility of local AD duration, latency to motor seizure and motor seizure duration for defining phenotypes. Preliminary examination of olfactory bulb AD durations among the BXD RI strains suggests a dissociation between this measure and the rate of kindled seizure development. For example, some strains which kindle relatively slowly (eg. strains 1 and 27) exhibit very long AD's which are more consistent with the profile of more rapid kindling strains (eg., D2 and strain 5). These data suggest that the genetic mechanisms which influence local excitability (as measured by AD duration) are independent of those which influence seizure generalization (as measured by stimulations to generalized motor seizure). The extent to which other measures of kindled seizure development and expression are dissociable will provide important information regarding the relative degree of genetic influence on these measures and may provide insights into the processes of neural reorganization underlying kindled seizure development.

## GENERAL DISCUSSION AND CONCLUSIONS

One major conclusion that can be drawn based on our preliminary studies is that susceptibility to kindling stimulation is a largely heritable trait. Traditionally, the standard measure used to define susceptibility in the kindling paradigm has been kindling rate. Our data are most complete for this measure. The near identity between the rate of kindled seizure development in D2 and $F_1$ hybrids indicates an almost complete dominance of the D2 phenotype for this measure. Data gathered within the context of the classical mendelian design suggest a complex mode of inheritance involving at least two genes. This interpretation is supported by preliminary analysis of the RI SDP indicating that at least two major genes influence the rate of kindled seizure development.

Our preliminary analyses indicate that other measures of kindled seizure development and expression also are heritable. Afterdischarge duration during initial seizure generalization and latency to motor seizure exhibit clear D2 dominance in the $F_1$ generation. However, measures of AD duration following 6 consecutive stage 5 seizures and motor seizure duration in $F_1$ hybrids are more consistent with the B6 phenotype. These data indicate that while these measures reflect heritable processes, the genetic influences on these processes may be different from those influencing kindling rate. Data gathered on AD durations in our preliminary studies of the BXD RI strains suggest a dissociation between this measure and kindling rate. If this dissociation is reliable, it will strongly suggest that the genetic influences on local afterdischarge duration are distinct from those influencing the rate of kindled seizure development. This finding is important and suggests that these measures may reflect fundamentally different processes and may not be used interchangeably.

It is clear from our data that other characteristics of kindling other than rate of seizure development are heritable. This outcome is not surprising. What is surprising, however, is the suggestion that the various aspects of kindling observed may not be inherited as a unit. That various characteristics of the kindling phenomenon can be dissociated during inheritance has implications for understanding the basic mechanisms underlying the kindling process. This observation suggests that the experimental approach outlined in this paper may prove useful for dissecting the kindling process and may provide insights into the underlying architecture of kindled seizure development and expression from the perspective of the genome. The preliminary data gathered within the context of this genetically based framework are sufficiently intriguing to warrant further experimentation. The ultimate validation of our approach must await the outcome of these efforts.

What mechanisms account for the differences in kindling rates in D2 and B6 mouse strains? At present, the available data do not allow for any clear, coherent conclusions regarding the underlying basis for the differing susceptibilities to kindled seizure development in these strains. Given the complex mode of inheritance suggested by our preliminary studies, it would not be surprising if several or even many neurochemical or neurophysiological factors were related to the observed phenotypic differences between these strains. However, conducting experiments within the genetic framework outlined here, allows one to both formulate and rigorously test hypotheses regarding mechanisms governing kindled seizure susceptibility. The strategy is simple, and we believe the outcome will be conclusive. The initial step is to conduct linkage analyses through a comparison of the phenotypic SDP with the genetic maps available for the BXD RI strains. Once linkage with a marker is established, the known genetic material traveling in that linkage group can be evaluated to generate hypotheses regarding underlying mechanisms. These hypotheses can then be initially tested by determining whether the D2 and B6 parental strains exhibit different phenotypes with respect to this variable. If the parents differ, then F1 hybrids can be evaluated to determine if the candidate mechanism is concordant with the F1 behavioral phenotype. Once a candidate mechanism has met these initial criteria, the hypothesis can be rigorously tested in the BXD RI strains. The distribution pattern across the BXD RI strains of any candidate mechanism must exhibit a high degree of correlation with the SDP of the RI strain series. Only those mechanisms which satisfy this final criterion can be considered for further experimentation.

In summary, the ability to search for mechanisms of susceptibility at the level of the genome adds a dimension to our strategy which is generally not available in other experimental contexts. Establishing an association between genetic loci and seizure phenotype can provide a strong base from which to generate and rigorously test hypotheses regarding underlying mechanisms of seizure susceptibility to kindling stimulation as well as to other seizure-inducing agents. This is a long range goal of our research employing the experimental strategy outlined in this paper.

## REFERENCES


1. Applegate, CD and Burchfiel, JL. Intra-amygdala muscimol attenuates the motor expression of seizures kindled from prepiriform cortex. Epilepsia. 1987; 28:594.

2. Applegate, CD, DiFazio, S, Spanknebel, K, Burchfiel, JL and Neumann, PE. A classical genetic analysis of kindled seizure susceptibility in DBA/2J and C57/6J mice. Soc. Neurosci. Abs. 1988; 14:253.

3. Applegate, CD, Burchfiel, JL and Neumann, PE. A classical genetic and recombinant inbred stain of seizure susceptibility in mice to pentylenetetrazol. Epilepsia. 1988; 29:676.

4. Applegate, CD, Samoriski, G, and Neumann, PE. Susceptibility to electroconvulsive shock in DBA/2J and C57BL/6J mice: A classical genetic and recombinant inbred strain analysis. Epilepsia. 1989; in press.

5. Bailey DW. Recombinant inbred strains. Transplantation. 1971; 11:325.

6. Burnham, WM, Racine, RJ and Okazaki, MO. Kindling mechanisms II. Biochemical studies. In: J. Wada (Ed). Kindling III. Raven Press, N.Y., 1986, 283.

7. Cain DP. Seizure development following repeated electrical stimulation of central olfactory structures. Ann. NY Acad. Sci. 1977; 290:200.

8. Carey G. Admixture analysis and the search for a major gene in recombinant inbred strains. Behav. Genet. 1986; 16:775.

9. Collins RL. A new genetic locus mapped from behavioral variation in mice. Behav. Genet. 1970; 1:99.

10. Engstrom, FL, Woodbury, DM. Seizure susceptibility in DBA and C57 mice: The effects of various convulsants. Epilepsia. 1988; 29:389.

11. Falconer DS. Introduction to Quantitative Genetics. Ronald Press, NY, 1960.

12. Freund RK, Marley RJ, and Wehner JM. Differential sensitivity to bicuculline in three inbred mouse strains. Brain Research Bull. 1987; 18:657.

13. Green, RC, Elizondo, BJ and Seyfried, TN. Accelerated kindling in the mutant epileptic mouse El. Soc. Neurosci. Abs. 1988; 14:252.

14. Goddard GV, McIntyre DC, and Leech CK. A permanent change in brain function resulting from daily electrical stimulation. Exp. Neurol. 1969; 25:295.

15. Henry KR, Bowman RE. Effects of acoustic priming on audiogenic, electroconvulsive and chemoconvulsive seizures. J. Comp. Physiol. Psychol. 1969; 67:401.
16. Klein TW. Analysis of major gene effects using recombinant inbred strains and related congenic lines. Behav. Genet. 1978; 8:261.
17. Leech CK. Rate of development of electrically kindled convulsions compared to audiogenic seizures and learning ability in six inbred mouse strains. Unpublished Dissertation. 1972.
18. Leech CK, McIntyre DC. Kindling rates in inbred mice: An analog of learning? Behav. Biol. 1976; 16:439.
19. Mather K, Jinks JL. Introduction to Biometrical Genetics. Cornell Univer. Press, Ithica NY, 1977.
20. Marks MJ, Burch JB, Collins AC. The genetics of nicotine response in four inbred strains of mice. J. Pharmacol. Exp. Ther. 1983; 226:291.
21. Marks MJ, Patinkin DM, Artman LD, Burch JB, Collins AC. Genetic influences on cholinergic drug response. Pharmacol. Biochem. Behav. 1981; 15:271.
22. Marley RJ, Gaffney D, and Wehner JM. Genetic influences on GABA-related seizures. Pharmacol. Biochem. Behav. 1986; 24:665.
23. Miner LL, Marks MJ, Collins AC. Classical genetic analysis of nicotine-induced seizures and nicotinic receptors. J. Pharmacol. Exp. Ther. 1984; 231:545.
24. McCall, RD and Frierson, D. Evidence that two loci predominantly determine the difference in susceptibility to high pressure neurological syndrome type 1 seizures in mice. Genetics. 1981; 99:285.
25. McNamara JO, Bonhaus D, Shin C, Craine B, Gellman R, Giacchino J.The kindling model of epilepsy: A critical review. CRC Crit. Rev. Clin. Neurobiol. 1985; 1:341.
26. Phillips, TJ and Dudek, BC. A diallel cross genetic analysis of bicuculline-induced seizures in mice. Soc. Neurosci. Abs. 1984; 10:649.
27. Racine R. Modification of seizure activity by electrical stimulation: II. Motor seizure. Electroenceph. Clin. Neurophysiol. 1972; 32:281.
28. Racine, R, Burnham, WM, Gilbert, M, and Kairiss, EW. Kindling mechanisms: I. Electrophysiological studies. In: Wada, J. (Ed). Kindling III. Raven Press, NY. 1986, 263.
29. Savage, DD, Reigel, CE and Jobe, PC. Angular bundle kindling is accelerated in rats with a genetic predisposition to acoustic stimulus-induced seizures. Brain Res. 1986; 376:412.
30. Seyfried TN, Yu RK, Glaser GH. Genetic analysis of audiogenic seizures in B6 x D2 recombinant inbred strains of mice. Genetics. 1980; 94:701.
31. Suzuki, D., Griffiths, A., Miller, J., Lewontin, R. An Introduction to Genetic Analysis. W.H. Freeman & Co., New York, 1986.
32. Swank RT, Bailey DW. Recombinant inbred lines: Value in the genetic analysis of biochemical variants. Science. 1973; 181:1249.
33. Taylor, BA. Genetic analysis of susceptibility to isoniazid-induced seizures in mice. Genetics. 1976; 83:373.
34. Taylor BA. Recombinant inbred strains: Use in gene mapping. In: Morse H. (ed). Origins of Inbred Mice. 1978, 423.
35. Tepper JM, Wilson JR, Schlesinger K. Relations between nicotine-induced convulsive behavior and blood and brain levels of nicotine as a function of sex and age in two inbred strains of mice. Pharmacol. Biochem. Behav. 1979; 10:349.
36. Torchiana, ML and Stone, CA. Post-seizure mortality following electroshock convulsions in certain strains of mice. Proc. Soc. Exp. Biol. Med. 1959; 100:290.

Discussion of Dr. Applegate's Presentation

DR. MCINTYRE: First, have you examined the catecholamine content in these strains, and second, I notice that your fastest developing strain, as we often see in any kindling, the faster the animal kindles, it seems the longer their latencies are to the actual provocation of that convulsion, and you saw that here, too?

DR. APPLEGATE: We have not at this point in time looked at catecholamine content. These strains have been examined for that and, in general, the DBA animals have lower norepinephrine concentrations. In terms of the latency to develop or to express a stage 5 seizure, my thinking is that what you're looking at is a sort of immature hook-up between the focus and the substrate which is necessary for expressing the seizure. I think that's what we're seeing here and in rats, for example, in their first stage 5 seizure, they seem to have a much longer latency to that seizure than they do later on after repeated seizures.

DR. MOSHE: Have you looked at other seizure models to see if this is specific for kindling, that there are 2 genes influencing the susceptibility? What happens to other seizure models? Is that a seizure predisposition?

DR. APPLEGATE: We've looked at pentylene tetrazol (PTZ) as well as electroconvulsive shock (ECS) in these strains and in all cases it seems like we're not going to see a single gene in any given model that's going to be critical for conferring susceptibility. In both ECS and PTZ, we're looking at 2 genes at least, and possibly 3, which are exerting a major influence on the basal susceptibility at this point. Because our data are so preliminary with the kindling model, I feel a little funny comparing the kindling to PTZ or ECS. PTZ and ECS, however, exhibit an amazing amount of concordance in the strain distribution patterns of the phenotypes and at this point I think we're very close to being able to link those phenotypes to loci on chromosome 1, which may be a common loci for ECS and PTZ.

# THE GABA HYPOTHESIS OF KINDLING

W. McIntyre Burnham and Georgia A. Cottrell

Department of Pharmacology, University of Toronto
Toronto, Ontario, M5S 1A8

## INTRODUCTION

The GABA hypothesis, as recently revised (6), may be stated as follows:

"The kindling procedure causes a permanent change in some part of the GABA-A inhibitory system. This leads to a chronic tendency toward GABAergic hypofunction, and to all (or part) of the physiological abnormalities which characterize the 'kindled state'."

The evidence related to the GABA hypothesis has been summarized in several recent reviews (5,6,9). The present discussion will concentrate on the hypothesis as a *theoretical entity*, attempting to assess the strengths and weaknesses of this approach to the basic mechanisms of kindling.

## DEFINITIONS/LIMITS TO DISCUSSION

The term "kindling" has been used to refer both to the acquisition of kindled seizures and to their persistence. In the present discussion, it will be used to refer only to *acquisition*. The permanent functional changes which result from kindling (e.g., lowered threshold, enhanced generalization) will be referred to as the "kindled state" (50). The permanent structural or biochemical brain abnormalities which give rise to the kindled state will be termed the "kindled change".

Given those definitions, we may note that the present discussion will be limited to biochemical hypotheses of the "kindled change". Due to space restrictions, it will further be limited to hypotheses relating the "change" to epileptiform abnormalities, and to those hypotheses which involve the traditional neurotransmitters and modulators.

## WHY STUDY THE BASIC MECHANISMS OF KINDLING?

From a critical perspective, the first question is, "Why study the basic mechanisms of kindling"? The kindling model - at age 20+ - is no longer "novel" or "cutting edge". Is it still worthy of our attention?

The kindling preparation was first presented as a model of "neural

*Kindling 4*
Edited by J. A. Wada
Plenum Press, New York, 1990

plasticity" [i.e. learning (23,24)]. It has become clear, however, that the plasticity in kindled brains differs in some ways from the plasticity observed in learning models (10). It was later argued that the kindled brain might offer clues to the "epileptic flaw" in human brains (5). This may or may not be true: despite the functional similarities observed in kindled and epileptic brains (5), there is no guarantee that the "kindled change" is identical to the "epileptic flaw".

Perhaps the strongest argument for the study of kindling is the argument that it will help us to understand the effects of seizures on the brain. It was once believed that repeated seizures had little effect on brain function - unless status epilepticus was involved (32). Recent clinical studies, however, suggest that repeated seizures exert long-lasting effects on both personality and cognition (e.g. 1,61). In light of the widespread incidence of epilepsy, and the fact that certain forms are poorly controlled, it is crucial to know how seizures affect the brain, and how these effects may be prevented. Kindling provides an excellent model for this study. If we also discover the basic mechanisms of epilepsy or learning, so much the better.

## WHY A BIOCHEMICAL APPROACH TO MECHANISMS?

A major factor in the development of biochemical kindling hypotheses was the failure of early studies to demonstrate a structural basis for the kindling phenomenon (23,54). Recent anatomical studies, however, have revealed several sorts of post-kindling change, including a loss of non-perforated axospinous synapses (20), sprouting (60), and the occurrence of astrocyte hypertrophy (Ivy, Racine, and Milgram, in preparation). If a structural basis is demonstrated for kindling, will the biochemical approach become obsolete?

The answer, of course, is "no". With a few exceptions, neurons communicate biochemically, and any meaningful structural abnormality is transformed, at the first synapse, into a biochemical abnormality. Structural theories of kindling - as they develop - will be integrated with biochemical theories. They will not replace them.

## CRITERIA: HOW TO ASSESS BIOCHEMICAL HYPOTHESES

Several years ago, Burnham et al. (9) proposed three criteria for evaluating biochemical hypotheses of kindling - or, more properly, of the "kindled change". In updated form, these suggest that:

1) Pharmacological induction of the proposed abnormality should simulate the kindled state, even in naive animals.

2) Pharmacological antagonism of the proposed abnormality should reverse the kindled state, returning "mature" focal-generalized seizures to their original focal form.

3) Biochemical assays should reveal the abnormality in animals sacrificed at least four weeks after kindling.

These criteria deserve some discussion, since they are basic to much of what follows.

<u>Why are Criteria Necessary?</u> Why formalize the study of kindling? The answer lies in the complexity of the biochemical/pharmacological data. At some point after kindling, an abnormality has been reported in nearly every transmitter system which has been studied (cf. 5,50). Likewise, a *wide* variety

of drugs have been shown to affect either the development or the occurrence of kindled seizures (cf. 50,64). Some of these biochemical/drug effects are probably relevant to the kindled state. Most of them, however, are probably non-specific or unrelated. Some sort of criteria are needed to evaluate the data.

Why These Criteria? How may data relevant to basic mechanisms be distinguished from unrelated or non-specific effects? There will be a constant association between the kindled state and the brain changes which cause it; there will not be a constant association between the "state" and other abnormalities. The three criteria outlined above provide a paradigm for seeking a constant association. They all relate to a single, basic criterion: "When the 'kindled change' is present, the kindled state will be present; when the 'kindled change' is absent, the kindled state will be absent." This criterion, dictated by simple logic, seems unarguable.

Why Update the Assay Criterion? In our original statement of criteria (9), we suggested that the proposed "change" must be demonstrable in kindled brains at least *two weeks* after the last seizure. Our present criteria suggest a delay of at least *four weeks*. This modification is based on findings which suggest that transitory changes caused by the kindling process (e.g., long-term potentiation) may take at least four weeks to return to baseline (10,53). Changes observed at four (or more) weeks are likely to be permanent.

Why Update the Pharmacological Criteria? Our original criteria suggested that pharmacological induction of the "kindled change" should accelerate kindling and that pharmacological antagonism should retard it. The updated criteria suggest that pharmacological induction of the "change" should simulate the kindled state - even in a naive animal - and that pharmacological antagonism should reverse it - even in a well-kindled animal. These new criteria are suggested both by logical and design considerations. Logically, if the "kindled change" can be created pharmacologically, a subject should be "kindled" - at least for the duration of the drug's action. Likewise, if the "kindled change" can be pharmacologically reversed, the kindled state should disappear - at least temporarily (cf. Mingo and Burnham, submitted). As yet, only a few experiments have attempted simulation/reversal, but *logically* both should work. In terms of design, it is clear that simulation and reversal offer a more direct test of biochemical hypotheses than acceleration and reversal. Many drugs might accelerate or retard kindling without relating, in any direct fashion, to the "kindled change". Many compounds do accelerate or retard kindling without simulating or reversing the kindled state (see below). Acceleration and retardation data will continue to be of interest in their own right, but simulation and retardation are superior for testing biochemical hypotheses.

Afterword: Critics may wish to point out that the criteria proposed above are "necessary" but not "sufficient"; that it would be possible (though not probable) for a given "change" to meet all of the criteria, but not be causal to the kindled state. This is true. Critics should also bear in mind, however, that while the criteria may not be sufficient, they are necessary. A hypothesis which cannot show simulation, reversal, and long-term assay changes must ultimately be rejected.

## HOW DO THE BIOCHEMICAL HYPOTHESES MEASURE UP TO THE PROPOSED CRITERIA?

How do the various biochemical kindling hypotheses measure up to the proposed criteria? A complete review is beyond the scope of this paper, but a summary of the evidence related to three major hypotheses - noradrenaline, glutamate and GABA - appears in Table 1. As indicated, the non-GABA hypotheses presently have some problems in meeting the proposed criteria.

In the case of the noradrenergic hypothesis (13,40), acceleration has frequently been demonstrated, but simulation does not seem to work. Retardation has been demonstrated, but reversal has not been shown - at least in whole animals. Relatively long-lasting changes have been demonstrated in receptor binding, but consistent changes in other noradrenergic parameters have not been found. (Details in Table 1.)

In the case of the glutamate hypothesis (3), neither acceleration nor simulation has been demonstrated. Retardation and reversal have been shown, but both are partial rather than complete. Long-term electrophysiological changes and changes in hippocampal slices have been reported, but few changes have been observed as yet in traditional biochemical assays. (Details in Table 1.)

Similar problems can be pointed out with regard to the cholinergic, dopaminergic and serotonergic hypotheses. Pharmacologically, the best supported non-GABA hypothesis is probably the adenosine hypothesis (14,15). Unfortunately, long-term assay data related to this hypothesis are lacking.

## HOW DOES THE GABA HYPOTHESIS MEASURE UP TO THE PROPOSED CRITERIA?

The pharmacological data have always been supportive of the GABA hypothesis (9). As indicated in Table 1, GABA antagonists accelerate kindling and GABA agonists retard it. GABA agonists also reverse the kindled state (4), and GABA antagonists, at least partially, simulate it (Mingo and Burnham, submitted). Recent experiments have reinforced this picture by demonstrating positive effects of GABA uptake blockers (25), direct infusion of GABA (19), etc.

Originally, however, it appeared that the GABA hypothesis could not meet the assay criterion. A series of assay studies, beginning in 1978, reported that GABA-related parameters were unchanged in kindled brains (cf. 6). These studies - which were generally conducted on whole-tissue ("metabolic" plus "synaptic") GABA - created some doubt as to the viability of the GABA hypothesis. A more recent series of studies, however, have largely reversed this negative impression. These have concentrated on "synaptic" GABA - less than half of the total pool (31). They have indicated that kindled brains show abnormalities in GAD (36,37,45), GABA levels (37), GABA-A, benzodiazepine and TBPS binding (e.g. 8,33,37), GABA re-uptake (12), GABA-T (26,45), GABA-stimulated chloride flux (7,33), and GABA immunoreactivity (30). (For summary, cf. Table 2; for review, cf. Ref. 6.) In general, the changes reported have been suggestive of GABAergic hypofunction, and several of them - changes in binding, uptake and chloride flux - have been demonstrated 28 or more days after the last kindled seizure. (Details in Table 2.)

Thus, the assay data - once a major problem for the GABA hypothesis - now represent one of its strong points.

## TOWARDS A GABAERGIC THEORY OF KINDLING

At present, the GABA hypothesis seems to be the best supported of the biochemical kindling hypotheses. It is still far from *proven*, however. What further data are needed to support (or disprove) this hypothesis?

1) <u>Biochemical Data:</u> Firstly, a great deal more work needs to be done in the biochemical field. As indicated by Table 2, there is still a great deal to learn about the nature, duration, and location of the GABAergic changes which occur. Particular attention should be paid to the receptor binding area. The present

*TABLE 1*

*SUMMARY OF CURRENT DATA RELATED TO THREE BIOCHEMICAL HYPOTHESES*

*(Note: No complete review has been attempted. Sample references are offered as a guide to the literature.)*

*GABA SYSTEM (GABA-A)*

*1) Acceleration/Simulation - Bicuculline and 3-mercaptopropionic acid accelerate kindling (29), and picrotoxin has recently been shown to do so as well (Mingo and Burnham, submitted). Bicuculline and picrotoxin partially simulate the kindled state (ibid).*

*2) Retardation/Reversal - Progabide and GABA-T blockers retard kindling (28,47), as do benzodiazepines (52,67) and barbiturates (29,64,67). Agonists, both specific and non-specific, fully reverse the kindled state, returning focal-generalized seizures to their prekindled "focal" form (4).*

*3) Long-Lasting Changes - Assays of "synaptic" GABA parameters have revealed a number of post-kindling changes (Table 2), several of which have been reported 28 or more days after the last seizure (6).*

*GLUTAMATE SYSTEM (NMDA)*

*1) Acceleration/Simulation - Apparently never attempted.*

*2) Retardation/Reversal - NMDA antagonists, both specific (APV) and non-specific (MK-801), partially retard the development of kindled seizures (11,43), and partially antagonize well-developed seizures. Retardation and reversal are not complete, however (ibid).*

*3) Long-Lasting Changes - Electrophysiological abnormalities, attributed to the NMDA system, have been reported one day (65), and six weeks (44) after the last seizure. Increased glutamate release (22) and inhibition of carbachol-stimulated phosphoinositide hydrolysis (46) have also been reported in slices at one month. Results from traditional biochemical assays, however, have been largely negative. Synthesis (55), release (34), levels (18), and uptake (55,58) of glutamate are reportedly unchanged by kindling. NMDA receptors are slightly decreased in the cortex and CA 1 of kindled rats at 28 days, and unchanged in other areas (49,58). (See, however, McNamara et al., this volume).*

*NORADRENERGIC SYSTEM*

*1) Acceleration/Simulation - Antagonism of the noradrenergic system accelerates kindling (13,50). Neither alpha nor beta blockade, however, seems to simulate the kindled state (Burnham, unpublished pilot data).*

*2) Reversal/Retardation - Enhancement of noradrenergic activity retards kindling (41). Elevating synaptic noradrenaline does not, however, antagonize well-established seizures (Okazaki, Cheng and Burnham, submitted). It has been suggested that selective stimulation of the alpha-2 receptors may antagonize well-established attacks (27). This has recently been questioned, however (21,41).*

*3) Long-Lasting Changes - Relatively long-lasting changes in receptor binding have been reported, but studies have not revealed consistent long-lasting abnormalities in release, levels, or metabolites (9); (see also Corcoran and Weiss, this volume.) The activity of locus coeruleus neurons does not seem to be changed by kindling (2).*

data related to GABA-A and benzodiazepine binding are confused and inconsistent, possibly due to variations in technical parameters (6).

2) Electrophysiological Data: A second area for investigation is the question of whether GABAergic changes are associated with areas of electrophysiological abnormality. The GABA hypothesis suggests not only that post-kindling GABAergic abnormalities exist, but also that they give rise to functional abnormalities (the kindled state).

Preliminary correlational data seem promising: Elevated binding in the fascia dentata appears to be associated with an increase in inhibition (48,62), whereas the loss of GABA immunoreactivity in CA 1 is accompanied by a loss of double-pulse inhibition (35). Loss of GAD, GABA and GABA binding in the amygdala is accompanied by hyperexcitation in that area, which has been detected in both gross-electrode and single-cell studies (42). An apparent exception, however, occurs in the substantia nigra (pars reticulata), where GAD and GABA binding are decreased (36,37), but electrophysiological function is normal (66). This exception deserves considerable attention; it is the exception which will test the rule.

Though generally promising, the present correlational data are few in number. Further work is needed. "Electrophysiological correlation" should probably become our fourth criterion in assessing biochemical hypotheses.

3) Data Related to Nature of the Cellular Change: A final area for investigation is the exact *nature* of the neuronal changes which give rise to GABAergic abnormalities. Research in this area is just beginning. We may speculate on some possibilities, however, and attempt to relate them to the biochemical data presently available.

## SPECULATIONS CONCERNING THE CELLULAR CHANGE

General Schema: A general schema, formulated by Lothman in his discussions of "rapid kindling", states that, "inhibitory interneurons are injured by seizures, irrespective of the cause" (38). With some necessary qualifications, this schema may be adopted as a working hypothesis for the entire kindling field. As one necessary qualification, we may note that neuronal damage probably extends beyond inhibitory neurons *per se*, since, in the GABA system, *post-* as well as *pre*-synaptic changes are observed. As a further qualification, we may propose that damage may be concentrated in those regions which suffer the severest stress: The focus would be fully stressed and should suffer a general loss in inhibitory function (and hence a drop in threshold). In extrafocal structures, only interneurons related to the focal-extrafocal pathway might be stressed - which would lead to an increased sensitivity to patterned input *without* a drop in threshold (cf. 51). This schema would explain a good deal of what we know about kindling.

Types of Neuronal Change: What sorts of neuronal change might contribute to a selective loss of inhibitory function? Basically, there are three possibilities - growth-like changes, cell death (or damage), and long-lasting dysfunction: 1) Growth: The case for growth-like changes, or "enhancement of synaptic efficiency", has been summed up by Morrell and colleagues (20). This approach finds some support in recent anatomical studies (60), although it faces the criticism that kindling seems to occur only when afterdischarge is present, and that growth-like changes have been easier to detect in other plasticity models (10). In the present context, "growth" would presumably occur in neurons that *inhibit* the GABA system. 2) Death: Cell death (or damage) is a possibility that has found little favor; most theorists argue against it (26,30,60). In many cases, however, the argument is simply that dead cells have *not yet been found.* This argument is countered, in part, by recent findings of post-kindling astrocyte hypertrophy (Ivy, Racine and Milgram, in preparation). Cell death is a hypothesis that still requires investigation. 3) Dysfunction: A third possibility is that "kindled" cells are alive and whole, but permanently

*TABLE 2. SUMMARY OF ASSAY DATA RELATED TO THE GABA HYPOTHESIS**

| *SITE* | *PARAMETER* | | | | | | | |
|---|---|---|---|---|---|---|---|---|
| | *GAD* | *Levels* | *Uptake* | *Binding*** *GABA-A* | *BZD* | *GABA-T* | *Cl- Flux* | *Immuno React.* |
| Amygdala | 0 | - | ? | 0/- | 0/+ | - | ? | ? |
| *Brain Stem**** | 0 | 0 | ? | ? | ? | - | - | ? |
| *Cere-bellum* | 0 | 0 | ? | ? | ? | ? | 0 | ? |
| *Cortex* | -/0 | 0 | ? | 0 | 0/- | - | 0 | ? |
| *Hippo-campus* | 0 | 0 | + | 0 | 0/+ | - | ? | ? |
| *Cornu Ammonis* | ? | ? | ? | 0 | 0 | ? | ? | - |
| *Fascia Dentata* | ? | ? | ? | 0/+ | 0/+ | ? | ? | ? |
| *Hypo-thalamus* | 0 | 0 | ? | ? | - | ? | ? | ? |
| *Olfactory Bulbs* | 0 | 0 | ? | ? | + | ? | ? | ? |
| *Substantia Nigra* | - | 0 | ? | 0/- | 0 | ? | ? | ? |
| *Striatum* | 0 | 0 | ? | 0/+ | 0/- | ? | ? | ? |
| *Thalamus* | 0 | 0 | ? | ? | ? | - | ? | ? |

** Table 2 is limited to studies using "synaptic" assays, conventional kindling parameters and sacrifice 24 or more hours after the last seizure. When groups have worked at 2 or more intervals, the long-interval data are reported. Abbreviations: "0" = No change; "+" = Increase; "-" = Decrease; "?" = Not Assayed. References: GAD activity (36,37,45); GABA levels (37); GABA-A binding (37,57,63); selected benzodiazepine binding data (8,56,57,63); GABA reuptake (12); GABA-T activity (26,45); Chloride flux (7,33); GABA immunoreactivity (30). For full review, cf. Ref. 6*

*** No attempt has been made to summarize the complete binding data. Selected references have been included to indicate: a) where changes have been reported, and b) that the data are often contradictory. (A full review appears in ref. 5).*

**** "Brain Stem" includes midbrain, pons and medulla. Thalamus, cerebellum and substantia nigra are listed separately.*

hypofunctional. The problem here is to explain how a non-structural dysfunction can be perpetuated for the life of the cell. One interesting possibility, suggested by recent studies of c-fos proteins (16), is that kindling may somehow alter gene transcription.

Present Data: Which of the above possibilities is favored by the existing data? In *the presynaptic context*, there are reports of lowered GAD, lowered GABA levels in synaptosomes, lessened GABA-T synthesis, etc. (Table 2). These might seem to suggest a loss of GABAergic neurons possibly resultant from hyperactivity in "upstream" excitatory neurons. (*In vitro* studies have indicated that glutamic hyperexcitation - and related calcium influx - can cause neural damage (39,59).) Both Itagaki and Kimura (26) and Kamphuis et al. (30), however, argue that GABAergic changes occur without GABA neuron loss. Present theorists, therefore, seem to favor the option of presynaptic *hypofunction.*

*In the postsynaptic context*, there are reports of binding changes, and of decreased GABA-stimulated chloride flux (Table 2). Discussion of the binding data may be deferred until consistent data are achieved. Long-lasting reductions in GABA-stimulated chloride flux, however, have been reported in synaptoneurosomal preparations derived from both electrically kindled (7) and chemically kindled animals (33). (In short-term studies, similar changes have also been observed after kainic acid seizures (17).) Recent studies from our own laboratory suggest that flux changes occur in the absence of changes in GABA-A or benzodiazepine binding (Nobrega, Burnham et al. submitted). Thus, the postsynaptic cells seem to be present, although less responsive to GABA. In the chemical kindling and kainic acid models, flux changes have been associated with changes in binding of TBPS, which suggests the site of dysfunction may lie in the GABA-related chloride ionophore. How such a change could occur, and how it could persist when the GABA-A complexes are re-synthesized every few days, is not yet clear. The present post-synaptic data, however, also seem to favor the option of hypofunction rather than cell death.

## LIMITS TO THE GABA HYPOTHESIS

In the present paper we have argued that the GABA hypothesis is the best supported of the current biochemical hypotheses of kindling. We do not mean to suggest, however, that GABAergic abnormalities are the only abnormalities which contribute to the kindled state. Two considerations make this unlikely:

Firstly, the different transmitters do not exist in isolation; they work in interactive partnerships. It is unlikely, therefore, that an abnormality in any system will exist alone. The "kindled change" is likely to be a complex pattern involving two or more transmitters. In the case of GABA, two of the other transmitters are likely to be noradrenaline and glutamate, since both are known to interact with the GABA system (59). In Lothman's studies of "rapid kindling" abnormalities are found in *both* GABA and glutamate (38).

*Secondly*, there are now experimental data which suggest that GABA may be responsible for part, but not all, of the kindled state. Pharmacological simulation data produced by our laboratory (Mingo and Burnham, submitted) indicate that blockade of the GABA system to "just subconvulsant" levels produces an animal which responds to a focal seizure with partial, but not complete generalization. The subjects look "half kindled". It may be that we have simply not been able to create - at subconvulsant doses - the complete "kindled change". Alternately, it may be that two or more systems contribute to the "kindled change". Certainly, at the present time, no single system should be considered the sole cause of the kindled state.

REFERENCES


1. Blumer, D., 1975, Temporal lobe epilepsy and its psychiatric significance, in: "Psychiatric Aspects of Neurologic Disease", D. Blumer, and D.F. Benson, eds., Grun and Stratton, New York.
2. Bonhaus, D.W., and McNamara, J.O., 1987, Activity of locus coeruleus neurons in amygdala kindled rats: role in suppression of after-discharge, Brain Res., 407:102.
3. Bonhaus, D.W., and McNamara, J.O., in press, TCP binding: a tool for studying NMDA receptor mediated neurotransmission in kindling, Neurosci. Behav. Rev.
4. Burnham, W.M., 1985, Progabide and other GABAmimetics in the kindling model, in: "Epilepsy and GABA Receptor Agonists. Basic and Therapeutic Research", G. Bartholini, L. Bossi, K. G. Lloyd, and P. L. Morselli, eds., Raven Press, New York.
5. Burnham, W.M., 1988, Receptor binding in the kindling model of epilepsy, in: "Receptors and Ligands in Neurological Disorders", A.K. Sen and T. Lee, eds., Cambridge University Press, Cambridge.
6. Burnham, W.M., in press, The GABA hypothesis of kindling: recent assay studies, Neurosci. Behav. Rev.
7. Burnham, W.M., Kish, S.J., and Sneddon, W.B., 1988, GABA-stimulated chloride flux in the kindling model of epilepsy. Neurosci. Absts., 14:1035.
8. Burnham, W.M., Niznik, H.B., Okazaki, M.M., and Kish, S.J., 1983, Binding of 3H-flunitrazepam and 3H-R0-4864 to crude homogenates of amygdala kindled rat brain: two months post-seizure, Brain Res., 279:259.
9. Burnham, W.M., Racine, R.J., and Okazaki, M.M., 1986, Kindling mechanisms: II. biochemical studies, in: "Kindling 3", J.A. Wada, ed., Raven Press, New York.
10. Cain, D.P., 1989, Long-term potentiation and kindling: how similar are the mechanisms? Trends Neurosci., 12:6.
11. Cain, D.P., Desborough, K.A., and McKitrick, D.J., 1988, Retardation of amygdala kindling by antagonism of NMD-aspartate and muscarinic cholinergic receptors: evidence for the summation of excitatory mechanisms in kindling, Exp. Neurol., 100:179.
12. Chaudieu, I., Rondouin, G., Chicheportiche, M., and Chicheportiche, R., 1987, Presynaptic alteration of (3H) GABA transport in hippocampus by amygdala kindling, Neurosci. Lett., 76:329.
13. Corcoran, M.E., 1981, Catecholamines and kindling, in: "Kindling 2", J.A. Wada, ed., Raven Press, New York.
14. Dragunow, M., and Goddard, G.V., 1984, Adenosine modulation of amygdala kindling, Exp. Neurol., 84:654.
15. Dragunow, M., Goddard, G.V., and Laverty, R., 1985, Is adenosine an endogenous anticonvulsant? Epilepsia, 26:480.
16. Dragunow, M., Robertson, H.A., and Robertson, G.S., 1988, Amygdala kindling and c-fos proteins, Exp. Neurol., 102:261.
17. Edgar, P., Bowe, M.A., and Schwartz, R.D., 1988, Regulation of the GABA-A receptor complex in brain following limbic seizures produced by kainic acid, Neurosci. Absts., 14:1036.
18. Fabisiak, J.P., and Schwark, W.S., 1982, Cerebral free amino acids in amygdaloid kindling model of epilepsy, Neuropharmacology, 21:179.
19. Fikuda, H., Brailowsky, S., Menini, C, Silva-Barrat, C., Riche, D., and Naquet, R., 1987, Anticonvulsant effect of intracortical, chronic infusion of GABA in kindled rats: Focal seizures upon withdrawal, Exp. Neurol., 98:120.
20. Geinisman, Y., Morrell, F. and deToledo-Morrell, L., 1988, Proc. Natl. Acad. Sci. USA, 85:3260.
21. Gellman, R.L., Kallianos, J.A., and McNamara, J.O., 1987, Alpha-2 receptors mediate an endogenous noradrenergic suppression of kindling development, J. Pharm. Exp. Ther., 241:891.
22. Geula, C., Jarvie, P.A., Logan, T.C., and Slevin, J.T., 1988, Long-term enhancement of K+-evoked release of L-glutamate in entorhinal kindled rats, Brain Res., 442:368.

23. Goddard, G.V. and Douglas, R.M., 1975, Does the engram of kindling model the engram of long term memory? Can. J. Neurol. Sci., 2:385.
24. Goddard, G.V., McIntyre, D., and Leech, C., 1969, A permanent change in brain function resulting from daily electrical stimulation, Exp. Neurol., 25:295.
25. Heit, M.C. and Schwark, W.S., 1988, Pharmacological studies with a GABA uptake inhibitor in rats with kindled seizures in the amygdala, Neuropharmacology, 27:367.
26. Itagaki, S. and Kimura, H., 1986, Retardation of resynthesis of GABA-transaminase in some brain regions of amygdala-kindled rats, Brain Res., 381:77.
27. Joy, R.M., Stark, L.G., and Albertson, T.E., 1983, Dose-dependent proconvulsant and anticonvulsant actions of the alpha 2 adrenergic agonist, xylazine, on kindled seizures in the rat, Pharmacol. Biochem. Behav., 19:345.
28. Joy, R.M., Albertson, T.E., and Stark, L.G., 1984, An analysis of the actions of progabide, a specific GABA receptor agonist, on kindling and kindled seizures, Exp. Neurol., 83:144.
29. Kalichman, M.W., Livingston, K.E., and Burnham, W.M., 1981, Pharmacological investigation of gamma-aminobutyric acid (GABA) and the development of amygdala-kindled seizures in the rat, Exp. Neurol., 74:829.
30. Kamphuis, W., Wadman, W.J., Buijs, R.M., and Lopes da Silva, F.H., 1986, Decrease in number of hippocampal gamma-aminobutyric acid (GABA) immunoreactive cells in the rat kindling model of epilepsy. Exp. Brain Res., 64:491.
31. Kuriyama, K. Subcellular localization of the GABA system in brain, 1976, in: "GABA in Nervous System Function", E. Roberts, T.N. Chase and D.B. Tower, eds., Raven Press, New York.
32. Lennox, W.G., 1941, "Science and Seizures" Harper and Brothers, New York.
33. Lewin, E., Peris, J., Bleck, V., Zahniser, N.R., and Harris, R.A., 1989, Chemical kindling decreases GABA-activated chloride channels of mouse brain, Eur. J. Pharmacol., 160:101.
34. Liebowitz, N.R., Pedley, T.A., and Cutler, R.W.P., 1978, Release of gamma-aminobutyric acid from hippocampal slices of the rat following generalized seizures induced by daily electrical stimulation of entorhinal cortex. Brain Res., 138:369.
35. Lopes da Silva, F.H., 1987, Hippocampal kindling: physiological evidence for progressive disinhibition, in: "Advances in Epileptology, Vol. 16", D. Wolf, M. Dam, D. Janz, and F.E. Dreifuss, eds., Raven Press, New York.
36. Loscher, W. and Schwark, W.S., 1985, Evidence for impaired GABAergic activity in the substantia nigra of amygdaloid kindled rats, Brain Res., 339:146.
37. Loscher, W. and Schwark, W.S., 1987, Further evidence for abnormal GABAergic circuits in amygdala-kindled rats, Brain Res., 420:385.
38. Lothman, E.W., Bennet, J.P., and Perlin, J.P., 1987, Alterations in neurotransmitter amino acids in hippocampal kindled seizures, Epilepsy Res., 1:313.
39. Mattson, M.P., and Kater, S.B., 1989, Excitatory and inhibitory neurotransmitters in the generation and degeneration of hippocampal neuroarchitecture, Brain Res., 478:337.
40. McIntyre, D.C., 1981, Catecholamine involvement in amygdala kindling of the rat, in: "Kindling 2", J.A. Wada, ed., Raven Press, New York.
41. McIntyre, D.C. and Guigno, L., 1988, Effect of clonidine on amygdala kindling in normal and 6-hydroxydopamine-pretreated rats. Exp. Neurol., 99:96.
42. McIntyre, D.C. and Racine, R.J., 1986, Kindling mechanisms: current progress on an experimental epilepsy model. Prog. in Neurobiol., 27:1.
43. McNamara, J.O., Russell, R.D., Rigsbee, L., and Bonhaus, D.W., 1988, Anticonvulsant and antiepileptogenic actions of MK-801 in the kindling and electroshock models, Neuropharmacology, 27:563.
44. Mody, I. and Heinemann, U., 1987, NMDA receptors of dentate gyrus granule cells participate in synaptic transmission following kindling, Nature, 326:701.
45. Moiseev, I.N., Shandra, A.A., and Godlevskii, L.S., 1984, Cytophotometric study of changes in glutamate dehydrogenase and GABA transaminase

in the cerebral cortex during metrazol kindling, Bull. Exp. Biol. Med., 97:422.
46. Morrisett, R.A., Nadler, J.V., and McNamara, J.O., 1987, Evidence for enhanced N-methyl-D-aspartate receptor mediated inhibition of carbachol-stimulated phosphoinositide hydrolyis from kindled rats, Neurosci. Absts., 13:946.
47. Myslobodsky, M.S., and Valenstein, E.S., 1980, Amygdaloid kindling and the GABA system, Epilepsia, 21:163.
48. Oliver, M.W., and Miller, J.J. 1985, Alterations of inhibitory processes in the dentate gyrus following kindling-induced epilepsy, Exp. Brain Res., 57:443.
49. Okazaki, M.M., McNamara, J.O., and Nadler, J.V., 1989, N-methyl-D-aspartate receptor autoradiography in rat brain after angular bundle kindling, Brain Res., 482:359.
50. Peterson, S.L. and Albertson, T.E., 1982, Neurotransmitter and neuromodulator function in the kindled seizure and state, Prog. Neurobiol., 19:227.
51. Racine, R.J., 1972. Modification of seizure activity by electrical stimulation: I. after-discharge threshold, EEG clin. Neurophysiol., 32:269.
52. Racine, R.J., Livingston, K., and Joaquin, A., 1975, Effects of procaine hydrochloride, diazepam, and diphenylhydantoin on seizure development in cortical and subcortical structures in rats, EEG clin. Neurophysiol., 38:355.
53. Racine, R.J., Milgram, N., and Hafner, S., 1983, Long-term potentiation phenomena in the rat limbic forebrain, Brain Res., 260:217.
54. Racine, R.J., Tuff, L., and Zaide, J., 1975, Kindling, unit discharge patterns and neural plasticity, Can. J. Neurol. Sci., 2:395.
55. Reedy, D.P., McGeer, E.G., Staines, W.A., and Corcoran, M.E., 1978, Amygdaloid kindling and central enzyme activity, Neurosci. Absts., 4:146.
56. Rondouin, G., Chicheportiche, M., Lerner-Natoli, M., Ben Attia, M., Privat, A. and Baldy-Moulinier, M., 1986, Inhibitory processes in limbic kindling, in: "Kindling 3", J.A. Wada, ed., Raven Press, New York.
57. Shin, C., Pedersen, H.B., and McNamara, J.O., 1985, Gamma- aminobutyric acid and benzodiazepine receptors in the kindling model of epilepsy: a quantitative radiohistochemical study, J. Neurosci., 5:2696.
58. Slevin, J.T., and Ferrara, L.P., 1985, Lack of effect of entorhinal kindling on L-(3H)glutamic acid presynaptic uptake and postsynaptic binding in hippocampus, Exp. Neurol., 89:48.
59. Stelzer, A., Slater, N.T., and ten Bruggencate, G., 1987, Activation of NMDA receptors blocks GABAergic inhibition in an in vitro model of epilepsy, Nature, 326:698.
60. Sutula, T., Xiao-Xian, H., Cavazos, J., and Scott, G., 1989, Synaptic reorganization in the hippocampus induced by abnormal functional activity, Science, 239:1147.
61. Trimble, M.R., 1988, Cognitive hazards of seizure disorders, Epilepsia, 29:S19.
62. Tuff, L.P., Racine, R.J., and Adamec, R., 1983, The effects of kindling on GABA-mediated inhibition in the dentate gyrus of the rat. I. paired-pulse depression, Brain Res., 277:79.
63. Tuff, L.P., Racine, R.J., and Mishra, R.K., 1983, The effects of kindling on GABA-mediated inhibition in the dentate gyrus of the rat: I. receptor binding, Brain Res., 277:91.
64. Wada, J.A., 1977, Pharmacological prophylaxis in the kindling model of epilepsy, Arch. Neurol., 34:389.
65. Wadman, W.J., Heinemann, U., Konnerth, A. and Newhaus, S., 1985, Hippocampal slices of kindled rats reveal calcium involvement in epileptogenesis, Exp. Brain Res., 57:404.
66. Waszczak, B.L., Applegate, C.D., and Burchfield, J.L., 1988, Kindling does not cause persistent changes in firing rates or transmitter sensitivities of substantia nigra pars reticulata neurons, Brain Res., 455:115-122.
67. Wise, R.A., and Chinerman, J., 1974, Effects of diazepam and phenobarbital on electrically-induced amygdaloid seizures and seizure development, Exp. Neurol., 45:355.

Discussion of Dr. Burnham's Presentation

DR. ADAMEC: Thank you, Mac. It was a very good presentation. There's an odd chance that someone will take your criteria seriously. With respect to simulation, I think you just touched on it in your final comments - I think it might not be possible to simulate kindling until you have a firm idea of what the state looks like. I raise this only because pharmacological and drug manipulations sometimes, it seems to me, treat the brain as if it were one gigantic receptor and that you should expect changes in only one direction, and some single gigantic neuron which is the kindled neuron. It's very possible, and certainly at least in terms of short-term changes it has been shown that, for example, in a GABA- mediated system the tri-synaptic circuit, dentate for example, you get an increase in inhibition. I'm going to talk tomorrow about increase inhibition in CA3, CA1 you get a relief of it. That is not the kindled state, but that is a case where even in the same neurotransmitter system you get a pattern change depending upon the piece of circuitry you're in. So, that criterion to my mind, would be the one you'd use last and not first. I mean, I wouldn't give up on a manipulation if you couldn't find that you could simulate it by a systemic manipulation. It would be better to track down the other two, get some idea of whether there is a pattern, and then try to simulate the pattern. Would you agree or disagree?

DR. BURNHAM: I think that's a good comment. Simulation's going to be hard, and when you drive up or down levels in the whole brain, it could be that the brainstem will cause generalized seizures before you cause an extra-brainstem change; could be the animal will die before you get to the right level. It's a difficult criterion; I think that people should try it though. Try it, you'll like it. Please don't put that in the book. One thing we're going to try is, of course cannulating GABA blockers into specific areas. It's been done before, but we wonder if we can cause site specific simulation so that the animal goes from focal to generalized seizure immediately without kindling. We're going to try it and see what happens. The criteria I hope will be discussed.

DR. CAIN: Mac, I'd like to encourage you to keep sort of stepping back and doing this philosophical sort of thing, because I really enjoy these talks. But, picking up a bit on what Bob was just saying, I do have a little problem with things you said at two different times. For example, the reversal and simulation criteria; if it's true, as I've argued and others have argued including I think, earlier today, that there is a multiplicity of neurotransmitter systems, excitatory ones for example, that contribute, you can't completely block kindled seizures with any single antagonist, then I don't see how logically you can expect a reversal experiment to work. I think you must always expect that there's going to be a failure of that, likely because of the presence of other excitatory systems that are going to be stimulated and working.

DR. BURNHAM: You know, I think you can reverse full kindled animals both with GABA agonists and with adenosine agonist. Those are the two systems where blockade, at least partially, simulates the kindled seizure. Likewise, phenonbarb or progabide in our hands will take whole focal-generalized seizure right back to the focus. I think that cannula-applied GABA in certain cases has accomplished similar things.

DR. CAIN: Well, I like the GABA story. I was thinking more of the excitatory end of things, because that's what I do work in, but I'll leave that. I was also going to ask you, since you're much more up on this than I am, whether there have been any demonstrations in epileptic human brain of enduring changes in GABA systems.

DR. BURNHAM: Yeah, that's an interesting story. Lloyd circa 1980 working with Bancaud in Paris produced some data which suggested that the human focus was deficient in GABA-A binding as well as GAD levels. Now this, just as a sideline, is an interesting point. Quite frequently, in different models and in this human tissue, you get both GAD changes which are presynaptic and GABA-A binding changes which are post-synaptic. In many cases they go together, it's not what you would predict, but it seems to be what is found. And I think that if we understood that, we'd understand an important thing. Anyway, Lloyd with Bancaud reported a deficiency in GAD and GABA-A. Now, a number of other people, including Sherwin in Montreal and our group in Toronto, did not find similar things in human tissue. Mostly we looked at the benzos. What Lloyd said (and he's right), is that his tissue, which was done with depth electrodes to locate the focus, was vastly better than ours, which was done on the basis of interictal spiking. He got the focus; I think mostly we got cortex near the focus but not necessarily in the focus. The recent evidence which supports Dr. Lloyd's findings is from a group with Savic, I expect his name is et al., and they have done PET scans in Scandinavia, epileptic people, and they have located the focus by PET scan. They have done benzo binding in it, and they find quite sizeable deficiencies in benzo binding in the awake human. I think that's the most interesting thing that's come out lately.

# NORADRENALINE AND KINDLING REVISITED

Michael E. Corcoran and Gerald K. Weiss*

Department of Psychology
University of Victoria
Victoria, BC
Canada V8W 2Y2

*Department of Physiology
University of New Mexico
Albuquerque, NM 87131
USA

One of the most consistent findings in the literature on kindling has been the observation that the process of seizure development is facilitated by treatments that interfere with the actions of noradrenaline (NA). This body of research has given rise to the hypothesis that part of the mechanism of kindling may be a decrease in the effects of NA, produced either by a progressive suppression of the presynaptic release of NA[1] or a decline in the postsynaptic response to NA[2]. In the present chapter we provide an update on recent evidence that NA may modulate kindling.

## EFFECTS OF NORADRENERGIC ANTAGONISTS

The original evidence that the spread of ictal discharge is restricted by NA came from studies of the effects on kindling of NA antagonists, particularly the neurotoxin 6-hydroxydopamine (6-OHDA). Depletion of NA produced by intracerebral or intraventricular infusion of 6-OHDA results in a tremendous acceleration in the rate of kindling provoked by stimulation of limbic sites[1,2] or anterior neocortex[3]. In some NA-depleted rats, generalized seizures can be triggered by the first amygdaloid AD[1]. A similar facilitation of amygdaloid kindling is produced by systemic injections of antagonists for α-2 but not α-1 or β noradrenergic receptors[4], suggesting that the antikindling effects of NA are produced at α-2 receptors. The antikindling actions of NA are apparently produced in the forebrain, because kindling is facilitated by infusions of 6-OHDA into either the dorsal NA bundle[1] or the amygdala itself[2].

The facilitation of kindling after depletion of NA is not due to effects at the stimulated site, as would be reflected in changes in the threshold or duration of the initial afterdischarge (AD); rather NA depletion accelerates the spread of AD to distant sites and the consequent development of behavioral seizures[1]. Because the duration and intensity of generalized seizures in NA-depleted rats are no different from those in controls, and because depletion of NA after seizure generalization does not potentiate seizures[5], it has long been recognized that NA dampens not kindled seizures but instead the kindling process itself. Only recently, however, has it become clear

*Kindling 4*
Edited by J. A. Wada
Plenum Press, New York, 1990

just where the antikindling effects of NA are exerted. Michelson and Buterbaugh[6] demonstrated in juvenile rats receiving amygdaloid stimulation that depletion of NA accelerated progression out the early stages of nonconvulsive or partial seizure but did not affect the amount of time spent in intermediate or later stages of seizure. As shown in Figure 1, Corcoran[7] has replicated their analysis in adult rats; and Uemura et al.[8] reported generally comparable findings in cats treated with intraamygdaloid infusions of 6-OHDA. Systemic administration of the α-2 antagonist idazoxan similarly facilitates kindling by truncating the time spent in the early stages of seizure development[4].

Finally, NA's involvement in kindling has also been demonstrated in several paradigms other than conventional kindling. McIntyre et al.[9] showed that the antagonistic effects of applying kindling stimulation at short (e.g., 10 min) interstimulation intervals were absent after depletion of NA. This of course suggests that the antikindling effects of stimulating at short intervals are due to actions of NA, presumably released at sites in the forebrain. Applegate et al.[10] have shown that NA is also responsible for "kindling antagonism," wherein sequential stimulation of two sites results in domination by one and suppression of kindling provoked from stimulation of the other. In contrast to the involvement of forebrain NA in regulating other aspects of kindling, however, kindling antagonism is apparently due to the effects of NA in the hindbrain.

## EFFECTS OF NORADRENERGIC AGONISTS

Compatible with the results obtained with noradrenergic antagonists, studies with agonists have shown that kindled seizures are relatively insensitive to noradrenergic manipulations[11]. However, the results have not been entirely consistent. For example, Gellman et al.[4] and McIntyre and Giugno[12] have both reported that the α-2 agonist clonidine had little if any effect on kindled amygdaloid seizures, whereas Loscher and Czuczwar[13] found small but significant antiepileptic effects of clonidine. We reexamined the antiepileptic effects of

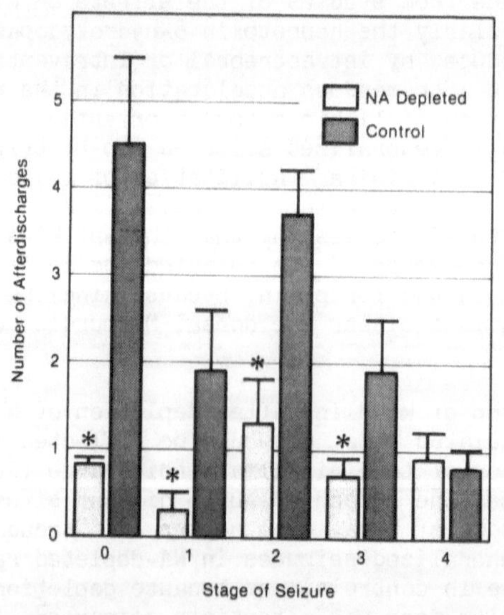


Figure 1. Mean number of ADs required for development of each stage of seizure in NA-depleted and control rats. *Significantly different from control, $p<0.05$.

clonidine by measuring the effects of ip injections at doses of 10, 100, 200, and 500 g/kg on kindled amygdaloid seizures (Corcoran, unpublished observations). As shown in Table 1, clonidine produced slight reductions in the mean duration and intensity of kindled seizures, but these changes were not statistically significant. Thus the bulk of the evidence indicates that kindled seizures are not reliably suppressed by clonidine.

In contrast to the relative insensitivity of established kindled seizures, the process of kindling is reliably suppressed by clonidine, as demonstrated by Gellman et al.[4] and McIntyre and Giugno[12]. Because the prophylactic effect of clonidine was evident in NA-depleted rats pretreated with 6-OHDA[12] or DSP-4[4], it seems that NA acts at postsynaptic α-2 receptors to delay kindling. Additional evidence that endogenous NA acts to delay kindling comes from an experiment performed by Barry et al.[14]. They found that transplantation of cell suspensions from fetal locus coeruleus (LC), rich in NA, could offset the facilitatory effects on hippocampal kindling of 6-OHDA-induced depletion of NA.

One likely site of the prophylactic effects of NA is the amygdala, and we have been examining the effects of intraamygdaloid infusions of clonidine during kindling (Pelletier & Corcoran, unpublished observations). Our preliminary findings are that rapid kindling with low-frequency stimulation (mean [±SEM] of 2.3 [± 0.3] ADs to generalized seizure in saline-infused control rats) can be slowed significantly by infusions of 0.5 μl of 1.0 mM clonidine (mean of 4.4 [± 0.9] ADs), a concentration chosen because it has been shown in other studies to affect learning. A similar slowing of low-frequency kindling was produced by intraamygdaloid infusions of 1.0 mM of the β antagonist propranolol (mean of 4.1 [±0.5] ADs).

Another line of evidence suggesting that endogenous NA acts in the intact brain to delay kindling is found in our observation that electrical stimulation of the locus coeruleus (LC) can delay amygdaloid kindling[15]. We exposed rats to 30 min of LC stimulation before each amygdaloid AD and found that there was a delay in the progression out of the early stages of partial seizure. LC-stimulated rats unexpectedly displayed significantly longer AD than controls in the early stages of kindling and a significant facilitation of generalization from stage 3 to later stages of seizure. This differs from the characteristics of the prophylactic effects of clonidine[4], which both shortens AD duration and slows development of behavioral seizures. It seems clear that the effects of clonidine are related to activation of receptors for NA[4], but it cannot be assumed without question that the effects of LC stimulation are necessarily related to the release of NA at critical forebrain sites. To address the question of whether forebrain NA mediates the effects of LC stimulation on kindling, we have examined the effects of LC stimulation on amygdaloid kindling after 6-OHDA-induced depletion of NA (Weiss, Lewis, Jiminez-Rivera, Laubert, Voltura, & Corcoran, unpublished observations). As shown in Figure 2, we found that LC stimulation prolonged the occurrence of the early stages of partial seizure, but that this did not occur in NA-depleted rats. Thus the delaying effects of LC stimulation on amygdaloid kindling depend on the integrity of noradrenergic fibers innervating the forebrain and midbrain, a finding that is compatible with the antikindling effects of noradrenergic agonists. In contrast, depletion of NA failed to affect the enhanced duration of AD and the significant reduction in number of ADs spent in stages 3 and 4 of kindling in rats receiving LC stimulation. Thus the facilitatory effects of LC stimulation on amygdaloid kindling are independent of the noradrenergic innervation of the forebrain and midbrain and must instead depend on activation either

Table 1. Effects of Clonidine on Kindled Amygdaloid Seizures[a]

| Group | Per cent having generalized seizure | Mean stage of seizure[b] | Mean duration of seizure[b] | Mean duration of AD[b] |
|---|---|---|---|---|
| Control | 100.0 | 4.7 (±0.2) | 39.8 (±1.6) | 39.2 (±1.6) |
| 10 μg/kg | 88.9 | 4.4 (±0.3) | 31.1 (±1.1) | 43.1 (±1.7) |
| 100 μg/kg | 62.5 | 3.5 (±0.5) | 21.0 (±1.3) | 44.5 (±1.8) |
| 200 μg/kg | 77.9 | 3.6 (±0.5) | 26.4 (±1.3) | 64.2 (±2.2) |
| 500 μg/kg | 80.0 | 3.6 (±0.5) | 20.3 (±1.5) | 27.0 (±1.9) |

[a]Data expressed as mean (±SEM)
[b]All ps>0.05

of noradrenergic fibers spared by dorsal bundle infusions of 6-OHDA or of non-noradrenergic systems in or near the LC.

Studies using agonists and antagonists of NA collectively suggest that endogenous NA acts in the intact brain to delay the emergence of the early features of kindling. The results indicate that NA acts at a postsynaptic α-2 receptor to delay kindling, either by reducing the progressive lengthening of AD duration, or by retarding the spread of AD from the stimulated site, or both. The data are in good agreement that the antikindling actions of NA are restricted to the earliest stages of seizure development, in which nonconvulsive or partial seizures are being expressed. Once kindling has progressed beyond these early stages, NA seems to lose its ability to suppress ictal manifestations, and noradrenergic manipulations have little if any effect on established kindled seizures. The results are therefore consistent with the hypothesis that the ability of seizures to evolve beyond the early stages of kindling might in part involve an erosion of the inhibitory effects of NA. Next we review evidence suggesting that the hypothesized decrease in NA's effects is due to declines both in

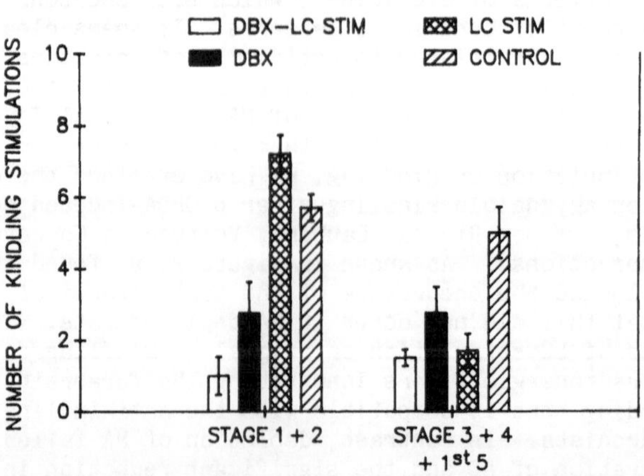


Figure 2. Mean number of ADs required for development of different stages of seizure in NA-depleted (DBX) and control rats receiving stimulation of the LC (LC STIM).

presynaptic activity in noradrenergic neurons[1] and in responsiveness of postsynaptic target neurons to NA[2].

NORADRENERGIC CORRELATES OF KINDLING: NEUROCHEMISTRY

Although a number of early studies suggested that alterations in NA concentrations might be detected after kindling (reviewed in ref. 1), recent studies have failed to document any long-lasting changes in NA under carefully controlled conditions. For example, Blackwood[16] used the radioenzymatic assay to measure concentrations and turnover of NA in the amygdala and hippocampus of amygdaloid kindled rats and unstimulated control rats killed one month after the last seizure. He was unable to find any significant differences between the groups.

Reasoning that variability in previous results might have been due to inconsistencies in procedures, including wide differences in the postkindling intervals at which assays were performed, we recently reexamined regional concentrations of NA in the brains of amygdaloid kindled rats killed at 2 intervals postkindling, 2 and 4 weeks[17]. We were very careful to yoke each kindled rat to a control that received one sec of low-frequency stimulation daily at an intensity that failed to evoke AD. We found minor fluctuations in the concentrations of NA after kindling, but no consistent or comprehensible pattern emerged. For example, there was a small decrease in NA in the ipsilateral frontal cortex 2 weeks after kindling, but this had reverted to control levels by 4 weeks. Concentrations of NA in the stimulated amygdala and ipsilateral hypothalamus were not different from control at 2 weeks, whereas they had dropped to significantly below control levels at 4 weeks. The very small magnitude of these changes and their lack of a coherent pattern highlight the instability of the kindled brain; but they cannot be considered to be a consistent alteration in noradrenergic markers. In a similar series of experiments, Okazaki et al.[18,19] have examined concentrations and turnover of NA and concentrations of its glycol metabolites MHPG and DHPG 2 months after completion of amygdaloid kindling. Okazaki et al. compared regional measures of NA and metabolites in kindled rats to yoked controls. They found no significant differences between the groups in any region examined. Collectively these results indicate that there are no long-lasting and stable alterations in presynaptic biochemical markers for NA that can be detected after limbic kindling. In view of evidence that NA regulates only the very early stages of kindling[4,6-8], however, we suggest that these correlative studies do not clarify the role of NA in kindling. A more appropriate test of the NA hypothesis would be to look for changes in markers for NA during the early stages of kindling, when NA is presumably most involved.

Considerable evidence suggests that kindling results in long-term alterations in binding of ligands for NA receptors. McIntyre and Roberts[20] measured regional binding of a ligand to β receptors for NA 3 weeks after kindling of generalized amygdaloid seizures produced by daily stimulation. They found that there was a significant decrease in the number of neocortical β receptors, with no change in affinity. Using hourly stimulation of the amygdala, Stanford and Jefferys[21] kindled seizures and measured binding to α-2 and βreceptors at various intervals after kindling. They found that the densities of α-2 and β receptors in olfactory cortex and of β receptors in neocortex were reduced 24 hr after kindling, but that only the decrease in neocortical β receptors was detectable 3 weeks later. A preliminary report by Chen and McNamara[22] has also described decreased binding of clonidine to α-2 receptors in pyriform cortex and the central amygdala 24 hr after amygdaloid kindling.

In a more complicated paradigm that varied aspects of the parameters of kindling, Michelson and Buterbaugh[23] kindled partial or generalized seizures in juvenile or adult rats using hourly or daily stimulation. Three weeks postkindling they measured β receptors in the whole forebrain. Michelson and Buterbaugh found decreased affinity of β receptors in juvenile rats in which partial seizures had been kindled by hourly stimulation but no changes after kindling of generalized seizures. Affinity was increased after kindling of partial seizures using hourly stimulation in adult rats, whereas kindling of generalized seizures resulted in no changes in β receptors, in contrast to the results of Stanford and Jefferys[21]. Using daily stimulation, however, kindling of generalized seizures in adult rats led to decreased numbers of β receptors, similar to the results of McIntyre and Roberts[20]. Michelson and Buterbaugh were therefore able to confirm that kindling can, under certain circumstances, lead to a long-lasting decrease in the density of β receptors for NA. Their results further indicate, however, that changes in β receptors are a function of a variety of factors, such as the age of the rats, the kindling paradigm used, and the degree of seizure development achieved.

There thus seems to be some agreement that limbic kindling can be followed by a short-lasting decrease in the number of α-2 receptors and a long-lasting decrease in the number of β receptors for NA in various regions of the forebrain. The studies all measured NA receptors after development of generalized seizures, however, and the results do not indicate whether binding of ligands for noradrenergic receptors can be detected during kindling itself. The results also do not discriminate between kindling-induced changes in autoreceptors for NA and changes in receptors located on postsynaptic target neurons.

In order to determine whether an identified population of NA receptors changes <u>during</u> kindling, we have used in vitro receptor autoradiography to examine the specific binding of $^3$H-idazoxan, a ligand for α-2 receptors, in the noradrenergic LC at three different stages of amygdaloid kindling[24]. Activity of LC neurons is partially regulated by recurrent inhibition from noradrenergic axon collaterals that release NA onto α-2 receptors located on LC cell bodies[25]. Activation of these α-2 receptors produces a reduction in terminal release of NA, similar to the effect of activation of presynaptic α-2 autoreceptors on NA axon terminals. In rats killed 90 min after the triggering of a kindled seizure, we found a mean increase of 26.5 per cent ($p<0.005$) in α-2 binding in the LC after 2 stage-1 seizures but no difference between kindled and control tissues after 1 stage-3 seizure or up to 3 stage-5 generalized seizures, as shown in Figure 3. Scatchard analysis revealed that there was an increase in the number and not affinity of receptors. Given the feedback inhibitory effects of NA on LC activity, this finding suggests that there is a decrease in the ability of the LC to sustain increased firing and continue to inhibit seizure development. The consequence is a disinhibition of the spread of AD and development of further stages of seizure. The finding also suggests that, as kindling progresses, the enhanced suppression of LC activity is reversed, so that normal levels of electrophysiological activity and reactivity to forebrain AD would be expected to be evident by the time generalized seizures have developed.

Transient elevation of α-2 receptors during early stages of kindling should also occur on LC axon terminals projecting to forebrain sites. We therefore labelled α-2 binding sites with $^3$H-idazoxan in 5 different areas of the forebrain at 2 different stages of kindling[26]. After 2 stage 1 seizures, the number but not affinity of α-2 binding sites was elevated by 10.5 per cent ($p=0.01$) in the amygdala but not in other forebrain regions (Figure 4). In contrast, there were no differences in any region after kindling of a single stage 5 seizure. This transient increase in the amygdala is qualitatively and temporally similar to the increase in binding observed in the LC, although the magnitude of the change is much smaller. The difference may be due to the fact that α-2 receptors in the forebrain are located both presynaptically and postsynaptically[26], whereas α-2 sites in the LC are equivalent to presynaptic sites[25]. If the number of presynaptic α-2 receptors increases in the early stages of kindling but postsynaptic α-2 receptors either do not change or decrease, the magnitude of the overall change would be a function of the ratio of presynaptic to postsynaptic receptors. Clarification of this issue awaits availability of a ligand that selectively binds to presynaptic or postsynaptic α-2 receptors.

In summary, kindling of partial seizures produces a transient increase in binding sites for NA in the LC and possibly on noradrenergic axon terminals in target regions. This change in receptors might lead to decreases in both noradrenergic unit activity and release of NA in target regions, both of which would disinhibit the further development of seizures. Kindling of generalized seizures is associated with a decrease in the density of α-2 and β receptors in target regions in the forebrain, although only the latter persists after a rest period. If these receptors are located on neurons postsynaptic to NA terminals, the decrease in α-2 receptors would be an additional disinhibitory mechanism contributing to seizure development, in view of evidence that activation of postsynaptic α-2 receptors retards kindling[4]. In contrast, since activation of β receptors can facilitate epileptiform discharge[27], the long-lasting decrease in β receptors after kindling might represent a compensatory response that reduces seizure susceptibility.

## NORADRENERGIC CORRELATES OF KINDLING: ELECTROPHYSIOLOGY

It has been shown in several other forms of experimental epilepsy that central NA neurons are activated during seizures. If central NA inhibits early stages of kindling it would be predicted that LC activity should also increase during early stages of kindling. Increased activity in response to forebrain AD would provide a negative feedback mechanism that reduces the further spread of AD. The binding studies discussed above suggest a mechanism that would then reduce the

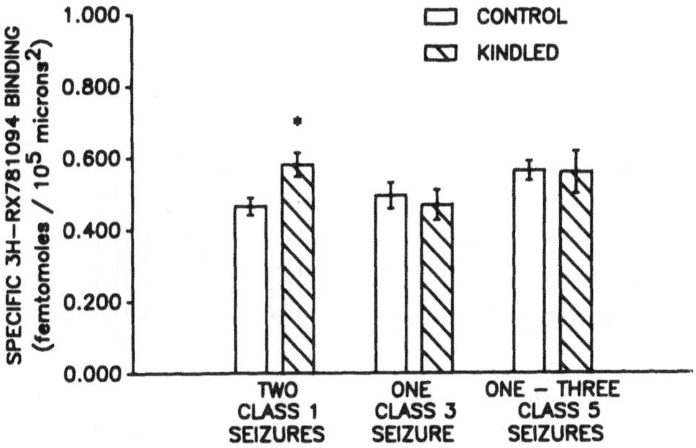


Figure 3. Specific [$^{3}$H]-idazoxan binding in the LC at various stages of kindling. *Significantly greater than in unstimulated controls, $p<0.01$.

effectiveness of this negative feedback. There are only limited electrophysiological data available that bear on these hypotheses, but some of the data are supportive.

Two experiments have examined firing rates of LC neurons during kindling. Jiminez-Rivera and Weiss[28] measured multiple unit activity from the LC in awake rats while they were being kindled. Early amygdaloid ADs were accompanied by bursting activity in the LC (figure 5). In a second study they used a procedure for inducing amygdaloid AD in rats anesthetized with gamma-butyrolactone. Single neurons in LC fired in rapid bursts time-locked to the epileptiform spikes of the first amygdaloid AD. Thus initial limbic ADs can excite neurons in the LC, as would be expected if LC activity is a negative feedback mechanism that dampens the early spread of AD. The data on α-2 binding in the LC suggest that LC neurons would not be activated as readily by forebrain AD as the early stages of kindling progress but that as seizures generalize their reactivity would recover. Thus LC activity would again be driven during generalized ADs, and the increased release of NA that occurs as a consequence might be responsible for the decreased number of postsynaptic α-2 receptors observed after generalization[21,22]. Unfortunately this hypothesis has not yet been tested. In the second experiment, Bonhaus and McNamara[29] recorded the activity of single neurons in the LC of rats maintained under paralysis and artificial ventilation. They compared activity in naive control rats and in rats in which generalized seizures had been kindled by amygdaloid stimulation. Bonhaus and McNamara found no differences between kindled and control rats in LC activity either between or during amygdaloid ADs,and they concluded that the LC is not the site at which attenuation of NA neurotransmission occurs during kindling. Their results aredifficult to interpret, however, because the spontaneous firing rates of LC neurons in their preparation was about 5 times higher than in freely moving or anesthetized rats, suggesting that their rats were severely stressed. Subtle differences between control and kindled rats could conceivably have been masked by the activating effects of stress. Electrophysiological data support the idea that the postsynaptic effects of NA are also reduced after kindling. McIntyre and Wong[30] have shown

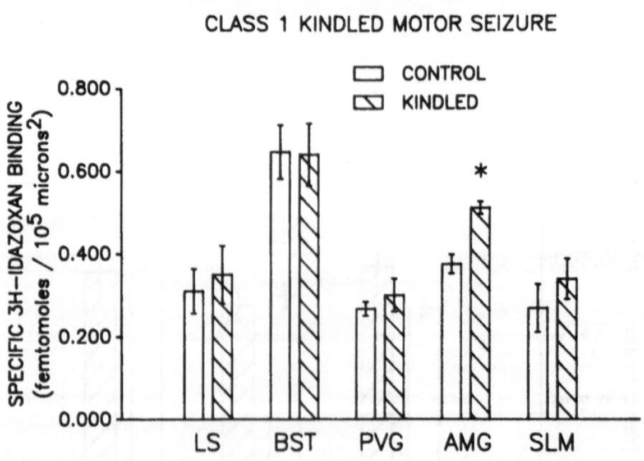


Figure 4. Specific [$^3$H]-idazoxan binding in the forebrain 90 min after 2 stage-1 seizures *Significantly greater than unstimulated controls, p=0.0027. Abbreviations: LS, lateral septum; BST, bed nucleus of stria terminalis; PVG, periventricular grey; AMG, amygdala; SLM, stratum lacunosum moleculare.

that NA and the α-2 agonist clonidine can suppress the epileptiform burst discharge of single neurons in the pyriform cortex of rats and that tissue from amygdaloid kindled rats was less sensitive to the suppressive effects of both NA and clonidine than was tissue from control rats. More recently, Stanton et al.[31] measured a variety of effects of NA on granule cells of the dentate gyrus. They found that NA acted at β receptors to induce depolarization and long-term potentiation (LTP) of synaptic potentials and to increase input resistance, firing, and influx of calcium. NA acted at α-1 receptors of granule cells to attenuate calcium-dependent regenerative potentials. All of these actions of NA were reduced in tissue taken from rats in which generalized seizures had been kindled, at various intervals beforehand, by repeated stimulation of the hippocampal commissure or the amygdala.

We suggest that the bulk of the evidence from the correlative studies of NA and kindling favors the following model: Initial limbic ADs activate firing of noradrenergic neurons in the LC, and the NA so released from the axon terminals of LC neurons retards the rate of spread of AD in the forebrain. Induction of repeated AD and coincident repeated activation of LC neurons (during the development of nonconvulsive or partial seizures) leads to an increase in the number of α-2 autoreceptors on LC neurons and their axon terminals, possibly in the subpopulation of LC neurons projecting specifically to the stimulated amygdala. The mechanisms of this "up-regulation" of α-2 receptors are unknown. The increase in presynaptic α-2 receptors would presumably lead to a decrease in the firing of LC neurons and in the release of NA, with a consequent disinhibiton of the spread of AD in target regions. Perhaps as a consequence of the recovery in LC activity that seemingly occurs by the time of seizure generalization, the density of α-2 and  receptors on postsynaptic target neurons decreases as well; however, these changes may also be induced by agonist-independent discharge[32]. The decreased density of postsynaptic α-2 receptors would contribute to disinhibition of the spread of AD in noradrenergic target regions, whereas the decrease in β receptors would result in a decrease in responsiveness to proepileptogenic effects of NA.

The increased density of LC α-2 receptors and of presumed presynaptic α-2 receptors in target regions is transient and restricted to the earlier stages of kindling. The density of presynaptic α-2

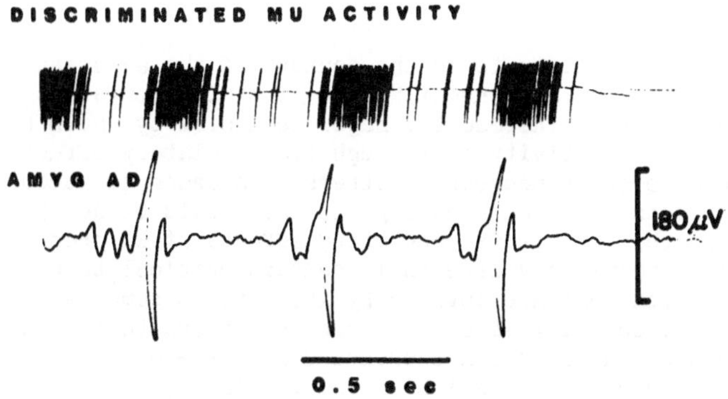


Figure 5. Multiple-unit activity (MUA) recorded from the LC of an awake rat during a stage-3 seizure. Top channel: output of discriminator triggered on MUA with amplitude twice the baseline noise. Lower channel: EEG recorded from the amygdala.

receptors reverts to control levels as seizures generalize. The decreases in presumed postsynaptic $\alpha$-2 receptors are also transient; although decreases are still evident after the appearance of stage-5 seizures, receptor density recovers to control levels within 3 weeks after kindling. The changes in $\beta$-receptor density are more long-lasting, and are still evident 3 weeks after completion of limbic kindling. The decreases in densities of $\alpha$-2 and $\beta$ receptors are both correlated with decreases in the functional effects of occupation of these receptors by appropriate ligands.

## PHYSIOLOGICAL MECHANISMS OF NA'S ANTIKINDLING EFFECTS

In addition to its ability to suppress spontaneous activity via hyperpolarizing effects on postsynaptic membranes, NA also exerts less time-locked and more indirect effects on postsynaptic neurons that have been referred to variously as "modulatory," "gating," "biasing," "enabling," enhancing signal-to-noise ratio," or "metabotropic." For example, in some situations iontophoretic application of NA, at currents below those that directly affect neuronal activity, can enhance the target neuron's responsiveness to other neurotransmitters that exert conventional excitatory or inhibitory effects, whereas in others NA potentiates evoked activation or suppression without affecting spontaneous background activity[33]. How can the inhibitory effects of NA on spread of ictal discharge, and the disinhibitory effects of depletion of NA, be interpreted in the context of these broad modulatory effects?

We suggest several possibilities. The first is based on evidence that NA generally produces 3 effects on its target neurons: a direct hyperpolarizing effect, an indirect modulatory effect that potentiates the target cells' responsiveness to inhibitory neurotransmitters such as GABA, and an indirect modulatory effect that potentiates the target cells' responsiveness to excitatory neurotransmitters such as glutamate[33]. Some evidence suggests that the modulatory effects have a lower threshold than the direct inhibitory effects of NA. If seizure discharge drives noradrenergic neurons sufficiently strongly so that all 3 effects are of equal strength, the net overall effect of the NA released during early stages of kindling would be inhibitory, resulting in a dampening of activity in target neurons and the spread of AD through neural circuits. It follows that, by this hypothesis, the net effect of depleting NA or decreasing the presynaptic release of NA or postsynaptic responsiveness to NA would be disinhibition of the spread of AD.

A related possibility is in the context of the argument[33] that the direct hyperpolarizing actions of NA are of minimal physiological significance, and that instead the major mechanism by which NA influences neuronal activity is through its modulatory effects on responsiveness to other neurotransmitters. In order to account for the inhibitory effects of NA on kindling, it then would be necessary to postulate that the inhibitory-modulatory effects of NA predominate over NA's excitatory-modulatory effects in certain critical target regions of the nervous system that are invaded by AD. If, as some evidence suggests, NA's modulatory effects on responsiveness to inhibitory and excitatory neurotransmitters are produced at different receptors[33], the relative predominance of NA's inhibitory-modulatory effects would then be accounted for by a greater dominance of the appropriate type of receptor. A final possibility is based on the observation[34] that NA reduces afterhyperpolarizations (AHPs) in hippocampal neurons. If AHPs contribute to the population synchrony underlying epileptiform discharge, a decrease in the availability or effectiveness of NA would facilitate development of seizures. There is at present no conclusive basis for determining which of these possibilities, or some other not considered here, is correct.

## FACILITATORY EFFECTS OF NA ON LONG-TERM POTENTIATION (LTP): A PARADOX?

Finally, we note that kindling and LTP are triggered by similar patterns of electrical stimulation applied to forebrain structures[35], and that it has been suggested that an important mechanism of kindling might be LTP of excitatory synaptic drive[36]. An apparent paradox therefore emerges when one considers the facts that NA suppresses kindling but facilitates or may even trigger LTP in the hippocampal formation[31,37,38]. How can these discrepancies be reconciled?

One possibility is that there might not be a paradox at all, because the antagonistic effects of NA on kindling have been demonstrated typically with amygdaloid stimulation whereas the facilitative effects of NA on LTP have been demonstrated with stimulation of the perforant path[37] or mossy fibers[38] but not of the Schaffer collaterals[37]. Thus it might be argued that NA's effects on LTP and kindling are regionally specific and that a critical test of the involvement of NA in the 2 phenomena should involve a comparison of kindling and LTP in the same region or pathway. We have attempted to provide this test by examining the effects of 6-OHDA-induced depletion of NA on kindling with stimulation of the perforant path (Popham and Corcoran, unpublished observations), so that this can be compared to the antagonistic effects of depletion of NA on LTP with stimulation of the perforant path[37]. We found that depletion of NA facilitates kindling of seizures with perforant path stimulation (mean [±SEM] ADs to generalization: control rats, 19.5 [±4.0]; NA-depleted rats, 9.6 [±1.5]). Thus it appears that even when the site of stimulation is held constant, NA exerts opposing effects on kindling and LTP.

Another possibility is that the involvement of NA in various forms of plasticity is not only regionally specific, but also receptor-specific. Thus there is evidence that activation of $\beta$receptors by NA exerts a facilitation of LTP[37] and some forms of epileptiform activity[27], whereas activation of $\alpha$-2 receptors by NA exerts a suppressive effect on epileptiform activity[27,30]. These data suggest that it would be useful to characterize further the effects of activation of $\beta$-noradrenergic receptors on kindling.

Whatever the correct explanation for the findings, they lead us to the conclusion that NA does indeed play different roles in LTP and kindling. The implications of this conclusion for the hypothesis that LTP is part of the mechanism of kindling remain to be explored further.

Acknowledgements. We thank Pergamon Press for permission to reproduce Figure 1 (from ref. 7) and Elsevier for permission to reproduce Figures 3 and 4 (from ref. 24 and 26, respectively).

## REFERENCES

1. M. E. Corcoran, Catecholamines and kindling, in: "Kindling 2," J. A. Wada, ed., Raven, New York (1981).
2. D. C. McIntyre, Catecholamine involvement in amygdala kindling of the rat, in: "Kindling 2," J. A. Wada, ed., Raven, New York (1981).

3. I. M. Altman and M. E. Corcoran, Facilitation of neocortical kindling by depletion of noradrenaline, Brain Res. 270:174 (1983).
4. R. L. Gellman, J. A. Kallianos, and J. O. McNamara, Alpha-2 adrenergic receptors mediate an endogenous noradrenergic suppression of kindling development, J. Pharmacol. exp. Therap 241:891 (1987).
5. V. S. Westerberg, J. Lewis, and M. E. Corcoran, Depletion of noradrenaline fails to affect kindled seizures, Exp. Neurol, 84:237 (1984).
6. H. B. Michelson and G. Buterbaugh, Amygdala kindling in juvenile rats following neonatal administration of 6-hydroxydopamine, Exp. Neurol. 90:588 (1985).
7. M. E. Corcoran, Characteristics of accelerated kindling after depletion of noradrenaline in adult rats, Neuropharmacology, 27:1081 (1988).
8. S. Uemura, H. Kimura, A. Kashia, H. Kumashiro, and J. A. Wada, Bifunctional roles of catecholamines in the development of amygdala kindling by continuous intra-amygdala infusion of 6-hydroxydopamine, Brain Res., 448:162 (1988).
9. D. C. McIntyre, J. Rajala, and N. Edson, Suppression of amygdala kindling with short interstimulus intervals: effect of norepinephrine depletion, Exp. Neurol. 95:391 (1987).
10. C. D. Applegate, R. J. Konkol, and J. L. Burchfiel, Kindling antagonism: a role for hindbrain norepinephrine in the development of site suppression following concurrent, alternate stimulation, Brain Res. 407:212 (1987).
11. W. M. Burnham, R. J. Racine, and M. Okazaki, Kindling mechanisms. II. Biochemical studies, in: "Kindling 3," J. A. Wada, ed., Raven, New York, 1986.
12. D. C. McIntyre and L. Giugno, Effect of clonidine on amygdala kindling in normal and 6-hydroxydopamine-pretreated rats, Exp. Neurol. 99:96 (1988).
13. W. Loscher and S. J. Czuczwar, Comparison of drugs with different selectivity for central α1- and α-2 adrenoceptors in animal models of epilepsy, Epilepsy Res. 1:1165 (1987).
14. D. I. Barry, I. Kikvadze, P. Brundin, T. Bolwig, A. Bjorklund, and O. Lindvall, Grafted noradrenergic neurons suppress seizure development in kindling-induced epilepsy, Proc. Natl. Acad. Sci. USA 84:8712 (1987).
15. C. Jimenez-Rivera, A. Voltura, and G. K. Weiss, Effect of locus ceruleus stimulation on the development of kindled seizures, Exp. Neurol. 95:13 (1987).
16. D. Blackwood, The role of noradrenaline and dopamine in amygdaloid kindling, in: "Neurotransmitters, Seizures, and Epilepsy," P. Morselli, K. G. Lloyd, W. Loscher, B. Meldrum, and E. H. Reynolds, eds., Raven, New York (1981).
17. J. Lewis, V. Westerberg, and M. E. Corcoran, Monoaminergic correlates of kindling, Brain Res. 403:205 (1987).
18. M. M. Okazaki, J. J. Warsh, and W. M. Burnham, Unchanged norepinephrine turnover and concentrations in amygdala-kindled rat brain regions 2 months postseizure, Exp. Neurol. 94:81 (1986).
19. M. M. Okazaki, J. J. Warsh, and W. M. Burnham, Unchanged regional norepinephrine glycol metabolite levels in rat brain two months after amygdala kindling, Epilepsy Res. 2:72 (1988).
20. D. C. McIntyre and D. C. S. Roberts, Long-term reduction in beta-adrenergic receptor binding after amygdala kindling in rats, Exp. Neurol. 82:17 (1983).
21. S. C. Stanford and J. G. R. Jefferys, Down-regulation of α-2 and β-

adrenoceptor binding sites in rat cortex caused by amygdalar kindling, Exp. Neurol. 90:108 (1985).

22. L. S. Chen and J. O. McNamara, Autoradiographic localization of reduced alpha2 adrenergic receptor binding in kindled rats, Soc. Neurosci. Abstr. 14:1148 (1988).
23. H. B. Michelson and G. G. Buterbaugh, Alterations in β-adrenergic receptor binding in partially and fully amygdala-kindled juvenile and adult rats, Exp. Neurol. 95:56 (1987).
24. C. Jimenez-Rivera. M. J. Chen, A. Vigil, D. D. Savage, and G. K. Weiss, Transient elevation of locus coeruleus α-2-adrenergic receptor binding during the early stages of amygdala kindling, Brain Res. 485:363 (1989).
25. S. Foote, F. Bloom, and G. Aston-Jones, Nucleus locus coeruleus: new evidence of anatomical and physiological specificity, Physiol. Rev. 63:884 (1983).
26. M. J. Chen, A. Vigil, D. D. Savage, and G. K. Weiss, Transient elevation of amygdala $alpha_2$ adrenergic receptor binding sites during the early stages of amygdala kindling, Epilepsy Res. in press.
27. A. L. Mueller and T. V. Dunwiddie, Anticonvulsant and proconvulsant actions of alpha- and beta-noradrenergic agonists on epileptiform activity in rat hippocampus in vitro, Epilepsia 24:57 (1983).
28. C. Jiminez-Rivera and G. K. Weiss, The effect of amygdala kindled seizures on locus coeruleus activity, Brain Res. Bull. in press.
29. D. W. Bonhaus and J. O. McNamara, Activity of locus coeruleus neurons in amygdala kindled rats: role in the suppression of afterdischarge, Brain Res. 407:102 (1987).
30. D. C. McIntyre and R. K. S. Wong, Cellular and synaptic properties of amygdala-kindled pyriform cortex in vitro, J. Neurophysiol 55:1295 (1986).
31. P. K. Stanton, I. Mody, and U. Heinemann, Down-regulation of norepinephrine sensitivity after induction of long-term neuronal plasticity (kindling) in the rat dentate gyrus, Brain Res. 476:367 (1988).
32. R. Dasheiff and J. O. McNamara, Evidence for an agonist independent down regulation of hippocampal muscarinic receptors in kindling, Brain Res. 195:345 (1980).
33. B. D. Waterhouse, F. M. Sessler, J. T. Cheng, D. J. Woodward, S. A. Azizi, and H. C. Moises, New evidence for a gating action of norepinephrine in central neuronal circuits of mammalian brain, Brain Res. Bull 21:425 (1988).
34. H. L. Haas and A. Konnerth, Histamine and noradrenaline decrease calcium-activated potassium conductance in hippocampal pyramidal cells, Nature 302:432 (1983).
35. R. J. Racine, N. W. Milgram, and S. Hafner, Long-term potentiation phenomena in the rat limbic forebrain, Brain Res 260:217 (1983).
36. R. J. Racine, J. G. Gartner, and W. M. Burnham, Epileptiform activity and neural plasticity in limbic structures, Brain Res. 47:262 (1972).
37. P. K. Stanton and J. M. Sarvey, Depletion of norepinephrine, but not serotonin, reduces long-term potentiation in the dentate gyrus of rat hippocampal slices, J. Neurosci 5:2169 (1985).
38. W. F. Hopkins and D. Johnston, Noradrenergic enhancement of long-term potentiation at mossy fiber synapses in the hippocampus, J. Neurophysiol 59:667 (1988).

Discussion of Dr. Corcoran's and Dr. Weiss's Presentation

DR. MCINTYRE: Why did you examine the alpha-2 receptors 90 minutes after stimulation rather than 24 hours? When you kindle at 90 minute rates your seizure genesis is really not nearly so good as it is at once a day. Therefore, that up-regulation might actually represent the slowness of kindling at that interval. Whereas if you had done it at once a day at a 24 hour interval for your assessment of the receptors you might have seen a down-regulation of those alpha-2 receptors. What do you think?

DR. WEISS: You're thinking of stopping at stage 1 and then over time ... Yes, we're doing that. We're looking at the time sequence now of this change, and we'll take it out at 24 hours until we see a change. I don't know what it will be. But certainly the animals continued into stages 2 and then 3. But they're experiencing more seizures, that's true, and they don't show it, in fact it starts to come down. Now, the other thing we need to do is to do more stage 5. Six, or 7, or 8 stage 5's and then look and see if they have been driven down, I don't know what the mechanism is. Why does it go up and then come back, I have no idea.

DR. MCNAMARA: I congratulate you on a technically difficult study with those single units in the LC during the after-discharge. It would have been really interesting if perhaps by the time the animal got to the stage in which we first see a loss, at the transition from limbic to generalized seizures, inactivation of LC firing at that point.

DR. WEISS: Right. That's going to take some doing. We tried to look at just the after-discharges, whether it responded more or less to the after-discharges, but that's not quantitative enough. What we need to do is to go in and look at the functioning-like squirting some clonidine and then pinching, doing a quantitative pinch to see whether they change to the response.

DR. MCNAMARA: Right. But it seems to me that assuming the effects of norepinephrine on seizure propagation and kindling development are exerted during the seizure, as opposed to between the seizures - it would seem to me that the electrophysiological experiments argue against a defect intrinsic to the locus coeruleus causing kindling development. OK? Because you've got equivalent LC firing throughout.

DR. WEISS: That's right, but I think by measuring in the LC like we did, and continuing to have seizures, I think it can be subtle enough. Remember we're not looking at big changes and they might be very specific to the area from which the seizure is coming. In other words is it in the amygdala? We should try kindling in the hippocampus and see whether the change in alpha-2 presynaptically in the hippocampus is changed as opposed to the amygdala, for instance. I think it can be a subtle enough change to allow that breakout of the early stages. What we visualize is that that driving of the LC is a feedback to keep the seizure from spreading. Now you're right, I think, if I go in at stage 4 and measure LC activity, that after-discharge is going to continue to drive it.

DR. MCNAMARA: So then it seems to me that there are three possibilities; either a defect intrinsic to the noradrenergic system does not occur in the kindled brain - that's causing kindling development and that's certainly a plausible explanation. A second possibility is that there is inactivation of release of norepinephrine from terminals, and as I'm sure you realize, there is work many years ago, demonstrating local control of norepinephrine release in the presence of seizures which all may be happening at the terminal, not at the LC itself. And a third possibility is that the changes intrinsic to the post-synaptic neurons - and along those lines, we won't be talking about it here, but we've submitted what Len Chin did in my laboratory - is that there is a reduction of pyramidal clonidine histochemically in the pyriform cortex of all places as well as in a few amygdaloid nuclei. But if I had to put my money on it, I bet if there is a change intrinsic to that system, it's intrinsic to noradrenergic receptive neurons that contain alpha 2-receptors.

DR. WEISS: You think that the driving of the LC is going to down-regulate or decrease the binding of post-synaptic?

DR. MCNAMARA: It could be. I don't know what the mechanism of the reduction is, but I suspect that there is a change intrinsic to noradrenergic receptive neurons that contain alpha-2 receptors. Whether it's in the receptor itself or in the transduction ...

DR. WEISS: I think, this is a typical mechanism that's going all the time in the central noradrenergic system. When it is overdriven I don't care for what, for stress, for anything, its initial response is to up-regulate, and I put quotes on that term, it's going to increase the number of alpha-2 receptors as an autoregulatory mechanism for reducing that activity. It just so happens that this is what happens initially then in a seizure driven event. It doesn't matter if it's driven from the sensory receptors, it doesn't matter if it was driven from anywhere. I think it's an intrinsic regulatory mechanism of the LC and after-discharges happen to be something that's driving it. So I think that both things are probably going to occur. I also think that the system is overwhelmed by continued activity of after-discharges and that eventually it does drive the LC, but there's been that period of time when it was autoregulating which allowed now the spread of the seizure. So that later on as it continues to drive it will then drive down the post-synaptic alpha-2's - that's where I'll put my money, that you'll only see it late.

DR. MCNAMARA: Yes, right, maybe after a few class 1 or 2 seizures.

DR. BURNHAM: I would like to second Dr. McNamara's suggestion that you look for these binding changes in the locus coeruleus at a prolonged period of time. In the GABA system there is quick up-regulation of both GABA binding and benzo binding that lasts for only about an hour after seizures, and then disappears. My question would be, in the locus coeruleus you're thinking that your binding changes alpha-2 are post-synaptic, is there also pre-synaptic alpha-2 binding in the locus coeruleus?

DR. WEISS: You're asking if there are alpha-2 receptors on the terminals of the feedback collaterals within the LC. Yes, these probably do exist but all of the data from Aghajanian, Cederbaum, in fact everybody who's looked at the LC in terms of any kind of input that alters alpha-2 receptors conclude that the effects are due to activation of post-synaptic receptors on the neuronal cell bodies in the LC. This is because if the pre-synaptic terminals were the culprits, the opposite results would have occurred. The reviewer of our recently published paper asked the same question. This idea is discussed in that paper. The conclusion is that the majority of the effects of alpha-2 activation in the LC are post-synaptic to the recurrent collaterals. We call them post-synaptic relative to the collaterals but they're really behaving, as Nakamura showed, the same as pre-synaptic receptors that exist on the terminals of the LC axon. They both inhibit the LC neurons. These are the receptors being studied.

# MECHANISMS OF KINDLING IN DEVELOPING ANIMALS

Ellen F. Sperber
Kurt Haas and
Solomon L. Moshé

Departments of Neurology
Neuroscience and Pediatrics
Laboratory of Developmental Epilepsy
Albert Einstein College of Medicine
Bronx, N.Y.

## INTRODUCTION

Kindling was first demonstrated by Goddard and his colleagues in the late 1960s (1,2) and has since become an important animal model of partial epilepsy with secondary generalization. Kindling is most commonly achieved with the administration of repeated local subconvulsive electrical stimulations which eventually lead to the development of generalized seizures. Chemical kindling can also be demonstrated with a variety of convulsant drugs (pentylenetetrazol, lidocaine, cocaine, penicillin and carbachol) administered at subconvulsive doses (3-10). Irrespective of the type of stimulus used for the kindling process, kindling progresses in a predictable manner. The initial stimulus results in a brief focal electrical seizure or afterdischarge (AD) in the absence of any behavioral manifestations. Gradually, as the behavioral manifestations become more apparent, the ADs intensify and increase in duration. Once established, the kindling effect persists for several months suggesting that it produces permanent changes in the brain (2).

Kindling has been demonstrated in rat pups as young as 15-16 days of age stimulated either in the amygdala or hippocampus (11-17). However, there are age-related differences both in local excitability as well as in the mechanisms involved in the propagation and suppression of seizures. An index of local excitability is the afterdischarge threshold, defined as the lowest intensity of current capable of inducing an AD. The amygdala AD threshold is raised in 15-day old rats compared to older rats (12). Depending on the stimulated site (for example in the hippocampus) the duration of the initial AD at the AD threshold may be significantly longer in pups than in adults (17).

Rat pups have a greater susceptibility to kindled seizures than older rats. In pups, kindling occurs when stimulations are

*Kindling 4*
Edited by J. A. Wada
Plenum Press, New York, 1990

delivered using a short interstimulus interval. Thus, pups can be stimulated as often as every 15 min (16), whereas in adults this type of stimulation inhibits or retards the development of kindling (2,16,18,19). This suggests that rat pups have a shorter refractory period than adult rats. Pups are also more prone to develop recurrent kindled seizures and status epilepticus than older rats (20).

As kindling progesses, pups display several different behaviors from adults (11,12,17) (Table 1). In general, in pups, there is a tendency for rapid seizure generalization. Pups spend proportionately less time in the early stages which are associated with focal seizures and quickly progress to bilateral asynchronous convulsions where they remain for a relatively longer period of time than adults. More specifically, a stage 1 seizure in pups is characterized by facial movements such as chewing, licking, sniffing and squinting. During stage 2, rhythmic head nodding and orienting the body toward the stimulated side are apparent. At stage 3, pups often demonstrate alternating forelimb clonus, intermittent hindlimb clonus and "wet dog shakes". The "wet dog shakes" are persistent and prolonged in pups, particularly with hippocampal kindling. Stage 4 consists of bilateral forelimb clonus and occasional rearing or of rotary movements of the tonically extended forelimbs. Stage 5 is characterized by a loss of balance and is not observed in every pup. Pinel and Rovner (21) have described additional kindling stages that they observed when adult rats were stimulated beyond stage 5 seizures. These stages, 6, 7 and 8, are shown in Table 1 as modified by Moshé and Ludvig (22).

Table 1. Behavioral Manifestations of Amygdala Kindled Seizures in Adults and Rat Pups.

| Stage | Adult Signs | Pup Signs |
|---|---|---|
| 1 | Chewing | Facial movements |
| 2 | Head nodding | Rhythmic head movements or turning of body toward stimulated side |
| 3 | Contralateral forelimb clonus | Unilateral or alternating forelimb clonus, hindlimb clonus, "wet dog" shakes" |
| 4 | Symmetrical forelimb clonus with rearing | Bilateral clonus or rotatory movements of tonically extended forelimbs; occasional rearing |
| 5 | Loss of balance | Loss of balance |
| 6 | Wild running, jumping, rolling and vocalizing | |
| 7 | Tonic posturing | |
| 8 | Spontaneous seizures | |

A variation of the typical kindling protocol called kindling antagonism has been described (23,24). It involves kindling two seperate structures simultaneously by applying subthreshold electrical current to each site on alternating trials. Gradually, one of the two sites (the site which results in the initial behavioral seizure manifestation) will become dominate for the development of kindling. The other site is usually suppressed so that kindling fails to develop despite the presence of prominent ADs. Kindling antagonism has been demonstrated between several limbic structures located ipsilaterally and contralaterally to each other. In addition, mutual antagonism has also been reported in which neither site kindles. This is particularly common when the two amygdala are stimulated.

Several studies have demonstrated that rat pups have an increased susceptibility to seizures induced by kainic acid, flurothyl, pentylenetetrazol and kindling (25-29). It is our hypothesis that pups have an inability to suppress seizures because they have not developed efficient systems which suppress seizures later in life. One such system revolves around the substantia nigra pars reticulata and its GABA-sensitive efferent outputs. In adult rats, infusions of GABA-A agonists into the substantia nigra pars reticulata can suppress several types of experimental seizures (30-36). While two-week old pups have approximately 60-80% of total brain GABA and normal GABA binding (37,38), they also have a site-specific paucity of high affinity GABA-A receptors in the substantia nigra (39). The relative lack of high affinity GABA-A receptors in the substantia nigra may account, in part, for the failure of the GABA-A agonist, muscimol, to suppress flurothyl seizures, when infused directly into the substantia nigra in pups (40). On the other hand, nigral or systemic infusions of the GABA-B agonist baclofen can suppress flurothyl seizures in pups (41), but not in adults (36). These data suggest that baclofen may act as an age-specific anticonvulsant, at least, in the flurothyl model.

Based on the age-related differences in seizure suppressing mechanisms we made the following predictions. Severe seizures (stage 6 and 7) should be elicited from kindled pups with fewer stimulations than is required for the occurrence of seizures of equal severity in adult rats. Secondly, the phenomenon of kindling antagonism may not occur in rat pups. In this paper, we report data supporting both predictions. In addition, we provide evidence that early in life, the GABA-B agonist baclofen may be an effective anticonvulsant in the kindling model too.

## METHODS

Rat pups were housed with their dam. At fourteen days of age (day of birth counts as day 0) the pups were implanted, under ketamine anesthesia, with stimulating/recording electrodes into the amygdala, hippocampus or both sites. When two electrodes were implanted, they were placed contralaterally to each other. Following a 2 day recuperative period, kindling was initiated. The kindling stimulus consisted of a 400 μA, 60 Hz sinusoidal current delivered for 1 sec. Following each stimulation, the duration of the ADs was recorded. Each pup was stimulated every 15-20 min. Following each experiment, the animals were sacrificed and standard histological techniques were used to verify electrode placement.

Experiment 1. Severe kindled seizures

For this experiment, rat pups were kindled from the amygdala or hippocampus, or from both sites on alternating trials. All pups were stimulated for a total of 30 stimulations, 20 the first day and 10 the second day. The severity of the seizures was classified as shown in Table 1 with stages 6 and 7 derived from Pinel and Rovner (21).

Experiment 2. Kindling antagonism

Pups were stimulated from the amygdala or the hippocampus on alternating trials. Two additional control groups were kindled from either the amygdala or the hippocampus. All animals were stimulated until they demonstrated stage 4 seizures (Table 1).

Experiment 3. Effect of baclofen on the development and severity of kindled seizures

To determine the effect of baclofen on kindling, amygdala kindled pups were administered (-) baclofen, 2 mg/kg, ip, or an equivalent volume of saline, thirty min prior to the first kindling trial. The pups were kindled until they reached a criterion of 3 consecutive stage 4 seizures.

In order to study the effect of baclofen on the severity of established kindled seizures, another group of pups were first kindled and then treated with either (-) baclofen (2 mg/kg, ip) or saline. These animals were then exposed to a single stimulation and the observed seizure stage and AD duration were recorded.

RESULTS

Experiment 1. There were no observed differences in seizure severity between pups that were kindled from alternating sites or from the same site. All 3 groups demonstrated their first Stage 6 seizure by stimulation 23 or 24. Several pups had stage 7 seizures by the end of the 30 stimulations. There was no significant difference in the number of stimulations required for the occurrence of severe seizures between the group receiving alternating stimulations and the groups receiving kindling stimuli from the amygdala or hippocampus only.

Table 2. Number of Stimulations Required for Rat Pups to Reach Each Stage of Kindling Following Baclofen (2mg/kg, ip) or Saline Administration.

| Seizure Stage | Baclofen | Saline |
|---|---|---|
| 1 | 5.2±.44 | 2.25±.21 * |
| 2 | 9.7±1.44 | 3.75±.42 * |
| 3 | 14.6±1.14 | 5.75±1.74 * |
| 4 | 18.2±1.23 | 9.75±1.46 * |
| Criterion | 19.0±8.0 | 11.75±1.1 * |

* p<.02

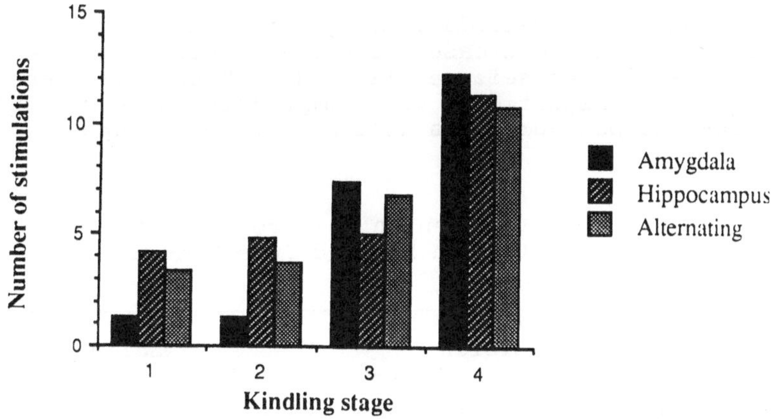


Fig 1. Development of kindled seizures following stimulation of the amygdala, hippocampus or alternating stimulations in pups. There was no significant difference in the rate of kindling regardless of whether the pups were stimulated from the amygdala, hippocampus or the stimulations were alternated between these two sites.

Experiment 2. Kindling antagonism was not observed in the pups (Fig 1). Alternating stimulations between the amygdala and hippocampus did not suppress kindled seizures from either site. These pups did not require signifcantly more stimulations to reach stage 4 seizures than control pups kindled from only one site.

Experiment 3. Systemic injections of baclofen significantly retarded the development of amygdala kindling in pups (Table 2). Baclofen-treated pups required significantly more stimulations to reach each seizure stage and the criterion of 3 consecutive stage 4 seizures compared to the control pups. In general, the pups infused with the baclofen, required approximately 70% more stimulations to reach criterion than the saline administered animals. One baclofen-treated pup failed to reach criterion after 24 trials. Although baclofen delayed the development of kindling, it prolonged the duration of the ADs. Table 3 compares the AD durations between baclofen and saline treated pups at each stage.

Table 3. Duration of ADs (sec) at Each Stage of Kindling in Pups Following Baclofen (2mg/kg, ip) or Saline Administration.

| Seizure stage | Baclofen | Saline | |
|---|---|---|---|
| 1 | 21.3±1.6 | 15.7±1.9 | |
| 2 | 43.0±9.5 | 19.0±9.0 | |
| 3 | 83.6±8.1 | 22.5±1.5 | * |
| 4 | 86.4±6.7 | 48.5±8.8 | * |
| Criterion | 89.6±6.29 | 71.2±4.6 | |

* $p<.02$

Once kindling was established, systemic injections of baclofen significantly decreased both the severity and the duration of the kindled seizure (Table 4). These differences were statistically significant when comparisons were made within a group (pre- and post-drug) as well as between groups.

Table 4. Effect of Baclofen on Seizure Severity and AD Duration (sec) in Kindled Pups Pre- and Post-Drug Infusion.

| | Baclofen | | Saline | |
|---|---|---|---|---|
| | Pre- | Post- | Pre- | Post- |
| Seizure stage | 3.8±.1 | 2.7±.3*,** | 3.7±.2 | 3.7±.1 |
| AD duration | 77.3±3.6 | 61.4±5*,** | 81.6±7 | 78.6±4 |

* p<.003 compared to pre-drug values; ** p<.002 compared to saline controls.

## DISCUSSION

The results of the present study indicate that rat pups can display severe seizures; that kindling antagonism does not occur in rat pups and that baclofen administered systemically suppresses both the development of kindling and severity of established kindled seizures. These results further demonstrate that rat pups have an increased susceptibility to seizures and this may be related to a delay in the development of efficient inhibitory-to-seizures mechanisms. There is, however, a system in the pup brain that is capable of suppressing seizures that is quite different from that present in the adult.

In the past, pups have been reported to have an increased susceptibility to seizures in a variety of seizure models including kindling, electroshock, kainic acid, pentylenetetrazol, picrotoxin, strychnine, bicuculline and pilocarpine induced seizures (25-29). The present study once again demonstrates pups have a faster kindling rate than adult rats. In addition, this is the first study reporting that rat pups can experience severe seizures (stage 6 and 7) after relatively few stimulations.

In the adult, severe seizures are characterized by wild running, jumping and vocalizations and later, by tonic posturing (21). At a glance, the severe seizures of the pups are similar to those of the adults. However, there are some characteristic differences between the two. In the adults, many more stimulations are required over several weeks of stimulating until stage 6 and 7 seizures are seen. In contrast, in the pups, severe seizures can be observed following fewer than 25 stimulations delivered within two days. As the seizures progress, there is also a different pattern of transition between the seizure stages as a function of age. Pups spend significantly less time in the early stages of kindling and rapidly progress to a stage 3. It was further observed that the kindled seizures in pups progressed from a stage 3 or 4 to a stage 6 seizure. On the other hand, in the adult rats, there is generally a more orderly development of kindling from a stage 3 to a stage 4 to a stage 5 to a stage 6 (21). In addition, although the severe seizures of both adult and rat pups are characterized by tonus, the tonus is more severe in the younger animals. In fact, some pups demonstrated such severe tonus in their hindlimbs and forelimbs simultaneously that they remained on their side. This was often followed by extensive hindlimb and forelimb clonus while remaining in the prone position. These seizures are similar to those observed in pups following flurothyl induced tonic seizures (29). Such severe tonic-clonic seizures are uncommon with kindling and difficult to demonstrate with flurothyl in adult rats.

While kindling antagonism has been observed in adult rats (23,24), our results indicate that kindling antagonism does not occur in the immature rat. One site did not develop kindled seizures preferentially over the other and thus, kindled seizures could be elicited from either the amygdala or the hippocampus at the same rate that was observed when each site was stimulated separately. There were no differences among the groups concerning seizure severity and duration. There was also no apparent difference in the morphology of the ADs.

What are the mechanisms underlying the differences in seizure susceptibility and lack of antagonism? Discrepancies in AD thresholds can not adequately explain these differences, since the AD threshold is often higher in pups than in adults (12). It is more likely that the delayed maturation of several neurotransmitter systems may have important consequences concerning the age-related expression of seizures. One such system is the norepinephrine system.

In adult rats, depletion of forebrain norepinephrine facilitates the development of kindling while an increase in norepinephrine activity inhibits kindling (42). Burchfiel and Applegate (43,44), in a series of elegant studies, have shown that depletion of total brain norepinephrine prevents the occurrence of kindling antagonism. A pattern of forebrain norepinephrine depletion associated with profound hypertrophy of hindbrain norepinephrine innervation maintains kindling antagonism implicating the crucial role of brain stem structures (43,44).

During development, norepinephrine levels do not reach adult values till approximately 3-8 weeks postnatally (12,45,46). Our pups are substantially younger (16 days of age). Thus, it is possible that their inability to demonstrate kindling antagonism is a result of a lack of maturity of the norepinephrine system at this age. In this respect, pups behave similarly to adults with total brain norepinephrine depletion. Norepinephrine depletion also markedly enhances the establishment of kindling which may occur after few stimulations are delivered (42). The relative low norepinephrine levels may be responsible, in part, for the rapid spread and generalization of seizures that occurs in rat pups.

Another important system may be the substantia nigra-based inhibitory-to-seizures circuit which is extremely sensitive to GABA. In adults, this system appears to be quite effective in controlling the propagation and termination of seizures (30-35). Pharmacological studies suggest that the nigral effects on seizures are mediated by the GABA-A receptors located on the

GABAergic neurons of the substantia nigra pars reticulata. Activation of the local GABA-B receptors does not alter the course of flurothyl seizures (36).

In rat pups, the situation is different. Activation of the nigral GABA-A receptors, by local application of the GABA-A agonist muscimol, not only fails to control flurothyl seizures but, instead, results in seizure facilitation (40). On the other hand, activation of the GABA-B receptors delays the development of flurothyl seizures (41).

To explain the age-related differences in the nigral-mediated seizure control, we have proposed the following. During a seizure, there is increased GABA release in the substantia nigra pars reticulata. GABA preferentially binds to GABA-A receptors. In adults, the GABA-A receptor activation leads to the suppression of seizures, while in pups, such activation has the opposite effect. Presumably, GABA will also bind to GABA-B receptors. In adults, this is inconsequential. In pups, the nigral GABA-B receptor-mediated seizure suppression is masked by the proconvulsant GABA-A system. However, nigral administration of the GABA-B receptor agonist baclofen can alter this course. We have further demonstrated that even systemic infusions of baclofen can suppress flurothyl seizures early in life.

In the present study, we examined the effects of baclofen on kindling. The data indicate that baclofen-treated pups require significantly more stimulations to reach each stage of seizure development. In addition, kindled pups that received baclofen were found to regress to an earlier seizure stage and their seizures were shortened. Thus, baclofen is an effective anticonvulsant early in life, able to curtail both generalized seizures induced by flurothyl and seizures of focal onset induced with kindling.

ACKNOWLEDGMENTS

Supported by NIH grant NS-20253 and NRSA training grant NS-07183 from the NINCDS and grant R-36986 from the United Cerebral Palsy Associations. The (-) baclofen was a gift from Ciba-Geigy.

REFERENCES

1. G. V. Goddard, Development of epileptic seizures through brain stimulation at low intensity. Nature 204:1020-1021 (1967).
2. G. V. Goddard, D. C. McIntyre, and C. K. Leech, A permanent change in brain function resulting from daily electrical stimulation. Exp. Neurol. 25:295-330 (1969).
3. H. Vosu, and R. A. Wise, Cholinergic seizure kindling in the rat: comparison of caudate, amygdala and hippocampus, Behav. Biol. 13:491-495 (1975).
4. R. M. Post, R. T. Kopanda, and A. Lee, Progressive behavioral changes during chronic lidocaine administration: relationship to kindling, Life Sci. 17:943-950 (1975).
5. R. M. Post, Progressive changes in behavior and seizures following chronic cocaine administration relationship to

kindling and psychosis, Advan. Behav. Biol. 21:353-372 (1977).

6. J. S. Stripling, and E. H. Ellinwood, Jr., Augmentation of the behavioural and electrophysiologic responses to cocaine by chronic administration in the rat, Exp. Neurol. 54:546-564 (1977).
7. R. C. Collins, Kindling of neuroanatomic pathways during recurrent focal penicillin seizures, Brain Res. 150:503-517 (1978).
8. J. P. Fabisiak, and W. S. Schwark, Aspects of the pentylenetetrazol kindling model of epileptogenesis in the rat, Exp. Neurol. 78:7-14 (1982).
9. D. P. Cain, Bidirectional transfer of electrical and carbachol kindling, Brain Res. 260:135-138 (1983).
10. C. G. Wasterlain, and V. Jonec, Chemical kindling by muscarinic amygdaloid stimulation in the rat, Brain Res. 271:311-323 (1983).
11. S. L. Moshé, The effects of age on the kindling phenomenon, Dev. Psychobiol. 14:75-81 (1981).
12. S. L. Moshé, N. S. Sharpless, and J. Kaplan, Kindling in developing rats: afterdischarge thresholds, Brain Res. 211:190-195 (1981).
13. M. E. Gilbert, and D. P. Cain A developmental study of kindling in the rat, Dev. Brain Res. 2:321-328 (1982).
14. G. L. Holmes, and D. A. Weber, Effects of ACTH on seizure susceptibility in the developing brain, Ann. Neurol. 20:82-88 (1986).
15. S. Lee, H. Kawawaki, O. Matsuoka, and R. Murata, Effect of Ca-antagonist (flunarizine) on kindling seizures in suckling rats, No To Hattatsu, 18:292-298 (1986).
16. S. L. Moshé, B. J. Albala, R. F. Ackermann, and J. Engel Jr., Increased seizure susceptibility of the immature brain, Dev. Brain Res. 7:81-85 (1983).
17. S. Lee, R. Murata, and S. Matsuura, Developmental study of hippocampal kindling, Epilepsia 30:266-270 (1987).
18. R. J. Racine, M. W. Burnham, J. G. Gartner, and D. Levitan, Rates of motor seizure development in rats subjected to electrical brain stimulation: strain and interstimulation interval effects, Electroencephal. Clin. Neurophysiol. 35:553-556 (1973).
19. S. L. Peterson, T. E. Albertson , and L. G. Stark, Intertrial intervals and kindled seizures, Exp. Neurol. 71:144-153 (1981).
20. S. L. Moshé, and B. J. Albala, Maturational changes in postictal refractoriness and seizure susceptibility in developing rats, Ann. Neurol. 13:552-557 (1983).
21. J. P. Pinel, and L. I. Rovner, Experimental epileptogenesis: kindling induced epilepsy in rats, Exp. Neurol. 58:190-202 (1978).
22. S. L. Moshé, and N. Ludvig, Kindling, in: "Recent Advances of Epilepsy 4" T. A. Pedley and B. S. Meldrum, eds., Churchill Livingstone, N. Y. (1988).
23. M. S. Duchowny, and J. F. Burchfiel, Facilitation and antagonism of kindled seizures development in the limbic system of the rat, Electroencephal. Clin. Neurophysiol. 51: 403-416 (1981).

24. J. L. Burchfiel, K. A. Serpa, and F. H. Duffy, Further studies of antagonism of seizure development between concurrently developing kindled limbic foci in the rat, Exp. Neurol. 75:476-489 (1982).
25. A. Vernadakis, and D. M. Woodbury, The developing animal as a model, Epilepsia 10:163-178 (1969).
26. A. Zouhar, P. Mares, and G. Brozek, Electrocorticographic activity elicited by metrazol during ontogenesis in rats, Arch. Int. Pharmacodyn. 248:280-288 (1980).
27. B. J. Albala, S. L. Moshé, and R. Okada, Kainic acid induced seizures: a developmental study, Dev. Brain Res. 13:139-148 (1984).
28. E. A. Cavalheiro, D. F. Silva, W. A. Turski, L. S. Calderazzo-Filho, Z. A. Bartolotto, and L. Turski, The susceptibility of rats to pilocarpine-induced seizures is age dependent, Dev. Brain Res. 37:43-58 (1987).
29. E. F. Sperber, and S. L. Moshé, Age-related differences in seizure susceptibility to flurothyl, Dev. Brain Res. 39:295-297 (1988).
30. M. J. Iadarola, and K. Gale, Substantia nigra: site of anticonvulsant activity mediated by gamma-aminobutyric acid, Science 218:1237-1240 (1982).
31. G. Le Gal La Salle, M. Kajima, and S. Feldblum, Abortive amygdaloid kindling seizures following microinjections of gamma-vinyl-GABA in the vicinity of substantia nigra in rats, Neurosci. Lett. 36:69-74 (1983).
32. L. P. Gonzalez, and M. K. Hettinger, Intranigral muscimol suppresses ethanol withdrawal seizures, Brain Res. 298: 163-166 (1984).
33. J. O. McNamara, M. T. Galloway, L. L. Rigsbee, and C. Shin. Evidence implicating substantia nigra in regulation of kindled seizure threshold, J. Neurosci. 4:2410-2417 (1984).
34. R. Okada, S. L. Moshé, B. Y. Wong, E. F. Sperber, and D. Zhao, Age related substantia nigra mediated seizure facilitation, Exp. Neurol. 93:180-187 (1986).
35. L. Turski, E. A. Cavalheiro, M. Schwarz, W. A. Turski, L. E. A. M. Mello , Z. A. Bartolotto, T. Klockgether, and K. H. Sontag, Susceptibility to seizures produced by pilocarpine in rats after microinjection of isoniazid or gamma-vinyl GABA into the SN, Brain Res. 370:294-309 (1986).
36. E. F. Sperber, J. N. D. Wurpel, D. Y. Zhao, and S. L. Moshé, Evidence for the involvement of nigral $GABA_A$ receptors in seizures of adult rats, Brain Res. 480:378-382 (1989).
37. J. T. Coyle, and S. J. Enna, Neurochemical aspects of the ontogenesis of gabanergic neurons in the rat brain. Brain Res. 111:119-133 (1976).
38. J. M. Palacios, D. L. Niehoff, and M. J. Kuhar, Ontogeny of GABA and benzodiazepine receptors: effect of Triton X-100, bromide and muscimol, Brain Res. 179:390-395 (1979).
39. J. N. D. Wurpel, A. Tempel, E. F. Sperber, and S. L. Moshé, Age-related changes of muscimol binding in the substantia nigra, Dev. Brain Res. 43:305-307 (1988).
40. E. F. Sperber, B. Y. Wong, J. N. D. Wurpel, and S. L. Moshé, Nigral infusions of muscimol or bicuculline facilitate seizures in developing rats, Dev. Brain Res. 37:243-250 (1987).

41. E. F. Sperber, J. N. D. Wurpel, and S. L. Moshé, Evidence for the involvement of nigral $GABA_B$ receptors in seizures of rat pups, Dev Brain Res.47:143-146 (1989).
42. C. McIntyre, Amygdala kindling in rats: facilitation after local amygdala norepinephrine depletion with 6-hydroxydopamine, Exp. Neurol. 69:395-407 (1979).
43. C. D. Applegate, J. L. Burchfiel, and R. J. Konkol, Kindling antagonism: effects of norepinephrine depletion on kindled seizure suppression after concurrent alternating stimulation in rats. Exp. Neurol. 94:379-390 (1986).
44. J. L. Burchfiel, C. D. Applegate, and R. J. Konkol, Kindling antagonism: A role for norepinephrine in seizure suppression, in: "Kindling 3", J. A. Wada, ed., Raven Press, N.Y. (1972).
45. L. P. Lanier, A. J. Dunn, and C. V. Hartesveldt, in: "Reviews of Neuroscience vol. 2", S. Ehrenpreis and I.J. Kopin, eds., Raven Press, N. Y. (1976).
46. R. J. Konkol, E. G. Bendeich, and G. R. Breese, A biochemical and morphological study of the altered growth pattern of central catecholamine neurons following 6-hydroxydopamine, Brain Res. 140:125-135 (1978).

Discussion of Dr. Moshe's Presentation

DR. SHOUSE: I think we saw everything that you did in the kittens except we got a lot of spontaneous seizures. Would you want to make some more comments about that?

DR. MOSHE: I'd hope you would say how you saw them, because you observe them much more closely than I do. I think that my problem is that after we put the rats in the animal institute we cannot observe them for a prolonged period of time. When we did the study of recycling with age I spent literally every day watching the animals without doing anything to them, and recording the EEG, and the only thing I would see was some fast EEG activity, but I'm not sure that I would call it spontaneous seizure. I did not see any good stage 3 or 4 seizures. If you observe the animals long enough you could see that they mouthed. If I'm going to say spontaneous seizure, I'd want to see a good generalized spontaneous seizure, and I've never seen that, but it could be that my time of observation was very limited. I think they would occur.

# THE KINDLING PROCESS AND VULNERABILITY TO STATUS EPILEPTICUS

**Gary G. Buterbaugh and Gail M. Hudson**

**Department of Pharmacology and Toxicology**
**University of Maryland School of Pharmacy**
**Baltimore, Maryland**

Several years ago, we developed an experimental model of status epilepticus (SE) that exploited the interaction of pilocarpine and seizure discharge upon a background of epileptogenesis resulting from amygdala kindling (1). We were initially interested in using the model to study the mechanisms by which the sustained seizures were initiated and maintained. However, we were struck by the extensive bilateral neuropathology resulting from a relatively short duration of SE in our model compared to other models of SE using pilocarpine (2, 11). We naturally wondered if the kindling process established an increased vulnerability to seizure-induced damage. If so, we reasoned that elucidation of the mechanisms responsible for the enhanced vulnerability might reveal information about the mechanisms underlying the acquisition of kindled seizures. Described here is some of the work that evolved from this idea, the results of which indicate that kindling does indeed alter vulnerability to SE, but also that the picture is more complex than we had originally envisioned.

## GENERAL METHODS

Bipolar electrodes for kindling stimulations were implanted into the amygdala (or other regions described later) of adult Sprague-Dawley rats under ketamine anesthesia. Electrodes were placed over the contralateral parietal-frontal cortices from which EEG was recorded on magnetic tape for later analysis. In some cases, an additional electrode was placed into a contralateral limbic region. After 10 days recovery, kindling stimulations were delivered daily or hourly (8 - 10 per day) at twice afterdischarge (AD) threshold until three consecutive stage 5 generalized convulsions were observed.

SE was induced in kindled rats by kindling site stimulation 20 min following pretreatment with pilocarpine (20 - 100 mg/kg depending on experimental design). Under these conditions, the stimulation-evoked seizure does not clearly end, but quickly develops into SE. The SE was terminated with diazepam (1 mg i.p.) and 2% halothane delivered by inhalation in 40% oxygen/60% nitrogen. Halothane anesthesia (1 - 2%) was maintained for 45 minutes and the animals allowed to recover. In most cases, 10% sucrose was administered p.o. for two or three days until the rats resumed intake of water and solid food. At least 10 days following SE, rats were deeply anesthesized with urethane and perfused intracardially with 0.9% saline followed by 4% buffered formaldehyde. Brains were removed, equilibrated in 30% sucrose and frozen sections stained with cresyl violet.

*Kindling 4*
Edited by J. A. Wada
Plenum Press, New York, 1990

## RESULTS AND DISCUSSION

Characteristics of the Model of SE. Two characteristics of the SE are important to point out. First, the SE is nonconvulsive to partial in nature. During the onset of SE, rats frequently showed brief episodes of rearing with forelimb clonus. Within about 10 minutes, however, all animals stood or crouched immobilized with vibrissae movements, ear twitching and mild body tremors. Although some rats did display a few brief episodes of jumping activity after about two hours of SE, and most rats developed weak forelimb clonus and head jerks during the third hour, severe sustained convulsive behavior was never observed.

Second, the SE consisted of high-frequency, high-amplitude, generalized seizure discharge. Power spectral analysis of the neocortical EEG during the early phase (first 30 min) of SE, evoked three weeks after Am kindling, revealed a wide frequency range centered on a mean frequency (MF) of 12.2 Hz with an edge frequency (EF) of 32.9 Hz (Figure 1). In fact, 47% of the total power in the EEG was above 15 Hz. Am and dorsal hippocampal (dH) discharge was analyzed in some rats and was slightly slower compared to the neocortex, but always above 10 Hz. The EEG seizure continued uninterrupted and the MF and EF of the cortical EEG showed a linear decrease with time during the SE until by three hours it appeared as a repetitive waveform with a considerable decline in MF (3.9 Hz) and EF (14.3 Hz) and an approximate 70% decrease in absolute power (Figure 1). EEG recorded from the dH and Am showed an almost identical waveform at this time. The gradual degeneration of the EEG during SE likely reflects loss of neuronal function during the sustained seizures.

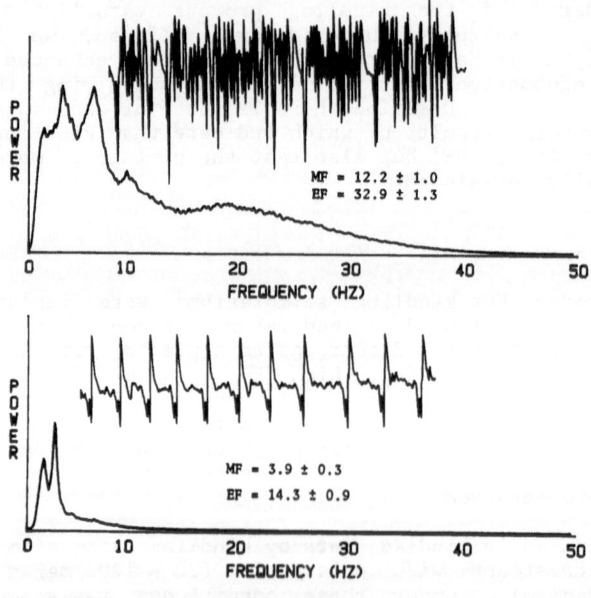


Figure 1. Power spectra of neocortical EEG during SE in amygdala kindled rats three weeks post-kindling. Upper panal is from the first 30 min and lower panal is from the end of the 3 hour SE episode. EEG was digitized with a Nicolet Pathfinder I computer, 100 Hz sampling rate, 1 Hz high-pass and 50 Hz low-pass filtering. For each rat, 112 consecutive 10 sec epochs (1024 data points each) of EEG were subjected to power spectral analysis and averaged. The spectra here are the grand average of the spectrum from 5 rats. The waveform inserts are each 10 seconds of representative digitized EEG.

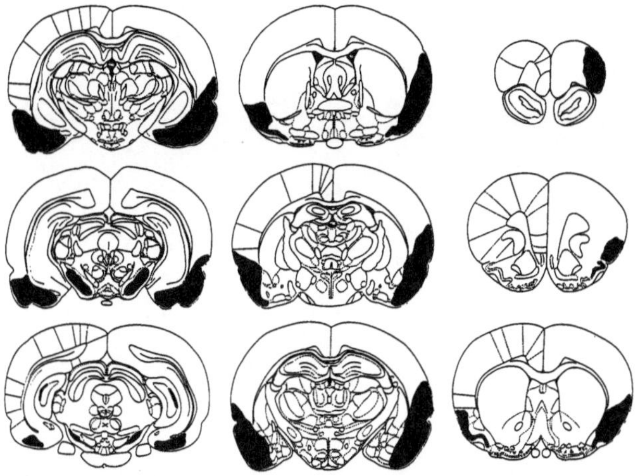

Figure 2. Schematic composite of the typical extent of the degeneration of the amygala and piriform and entorhinal cortices after two hours of SE. Damaged areas are blackened. Ipsilateral hemisphere is on left.

Neuropathology Resulting from SE in Amygdala Kindled Rats. Three hours of SE (N = 5) resulted in widespread, bilateral damage including thalamic regions, CA1 and CA3 pyramidal cells of the ventral hippocampus (vH) and the substantia nigra reticulata (SNr). Compared to the vH, dH regions were relatively spared from damage. The most remarkable damage in these animals was extensive, almost complete degeneration of the piriform (Pir) and entorhinal (Ent) cortices and most of the Am complex (Figure 3A). This damage extended from the anterior Pir caudally into the Ent, dorsally into perirhinal and insular cortices and medial to include the claustram and dorsal endopiriform nucleus. Extensive bilateral areas of damage were also found within anterior frontal cortex.

Two hours of SE (N = 8) resulted in considerable damage. However, there was little damage to frontal cortex and the damage to Pir, Ent and Am regions was not as spatially extensive, especially within the ipsilateral hemisphere where it did not extend anterior beyond the genu corpus callosum or above the rhinal fissure. Figure 2 is a composite of the damage within these regions and clearly shows the more extensive damage within the contralateral hemisphere.

Dan McIntyre (5) has shown in a different model of SE using Am-kindled rats that prolonged partial SE (10 - 24 hrs duration) results in severe damage to the ipsilateral Am and Pir cortex. Moreover, 4 hours of SE results in moderate to severe damage restricted to ipsilateral Am-Pir regions, certainly unlike the extensive bilateral damage shown here after only two hours of SE. One major difference between the two models using Am kindled rats is the higher frequency of the seizure discharge of our model which may more actively evoke excitotoxic mechanisms (10) of irreversible cell damage compared to the lower (< 3 Hz) frequency in McIntyre's model. This suggests that an important factor responsible for the contrasting damage patterns in the two SE models using Am kindled rats may be different frequency/seizure duration interactions. However, evidence presented later indicates that this is not the sole explanation.

SE limited to only one hour duration (N = 6) resulted in damage confined to bilateral degeneration of the SNr, clearly dissociating SNr

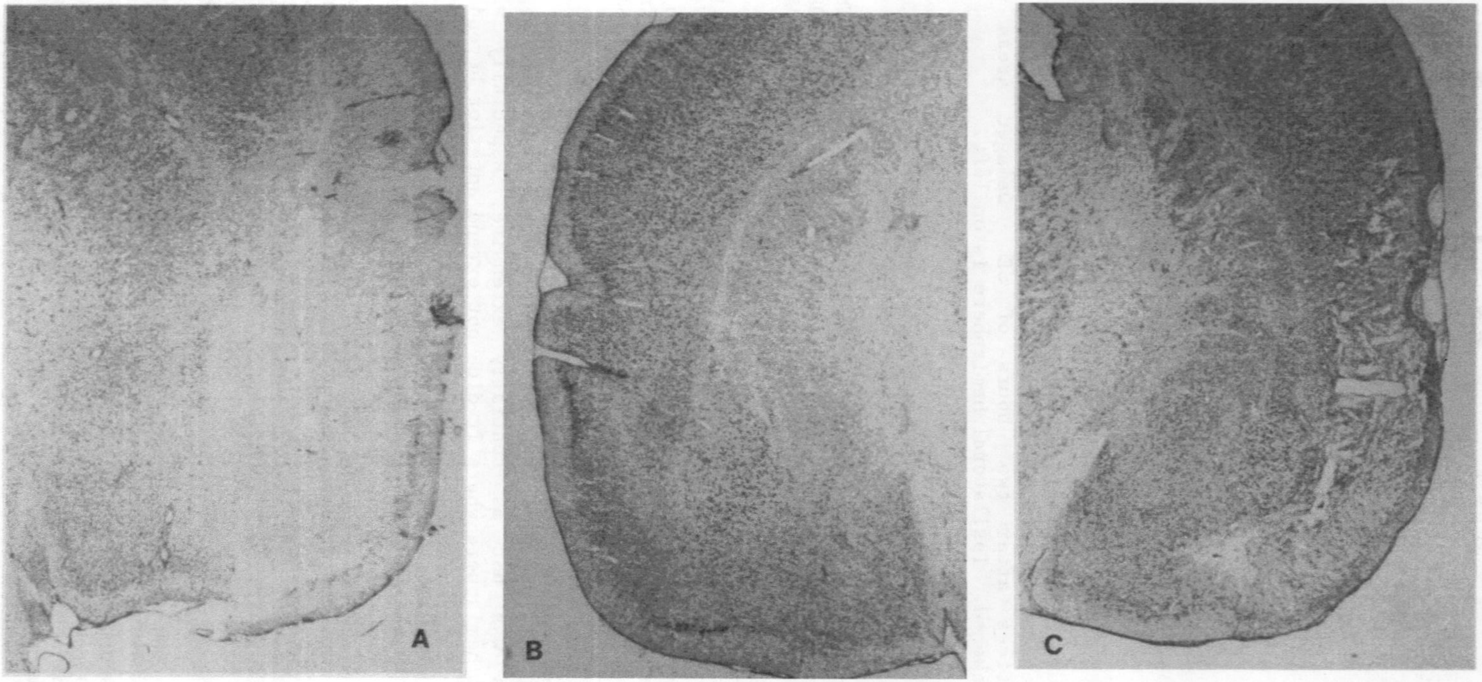


Figure 3. Coronal sections at the level of the basolateral amygdala comparing damage produced by three hours of SE. A. Amygdala kindled. B. Piriform cortex kindled. C. SE induced by pilocarpine (400 mg/kg) in unkindled rats.

damage from damage in forebrain regions such as the Am and Pir, and confirming other evidence that the SNr is highly sensitive to sustained seizure discharge (4, 9). These rats responded to Am stimulation with stage 4 or 5 generalized limbic seizures but with a 4 - 6 fold elevation of seizure threshold, compatible with evidence that the SNr participates with evidence that the SNr participates in the regulation of kindled seizure thresholds (McNamara et al., 1984). Overall, these results indicate that one to two hours of sustained seizure activity in our SE model in Am kindled rats is inclusive of a critical duration heralding the onset of damage within many vulnerable structures. However, if SE continues for three hours, damage to the Am, Pir, Ent and frontal cortex becomes progressively worse.

Other Kindling Sites. We had the opportunity to examine the pattern of brain damage resulting from three hours of SE in rats kindled from other anatomical sites in the brain, including the ventral and dorsal hippocampus, the piriform cortex and the anterior neocortex (Anc). At least 6 rats/site were studied except for the Anc where N = 2 and all rats had completed kindling at least four weeks before SE. The behavioral characteristics during SE in these rats resembled those described for Am kindled rats. Spectral analysis of cortical EEG during early and late SE showed frequency characteristics similar to three week post-kindling Am kindled rats (see Figure 1), including the progressive decline in MF and EF by three hours duration.

Based upon the similarities in the behavioral and electrographic components of the SE in these animals compared to Am kindled rats, we expected similar patterns of brain damage. Indeed, damage within posterior vH and thalamic regions, although variable, was much like that described for Am kindled rats. Remarkably, however, the AM and Ent were little effected and damage within the Pir was much less extensive in these rats compared to Am kindled rats (Figures 3B and 4). Of the four kindling sites, dH kindled rats had the most severe damage within these regions, but was still less extensive compared to Am kindled rats after only two hours of SE. This observation is interesting in that there is evidence that the amygdala plays a role in hippocampal kindling (Le Gal La Salle and Feldblum, 1983).

The remarkable contrast in damage between Am and Pir kindled rats is especially intriguing because of the proximity of these regions, the likelihood that the Pir is within downstream circuitry established during Am kindling (Morimoto et al., 1986) and in vitro evidence of the sensitivity of the amygdala/piriform region to kindling (McIntyre and Wong, 1986). Although the limited Pir damage did not extend rostrally beyond the ventral hippocampal commissure into the region of the electrode site, some Pir kindled rats had AD/seizure thresholds 2 - 3 times greater than pre-SE thresholds. Also, 5 of 7 rats displayed stage 5 responses within 3 Pir stimulations. The two exceptions did not progress beyond stage 1 responses after 10 - 12 stimulations; one rat had marked bilateral damage within the ventral pallidum and globis pallidus, while considerable damage to granule cells within the ventral posterior hippocampus was found in the other. These regions may have special importance to the Pir kindled seizure network.

Based upon a three hour duration of SE with comparable frequency characteristics in rats kindled from five different stimulation sites, Am kindled rats, and to a lesser extent, dH kindled rats, stand out as being much more sensitive to seizure-induced damage within Pir, Ent and Am regions of the brain. We leave it to the reader to speculate on the reasons for this kindling-site dependent sensitivity to damage. It is unlikely due to differences in kindling rates since the vH kindles considerably faster than the dH. We do suggest that the underlying basis may relate to different kindled seizure networks established from the various sites of kindling stimulations. While these different networks may not be completely anatomically distinct, they may differ appreciably in the manner in which seizure discharge converges upon certain regions

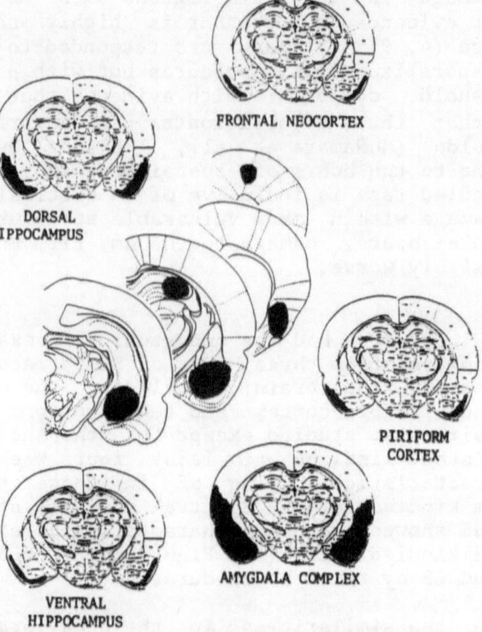


Figure 4. Schematic comparison of the damage to the amygdala and the piriform cortex following 3 hours of SE three weeks post-kindling in rats kindled by stimulation of one of the five labelled anatomically distinct sites. The center composite illustrates the stimulation sites. The darkened areas in the surrounding sections indicate comparative damage.

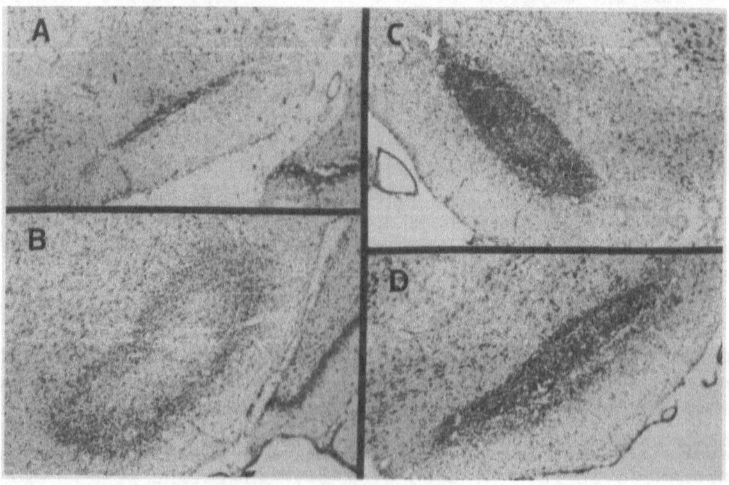


Figure 5. Examples of different patterns of glial infiltration within the substantia nigra reticulata following three hours of SE in rats kindled from different regions. A. Dorsal hippocampus. B. Ventral hippocampus. C. Piriform cortex. D. Anterior neocortex. Amygdala kindled rats (not shown) presented a pattern intermediate between the dorsal hippocampus and piriform cortex.

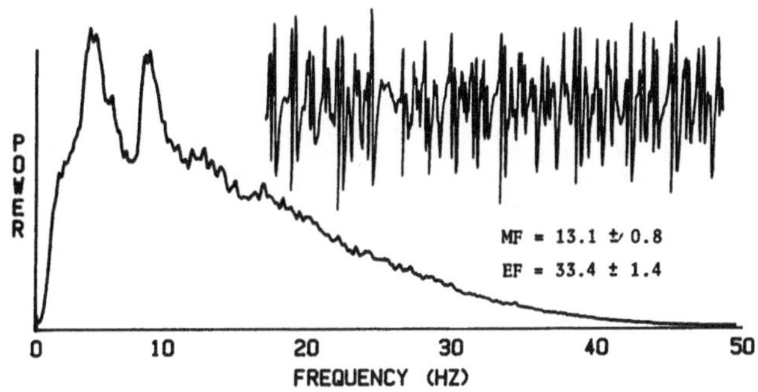


Figure 6. Power spectrum of the early phase of SE produced by pilocarpine (400 mg/kg) administered to unkindled rats. N = 6. See legend to Fig. 1 for details of method.

such as the Pir, Ent and Am. In this regard, the one common denominator of damage was the SNr which was always bilaterally damaged by SE regardless of the kindling site. However, several distinct kindling-site dependent patterns of gliotic reaction within the SNr were found (Figure 5). These different patterns may reflect subtle but important differences in the way in which the SNr is recruited into kindling-site dependent seizure networks, resulting in temporal or spatial differences in SNr activation during SE and different patterns of damage.

<u>Pilocarpine SE in Unkindled Rats</u>. Because of the marked kindling-site dependent differences in sensitivity to seizure-induced neuropathology, we were interested in determining how the neuropathology resulting from the seizures of a different SE model in unkindled rats would compare. Six unkindled rats were treated with pilocarpine (400 mg/kg) to evoke SE which was allowed to continue for three hours. Although the damage to the Am, Pir and Ent regions showed considerable variation among rats, it was always less extensive than that in Am kindled rats and was subjectively different in appearance (Figure 3C). Interestingly, the power spectra of neocortical EEG during the early phase of SE in these animals (Figure 6) showed frequency characteristics quite similar to those of Am kindled rats. Thus, the different extent of neuropathology produced by equal SE durations in this model compared to our model in Am kindled rats cannot be explained on the basis of marked differences in seizure discharge.

<u>Immediate Post-Kindling SE</u>. We noticed that the neuropathology resulting from SE immediately following completion of Am kindling was not as extensive as that observed in rats which experienced SE more than three weeks post-kindling, especially in the Am and the Pir and Ent. We speculated that the kindling process generated a neuroprotective condition that was maximum immediately following kindling, but which decayed with post-kindling time. To test this possibility, 8 rats were kindled by hourly Am stimulation and within two days after kindling completion, subjected to three hours of SE (pilocarpine 100 mg/kg). Our first clue that our suspicions were right about the relative resistance of these animals to seizure-induced brain damage was the retention of a relatively high MF (7.9 Hz) and EF (25.6 Hz) in the cortical EEG at the end of three hours of SE (Figure 7). The end-SE EEG of these rats contrasts sharply with the end-SE waveform of Figure 1 from rats experiencing SE three weeks post-kindling. These animals were often observed sleeping in a normal curled-up position within 24 hours post-SE indicating rapid recovery from the SE.

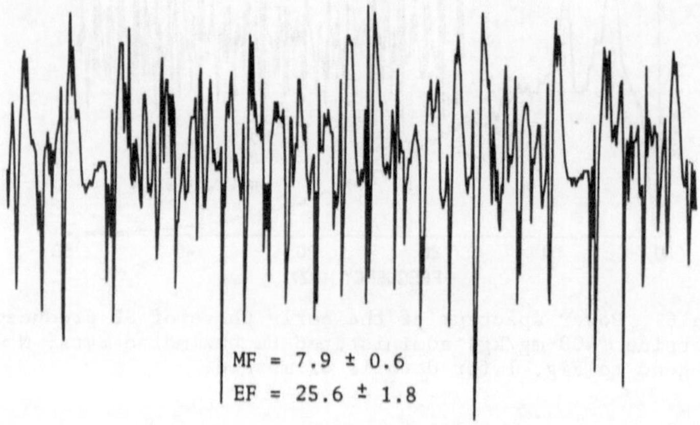


Figure 7. Representative 10 second samaple of digitized neocortical EEG at the end of three hours of SE evoked two days after amygdala kindling (Hourly stimulation).

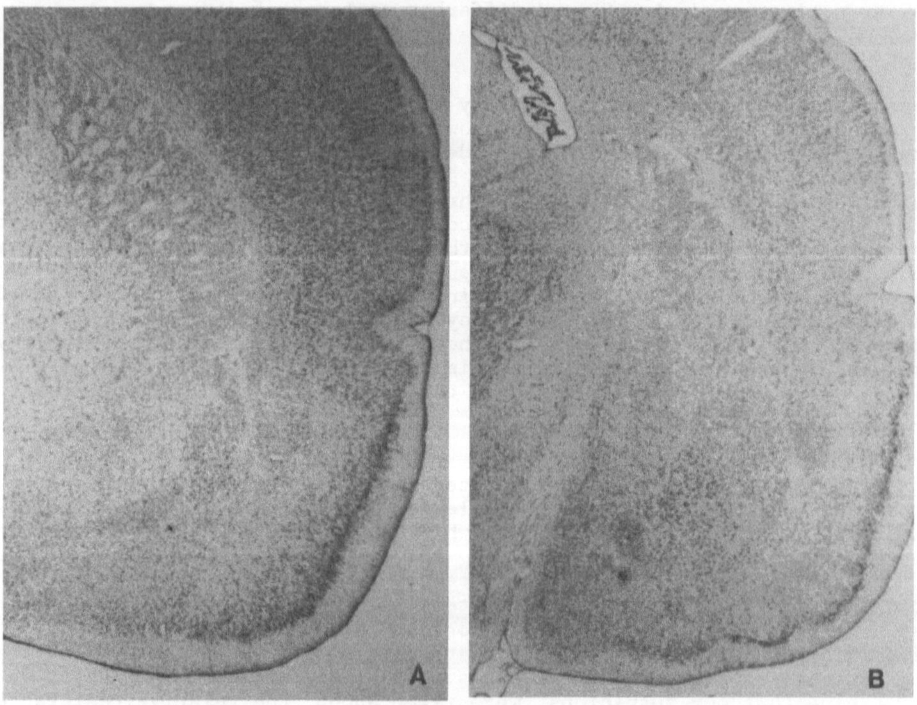


Figure 8. Representative coronal sections at the level of the basolateral amygdala from rats after three hours of SE. A. One episode of SE two days after amygdala kindling. B. Like A, but a second three hour SE episode five weeks after the first episode.

Two weeks after SE, the Am AD threshold in these rats was 103 ± 18 uA (mean ± s.e.m.) compared to 78.6 ± 7 uA before SE. All rats responded with a stage 5 generalized response within one to three stimulations. The dH AD threshold of these rats had not changed compared to pre-SE thresholds and fully generalized stage 5 seizures were evoked within six dH stimulations. Since in our experience, 40 - 50 AD trials are ordinarily required for dH kindling, this result represents a remarkable degree of transfer from Am to dH kindling in these rats suggesting a relatively healthy brain in these animals. Histological examination confirmed this in that little or no damage was found in the Am, Pir (Figure 8A), Ent and posterior vH. The lack of damage was not global, however, in that thalamic regions showed damage which was similar to that found in the three week post-kindling rats. Damage to the SNr was variable, ranging from a few scattered glial cells to moderate glial infiltration, always less severe than three week post-kindling rats. There was no correlation between SNr damage and AD thresholds; in fact, the AD threshold of the rat with most extensive SNr damage was identical to the pre-SE threshold.

In light of the resistance of these animals to the damage ordinarily produced by three hours of SE in the Pir, Ent, Am and to a lesser extent the SNr, it is interesting that power spectral analysis of cortical EEG recorded soon after the initiation of SE (Figure 10) showed a higher MF (15.6 Hz) and EF (37.3 Hz) compared to the same phase of SE in three week post-kindling rats and 58% of total power in the EEG was above 15 Hz.

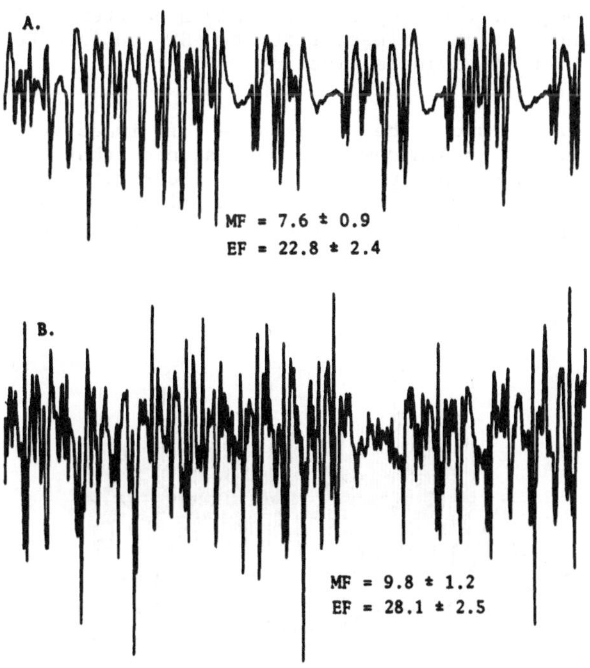


Figure 9. Representative 10 second samples of neocortical EEG at the end of three hours of SE in amygdala kindled rats. A. First SE episode within two days after amygdala kindling. B. Second SE episode in same rat four weeks later.

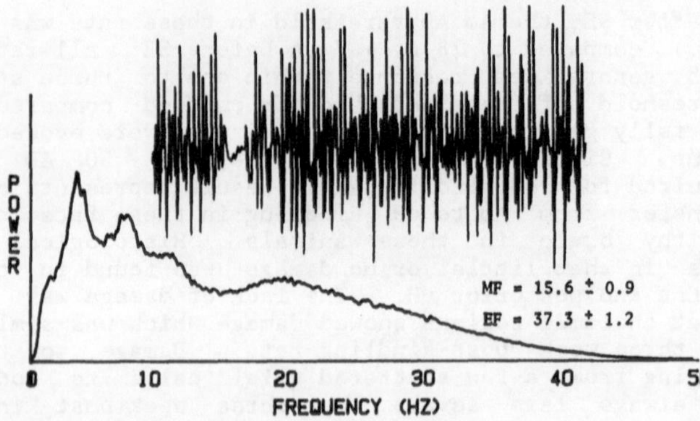


Figure 10. Power spectrum of the early phase of SE evoked in rats within two days after amygdala kindling. N = 8. See legend to Fig. 1 for details of method.

<u>SE in Amygdala "Long-Kindled" Rats</u>. We had the occasion to induce SE in three rats which had rested 4 - 5 months since completion of amygdala kindling. The behavioral characteristics of the SE in these rats resembled rats in SE a few weeks post-kindling. The cortical EEG was quite different, however, with 40% of total power centered on a MF of 10.6 Hz (Figure 11). The prominent peak in the power spectrum is reflected by the regular spiking activity evident in the waveform. EF was 27.1 Hz and only 27% of the total power was above 15 Hz. These animals were dying within 24 hours of SE and therefore histology was inconclusive. They were, however, beginning to show massive bilateral degeneration throughout the temporal lobe and neocortex. These findings suggest that changes in the vulnerability to SE continue to evolve long after the completion of amygdala kindling.

<u>Repeated SE</u>. Based upon the resistance of immediate post-kindling animals to SE-induced brain damage, we reasoned that similar animals might fully support a second episode of SE. Four rats were subjected to three hours of SE within two days after Am kindling. Three to five weeks later, a second episode of SE was induced. Power spectral analysis of

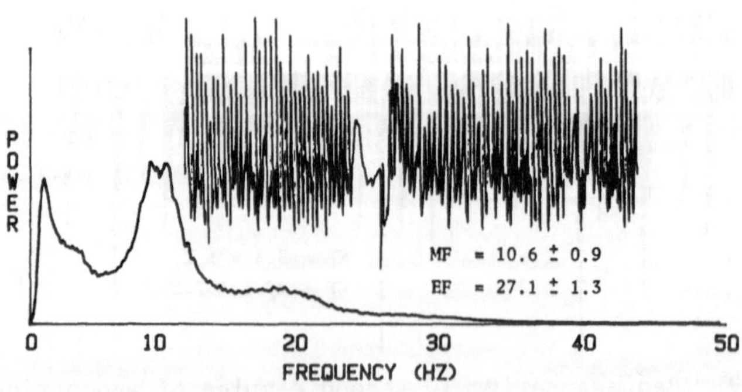


Figure 11. Power spectrum of the early phase of SE evoked in rats 4 - 5 months after amygdala kindling. N = 3. See legend to Fig. 1 for details of method.

the cortical EEG at the end of three hours of SE resulted in a MF and EF of 9.8 and 28.1 Hz, respectively (Figure 9), values greater than the corresponding frequencies of 7.6 Hz and 22.8 Hz at the end of the first SE episode. Ten days later, histological analysis revealed little damage to the Am, Pir (Figure 8B), Ent and posterior vH. These results suggest that exposure of animals to prolonged SE during the immediate post-kindling "neuroprotective" time period results in retention of the protection for as long as five weeks. As a step in understanding this phenomenon, we did comparative studies with the putative neuroprotective compound MK-801.

MK-801 Protection. We had previously found that MK-801 protected against the brain damage caused by SE in our model. Thus, four rats were Am kindled (daily stimulation) and rested four weeks before SE. Within 20 minutes of the onset of SE, the rats were treated with MK-801 (1 mg/kg i.p.) and the SE allowed to continue for three hours. In spite of the generalized seizure activity which was not overtly altered by MK-801, the rats displayed the behavior typically observed with MK-801 in naive rats not in SE. Within 10 minutes of the injection, the rats progressed through loss of posture and head weaving to finally remain through the duration of SE motionless on the cage floor with splayed hindlimbs. After a two week rest, these animals were subjected to a second three hour duration of SE (drug-free). The end-SE frequency characteristics of the two episodes of SE were comparable (Figure 12) and histological assessment was similar to that of the repeated SE animals described above.

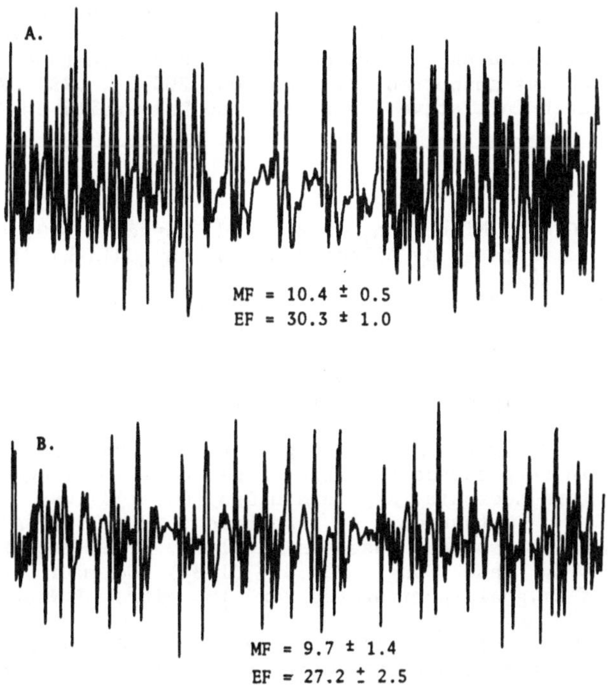


Figure 12. Representative 10 second samples of neocortical EEG at the end of three hours of SE in amygdala kindled rats. A. First SE episode four weeks after kindling during treatment with MK-801 (1 mg/kg). B. Second SE episode in the same rat two weeks later (drug-free).

## CONCLUSION

The results establish a definite relationship between the kindling process and the underlying vulnerability to the neuropathology resulting from SE. In particular, there appears to be something very unique about amygdala kindling in establishing conditions of sensitivity to seizure-induced brain damage, especially within temporal lobe regions and the amygdala. In spite of the differential, kindling-site dependent sensitivity to damage at least three weeks following kindling, the frequency characteristics of the the neocortical EEG during SE were quite comparable for the kindling sites studied. This suggests that the acute response of the brain during three hours of SE is similar for the various kindling sites. Amygdala kindling may thus establish conditions which allow the acute insult during SE to progress to a chronic degenerative process and cell death. The protective effects of MK-801 in amygdala kindled rats suggests that NMDA-preferring glutamate receptor coupled events may have importance to the results. In particular, the remarkable post-kindling change in the sensitivity of amygdala kindled rats to seizure-induced damage may reflect dynamic changes in neuronal functions related to the NMDA receptor, changes which may continue to evolve for many months post-kindling. The ability of amygdala kindled rats to retain resistance to SE induced brain damage for several weeks following a previous SE episode, experienced during the neuroprotective influence of immediate post-kindling conditions or MK-801, although inexplicable, certainly underscores the complexity of the kindling process.

## ACNOWLEDGEMENTS

Many people played some role in this project including Michael Gentry, Bruce Jones, Jennifer Kiefer, Lieser Mayo, Alison Dion and Rik Kline, and Drs. Rick Gussio, Karen Marquis, Hillary Michelson and Gerald Young. Thank you for your assistance and provocative discussion. The MK-801 was a generous gift from Merck, Sharp and Dohme Research Laboratories. Some of the research was supported by PHS NS 20670, BRSG RR 05770 and Designated Research Initiative Funds from the University of Maryland.

## REFERENCES

1. Buterbaugh, G.G., Michelson, H.B. and Keyser, D.O. (1986) Status epilepticus facilitated by pilocarpine in amygdala-kindled rats. Exp. Neurol. 94: 94-102.
2. Honchar, M.P., Olney, J.W. and Sherman, W.R. (1983) Systemic cholinergic agents induce seizures and brain damage in lithium-treated rats. Science 220: 323-325.
3. Le Gal La Salle, G and Feldblum, S. (1983) Role of the Amygdala in Development of Hippocampal Kindling in the Rat. Exp. Neurol. 82: 447-455.
4. Lindvall, O., Ingvar, M. and Gage, F.H. (1986) Short term status epilepticus in rats causes specific behavioral impairments related to substantia nigra necrosis. Exp. Brain Res. 64: 143-148.
5. McIntyre, D.C., Nathanson, D. and Edson, N. (1982) A new model of partial status epilepticus based on kindling. Brain Res. 250: 53-63.
6. McIntyre, D.C. and Wong, R.K.S. (1986) Cellular and synaptic properties of amygdala-pyriform cortex in vitro. J Neurophysiol. 55: 1295-1307.
7. McNamara, J.O., Galloway, M.T., Rigsbee, L.C. and Chin, C. (1984) Evidence implicating substantia nigra in regulation of kindled seizure threshold. J. Neurosci. 4: 2410-2417.
8. Morimoto, K., M. Dragunow and G.V. Goddard (1986) Deep prepyriform kindling and its relation to amygdala kindling in the rat. Exp. Neurol. 94: 637-648.
9. Nevander, G., Ingvar, M., Auer, R. and Siesjo, B.K. (1984) Irreversible brain cell damage after short periods of status epilepticus, Acta. Physiol. Scand., 120: 155-157.

10. Olney, J.W., Collins, R.C. and Sloviter, R.S. (1986) Excitotoxic mechanisms of epileptic brain damage. In Delgado-Escueta, A.V., Ward, A., Woodbury, D.M. and Porter, R.J. (eds): Advances in Neurology, Volume 44. Raven Press, New York, pp. 857-877.
11. Turski, W.A., Cavalheiro, E.A., Schwarz, M., Czuczwar, S.J., Kleinrok, Z. and Turski, L. (1983) Limbic seizures produced by pilocarpine in rats: Behavioral, electroencephalographic and neuropathological study. Behav. Brain Res. 9: 315-335.

Discussion of Dr. Buterbaugh's Presentation

DR. MCINTYRE: Gary, this is wonderful stuff. You obviously did quite a lot since we spoke 3 weeks ago. Is that why you did the speed kindling, trying to make this deadline? Did you, again after the second Herculean experience, stimulate these animals?

DR. BUTERBAUGH: Right, I'm glad you brought that up. I left that out. Let's take a speed/kindled, push condition. If we let these animals survive for a couple of weeks, and then their thresholds are normal in the amygdala, there's very little change; some animals may go up 20,30,40 microamps. We also put bipolar electrodes into the ipsilateral dorsal hippocampus, and those rats reached stage 5 convulsions within 5 or 6 stimulations, which is really fast - it's tremendous transfer. So they really had a healthy brain. The Hercules animals - Oh, gosh, after my 20 hours trying to get here yesterday, and three airlines, I'm not sure I can recall the numbers - their amygdala thresholds were relatively normal and they would support stage 5 seizures. So, with regard to thresholds and ability to have a convulsion, and histology, they looked pretty good.

DR. MOSHE: In the non-Herculean rats, the ones that have all the lesions, have you allowed any of them to survive, or stopped the status with some drugs, and see if you can expose them to another convulsant, if they're more susceptible to seizures?

DR. BUTERBAUGH: No, we haven't done that.

DR. MOSHE: It may be interesting because you have all these nigra lesions there exchanging.

DR. BUTERBAUGH: Right. The 3-hour amygdala kindled rats, of course, died. The 2-hour rats couldn't get out of the amygdala after-discharge, there's nothing to support it. The 1-hour animals, their amygdala threshold was up about threefold and they would support a full stage 5 seizure. Maybe that threefold increase is due to the nigral damage, could very well be.

DR. CAIN: Gary, your speed pushed rats suggest a throwaway comment or suggestion that I would like to make to kindlers in general. Mike brought up earlier, actually it was his discovery, the low frequency kindling effect, and I'd just like to mention that I've had some data for some years which I'll eventually publish, probably at least within the next decade, of some status with low frequency kindling. And I'd like to say that if you're interested in fast status induction it's fantastic. You just give 3 Hz stimulation to a naive, unkindled rat, for as long as it takes, and they go right into status - it might take 10 or 20 minutes, not terribly long, and they carry on unless you terminate the status. So it's a rapid and reliable way to develop status epilepticus. Bill Milgram who is in Toronto, not with us today, has done a similar kind of thing with slightly different stimulation, but that requires prior kindling. This is with naive rats. It's a very powerful technique, and I'd suggest it to anybody who's interested.

DR. BUTERBAUGH: I think the neuroprotective influence here is dependent upon the kindling process. We compacted it down to a 3-day period of time, but it's still a one hour space between stimulations, and I think that's the key. It's the kindling process that's generating that neuroprotective condition.

DR. CAIN: If I could just add, we had previously shown that low frequency kindling is genuine kindling and if you do that to them with low frequency, but you don't drive them into status, and then you leave them for about a week or so, and then hit them with a regular 1-second train of 60 Hz pulses, they show immediate stage 5s. So I liken this to a real kindling that just carried further within one and the same session.

DR. ADAMEC: I was wondering whether you had measured the core temperature of your rats during the status in the animals who do and do not show damage. The reason why I raise this point is because Corbett has recently found that in studies of ischemic damage where he was examining MK801 protection, there is a hypothermia during the ischemia. And if you prevent them by simply warming the animals up, you can prevent the damage. I was wondering if you have some more hypothermic process going on in your two groups of animals?

DR. BUTERBAUGH: How does that work? The MK801 produced the hypothermia?

DR. ADAMEC: No, no. If you warm the animals up during ischemia you get no damage, and you get protection as good or better than with treating with MK801.

DR. BUTERBAUGH: Well, we have measured the core temperatures in some of these animals and it certainly doesn't go down. If anything it goes up a little bit. We don't routinely do that, somebody happened to think of it. We should keep track of that. There isn't much of a change in temperature when we measured it.

DR. WADA: Is the neuroprotective action offered only when the animal has experienced a convulsive stage? A partially kindled stage wouldn't do?

DR. BUTERBAUGH: That's a good question. The partial kindled rats, we tried to induce the status within those animals and they wouldn't support it. We didn't push it very hard. I guess I should not use that term right now. But I rather think that the animals have to go through the full process of the acquisition of the kindled seizure.

# FREQUENCY- AND TIME-DOMAIN EEG TOPOGRAPHIC ANALYSIS OF THE AMYGDALA KINDLING EVOLUTION IN THE CAT

Augusto Fernández-Guardiola, Rodrigo Fernández-Mas, Adrián Martínez, Luisa Rocha, and Rafael Gutiérrez

División de Neurociencias, Instituto Mexicano de Psiquiatría
Calz. México-Xochimilco 101, CP 14370, México, D.F.

## INTRODUCTION

The kindling process has well studied short term (e.g. the amygdaloid after-discharge; AM/AD) and long lasting effects, that have been more difficult to evaluate. A great number of kindling experiments are performed in the rat, recording only the amygdaline or hippocampal ADs, with few if any, cortical recordings (a notorious exception are the works of Wada and his associates in sub-human primates), and correlating the ADs' duration (rarely the frequency) with the behavioral stages. Very often, this correlation fails to explain the dramatic motor and vegetative display of the tardive behavioral stages (4, 5 and 6) of kindling.

This failure is due to the poor understanding we have about the transfer effect from the primary stimulated site to the cerebral cortex. The involvement of the sensorimotor cortex (areas 4 and 6) and the prefrontal cortex (area 8) was elucidated when rabbits[1], cats[2,3], and monkeys[4,5] were used for kindling.

Split brain experiments (by callosal section and commissural lesions) as well as cortical ablations in these species were fruitful in proving a relationship between the motor and vegetative expression of the kindled animals and the functional integrity of the rostral cortex and of the amygdala or hippocampus rostral efferent connections.

But the lesion method, useful as it is, does not tell us about the orderly day to day progression of the amygdala-generated paroxisms, culminating in a generalized tonic-clonic seizure.

The initial work of Racine[6] provided a general description of the generalization of the ADs and the motor involvement, signaling the outstanding excitability changes occurring in the cortex, measured by the amplitude increment of the transcallosal and sensory evoked potentials. However, the timing and topography of the propagation of the AM/ADs to the cerebral cortex needs to be on-line recorded and analyzed.

Since this cortical activation is a dynamic and wide spread phenomenum from the beginning of the AM/AD generation, the analysis of the topographic features of the propagation has theoretical and practical implications.

*Kindling 4*
Edited by J. A. Wada
Plenum Press, New York, 1990

Traditionally, these electrical phenomena have been analyzed in regard to their time evolution, not considering their space-related integration. The EEG-AD has specific frequency patterns[7] and these must be correlated to the activity of specific cerebral areas, so the topographic localization of these frequencies can be easily accomplished by using band-specific topographic mapping techniques.

METHODS

The experiments were performed in 4 adult cats of either sex, weighing 3-5 kg. Sixteen stainless steel nail electrodes were epidurally implanted, forming an isometric matrix (modified 10-20 international standard; vertex electrodes were suppressed). This array covers the dorsal and lateral areas of the brain, mainly the gyrus lateralis or marginalis, the ectosylvian and suprasylvian gyri, and the rostral part of both hemispheres, including areas around the gyrus cruciatus. The posterior electrodes were placed over the specific visual projection areas in the posterior part of the gyrus marginalis (Fig. 1).

Two bipolar stainless steel electrodes were stereotaxically placed into the basolateral nucleus of the AM (2 cats in the right AM and 2 in the left AM), one for electrical stimulation and the other for recording. At least two weeks after the surgical procedure, the AM/AD threshold was determined by delivering a 1 s train of 1 ms monopolar rectangular pulses at 60 Hz. The stimulus' intensity was increased every 10 min until an AD of at least 2 s could be recorded or a twitching reaction of the ipsilateral eye could be observed. The animals were then kindled with this threshold intensity every 24 h (circa 10 am).

The behavior during each kindling trial was video tape recorded and the 16 channel signals plus the AD signal were simultaneously led to a Grass polygraph (78D) and to a computer (HP 9000) through a 12 bit A/D converter.

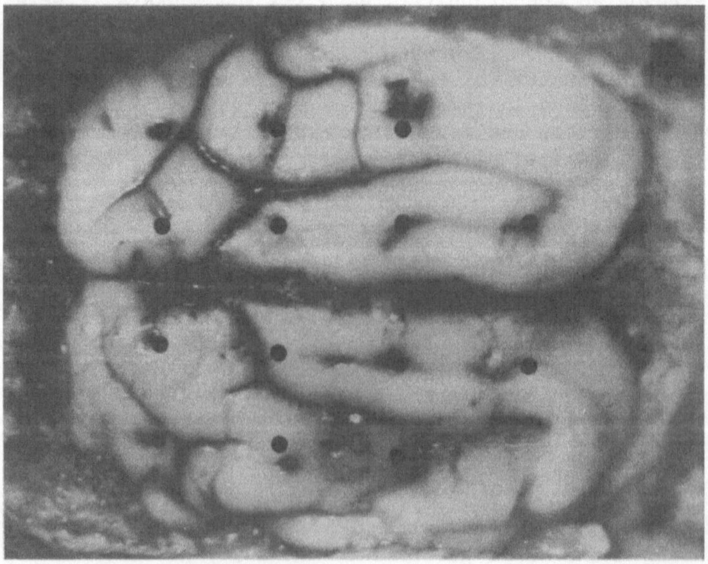

Fig. 1. Position of each electrode (black circles) of the 16-electrode isometric matrix in the cats' brain.

The signal acquisition was triggered 3-5 s (depending on the presence of artifacts) after the onset of the amygdaloid stimulation. Five 4-s sweeps (256 points) were acquired every 5 s, covering from the AM/AD through the postictal period. The sampling rate was selected following the Nyquist criteria. These signals were digitally stored in a hard disk for subsequent analysis (Fig. 2).

Time- and frequency-domain topographic color maps were daily on-line computed by means of a specifically designed software (RBEAM). The frequency-domain maps (FFT maps) were computed for a specific frequency band (8-16 Hz) (Fig. 3-I). The historical layout (time-domain maps) of the cortical projection of single AM/AD spikes was computed for a specific AD latency point during different behavioral stages. The selected spikes had a mean peak to peak duration of 86 ms. Each was divided into 6 intervals (Δt = 15.6 ms). A time-domain map was constructed for each interval, thus, the 6 maps show the spatiotemporal evolution of the spike.

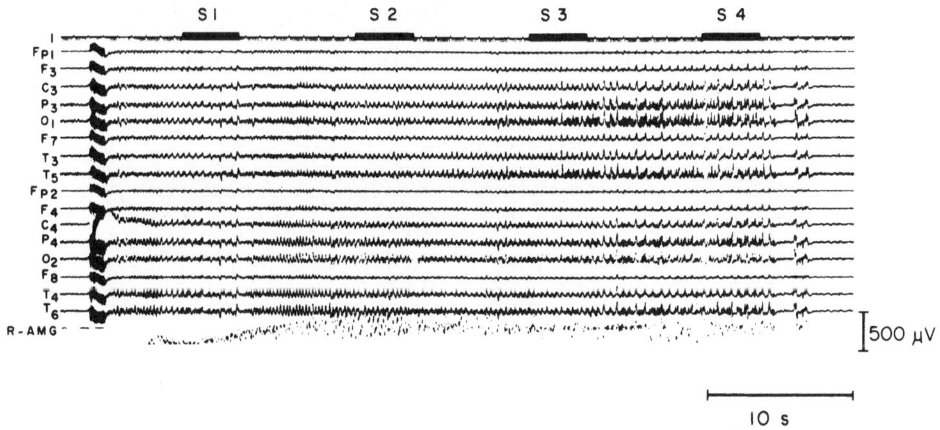


Fig. 2. 16-Channel EEG recording of the right AM/AD during the kindling trial 18 of cat 4. S1-S4 represent the acquisition sweeps. The first 8 channels correspond to the left hemisphere. Notice the different apparent frequencies at the beginning and at the end of the AD.

After the kindling was established, seven frequency-domain maps covering behavioral stages 4 and 5, were averaged for each cat (see Fig. 3-II). The video tapes were analyzed to establish the correlation between the behavior and the corresponding brain maps.

## RESULTS

Table I shows the threshold intensities for the first AM/AD of the 4 cats and the mean number of the daily kindling trials needed to elicit the first generalized convulsive seizure (GCS).

LEFT AMYGDALA

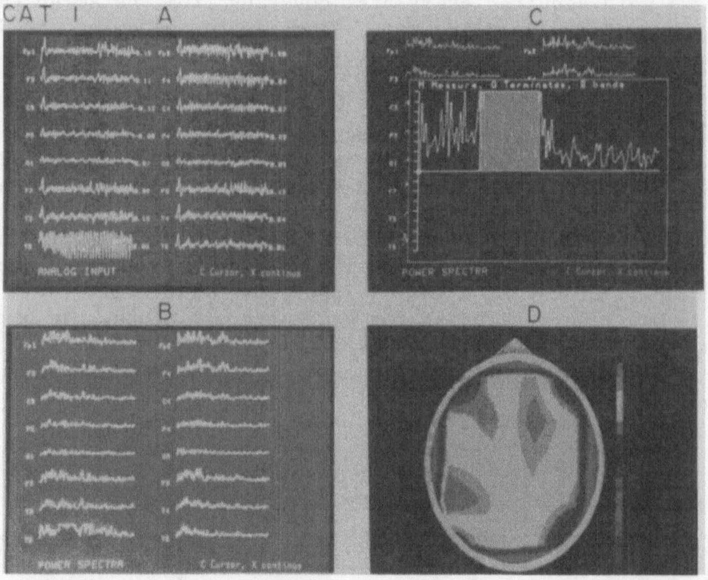


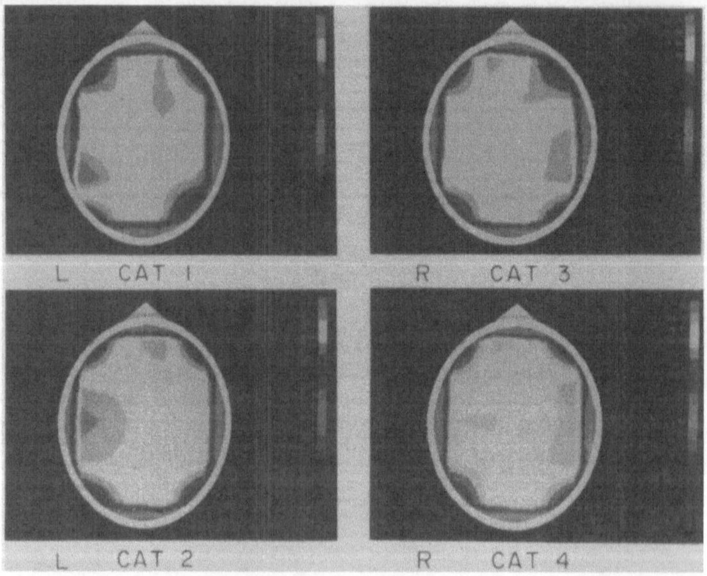


Fig. 3-I. Complete frequency-domain analysis sequence. Pannel A shows the raw data acquisition; pannel B shows the frequency-domain transformation (power spectra); pannel C shows the frequency band selection; pannel D shows its corresponding relative percentage topographic distribution. This sample was taken from the 21st kindling trial of cat 1, first acquisition sweep.

3-II. Frequency domain average (n=7) EEG maps during behavioral stages IV and V in the 4 cats. Notice the evident ipsilateral ectosylvian and contralateral frontal propagation with both left and right amygdala stimulation.

Table I. AM/AD threshold currents and trials needed to produce the first GCS (± SD)

| | μA | Kindling trials |
|---|---|---|
| Cat 1 | 320 | 40 |
| Cat 2 | 600 | 39 |
| Cat 3 | 400 | 33 |
| Cat 4 | 500 | 27 |
| | $\bar{X}$ = 566 ± 106 | $\bar{X}$ = 34.75 ± 5.2 |

The evolution of the duration and the frequency of the AM/AD of each animal are plotted in figure 5. Notice that the AD of cat 2 underwent a slow duration progression until it reached K33, though its frequency enhancement was similar to those of the other cats.

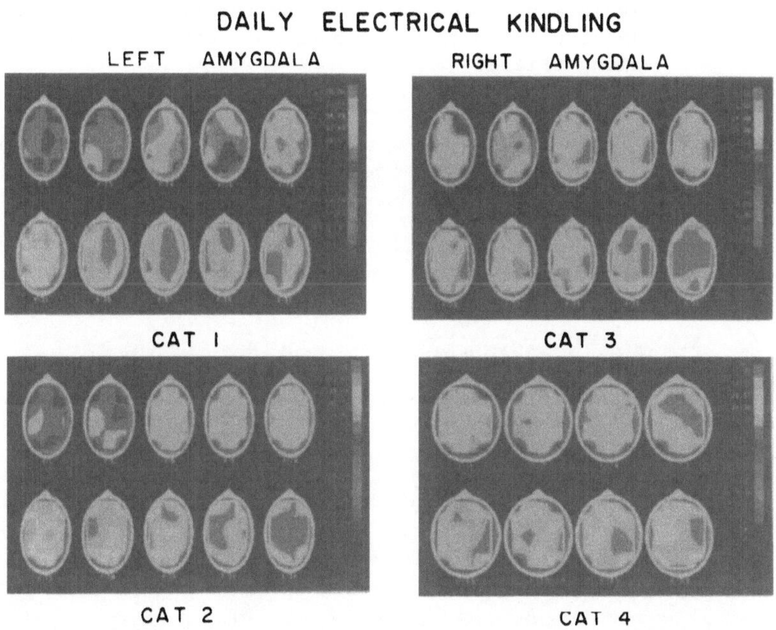


Fig. 4. Band-specific (8-16 Hz) frequency-domain maps of the evolution of the AM/AD cortical projection during the kindling process in the 4 cats. Notice the initial topographic specificity in the ipsilateral ectosylvian projection and in the contralateral frontal lobe and the subsequent widening of the spectral density areas. Also, notice that the propagation to the contralateral sylvian region occurs after the appearance of the contralateral prefrontal projection. This is evident in K2, K4 and K10 of cat 3; in K14-27 of cat 1; and in K10-32 of cat 2. In cat 4, the activation of the contralateral sylvian area is not evident until behavioral stage 5 is reached.

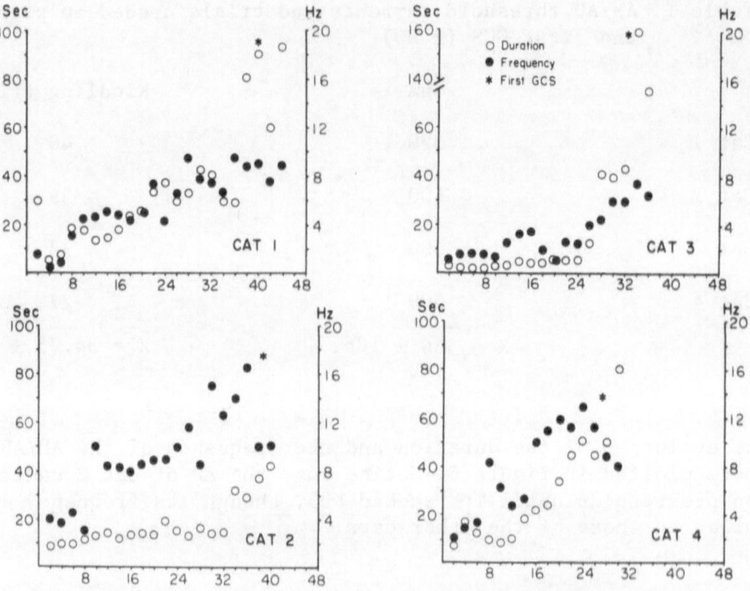


Fig. 5. Evolution of the duration and the frequency of the AM/AD of each animal. Abscissae represent the kindling trials.

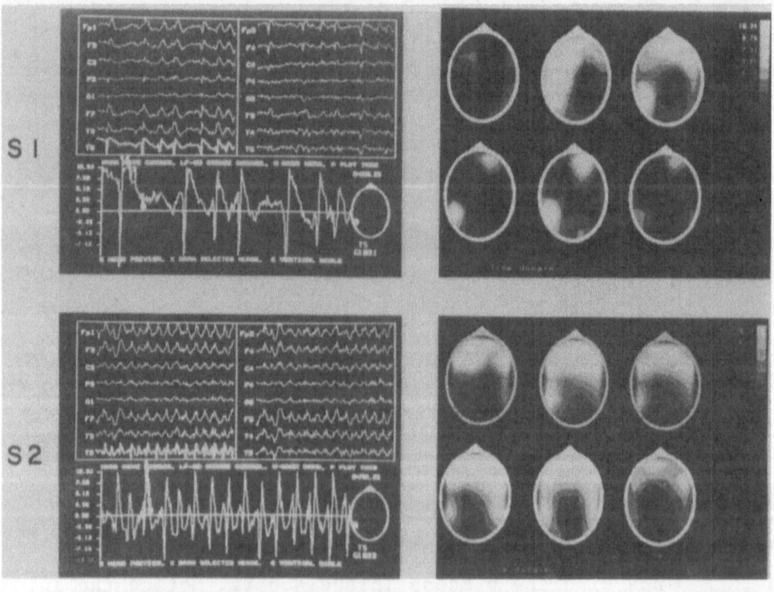


Fig. 6. Time evolution (historical layout) of the cortical projection of a single AM/AD spike of cat 1 during K2. It shows the raw data and the selection of the spike to be analyzed (magnified). Each map corresponds, from upper left to lower right, to a latency point selected within the AD by a cursor for sweeps 1 (S1) and 2 (S2). Notice the punctual localization of the spike in S1 (at the beginning of the AD) and the subsequent frontalization and the late invasion of the contralateral temporal lobe in S2.

Visual inspection of the 16 channel-EEG recordings revealed the presence of high frequency activity intermingled with the low frequency spiking. This activity was localized in the middle and posterior regions of the ectosylvian and suprasylvian gyri and it was simultaneously present in the forebrain (Fig. 2).

The frequency-domain EEG analysis during the AM/AD revealed a lack of topographic specificity of the low frequencies (1-8 Hz) which correspond to the AM spiking. But when higher frequencies were tested (8-16 Hz) a well defined topographic localization was evident (Fig. 4).

The band-specific frequency-domain maps show that the high-frequency component (8-16 Hz) of the AM/AD is first evident in a restricted area of the ipsilateral ectosylvian gyrus. As the kindling progressed high spectral density areas became prominent; i.e. its relative percentage increased and its topographic projection widens covering, at the beginning, the contralateral prefrontal areas and then propagates to the contralateral frontorolandic region (areas 4 and 6). Simultaneously, a contralateral ectosylvian focus appears. At this moment (behavioral stages 4 and 5), ipsilateral prefrontal and frontal activation is evident (see Fig. 4).

In the 4 animals, circling contralateral to the stimulated amygdala and mioclonias of the 4 limbs coincided with the propagation of high frequency activity from the prefrontal to the frontorolandic contralateral regions.

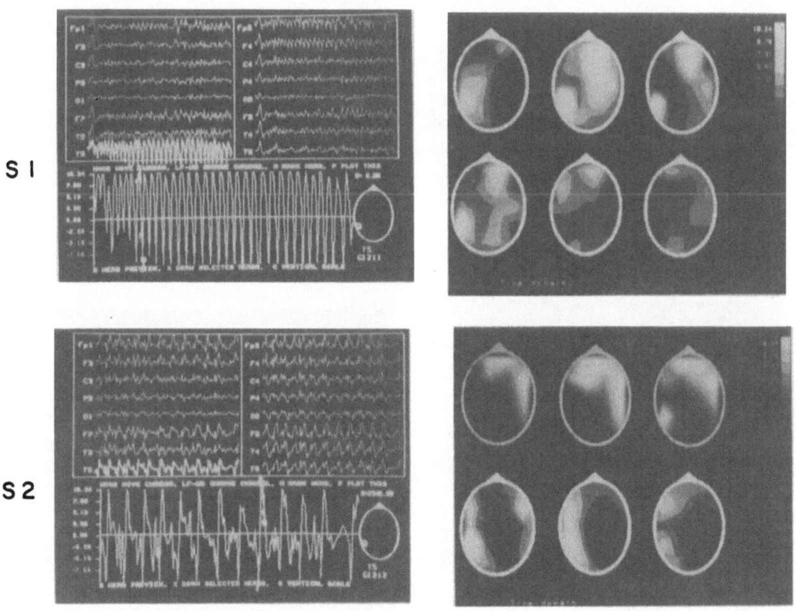


Fig. 7. Historical layout of the cortical projection of a single AM/AD spike of cat 1 during K21. Notice the frequency enhancement of the AM/AD in T5. The historical layout of the selected spike in S1 shows the same punctual projection in the contralateral frontal lobe with a higher density and covering a wider area plus a contralateral sylvian focus. In S2, a projection to both frontal lobes and an enhancement of the contralateral sylvian area are observed.

The time evolution of the cortical projection of single AM/AD spikes (historical layout) was computed for different latency points. This analysis shows that the cortical projections differ depending on the latency of the chosen spike within the AM/AD as well as on the kindling trial from which it is obtained. When the spike corresponded to the beginning of the AM/AD during the first days of kindling, the maps show a punctual focalization of the positive component of the spike in the posterior ipsilateral ectosylvian gyrus and in the contralateral prefrontal area. But when the spike corresponded to the second acquisition (17-22 s after the stimulus), the propagation of the positivity to the frontal areas is bilateral and a secondary activation focus is projected to the contralateral ectosylvian gyrus (Fig. 7).

If this analysis is carried out at the end of the kindling process (behavioral stage 5) it shows the same contralateral prefrontal activation plus a secondary focus in the contralateral ectosylvian gyrus. At this moment (see Fig. 8) this secondary focus coincides with the apparition of a frontal activation ipsilateral to the stimulated amygdala (for comparison see figures 6, 7 and 8). When the spike corresponded to the last acquisitions (K21 and K39), a noticeable activation of the occipital region could be observed.

The prefrontal contralateral propagation of the AM/AD of both, the primary and secondary AM focus, proved to be a robust phenomenum. Indeed, it was present in the 4 animals during different moments of the kindling process evolution. This is documented in Fig. 3-II which shows frequency-domain

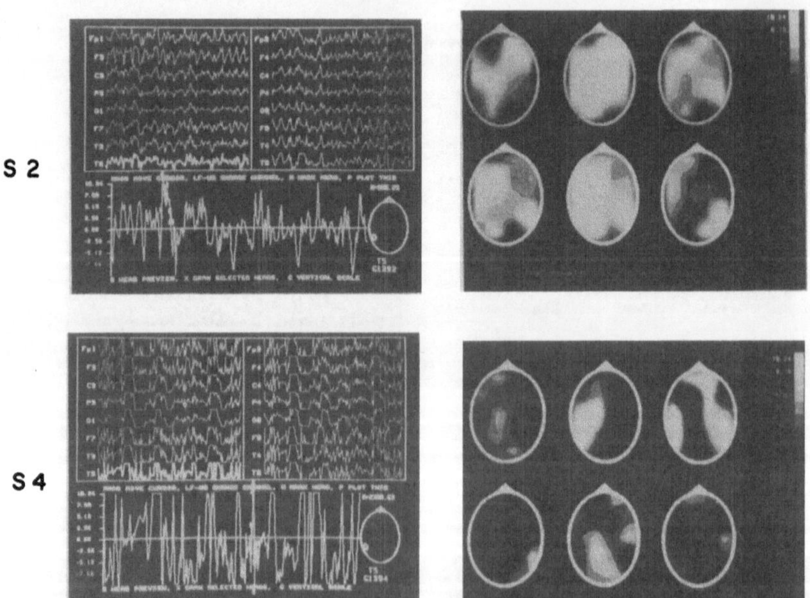


Fig. 8. Historical layout of the cortical projection of a single AM/AD spike of cat 1 during K31. During the behavioral stage 5 (the day before the first GCS), the contralateral frontal projection of the selected spike in S2 is evident at the beginning of the analysis, but after 32 ms (third map in S2 and S4), a contralateral ectosylvian focus appeared propagating to the left frontorolandic area. Notice that the frontorolandic area, ipsilateral to the stimulated amygdala, is activated by the contralateral temporal lobe.

average EEG maps. The standard deviation analysis revealed that the topographic distribution variance of the selected band was negligible.

## DISCUSSION

The cortical propagation of the amygdaline afterdischarge in rabbits, cats and monkeys has been subject of several studies in intact[1,2] and in split brain preparations[3,4,5,8].

The involvement of the frontal and frontorolandic areas was described by Wada et al.[4,5] both in Papio papio and Rhesus monkeys, nevertheless they do not make clear descriptions of cortical lateralization.

These initial works emphasize the final stages of kindling. They all consider that the propagation of the AM/AD to distant structures, mainly the subcortical ones, is a tardive phenomenum. There are few evidences regarding the propagation of the AM/AD during the first stages (1, 2 and 3) of kindling. The frequency-domain cortical mapping allowed us to detect very early frontal cortical changes. The most striking fact of our work is that this prefrontal focus was always contralateral to the stimulated amygdala. This frontal activation always precedes the cortical activation of the contralateral temporal lobe. The prefrontal ipsilateral activation was always present when the secondary temporal focus strengthened. This fact explains the changes in the direction of the head turning, the circling, and the forelimb mioclonic movements that suddenly appear during stages 4 and 5 in some animals.

These orderly specific changes are not likely to support the idea of the seizure generalization of reticular formation origin that acts as a diffuse system since our work prove that even in the last stages of kindling no simetrical cortical activation is observed.

The analysis of the historical layout reveals that the AD cortical projection is not uniform throughout the AM/AD itself, it is first punctual contralateral and then propagates bilaterally to the frontal regions during the last part of the afterdischarge and as the process continues. This phenomenum accentuates and the projection to occipital areas appears during the last behavioral stages.

In regard to the secondary focus in the contralateral ectosylvian gyrus, it is noteworthy to say that the spike propagation, 3-5 s after the stimulus, is not evident in this region. A few seconds later, within the same AD, spikes of the same amplitude clearly project their positivity on this area. In all the cases, this phenomenum follows the initial frontalization.

The AM/AD produced by the electrical kindling is a long-lasting artificially induced phenomenum. In the case of temporal lobe human epilepsy the AM/ADs are very short[9] and are more similar to the images of the first part of the AM/AD in the cat, thus they fail to propagate with the same intensity and they rarely produce generalized seizures. Nevertheless, many of the components of the psychomotor seizures are clearly of frontal lobe origin[10]. Thus, in these patients it is likely that the punctual contralateral frontal projections can also occur.

There are anatomical evidences that support the physiological findings regarding the amygdaloid frontal projection. Two pathways have been proposed. One is a projection from the basolateral amygdala to the medio dorsal thalamic nucelus, which projects to the medial prefrontal areas and less densely to the agranular insular area[11] in the rat. These authors and previously Llamas et al.[12] described, using the HRP labeling technique, a second pathway that comprises direct monosynaptic projections from the amygdala to the entire frontal cortex in the cat.

REFERENCES


1. Tanaka, A., 1972, Progressive changes of behavioral and electroencephalographic responses to daily amygdaloid stimulations in rabbit. Fukuoka Acta Med., 63: 152-164.
2. Wada, J. A., and Sato, M., 1974, Generalized convulsive seizures induced by daily electrical stimulation of the amygdala in the cat, Neurology, 24: 565-574.
3. Wada, J. A., and Sato, M., 1975, The generalized convulsive seizure state induced by daily electrical stimulation of the amygdala in split brain cats. Epilepsia, 16: 417-430.
4. Wada, J. A., Mizoguchi, T., and Komai, S., 1981, Cortical motor activation in amygdaloid kindling: observations in nonepileptic rhesus monkeys with anterior two-thirds callosal bisection, in: Kindling 2, J. A. Wada, ed., Raven Press, New York.
5. Wada, J. A., and Mizoguchi, T., 1984, Limbic kindling in the forebrain bisected photosensitive baboon, Papio papio, Epilepsia, 25: 278-287.
6. Racine, R., Tuff, L., and Zaide, J., 1976, Unit discharge patterns and neural plasticity, in: Kindling 1, J. A. Wada, ed., Raven Press, New York.
7. Goddard, G. V., and Maru, E., 1986, Forces for and against the kindling state as revealed by EEG and field potential analysis in the hippocampal dentate area of perforant path kindled rats, in: Kindling 3, J. A. Wada, ed., Raven Press, New York.
8. McCaughran, J. A., Corcoran, M. E., and Wada, J. A., 1977, Facilitation of secondary site amygdaloid kindling following bisection of the corpus callosum and hippocampal commissure in rats. Exp. Neurol. 57: 132-141.
9. Fernández-Guardiola, A., 1977, Reminiscences elicited by electrical stimulation of the temporal lobe in humans, in: Neurobiology of Sleep and Memory, J. R. McCough and R. R. Drucker-Colín, eds., Academic Press, New York.
10. Bancaud, J., Talairach, J., Morel, P., Bresson, M., Bonis, A., Geier, S., Hemon, E., and Buser, P., 1974, Generalized epileptic seizures elicited by electrical stimulation of the frontal lobe in man, Electroenceph. clin. Neurophysiol., 37: 275-282.
11. Sarter, M., and Markowitsch, H. J., 1984, Collateral innervation of the medial and lateral prefrontal cortex by amygdaloid, thalamic, and brain-stem neurons, J. Comp. Neurol., 224: 445-460.
12. Llamas, A., Avendaño, C., and Reinoso-Suárez, F., 1977, Amygdaloid projections to prefrontal and motor cortex. Science, 195: 794-796.

Discussion of Dr. Fernandez-Guardiola's Presentation

DR. WADA: Do you have any idea why it has to be contralateral frontal instead of ipsilateral frontal?

FERNANDEZ-GUARDIOLA: Well, all the ideas I have come from your work on the split brain. We are examining the diameter of the fibres of the anterior commissure from the amygdala to see if the reason is anatomical and the biggest fibers are conducting more rapidly to the contralateral lobe. That would be an explanation, but I have no idea why not ipsilateral.

DR. WADA: But the anterior commissure does not really directly connect the amygdalae. Then, whatever frequency band activity has to go to the opposite temporal lobe and then travel back to the frontal lobe, which seems to me an extraordinary route to take. If there is a direct connection, as you mentioned in the Science paper of some years ago, of ipsilateral amygdala projection to frontal, then why not ipsilateral frontal?

DR. FERNANDEZ-GUARDIOLA: As a matter of fact this is a fact. I cannot speculate but it is so constant that I think we are wrong if we think that in kindling there is activation of one amygdala, then the contralateral amygdala, then the propagation. This is not true. From the first day of kindling there is amygdala frontal. We are not kindling the amygdala. We are kindling a system, a limbic cortical system. From the first stimulation, this is my idea now.

DR. WADA: Does the EEG signature of seizure activity immediately connect up with clinical seizure development? In this particular frequency component, I wonder if what one is looking at is not necessarily a process representing seizure facilitation but rather suppression? Could it be a compensatory response to whatever may be taking place in the ipsilateral frontal? This is entirely speculative, but what do you think?

DR. FERNANDEZ-GUARDIOLA: You know in the cat the pyramidal system is not like in humans; cats have a strong direct pyramidal pathway, not only cross... In man it is different, and then I think the change in circling. When you are kindling a cat, I don't know about a rat, but the cat is circling in one sense and then suddenly it begins to circle in the other direction, and in that moment it is stage 4 and 5. You can see with this program the change in the participation of contralateral temporal lobe with the ipsilateral frontal. Ipsilateral will be stimulated, the contralateral in that moment to the maximal density of the chosen frequency. I think that the novelty of this program is that we don't use alpha, beta, delta classification - looking at all the frequencies because in that moment you see the slow frequencies. The beautiful thing about this program is to choose the frequencies and to analyze, to have time, because everything is kept in a hard-disc, or floppy disc, and you have time to analyze day by day all the frequencies.

DR. BERMAN: I am interested in the source of the interictal spiking, and I wonder if your data always show the interictal spikes being generated in the seizure focus, and whether they appear in other areas? Could you comment on how they spread.

DR. FERNANDEZ-GUARDIOLA: Well, our results confirm indirectly the idea that interictal spikes are different from the spike in the after-discharge. We don't have many studies, because this is a very recent subject of study. We have seen about a dozen interictal spikes always going differently, they are more occipital than the spikes of the after-discharge, and maybe they inhibit instead of activate. But they are different in the topographical projections.

DR. BERMAN: Were they always generated, as far as you could tell, at the focus?

DR. FERNANDEZ-GUARDIOLA. No, they are randomly generated in any place, not even in the temporal lobe, they would not necessarily appear. But they are more frequently occipital in localization.

DR. ADAMEC: I was wondering if you have tried or have the capability of computing coherence functions between your electrodes because I'm a little concerned about using the term propagation from the way you analyze the data. I think if you could compute a coherence function and also look at the phase diagrams, you could determine whether or not you are getting actual propagation from the amygdala discharge to the contralateral. You could presumably also estimate, or potentially estimate, the synaptic delay involved. I mean, you're concentrating particular frequency bands, but you could calculate the coherence function over a wide range of frequency bands, and determine whether or not you are really getting a contralateral projection, or an ipsilateral and a contralateral within different frequency domains.

DR. FERNANDEZ-GUARDIOLA: You want to answer that, Rodrigo?

DR. FERNANDEZ-MAS: Actually, we are interested in the spatio-temporal distribution of the after-discharge. If we compute the coherence we kill that evolution in the time; we cannot establish relationships between the coherence in this time and the coherence in the time of those 2 seconds, for example. All the reason for this program is to analyze the evolution in the time and in the space of the after-discharge. We think that mathematically the after-discharge can be considered as a function of the time, but this has inherent integration in the space.

DR. ADAMEC: Yes, you lose some time resolution, certainly, but you could compute coherence function with fixed epochs of let's say, a second, and then you could do second by second, if you had the computing power, changes in the coherence functions. It still might be worth doing simply to say whether or not the frequency bands you're looking at do in fact represent a propagation from the amygdala, and you can certainly use the phase diagrams to determine the phase relationships, to determine whether or not you're getting propagation of a complex wave from the amygdala to another cortical location. I'm just thinking, in terms of using the term propagation to the contralateral, particularly given the anatomy, it would be more comfortable in accepting the notion that propagation if you could have some coherence verification if that's what was going on.

# N-METHYL-D-ASPARTATE (NMDA) RECEPTORS AND THE KINDLING MODEL

James O. McNamara[1,2,3], Douglas W. Bonhaus[1,3],
J. Victor Nadler[2], and Geng-Chang Yeh[2]

Departments of [1]Medicine and [2]Pharmacology
Duke University Medical Center and
[3]Veterans Administration Medical Center
Durham, North Carolina

## INTRODUCTION

This chapter will consider recent pharmacologic, electrophysiologic, and biochemical studies examining the role of a subtype of excitatory amino receptor, the NMDA receptor, in the development of kindling and expression of kindled seizures.

## RATIONALE FOR STUDY OF NMDA RECEPTORS IN THE KINDLING MODEL

The rationale for study of NMDA receptors in the expression of kindled seizures is straightforward. An alteration in synaptic function is one way in which brain excitability could be regulated. An enhancement of excitatory and/or an attenuation of inhibitory synaptic function could account for the abnormal hyperexcitability of the kindled brain. Excitatory amino acids (EAA) have emerged as the principal excitatory neurotransmitter in the brain in the past several years. Among the receptors with which a synaptically released EAA might interact to effect a biologic response, the NMDA subtype is a particularly attractive candidate for regulation in kindling for two reasons. 1) The distinctive voltage-dependence of NMDA receptor-mediated synaptic transmission is associated with regenerative properties, which results in burst firing similar to that observed in epileptiform discharge (1, 2, 3, 4). ,2) NMDA receptor antagonists exhibit anticonvulsant properties in several seizure models (5, 6).

The rationale for study of NMDA receptors in the development of kindling is also straightforward. The NMDA subtype of EAA receptor has been linked to the development of plastic processes including synaptogenesis (7), long term potentiation (8) and a form of learning and memory (9). Since kindling is a robust form of neuronal plasticity, it seems reasonable to postulate a role for NMDA receptors in the development of kindling.

## PHARMACOLOGIC STUDIES

### Development of Kindling

These studies have used competitive antagonists of the NMDA receptor itself such as amino phosphonovaleric acid (AP5) and amino phosphonoheptanoic acid (AP7). Such

*Kindling 4*
Edited by J. A. Wada
Plenum Press, New York, 1990

studies have also used uncompetitive antagonists of NMDA responses such as phencyclidine, the phencyclidine analog N-(1-[thienyl]cyclohexyl)-piperidine (TCP), ketamine, and MK-801; these compounds appear to effect the uncompetitive blockade of NMDA responses by gaining access to the open state of the NMDA channel, binding to a site within the channel, and thereby blocking the flow of current.

We examined the effect of systemic administration of the uncompetitive antagonist, MK-801, on the rate of kindling development. This treatment inhibited the progression of behavioral seizures elicited by amygdaloid stimulation in a dose dependent manner (Fig. 1). Little or no progression of rank of behavioral seizure occurred in the presence of 0.33 mg/kg of MK-801, while class 1 and 2 seizures eventually appeared during treatment with 0.10 mg/kg (10). The suppressant effects of MK-801 could be due to an inhibition of the processes underlying the development of kindling ("anti-epileptogenic") or merely due to inhibition of the seizures without otherwise affecting the development of kindling per se ("anti-convulsant"). These possibilities were differentiated by permitting 48 hr to elapse after the thirteenth stimulation to permit clearance of the MK-801; subsequent stimulations were administered in the absence of the drug. Behavioral seizures remained significantly suppressed in response to the first afterdischarge elicited in the absence of treatment in the 0.33 mg/kg group, indicating the presence of anti-epileptogenic, not simply anti-convulsant, effects. These inhibitory effects of MK-801 were also observed in measurements of the duration of afterdischarge (Fig. 2), although the magnitude of effects on behavioral seizure were more marked.

The above observations raise two important questions. One is whether the effects of MK-801 are due to inhibition of NMDA-evoked current or some other effect of the drug. The effects are almost certainly due to inhibition of NMDA-evoked current since similar effects have been described with other drugs which are structurally distinct yet also inhibit NMDA-evoked currents including phencyclidine and ketamine (11, 12, 13, 14). Moreover, competitive antagonists of the NMDA receptor itself inhibit kindling development including AP5 (15, 16), AP7 (17), and CPP (Sato et al., this volume).

Another key question is whether activation of NMDA receptors is an absolute prerequisite for kindling to develop. The report by Rondouin et al. (this volume) may shed some light on this issue. These investigators tested the effect of a high dose of the NMDA channel blocker, TCP, on kindling development. Lower doses of TCP (0.5 and 1.0 mg/kg respectively) produced effects equivalent to that described above with 0.1 and 0.33 mg/kg of MK-801. By contrast, the high dose of TCP (5 mg/kg) eliminated all vestiges of development of kindling by behavioral indices; the number of stimulations required to elicit the development of kindling after cessation of drug treatment was not significantly different from vehicle treated animals. However, a slight trend towards lengthening of afterdischarge duration was present after many stimulations. This could reflect an NMDA receptor independent mechanism which contributes a tiny component to the development of kindling. Alternatively, this could reflect the fact that the onset of the TCP blockade of the channel is presumably use-dependent; a small amount of current may flow through the channel prior to full establishment of the channel block. The data do not permit distinguishing these possibilities.

<u>Expression of Kindled Seizures</u>

The powerful anticonvulsant effects of NMDA antagonists in a diversity of seizure models led to the expectation that NMDA antagonists would exert equivalent effects in the kindling model. Surprisingly, the anticonvulsant effects have been much less robust than expected. A dose of MK-801 capable of significant inhibition of kindling development (0.1 mg/kg) was devoid of significant anticonvulsant effects against kindled seizures (10). A larger dose (0.33 mg/kg) only partially inhibited afterdischarge and forelimb clonus (38% and 66% respectively). Similar results have been obtained with both channel blockers, ketamine and phencyclidine (11, 12, 13, 14) and TCP (Rondouin et al. this volume) as well as competitive antagonists of the receptor administered systemically or by an intra-ventricular route (15; Sato et al. this volume).

Identification of more marked inhibitory effects of these drugs on kindling development than on kindled seizures has been a uniform finding in a number of laboratories. Some variability of results has nonetheless been present in studies of the effects of MK-801 on kindled seizures. Sato et al. (18) observed more robust anticonvulsant effects of MK-801 against amygdaloid kindled seizures compared to Gilbert (19). Although these discrepancies are relatively minor, the greater efficacy observed by Sato et al. is likely due to use of a 2 hour interval between drug administration and test stimulus (the optimal interval) and a test stimulus at (not exceeding) the generalized seizure threshold. We obtained effects intermediate between the other two laboratories (10). Taken together, it seems that MK-801 and other NMDA antagonists elevate the generalized seizure threshold but have little or no effects on the afterdischarge threshold. The efficacy of the anticonvulsant effects of NMDA antagonists in the kindling model is minimal in comparison to standard anticonvulsants such as phenytoin (20), carbamazepine, phenobarbital, etc. (21, 22).

Interpretation of Pharmacologic Studies

The data are consistent with the idea that activation of NMDA receptors is an absolute prerequisite for the development of kindling. The profound inhibitory effects of a diversity of structurally distinct compounds acting at different sites in this receptor channel complex support this conclusion. Even if not an absolute prerequisite, activation of NMDA receptors is at the very least an extraordinarily important determinant. Whether activation of NMDA receptors is sufficient to induce kindling development is more difficult to assess. NMDA receptor activation almost certainly occurs under physiologic conditions such as some learning and memory behaviors, yet kindling as defined here does not occur. Perhaps the synchronous activation of NMDA receptors as presumably occurs during an afterdischarge is sufficient to induce kindling.

The data are also consistent with the idea that activation of NMDA receptors may contribute to the expression of a kindled seizure, since NMDA antagonists elevate the generalized seizure threshold of a kindled rat. If NMDA receptor and channel blockers successfully inhibit receptor and channel activation during a kindled seizure, then activation of this receptor channel complex is not necessary for the initiation or propagation of a kindled seizure; this is true because the anticonvulsant effects of NMDA antagonists can be overcome by high intensity stimulations. The presence of some anticonvulsant effects of NMDA antagonists raise the possibility that enhanced NMDA receptor-mediated synaptic transmission intrinsic to a kindled brain may contribute to the enhanced seizure propagation characteristic of these animals. This possibility was directly tested in the electrophysiologic experiments described below.

## ELECTROPHYSIOLOGIC STUDIES

Mody and Heinemann (23) addressed this question with intracellular recordings from dentate granule cells activated by perforant path stimulation in hippocampal slices isolated from electrode implanted unstimulated control animals and kindled animals. These investigators identified an APV reversible component to the EPSP in the granule cells of the kindled but not the control animals. The amplitude and duration of the EPSP increased with depolarization of the granule cells in the kindled animals, a result consistent with relief of the voltage regulated block of the NMDA channel by extracellular magnesium. By contrast, the EPSP diminished in response to partial depolarization in the granule cells of the control animals. These effects were described in slices from animals kindled from different sites in the limbic system and sacrificed at intervals ranging from one day to more than a month after the last kindled seizure.

These findings are remarkable insofar as they document the presence of enhanced function of excitatory synapses in the kindled brain and identify the responsible receptor. Some of our recent studies have focussed on one question raised by these findings, namely what is the molecular mechanism(s) responsible for these electrophysiologic findings?

## BIOCHEMICAL STUDIES

In theory, increased NMDA receptor mediated neurotransmission could be due to an increase in the concentration of neurotransmitter in the synapse, an increase in the response of postsynaptic neurons to a given concentration of neurotransmitter, or both.

An increase in intrasynaptic concentrations of excitatory amino acid (EAA) could increase NMDA postsynaptic responses either by directly activating more NMDA receptors or indirectly by activation of non-NMDA receptors, thereby depolarizing the target neuron, and relieving the magnesium blockade of NMDA receptors. Increases in intrasynaptic concentration of EAA could conceivably result from: 1) decreased uptake of EAA; 2) increased release of EAA; 3) shrinkage of the intrasynaptic volume; or 4) reduced diffusion of EAA from the synapse.

An increase in the response of postsynaptic neurons to a given concentration of neurotransmitter could be the result of a redistribution of NMDA receptors from extrasynaptic to synaptic sites, to any of a constellation of alterations intrinsic to the NMDA receptor channel complex, etc.

Our initial approach to address these possibilities was to determine whether kindling was associated with an increased sensitivity of NMDA-receptive neurons to given concentrations of NMDA. One approach to this question is to examine a molecular event controlled by NMDA receptors. In the hippocampus of adult rats, NMDA receptor agonists have been demonstrated to inhibit phosphoinositide (PI) hydrolysis stimulated by several agonists including muscarinic cholinergic agonists such as carbachol (24). We sought to understand the mechanism by which NMDA effects this inhibition and to determine whether the sensitivity to NMDA was increased in hippocampal slices of kindled rats.

To elucidate the mechanism by which NMDA inhibits PI hydrolysis, we measured PI hydrolysis in transverse slices of rat hippocampus by preincubation with [$^3$H] inositol and subsequent isolation of inositol phosphates (IP) by anion exchange chromatography (25). We also performed correlative electrophysiologic measurements of responses of populations of CA1 neurons evoked by stimulation of Schaffer-collateral afferents.

Our data support the conclusion that NMDA receptor activation inhibits PI hydrolysis by depolarization mediated by sodium flux through NMDA channels. We initially found that NMDA inhibits carbachol stimulated PI hydrolysis in a dose and time dependent manner. The maximal inhibition was 80% and the approximate $IC_{50}$ of NMDA was 25 uM. NMDA exerts this effect indirectly through channel activation because two different channel blockers, MK-801 and TCP, prevented this action. Prevention of the NMDA effect by removal of sodium, but not calcium, from the incubation buffer suggested that depolarization may be the responsible mechanism, because sodium carries the majority of current through the NMDA channel (26). Depolarization alone appears sufficient to inhibit cholinergic activation of PI hydrolysis, since both veratridine and elevated potassium inhibited cholinergic stimulation of PI hydrolysis. The effect of NMDA appears to require sodium flux through NMDA channels, not through voltage dependent sodium channels, because tetrodotoxin did not inhibit the effect of NMDA. Our correlative electrophysiologic experiments disclosed that NMDA profoundly inhibited evoked EPSPs and population action potentials of CA1 neurons, an effect almost certainly due to depolarization. Together the data are consistent with the idea that NMDA receptor activation inhibits PI hydrolysis by depolarization mediated by sodium flux through

NMDA channels. We suspect that the excitatory effects of NMDA reflected in the depolarization induce a resistance of these neurons to a heterologous excitatory (cholinergic) input. Such resistance may reflect a protective mechanism aimed at maintaining cell viability at the expense of cell responsiveness. Importantly, NMDA receptor control of this second messenger system provides a measure of the sensitivity of NMDA receptive neurons to NMDA.

To determine whether kindling modified the sensitivity of hippocampal neurons to NMDA, we measured the effects of multiple concentrations of NMDA on carbachol stimulated hydrolysis in hippocampal slices isolated from control animals and animals sacrificed 24 hours after the last kindled seizure (27). The response to NMDA (10 uM) was increased from 23 +/- 4% in control to 54 +/- 4% in slices from kindled rats (mean $\pm$ S.E.M., $p < 0.025$, Wilcoxon signed rank test). This increase in sensitivity to NMDA resulted in steepening of the dose-response relationship in slices from kindled rats (Fig. 3). To ascertain whether this enhanced sensitivity to NMDA was long-lasting, the inhibitory effects of NMDA (10 uM) were measured in slices isolated from rats sacrificed 28-35 days after the last class 5 kindled seizure. Persistence of the enhanced sensitivity to NMDA was detected in slices isolated from these animals (kindled 46 +/- 4%; control 20 +/- 2%; Wilcoxon signed rank test; $p< 0.05$). These were the only effects of kindling observed. Neither basal nor carbachol-stimulated PI hydrolysis was significantly changed. As in slices from control rats, NMDA by itself had only minimal effects on PI hydrolysis in slices from kindled rats.

To determine whether the effects of NMDA were selective, we examined two other compounds which inhibit carbachol-stimulated PI hydrolysis, kainic acid and phorbol-12,13-diacetate (PDA). The concentrations of these two compounds were selected from the midpoint of the concentration response curves. Kindling did not significantly alter the response to either KA or PDA, indicating that the effects of NMDA were selective.

This increased sensitivity to NMDA is consistent with a change intrinsic to the NMDA receptor channel complex or with a change in part of the molecular machinery coupling the NMDA receptor channel to PI hydrolysis. Since KA, PDA, and NMDA all inhibit carbachol stimulated PI hydrolysis but only the response to NMDA was increased in kindling, this suggests that an alteration intrinsic to the NMDA receptor channel complex is responsible.

To determine whether an increase in affinity or number of NMDA receptor recognition sites were present in the hippocampus of kindled animals, quantitative radiohistochemical measurements of $[^3H]$ glutamate binding were performed under conditions selective for NMDA receptors (28). These experiments were conducted with slide-mounted sections obtained from kindled and control animals. To our surprise, no increase in NMDA receptor binding was detected over any of the three principal neuronal populations of hippocampus. In fact, a small (7-11%) but statistically significant reduction was found in stratum radiatum of hippocampal area CA1. Therefore, an increase in the affinity or number of the agonist form of NMDA receptor recognition sites themselves cannot account for the electrophysiologic and biochemical studies indicative of enhanced NMDA receptor activation and sensitivity.

## UNANSWERED QUESTIONS

Both together and individually, the pharmacologic, electrophysiologic, and biochemical findings raise a number of intriguing questions. With respect to the electrophysiology, how widespread is this change in synaptic physiology within the kindled brain? What accounts for the apparent qualitative difference between control and kindled animals with respect to NMDA receptor participation in synaptic transmission? Indeed a quantitative difference would intuitively seem more plausible. A recent report (29) claimed that NMDA receptors contribute to the EPSP at perforant path synapses on dentate granule cells from normal rats; whether this finding was due to inclusion of the quisqualate/kainate receptor antagonist (CNQX) in the bath is uncertain. The CNQX could inhibit tonic activation of basket cells by dentate granule cells and thereby lead to a disinhibition and permit detection of an NMDA component of the EPSP in normals.

With respect to the biochemical findings, the molecular explanation of the enhanced sensitivity to NMDA remains unknown. Regulation of other components of the NMDA receptor complex such as the glycine receptor or regulatory sites for zinc or magnesium seems plausible. If the NMDA receptor is a multiple subunit receptor like the GABA A or nicotinic cholinergic receptor, variation in the subunit composition of the receptor could have profound implications on the agonist potency and/or efficacy (30, 31, 32) and yet disclose no change in the radiohistochemical measurements performed to date. A third possibility is that a post-translational modification of the receptor could lead to enhanced function of the receptor and yet appear unchanged in our measurements thus far. Apart from the changes intrinsic to NMDA receptive neurons, potential alterations at other sites in these synapses such as transmitter release (33), reuptake, etc. warrant further investigation.

Finally, the issue arises as to how the changes in synaptic physiology and biochemistry described in the kindled brain develop and are maintained. One intriguing possibility is that NMDA receptor activation occurring during the stimulus-evoked afterdischarges with kindling development somehow leads to an enhanced function of NMDA synapses which then maintains the abnormal hyperexcitability manifest as kindled seizures.

## ACKNOWLEDGMENTS

This work was supported by N.I.H. grants NS24448, NS17771, NS27311, and by two Merit Review Grants awarded by the Veterans Administration. Appreciation is extended to Ms. Rena Wethington for her assistance in the preparation of this manuscript.

## REFERENCES

1. Nowak, L., Bregestovski, P., Ascher, P., Herbert, A., and Prochiantz, A., 1984, Magnesium gates glutamate-activated channels in mouse central neurons, Nature, 307:462.
2. Mayer, M.L., Westbrook, G.L., and Gutherie, P.B., 1984, Voltage-dependent block by magnesium of NMDA responses in spinal cord neurons. Nature, 309:261.
3. Herron, C.E., Lester, R.A., Coan, E.J., and Collingridge, J.L., 1985, Intracellular demonstration of an N-methyl-D-aspartate receptor mediated component of synaptic transmission in the rat hippocampus, Neurosci.Lett. 60:19.
4. Hynes, M.A. and Dingledine, R., 1984, Attenuation of epileptiform burst firing in the rat hippocampal slice by antagonists of N-methyl-D-aspartate receptors, Soc.Neurosci. Abst. 10:229.
5. Croucher, M.J., Collins, J.F., and Meldrum, B.S., 1982, Anticonvulsant action of excitatory amino acid antagonists, Science 216:899.
6. Czuczwar, S.J. and Meldrum, B.S., 1982, Protection against chemically induced seizures by 2-amino-7-phosphonoheptanoic acid, Eur. J. Pharmacol. 83:335.
7. Cline, H.T., Debski, E.A., and Constantine-Paton, M., 1987, N-methyl-D-aspartate receptor antagonist desegregates eye-specific stripes, Proc.Natl.Acad.Sci. U.S.A. 84:4342.
8. Harris, E.W., Ganong, A.H., and Cotman, C.W., 1984, Long-term potentiation in the hippocampus involves activation of N-methyl-D-aspartate receptors. Brain Res. 323:132.
9. Morris, R.G.M., Anderson, E., Lynch. G.S., and Baudry, M., 1986, Selective impairment of learning and blockage of long-term potentiation by an N-methyl-D-aspartate receptor antagonist, AP5, Nature 319:774.
10. McNamara, J.O., Russell, R.D., Rigsbee, L., and Bonhaus, D.W., 1988, Anticonvulsant

and antiepileptogenic actions of MK-801 in the kindling and electroshock models, Neuropharmacology 27:563.
11. Callaghan, D.A. and Schwark, W.S., 1980, Pharmacological modification of amygdaloid-kindled seizures, Neuropharmacology 19:1131.
12. Bowyer, J.F., and Winters, E.D., 1981, The effects of various anesthetics on amygdaloid kindled seizures, Neuropharmacology 20:199.
13. Bowyer, J.F., 1982, Phencyclidine inhibition of the rate of kindling development. Exp. Neurol. 75:173.
14. Bowyer, J.F., Albertson, T.E., Winters, W.D. and Baselt, R.C., 1983, Ketamine-induced changes in kindled amygdaloid seizures, Neuropharmacology 22:887.
15. Holmes, K.H. and Goddard, G.B., 1986, A role for the N-methyl-D-aspartate receptor in kindling, Proc. Univ.Otago Med. Sch. 64:37.
16. Cain, D.P., Desborough, K.A., McKitrick, D.J., 1988, Retardation of amygdaloid kindling by antagonism of NMD-aspartate and muscarinic cholinergic receptors: evidence for the summation of excitatory mechanisms in kindling, Exp. Neurol. 100:203.
17. Vezzanni, A., Wu, H.Q., Moneta, E., and Samanin, R., 1988, Role of the N-methyl-D-aspartate-type receptors in the development and maintenance of hippocampal kindling in rats, Neurosci.Lett. 87:63.
18. Sato, K., Morimoto, K., and Okamoto, M., 1988, Anticonvulsant action of a non-competitive antagonist of NMDA receptors (MK-801) in the kindling model of epilepsy, Brain Res. 463:12.
19. Gilbert, M.E., 1988, The NMDA-receptor antagonist, MK-801, suppresses limbic kindling and kindled seizures, Brain Res. 463:90.
20. McNamara, J.O., Shin, C., Butler, L., and Rigsbee, L.C., 1989, Intravenous phenytoin is an effective anticonvulsant in the kindling model, Annals of Neurol., in press.
21. Albright, P.S., and Burnham, W.M., 1980, Development of a new pharmacological seizure model: effects of anticonvulsants on cortical- and amygdala-kindled seizures in the rat, Epilepsia 21:681.
22. Albertson, T.E., Peterson, S.L., Stark, L.G., 1980, Anticonvulsant drugs and their antagonism of kindled amygdaloid seizures in rats, Neuropharmacology 19:643.
23. Mody, I. and Heinemann, U., 1987, NMDA receptors of dentate gyrus granule cells participate in synaptic transmission following kindling, Nature 326:701.
24. Baudry, M., Evans, M., Lynch, G., 1986, Excitatory amino acids inhibit stimulation of phosphoinositide metabolism by aminergic agonists in hippocampus, Nature 319:329.
25. Morrisett, R.A., Chow, C.C., Sakaguchi, T., Shin, C., and McNamara, J.O., 1989, Inhibition of muscarinic-coupled phosphoinositide hydrolysis by N-methyl-D-aspartate is dependent upon depolarization via channel activation. J. Neurochem., in press.
26. Mayer, M.L., and Westbrook, G.L., 1987, The physiology of excitatory amino acids in the vertebrate central nervous system, Prof.Neurobiol. 28:197.
27. Morrisett, R.A., Chow, C., Nadler, J.V., and McNamara, J.O., 1989, Biochemical evidence for enhanced sensitivity to N-methyl-D-aspartate in the hippocampal formation of kindled rats, Brain Res., in press.
28. Okazaki, M.M., McNamara, J.O., and Nadler, J.V., 1989, N-methyl-D-aspartate receptor autoradiography in rat brain after angular bundle kindling, Brain Res. 482:359.
29. Lambert, J.D.C., and Jones, R.S.G., 1989, Activation of N-methyl-D-aspartate receptors contributes to the EPSP at perforant path synapses in the rat dentate gyrus in vitro, Neurosci. Lett. 97:323.
30. Sakmann, B., Methfessel, C., Mishina, M., Takahashi, T., Takai, T., Kurasaki, M., Fukuda, K., Numa, S., 1985, Role of acetylcholine receptor subunits in fazing of the channel, Nature 318:538.
31. Mishina, M., Takai, T., Imoto, K., Takahashi, T., Numa, S., Methsessel, C. and Sakmann, B., 1986, Molecular distinction between fetal and adult form of muscle acetylcholine receptor, Nature 321:406.
32. Levitan, E.S., Schofield, P.R., Burt, D.R., Rhee, L.M., Wisden, W., Kohler, M., Fujita, N., Rodriguez, H.F., Stephenson, A., Darlison, M.D., Barnard, E.A., and Seebug, P.H.

1988, Structural and functional basis of $GABA_A$ receptor heretogeneity, Nature 335:76.

33. Geula, C., Harvie, P.A., Logan, T.C. and Slevin, J.R., 1988, Long-term enhancement of $K^+$-evoked release of L-glutamate in entorhinal kindled rats, Brain Res., 442:368.

Discussion of Dr. McNamara's Presentation

DR. ACKERMAN: From what you just said, Jim, if you did the following experiment, which I don't suggest you do, if you space your stimulations far enough apart so that it's within the time that these effects you're talking about occur, would you say that kindling would progress more rapidly in terms of stimulations to kindle than if you stimulate them once a day?

DR. MCNAMARA: I don't know. We obviously have not worked out the time course of this yet. But it is a possibility. I don't think that it is a direct prediction. Well, look, we kindled the animals and they were stimulated twice a day, 5 days a week. It wasn't rapid kindling, but it was twice a day 5 days a week; it's entirely possible that the absence of this effect at 24 hours could be due to adaptive responses of some sort. But alternatively, it may well be that the same net effect of that NMDA receptor function is present at 24 hours and there's simply a post-translational modification of that receptor that accounts for, or an alteration in subunits that are expressed there, different subunits, if indeed there are different subunits of this thing which we suspect there would be, can have profound implications on, for example, GABA-A receptor responses. I don't know enough about this yet, Bob, that if we use different intervals we would see that it might go faster. It's not clear to me how these data lead to that prediction. Maybe you can see something that I don't see.

DR. BURCHFIEL: Have you looked at earlier stages of kindling? When did these changes begin to occur?

DR. MCNAMARA: We don't know that yet.

DR. BURCHFIEL: Are you going to look at that?

DR. MCNAMARA: Yes.

DR. RACINE: That was a nice presentation, Jim. There are a couple of things that don't fit in very well. One thing is, if you kindle animals in the dentate gyrus and test them 24 hours after reaching the criterion, they all show kindled convulsion. If you wait 28 days, same animals, none of them will show kindled convulsion on the first test. We have done that several times now. Also, the fact is that 24 hours after completion of kindling your animals are all showing kindled convulsions and you're not seeing this in NMDA.

DR. MCNAMARA: That's exactly what I was trying to say. This alone - the increased density of glycine receptors - obviously cannot explain kindling. Alone, it can't, because it's not there at 24 hours, it is at 28 days. Now, I think that it may well contribute to some of the biochemical and electrophysiological changes identified at 28 days, but if there is a change intrinsic to that NMDA receptor channel complex at 24 hours, which I suspect there is, based upon their electrophysiological studies and our biochemical studies, it is not this, and there's got to be some different mechanism responsible for it.

DR. RACINE: But you did say that it was involved in the expression of kindled seizures.

DR. MCNAMARA: Their studies electrophysiologically and our studies biochemically are demonstrating enhanced - well, they're finding enhanced NMDA receptor-mediated synaptic transmission at multiple times after completion of kindling. We found enhanced NMDA receptor-mediated responses biochemically at 24 hours and at 28 days. So I believe those things are present at both points in time. And for that reason I do think that it's entirely possible that enhanced NMDA receptor-mediated responses may contribute, I think they do contribute to the expression of kindled seizures. I don't think they're the whole explanation, but I do think they're part of the explanation. And in turn, I think that this increase in glycine receptor number may be part of the explanation of the enhanced NMDA receptor function. I don't think it's the whole explanation.

DR. RACINE: My other point just came to me. Both Graham Goddard's lab and our lab have shown that if you kindle in the dentate gyrus the granule cells are actually less excitable rather than more excitable. Perhaps it's NMDA changes that act as a compensatory response to the decrease in excitability.

DR. MCNAMARA: The thing is that we didn't kindle in the dentate gyrus; we kindled in the amygdala, and if you look at it either electrophysiologically or biochemically in slices, there's an enhanced function, and so it may be a function of the site from which you kindle.

DR. MCINTYRE: Jim, these were amygdala kindled animals, and you're looking at hippocampus, right?

DR. MCNAMARA: Right, for the glycine binding and for the NMDA stimulated inhibition. The NMDA receptor binding itself was done in angular bundle kindled animals.

DR. MCINTYRE: OK. I'm wondering, one, if you'd expect to see this somewhere else, perhaps in the pyriform cortex? The second thing is, very often when we've done amygdala kindling we find there is quite a lot of negative influence between the amygdala and the hippocampus, and indeed maybe these things are really in hippocampus. What you're looking at is kind of an anti-epileptic effect rather than an epileptic effect.

DR. MCNAMARA: Well, all I can say is that Istvan found increased excitatory synaptic transmission there in the kindled animals. We don't know yet, and we're obviously going to look at multiple brain regions, including the pyriform cortex. My suspicion is that we'll find the same thing in multiple different places, but we used this as a model to study it first and once we've got it we're going to go and figure out its distribution. I don't know the answer to your question; I don't know if we kindled in the dentate gyrus if we'd see it. I don't like to put an electrode in the same place that you do the biochemistry because now I've got a confounding variable.

DR. BUTERBAUGH: I concur with what you said at the end, and I didn't say too much about it yesterday, because I knew you could explain the whole thing better than I could. But I just wanted to point out again, what Dan said, that most of that has been shown on the hippocampus, and the big difference in the sensitivity to the seizure induced damage we found is in the temporal lobes.

DR. MCNAMARA: Well, we will look there. Rest assured we will look there.

DR. BURNHAM: Jim, thank you for some very interesting biochemical data. I just wanted to comment on an interesting parallel between what you found in the glutamate system and what we found in the GABA system. You seem to have found that the receptors are OK, but that the iononophore is changed in some long-lasting way, or at least binding sites on the ionophore. In the GABA system, we have also found that the GABA-A and benzo receptors are OK, but we find a change in GABA-stimulated chloride flux. Now, we haven't found a change in TB PS binding, but Edgar et al. and Lewin et al. have found both the change in flux and the change in binding. It looks as if kindling might do something to the ionophores rather than to the receptor sites. And I think that it is interesting that you're finding it and we're finding it. What puzzles us, and we don't know how to explain it, how does this change prolong itself for the life of the cell?

DR. MCNAMARA: I agree completely. The key question: how does it develop and how is it maintained?

DR. MODY: I just want to make a comment which probably relates to many of these events, especially to GABA and this last question raised. It's more and more apparent that we're dealing with a super-family of receptor channels here, and I bet that when the NMDA receptor channel is cloned, it will be very similar to the GABA receptor channel structure. It has been shown that the GABA channel can be phosphorylated from the inside. Now we have shown that the NMDA channel can be phosphorylated from the inside, so it is possible that phosphorylation events can regulate in fact both of these receptors, and both respond in the same way because both are part of the same family and both will be upregulated by phosphorylation. This brings me to your point that in the early stages of kindling, maybe such a mechanism can be effective, and later maybe the glycine site changes and takes over and maintains the upregulation function. In the same study where we have shown the intracellular responses to NMDA or synaptic components to NMDA, we have shown that population spikes on the other hand are inhibited in the same way that Dr. Racine has measured, and Graham Goddard and Mike Oliver and everybody else has done. But it just means that the second pulse is inhibited, not really the first one. The first one may be a lot more excited than the second, and of course through the calcium dependent mechanism, so this conductance changes which are certainly very different under these circumstances and I think it's very hard to compare extracellular studies to intracellular studies in terms of measuring inhibition. But I have a question, actually two. Was there any difference between the basal carbachol stimulated PI turnover between control and kindled? and second, when you measure the glycine effect and TCP binding, now the TCP binding is a use-depending thing, so do you have any glutamate in the medium, and could it be that just basal levels of glutamate are different in kindled versus control animals, and that's what you're picking up?

DR. MCNAMARA: To answer the first question, there were no significant differences. Second, those are extraordinarily well washed membranes. We add glutamate in excess; the glutamate concentration in all those things is about EC50, which I believe is about 300 nanomolar in this preparation, and these things had been through a lot of washes. So I don't think that's the explanation, I don't want to leave everybody with the impression that this is unambiguously a selective alteration in glycine receptors, because let's remember (a) the animals were kindled from different sites, (b) one's done in radio histochemically the other's done in membranes. If that's a selective alteration in a glycine receptors, that's a very important finding, apart from kindling, because it would demonstrate the potential for selective regulation. I don't know that that's occurring. We're obviously in the process of working it out. And I want to emphasize too that as far as the TCP binding is concerned, that's at a single non-equilibrium time point. It could be explained simply by increased numbers of TCP binding sites, the difference in the kinetics, it could be a lot of things.

# DOES ELECTRICAL AND EXCITATORY AMINO ACID KINDLING SHARE A COMMON NEUROBIOLOGICAL MECHANISM?

Norio Mori[1] and Juhn A. Wada[2]

[1]Department of Neuropsychiatry, Fukushima Medical College, Japan
[2]Division of Neurosciences and Neurology, University of B.C., Canada

## INTRODUCTION

There is convincing evidence that excitatory amino acids, particularly glutamate (GLU) and aspartate (ASP), are involved in basic mechanism of epilepsy (for reviews see refs. 1,2,26). Thus, enhanced release, and reduced tissue levels, of excitatory amino acids have been reported in various animal models of epilepsy (9,13,20,24,34,41,44). Reduced levels of both GLU and ASP have also been demonstrated in tissue excised from human epileptic foci (42,43). In addition, excitatory amino acid antagonists have been shown to block epileptic seizures and epileptiform activity in both rodent and primate models of epilepsy (4,5,6,26,34,39,40).

The amygdala (AM), which is known to be the most sensitive site for kindling manipulation among many brain areas (14), containes various neurotransmitters including GLU and ASP (17,47). In order to elucidate the role of excitatory amino acids in AM kindling, we carried out several experiments using chemical kindling by means of repeated intra-AM injection of GLU and/or ASP, and intracerebral injection of excitatory amino acid antagonists.

## MATERIALS AND METHODS

All the animals used in the following experiments were male adult rats of the Royal Victoria Hooded or Wistar strain. Chronic electrodes, made of twisted nichrome or stainless steel wire with a tip separation of 0.5 mm, were implanted into the bilateral AM and the substantia innominata (SI). For injecting drugs, a 23 gauge stainless steel cannula was attached to the AM or SI electrode. The experiments were started 2 weeks following surgery. For electrical kindling, daily AM stimulation was delivered bipolarly in a 1-sec train of constant current 60-Hz sine wave at an afterdischarge (AD) threshold (ADT). The pattern of seizure development was classified into five stages (35).

Histological examination confirmed that all electrode and cannula tips were located in the intended structure. There was no evidence of neuronal damage in the site of drug injection.

## EXPERIMENT I

Recently, Sato et al. (37) compared the electroclinical responses induced

*Kindling 4*
Edited by J. A. Wada
Plenum Press, New York, 1990

by single intra-AM injections of 3 μmol of GLU or ASP alone, or a range of different molar ratios of these amino acids in rats. A mixture of GLU/ASP combined in a molar ratio of 1:3 (GLU-0.75 μmol: ASP-2.25 μmol) was by far the most potent, evoking a generalized convulsion similar to electrically-kindled AM seizure. In Expt. I, the kindling effect produced by repeated intra-AM injection of an initially ineffective dose of the same combination, and bidirectional transferability between GLU/ASP and electrical kindling at the same AM site, in addition to the persistent nature of the GLU/ASP-kindled seizure susceptibility are described.

## Materials and Methods

Twenty-two male hooded rats of the Royal Victoria Hospital strain were used. For GLU/ASP kindling, monosodium-GLU 0.375 μmol and monosodium-ASP 1.125 μmol were dissolved in a 0.5 μl vehicle, i.e. 0.1 mol phosphate buffer (pH= 7.0). Using Hamilton syringe, intra-AM injection of either GLU/ASP or vehicle was made through the implanted cannula every second day.

The animals were divided into three groups, i.e. the I-A (n=6), I-B (n=10) and I-C (n=6) groups. Group I-A received GLU/ASP injection until five Stage 5 seizures occured. Subsequently, the same AM site was subjected to electrical kindling. Group I-B received 20 vehicle injections, followed by electrical kindling of the same AM site. Group I-C underwent AM electrical kindling, followed by GLU/ASP kindling of the same AM site. In order to determine whether the sensitization to GLU/ASP was sustained, 30 days after the last injection of GLU/ASP, all the group I-C animals received a single injection of GLU/ASP at half the dose used for GLU/ASP kindling.

## Results

### 1. Electroclinical consequences of GLU/ASP kindling

Results are summarized in Fig. 1. On the first day of GLU/ASP injection, two animals in group I-A showed sporadic spikes about 30 min postinjection, a feature which appeared in all the animals after repeated injections. With a mean of 3 injections, overtly apparent clinical seizure began to develop and progressed to Stage 5 seizure, identical to electrically-kindled AM convulsion. With a mean of 12.8 injections (range 9-16), all the animals in group I-A reached the first Stage 5 seizure ( Fig. 1 and Table 1). On the other hand, the majority (7/10) in group I-B showed no evidence of electroclinical ictal

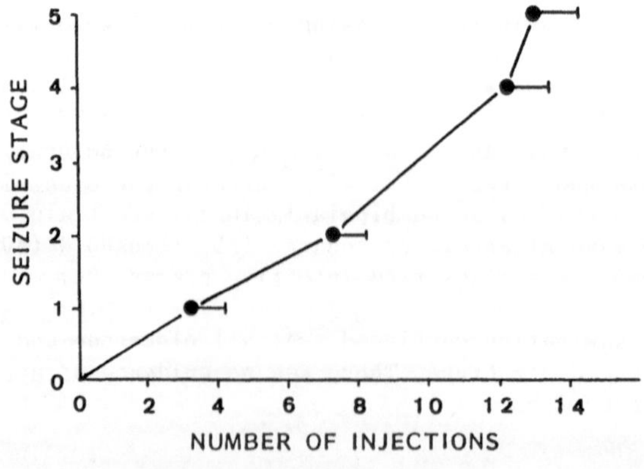


Fig .1. The rate of chemical kindling induced by GLU/ASP injection for the first seizure of the class specified on the ordinate. The values are mean±S.E.M.

Table 1. Bidirectional transfer between GLU/ASP and electrical kindling in the same AM site.
Values are mean±S.E.M. Numbers in parentheses=range.

| Group | n | Times to first Stage 5: GLU/ASP | | Times to first Stage 5: Electrical |
|---|---|---|---|---|
| I-A | 6 | [b] 12.8±1.0(9-16) | ⟶ | 2.7±0.6(1-5) [a] |
| I-B | 10 | (Vehicle) | ⟶ | 8.3±0.9(3-14) |
| I-C | 6 | 3.7±0.7(1-5) | ⟵ | 8.0±1.0(5-12) |

[a]P<0.005 and [b]P<0.001 by Student's t-test.

response. Only 3/10 of this group showed Stage 1 manifestation after 17-20 injections and only one of them eventually developed Stage 5 seizure.

2. Transferability between GLU/ASP and electrical kindling and the persistent nature of GLU/ASP-kindled seizure

Results are summarized in Table 1. When group I-A was subjected to electrical kindling, all the animals responded with Stage 5 seizure with a mean of 2.7 electrical stimulations (range 1-5). The animals in group I-B, which had received the vehicle injection, required a mean of 8.3 electrical stimulations (range 3-14) for development of the first electrically-kindled Stage 5 seizure. Therefore, the rate of electrical kindling was significantly faster in the I-A group than in the I-B group ($p<0.005$, Table 1). Group I-C required a mean of 8.0 electrical stimulations (range 5-12) for development of the first Stage 5 seizure. When group I-C was subjected to GLU/ASP kindling, the mean number of GLU/ASP injections required for the first Stage 5 seizure was 3.7 (range 1-5). Therefore, the rate of chemical kindling of the electrically-kindled animals was significantly faster than that of chemical kindling without prior electrical kindling ($p<0.001$, Table 1).

EXPERIMENT II

In this experiment, the relationship between GLU/ASP and electrical kindling at the AM using trans-hemispheric transfer and post-transfer interference as specific measures (3,14,25) is described.

Materials and Methods

Fifty-eight hooded rats of the Royal Victoria Hospital strain were used. Technical details of GLU/ASP kindling have been described in Expt. I. For testing trans-hemispheric transfer between GLU/ASP and electrical kindling, 39 animals were used, divided into four groups: II-A (n=9): upon completion of GLU/ASP kindling at the primary site AM, the secondary site AM was subjected to electrical kindling: II-B (n=10): vehicle injection into the primary site AM was administered for the identical number of times required for GLU/ASP kindling in group II-A, followed by electrical kindling at the secondary site AM: II-C (n=10): were given electrical kindling at the primary site AM followed by GLU/ASP kindling at secondary site AM: II-D (n=10): received daily electrical stimulation at the primary site AM at the intensity below the ADT (sub-ADT) for the identical number of times required for electrical kindling in group II-C, followed by GLU/ASP kindling at the secondary site AM.

For testing the post-transfer interference between GLU/ASP and electrical kindling, groups II-A (n=9) and II-C (n=10) were used, in addition to two new groups of animals: II-E (n=9) and II-F (n=10). In group II-E, as in group II-A, the primary site AM was subjected to GLU/ASP kindling, but for the secondary site AM, the sub-ADT stimulation was delivered for the identical number of times required for secondary site electrical kindling in group II-A. The animals in group II-F (as did those in group II-C) underwent electrical kindling at the primary site AM. Subsequently, the secondary site AM received the vehicle only for the identical number of times required for secondary site GLU/ASP kindling in group II-C. For primary site retest, animals in groups II-A and II-E received GLU/ASP injections, while those in groups II-C and II-F were stimulated at the previously established ADT.

## Results

### 1. Trans-hemispheric transfer between GLU/ASP and electrical AM kindling

Results are summarized in Table 2. After repeated intra-AM injections of GLU/ASP, the animals in group II-A showed progressive seizure development culminating in Stage 5 seizure. The mean number of GLU/ASP injections required for development of the first Stage 5 seizure was 9.1 (range 5-13). When the secondary site AM was subjected to electrical kindling, a mean of 4.4 stimulations (range 1-8) resulted in the first Stage 5 seizure. In the group II-B, no electroclinical seizure development occured when injecting the vehicle at the primary site AM. At the secondary site AM, a mean of 11.7 electrical stimulations (range 9-19) was required to reach the first Stage 5 seizure. The rate of electrical kindling at the secondary site AM of group II-A was significantly faster than that of group II-B ($p<0.001$, Table 2).

The animals in group II-C required a mean number of 11.1 stimulations (range 6-18) to reach the first Stage 5 seizure at the primary site AM. When the secondary site AM was subjected to GLU/ASP kindling, a mean of 3.6 injections (range 2-6) was required to develop the first Stage 5 seizure. In group II-D, no electroclinical seizure developed with sub-ADT stimulations at the primary site AM. At the secondary site AM, a mean of 9.9 injections (range 4-15) of GLU/ASP was required for the first Stage 5 seizure. The difference in the rate of GLU/ASP kindling at the secondary site AM between the II-C and II-D groups was significant ($p<0.001$, Table 2).

### 2. Post-transfer interference between GLU/ASP and electrical AM kindling

Results are summarized in Table 2. In group II-A, a mean of 3.7 injections (range 1-6) of GLU/ASP was required to re-establish a Stage 5 seizure at the primary site AM. In group II-E, primary site GLU/ASP kindling required a mean of 10.1 injections (range 4-17) to develop the first Stage 5 seizure. At the

Table 2. Bidirectional trans-hemispheric transfer and post-transfer interference between GLU/ASP and electrical kindling.
Values are mean±S.E.M. Numbers in parentheses=range.

| Group | n | Time to first Stage 5 | | |
|---|---|---|---|---|
| | | Primary site | Secondary site | Primary site retest |
| | | GLU/ASP | Electrical | GLU/ASP |
| II-A | 9 | 9.1±0.9(5-13) | 4.4±0.8(1-8) [a] | 3.7±0.7(1-6) [b] |
| II-B | 10 | (Vehhicle) | 11.7±1.1(9-19) | |
| II-E | 9 | 10.1±1.3(4-17) | (sub-ADT) | 1.4±0.4(1-5) |
| | | Electrical | GLU/ASP | Electrical |
| II-C | 10 | 11.1±1.1(6-18) | 3.6±0.5(2-6) [a] | 2.8±0.3(2-4) [a] |
| II-D | 10 | (sub-AD) | 9.9±1.0(4-15) | |
| II-F | 10 | 12.9±0.8(10-16) | (Vehicle) | 1.0±0 |

[a] $P<0.001$ and [b] $P<0.02$ by Student's t-test.

secondary site AM, sub-ADT stimulations produced no electroclinical change. At the primary site retest, a mean of 1.4 injections (range 1-5) was needed for the re-establishment of the Stage 5 seizure. Thus, the degree of post-transfer interferrence at the primary site AM was more marked in group II-A than in group II-E ($p<0.02$, Table 2).

In group II-C, the re-establishment of the Stage 5 seizure at the primary site AM needed a mean of 2.8 electrical stimulations (range 2-4). The animals in group II-F required a mean of 12.9 electrical stimulations (range 10-16) to reach the first Stage 5 seizure at the primary site AM. At the secondary site AM, vehicle injection produced no electroclinical developement. When the primary site AM was re-stimulated at the ADT, all the animals responded with a Stage 5 seizure upon the first stimulation. Therefore, group II-C required a much larger number of electrical re-stimulations than group II-F to re-establish the kindled Stage 5 seizure ($p<0.001$, Table 1).

## EXPERIMENT III

In this experiment, the comparative kindling effect by means of repeated intra-AM injection of an initially ineffective dose of either GLU or ASP in addition to bidirectional transferability between kindling produced by either GLU or ASP injection and electrical stimulation at the same AM site are described. We also report the effect of kynurenic acid (KYA), an antagonist for N-methyl-D-aspartate (NMDA), kainate (KA) and quisqualate (QUIS) receptors (18, 33) on the GLU- or ASP-kindled seizure development.

### Materials and Methods

Fifty-two male Wistar rats were used. In Experiments I and II, a combination of 0.375 $\mu$mol GLU and 1.125 $\mu$mol ASP was used for GLU/ASP kindling. Therefore, an identical dose of either GLU or ASP, i.e. 1.5 $\mu$mol GLU or 1.5 $\mu$mol ASP, was used for GLU or ASP kindling.

To elucidate the kindling effect induced by either GLU or ASP, and the bidirectional transfer between chemical kindling induced by GLU or ASP and electrical kindling, 33 animals were used. They were divided into five groups: III-A (n=6), III-B (n=7), III-C (n=7), III-D (n=6) and III-E (n=7). Group III-A continued to receive GLU injection and group III-B was injected with ASP every second day, until five Stage 5 seizures were produced. Group III-C group received 25 vehicle injections. Upon completion of chemical kindling, these three groups were subjected to electrical kindling at the same AM site. Groups III-D and III-E underwent electrical kindling. Upon development of five Stage 5 seizures, group III-D received GLU injection and group III-E received ASP injection until Stage 5 seizure was induced.

To examine the effect of KYA on GLU or ASP kindling, three new groups of animals, i.e. III-F (n=6), III-G (n=6) and III-H (n=7) were used. Group III-F received a combination of 0.1 $\mu$mol KYA and 1.5 $\mu$mol GLU for the identical number of times required for GLU kindling in group III-A. Group III-G received a combined KYA-0.1 $\mu$mol/ASP-1.5 $\mu$mol for the identical number of times required for ASP kindling in group III-B. Group III-H was given with 0.1 $\mu$mol KYA alone.

### Results

#### 1. Electroclinical consequences of GLU or ASP kindling

The results are summarized in Fig. 2. Animals in group III-A showed progressive seizure development in accordance with repeated GLU injections. This group reached the first Stage 1 seizure with a mean of 12.5 injections (range 8-20) and required a mean of 17.5 injections (range 10-30) to develop the

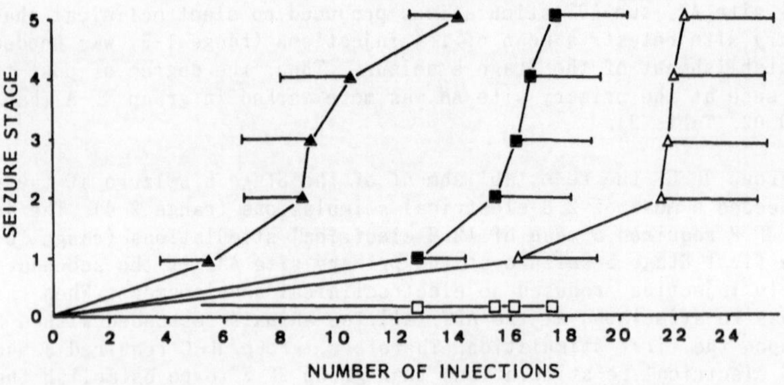


Fig. 2. The rate of chemical kindling induced by ASP (dark triangle), GLU (dark square), KYA/ASP (open triangle) or KYA/GLU (open square) for the first seizure of the class specified on the ordinate. The values are mean ± S.E.M.

first Stage 5 seizure (Fig. 2). Group III-B animals began to develop overtly apparent clinical seizure with a mean of 5.6 injections (range 2-17) of ASP and progressed to Stage 5 seizure identical to electrically-kindled AM convulsion (Fig. 2). With a mean of 14.3 injections (range 7-22) of ASP, this group reached the first Stage 5 seizure. The mean number of injections required for the development of the first Stage 1 seizure was significantly smaller in group III-B than in group III-A ($p<0.05$, Fig. 2), but there was no significant difference in the mean number of injections required for the first clinical manifestation of Stages 2-5 between the two groups. Group III-C showed no evidence of electroclinical response.

Table 3. Bidirectional transfer between chemical kindling induced either by GLU or by ASP and electrical kindling in the same AM site. Values are mean ± S.E.M. Mumbers in parentheses=range.

| Group | n | Time to first Stage 5 | | |
|---|---|---|---|---|
| | | GLU | | Electrical |
| III-A | 6 | [a] 17.5±2.9(10-30) | ⟶ | 3.5±0.3(2-4) [a] |
| III-D | 6 | 1.2±0.2(1-2) | ⟵ | 16.0±1.8(13-24) |
| | | ASP | | Electrical |
| III-B | 7 | 14.3±2.1(7-22) | ⟶ | 2.9±0.6(1-5) [a] |
| III-E | 7 | [a] [a] 2.0±0.5(1-4) | ⟵ | 15.0±1.7(11-23) |
| | | Vehicle | | Electrical |
| III-C | 7 | (-) | ⟶ | 13.4±1.3(9-19) |
| | | KYA/GLU | | Electrical |
| III-F | 6 | (-) | ⟶ | 13.1±2.5(7-24) |
| | | KYA/ASP | | Electrical |
| III-G | 6 | 21.7±1.9(16-28) | ⟶ | 3.8±1.5(2-6) |
| | | KYA | | Electrical |
| III-H | 7 | (-) | ⟶ | 11.4±2.0(8-23) |

[a] $P<0.001$ by Student's t-test.

2. Transferability between kindling induced by either GLU or ASP injection and electrical stimulation

Results are summarized in Table 3. When group Ⅲ-A was subjected to electrical kindling following the completion of GLU kindling, all the animals of this group responded with Stage 5 seizure with a mean of 3.5 stimulations (range 2-4). When the animals in group Ⅲ-B underwent electrical kindling upon the completion of ASP kindling, all the animals of this group responded with Stage 5 seizure with a mean of 2.9 stimulations (range 1-5). Group Ⅲ-C, which received the vehicle injections, required a mean of 13.4 electrical stimulations (range 9-19) for the first Stage 5 seizure. Therefore, the rate of electrical kindling of groups Ⅲ-A and Ⅲ-B was significantly faster than that of group Ⅲ-C ($p<0.001$, Table 3).

Group Ⅲ-D required a mean of 16.0 electrical stimulations (range 13-24) to reach the first Stage 5 seizure. When GLU was injected, this group developed the first Stage 5 seizure with a mean of 1.2 injections (range 1-2). Therefore, the rate of GLU kindling of the electrically-kindled animals was significantly faster than that of GLU kindling without prior electrical kindling ($p<0.001$, Table 3). Group Ⅲ-E reached the first Stage 5 seizure with a mean of 15.0 electrical stimulations (range 11-23). When subjected to ASP kindling, all the animals in group Ⅲ-E responded with Stage 5 seizure with a mean of 2.0 injections (range 1-4). Therefore, the degree of ASP kindling of electrically-kindled animals was more marked than that of ASP kindling without prior electrical kindling ($p<0.001$, Table 3).

3. Effect of KYA on the seizure development induced by GLU or ASP injection

Results are summarized in Fig. 2 and Table 3. None of the animals in group Ⅲ-F showed any electroclinical response after repeated KYA/GLU injections. In group Ⅲ-G, 1/6 of the animals showed sporadic spikes on the third day of KYA/ASP injection, and after 9 injections all the animals began to show this feature. Therefore, this group of animals continued to receive KYA/ASP injection until five Stage 5 seizures occured. Eventually, with a mean of 16.2 injections (range 13-23), overtly apparent clinical seizure began to develop and progressed to Stage 5 seizure typical of electrically-kindled AM convulsion. With a mean of 21.7 injections (range 16-28), all the Ⅲ-G animals reached the first Stage 5 seizure (Fig. 2 and Table 3). The rate of KYA/ASP kindling of group Ⅲ-G was significantly greater than that of ASP kindling of group Ⅲ-B ($p<0.001$, Table 3). Group Ⅲ-H was given with KYA alone and showed no evidence of electroclinical response.

When group Ⅲ-F was subjected to electrical kindling at the same AM site, the first Stage 5 seizure developed with a mean of 13.1 stimulations (range 7-24). On the other hand, group Ⅲ-G animals reached the first Stage 5 seizure with a mean of 3.8 electrical stimulations (range 2-6). The animals in group Ⅲ-H developed their first Stage 5 seizure with a mean of 11.4 stimulations (range 8-23) when subjected to electrical kindling.

EXPERIMENT Ⅳ

In this experiment, the comparative effect of intracerebral injection of 2-amino-7-phosphonoheptanoic acid (2-APH), a specific antagonist for NMDA receptors (7,48), into the AM and the SI of AM-kindled rats was studied.

Materials and Methods

Sixteen male hooded rats of the Royal Victoria Hospital strain were used. All the animals were electrically kindled at the left AM until a stable Stage 5 seizure was induced for 5 successive days. Subsequently, the stimulus intensity

Table 4. Sustained GST elevation following 2-APH injection into the kindled AM.

| Rat No. | Time following injection | | | | | | | | | | | | | | | | | | | | | | | |
|---|---|---|---|---|---|---|---|---|---|---|---|---|---|---|---|---|---|---|---|---|---|---|---|
| | 45min | 1 | 2 | 3 | 4 | 5 | 6 | 7 | 8 | 9 | 10 | 11 | 12 | 13 | 14 | 15 | 16 | 17 | 18 | 19 | 20 | 21 | 22 | 23(day) |
| 0.05 $\mu$M | | | | | | | | | | | | | | | | | | | | | | | | |
| 1 | − | − | − | − | − | | | | | | | | | | | | | | | | | | | |
| 2 | − | − | − | − | − | | | | | | | | | | | | | | | | | | | |
| 3 | + | + | + | + | + | + | + | + | + | + | + | + | − | − | − | − | − | | | | | | | |
| 4 | + | + | + | + | + | + | + | + | + | + | + | + | + | − | + | + | + | − | − | − | − | − | | |
| 5 | + | + | + | + | + | + | − | − | − | − | − | | | | | | | | | | | | | |
| 6 | − | − | − | − | − | | | | | | | | | | | | | | | | | | | |
| 0.1 $\mu$M | | | | | | | | | | | | | | | | | | | | | | | | |
| 7 | + | ++ | + | + | + | + | + | + | − | + | + | − | + | − | + | − | − | + | + | − | − | − | − | − |
| 8 | + | + | + | + | + | + | + | + | + | − | + | − | − | − | − | − | | | | | | | | |
| 9 | + | ++ | + | + | + | + | − | − | − | − | − | | | | | | | | | | | | | |
| 10 | + | + | + | + | + | + | + | − | − | − | − | − | | | | | | | | | | | | |
| 11 | − | ++ | − | − | − | − | − | | | | | | | | | | | | | | | | | |
| 12 | + | + | + | − | − | − | − | − | | | | | | | | | | | | | | | | |

When AD generation was suppressed at the previously established GST (−), the stimulaus intensity was increased 50 $\mu$A above the GST (+) and 100 $\mu$A above the GST (++) until Stage 5 seizure was elicited. Electrical stimulation was continued until five consecutive Stage 5 seizures were evoked at the GST.

was gradually reduced and the last intensity to induce Stage 5 seizure was designated as the generalized seizure triggering threshold (GST). The animals were divided into two groups, i.e. the IV-A (n=12) and IV-B (n=4) groups. The animals in group IV-A received either 0.05 μmol (n=6) or 0.1 μmol (n=6) injections of 2-APH into the kindled AM. Group IV-B received 0.1 μmol initially, and 0.2 μmol on the second occasion, into the SI ipsilateral to the kindled AM. Forty-five min after the injection, the kindled AM was stimulated at the previously established GST. If AD generation was suppressed, the stimulus intensity was increased by 50 μA steps until AD was elicited. Subsequently, the animals were tested every 24 hrs at the GST or the increased ADT. Each animal received the vehicle (0.01 mol phosphate-buffered saline, pH=7.0) injection prior to the drug treatment. The injection of the vehicle alone neither changed the behavior nor altered the effect of AM stimulation.

## Results

Results are summarized in Table 4 and Fig. 3. In contrast to intra-SI injection, which produced a transient behavioral change characterized by a reduction of postural muscle tone associated with incoordination, intra-AM injection of 2-APH produced no behavioral change. When the animals were stimulated at the GST, the kindled seizure was completely suppressed in a dose-dependent manner (Table 4). Kindled seizure could be elicited, however, when

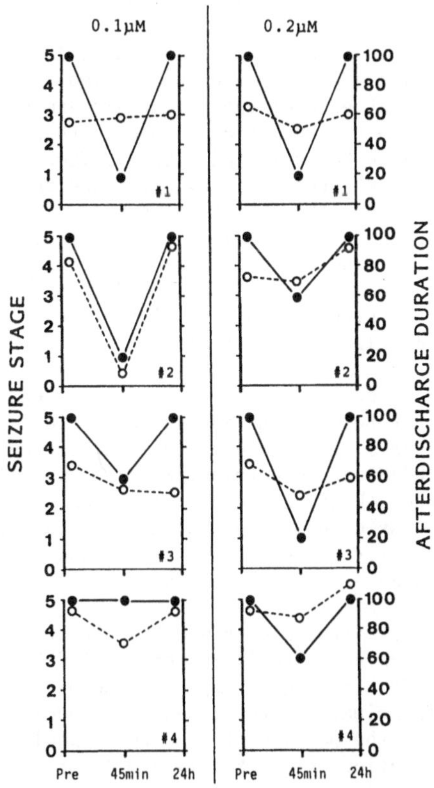


Fig. 3. Effects of intra-SI injection of 2-APH on seizure stage (dark circle) and after-discharge duration (open circle) induced by AM stimulation. 'Pre' represents the mean values of 3 kindled seizures.

the stimulus intensity was increased. This elevation of GST was maximal at 24 hrs and lasted for 1-18 days.

Intra-SI injection of 2-APH induced a transient incoordination followed by immobility with loss of the rightning reflex, beginning at about 5 min following the injection and lasting for about 3 hrs. In group IV-B, 3/4 of the animals injected with 0.1 $\mu$mol, and 4/4 of the animals injected with 0.2 $\mu$mol developed this behavioral change. Intra-SI injection of 0.1 $\mu$mol resulted in a regression of kindled seizure to Stages 1-3 in 3 animals , while the remaining animal, which showed no behavioral change, responded with kindled seizure. In all the animals, AD was readily produced and only 1 animal showed a significant reduction of its duration (Fig. 3). With 0.2 $\mu$mol, 4/4 of the animals showed a regression of kindled seizure to earlier stages with a slight reduction of AD duration (Fig. 3). At 24 hrs, all the animals responded to GST stimulation.

## DISCUSSION

GLU or ASP alone is only minimally effective as a neurotoxin and they both appear to be much less potent than expected (32,37). In an earlier report by Freeman (12), the combined release of GLU and ASP from the excitatory nerve terminals in a molar ratio of 1:3 was postulated to cause a potent excitation of the lobster neuromuscular synapse. Recently, using this formula, Sato et al. (37) demonstrated that intra-AM injection of GLU and ASP combined in a molar ratio of 1:3, i.e. 0.75 $\mu$mol GLU and 2.25 $\mu$mol ASP, resulted in a generalized convulsion similar to electrically-kindled AM seizure in rats.

In Expt. I, we found that repeated spaced intra-AM injections of the same combination, but at a lower quantity, i.e. 0.375 $\mu$mol GLU and 1.125 $\mu$mol ASP, can kindle animals in the absence of specific brain damage. GLU/ASP applied intracerebrally acts on the postsynaptic neurons, although the terminals of presynaptic neurons may also be activated (15,48). Therefore, lack of neuronal damage at the injected AM is important for understanding the mechanisms of GLU/ASP kindling. In addition, the results of Expt. I show not only a strong bidirectional transfer between GLU/ASP and electrical kindling at the same AM site but also that GLU/ASP-kindled seizure can be triggered 30 days after the last injection with half the dose used for GLU/ASP kindling. These findings suggest that the GLU/ASP system participates in the development and the persistence of increased seizure susceptibility.

In Expt. II, clear evidence was obtained of reciprocal and bidirectional trans-hemispheric transfer between GLU/ASP and electrical kindling at the contralateral homotopic AM site. The result of bidirectional post-transfer interference between GLU/ASP and electrical kindling was also obtained. These findings suggest that the GLU/ASP system participates actively in the trans-synaptic changes considered to be the underpinning of the kindling phenomenon, including positive transfer at the secondary site and post-transfer interference which accompanies AM kindling in rats (3,14,25).

In Expt. III, the respective roles played by GLU or ASP alone in AM kidling were examined. Repeated spaced intra-AM injection of GLU or ASP alone at the same dose used for GLU/ASP kindling in Expt. I and II, i.e. 1.5 $\mu$mol GLU or 1.5 $\mu$mol ASP can kindle animals in the absence of specific neuronal damage at the injection site. Since the pattern of electroclinical seizure development caused by GLU or ASP injection is strikingly similar to electrical kindling, it is suggested that kindling induced either by GLU or by ASP is supported by trans-synaptic changes. Evidence of strong bidirectional transfer between electrical kindling and chemical kindling induced either by GLU or by ASP at the same AM site suggests that each amino acid participates in the seizure development of electrically-kindled AM seizure.

In contrast to repeated intra-AM injection of GLU, which could kindle animals, repeated intra-AM injection of KYA/GLU failed to producc electroclical ictal reponse. In addition, intra-AM injections of KYA/ASP produced very slow seizure development, although all the animals tested developed Stage 5 seizure. KYA is known to reduce excitation elicited by NMDA, KA and QUIS (18,33). NMDA appears to excite ASP-preferring receptors, and QUIS appears to excite GLU-preferring receptors (7,48). Therefore, it is reasonable to assume that KYA diminishes excitation induced by intra-AM injection of GLU or ASP, resulting in a negative kindling effect with KYA/GLU or a significant retardation of seizure development with KYA/ASP. A similar result is obtained with electrical kindling, i.e. daily intracerebroventricular injection of KYA retards significantly the seizure development of electrically-kindled AM seizure in rats (39). Therefore, it is suggested further that excitatory amino acid and electrical kindling share common neurobiological mechanisms.

In Expt. IV, we found that intra-AM injection of 2-APH, a specific and potent antagonist for NMDA receptors (7,36), suppressed AD generation completely at the GST, although kindled seizure can be recalled by increasing the stimulus inteansity. This is comparable to a recent finding with another antagonist, 2-amino-5-phosphonovaleric acid (2-APV) (20). When 2-APV was injected into the kindled AM in rats, AD generation was totally or significantly suppressed at the GST. The elevation of the GST following intra-AM injection of 2-APH was maximal at 24 hrs and lasted for 1-18 days. It has been reported that a similar effects is obtained with intra-AM injection of 2-APV, i.e. it does not suppress AD generation 1 hr after the injection, but subsequenly does so from 24 to 72 hrs (20). Since both 2-APH and 2-APV were injected directly into the kindled site, it is conceivable that mechanisms of action other than a simple receptor blockade are involved.

In contrast to intra-AM injection, which did not cause any behavioral change, intra-SI injection of 2-APH produced a transient behavioral change characterized by a reduction of postural muscle tone, beginning at about 5 min and lasting for about 3 hrs following the injection. When the kindled AM was stimulated at the GST, 45 min after the injection, kindled seizure regressed to earlier stages, although AD duration remained largely unchanged. This finding indicates that intra-SI injection of 2-APH suppressed kindled seizure without affecting AM excitability. Regression of kindled seizure was observed in all the animals showing behavioral change, but not in the one which did not develop behavioral change. At 24 hrs, when the animals had completely recovered, kindled seizure was readily activated at the GST. It has been reported that the SI has reciprocal projections with the AM (10,11,30). Its cholinergic projections, believed to be excitatory (21,23), innervate wide cortical areas, including the sensory motor cortex (8,16,22,27). Therefore, it is reasonable to assume that intra-SI injection of 2-APH caused functional alteration maximal in the motor cortical area resulting in a limp paretic state. Since development of kindled AM seizure requires participation of the motor cortical area (38,45,46), inactivation of the latter by intra-SI injection of 2-APH was the probable cause of elimination of the convulsive component. The result of intra-SI injection of 2-APH is comparable to that in our previous studies, in which electrolytic lesioning of the SI or intra-SI injection of a GABAergic agent suppresses kindled seizure despite an intact AD generation at the GST (19,29,31, 36), and therefore, provides further support for the roles played by the SI in the generalization of limbic seizure.

In conclusion, the present study clearly suggests that (1) excitatory amino acids, GLU and/or ASP can kindle the AM, (2) excitatory amino acid kindling shares common neurobiological mechanisms with electrical kindling and (3) NMDA receptors in the AM and SI play a differential role in AM seizure, possibly for initiation and propagation, respectively.

ACKNOWLEDGMENT

This study was supported by grants from the Medical Research Council of Canada to J.A.W.

REFERENCES

1. Bradford, H.F. and Dodd, P.R. (1976) Biochemistry and basic mechanisms in epilepsy. In: Biochemistry and neurological disease, A.N. Davison (ed), Blackwell, Oxford, pp 114-168.
2. Bradford, H.F. and Peterson, D.W. (1987) Current views of the pathobiochemistry of epilepsy. Mol. Aspects Med., 9: 119-172.
3. Burnham, W.M. (1975) Primary and 'transfer' seizure development in the kindled rat. Can. J. Neurol. Sci., 2: 417-428.
4. Coutinho-Netto, J., Abdul-Ghani, A.S., Collins, J.F. and Bradford, H.F. (1981) Is glutamate a trigger factor in epileptic hyperactivity. Epilepsia, 22: 289-296.
5. Croucher, M.J., Collins, J.F. and Meldrum, B.S. (1982) Anticonvulsant action of excitatory amino acid antagonists. Science, 216:889-901.
6. Croucher, M.J. and Meldrum, B.S. (1984) The role of dicarboxylic amino acis in epilepsy and the use of antagonists as antiepileptic agents. In: Neurotransmitters, Seizures and Epilepsy II, Fariello et al (ed), Raven, York, pp 227-236.
7. Curtis, S.T. and Johnstone, G.A.R. (1974) Amino acid transmitters in mammalian central nervous system. Ergbn. Physiol., 69: 97-188.
8. Divac, I. (1975) Magnocellular nuclei of the basal forebrain project to neocortex, brain stem and olfactory bulb. Review of some functional correlates. Brain Research, 93: 385-398.
9. Dodd, P.R. and Bradford, H.F. (1976) Release of amino acids from the maturing cobalt-epileptic focus. Brain Research, 111:377-388.
10. Emson, P.C., Paxinos, G., Le Gal la Salle, G., Ben-Ari, Y. and Silver, A. (1979) Choline acetyltransferase and acetylcholinesterase-containing projections from the basal forebrain to the amygdaloid complex of the rat. Brain Research, 165: 271-282.
11. Femano, P.A., Edinger, H.M. and Siegel, A. (1983) The effects of stimulation of substantia innominata and sensory receiving areas of the forebrain upon the activity of neurons within the amygdala of anesthetized cat. Brain Research, 269: 119-132.
12. Freeman, A.R. (1976) Polyfunctional role of glutamic acid in excitatory synapstic transmission. Prog. Neurobiol., 6: 137-153.
13. Geula, C., Jarvie, P.A., Logan, T.C. and Slevin, J.T. (1988) Long-term enhancement of $K^+$-evoked release of L-glutamate in entorhinal kindled rats. Brain Research, 442:368-372.
14. Goddard, G.V., McIntyre, D.C. and Leach, C.K. (1969) A permanent change in brain function resulting from daily electrical stimulation. Exp. Neurol., 25: 295-330.
15. Goodchild, A.K., Dampney, R.A.L. and Bandler, R. (1982) A method for evoking physiological responses by stimulation of the cell body, but not the axons of passage, within localized regions of the central nervous system. J. Neurosci. Meth., 6: 351-363.
16. Gorry, J.D. (1963) Studies on the comparative anatomy of the ganglion basale of Mynert. Acta Anat., 55: 51-104.
17. Greenamyre, J.T., Young, A.B. and Penny J.B. (1984) Quantitative autoradiographic distribution of L-[$^3$H]glutamate-binding sites in rat central nervous system. J. Neurosci., 4: 2133-2144.
18. Herrling, P.L. (1984) Evidence that the cortically evoked e.p.s.p. in cat caudate neurones is mediated by non-NMDA excitatory amino acid receptors. J. Physiol. (Lond.), 353: 98P.

19. Kimura, H., Kaneko, Y. and Wada, J.A. (1981) Catecholamine and cholinergic systems and amygdaloid kindling. In: Kindling 2, J.A. Wada (ed), Raven, New York, pp 268-287.
20. Koyama, I. (1972) Amino acids in the cobalt-induced epileptogenic and non-epileptogenic cat's cortex. Can. J. Physiol. Pharmacol., 50:740-752.
21. Krnjevic, K., Pumain, R. and Renaud, L. (1978) The mechanisms of excitation in the cerebral cortex. J. Physiol. (Lond.), 215: 247-268.
22. Krnjevic, K. and Silver, A. (1965) A histochemical study of cholinergic fibers in the cerebral cortex. J. Anat., 99: 711-759.
23. Lamour, Y., Dutar, P. and Jobert, A. (1982) Excitatory effect of acetylcholine on different types of neurons in the first somatosensory neocortex of the rat: Laminar distribution and pharmacological characteristics. Neuroscience, 7: 1483-1494.
24. Leach, M.J., Marden, C.M., Miller, A.A., O'Donnel, R.A. and Weston, S.B. (1985) Changes in cortical amino acids during electrical kindling in rats. Neuropharmacology, 24: 937-940.
25. McIntyre, D.C. and Goddard, G.V. (1973) Transfer, interference and spontaneous recovery of convulsions kindled from the rat amygdala. Electroencephalogr. Cli. Neurophysiol., 35:535-543.
26. Meldrum, B.S. (1984) Amino aci transmitters and new approaches to anticonvulsant drug action. Epilepsia, 25:379-393.
27. Mesulam, M.H. and Van Hoesen, G.W. (1980) Acetylcholinesterase-rich projections from the basal forebrain of the rhesus monkey to neocortex. Brain Research, 109: 152-157.
28. Morimoto, N., Holmes, K.H. and Goddard, G.V. (1987) Kindling-induced changes in EEG recorded during stimulation at the site of stimulation. Ⅲ. Direct pharmacological manipulation of kindled amygdala. Exp. Neurol., 97: 17-34.
29. Morita, K., Okamoto, M., Seki, K. and Wada, J.A. (1985) Suppression of amygdala-kindled seizure in cats by enhanced GABAergic transmission in sustantia innominata. Exp. Neurol., 89: 225-236.
30. Nagai, R. and Kimura, H. (1982) Cholinergic projections from the basal forebrain of the rat amygdala. J. Neurosci., 2: 512-520.
31. Okamoto, M. and Wada, J.A. (1984) Reversible suppression of amygdaloid-kindled convulsion following unilateral gabaculline injection into the substantia innominata. Brain Research, 305: 389-392.
32. Olney, J.W. and Price, M.T. (1983) Excitotoxic amino acids as neuroendocrine research tools. Meth. Enzymol., 103: 379-393.
33. Perkins, M.N. and Stone, T.W. (1982) An iontophoretic investigation of the actions of convulsant kynurenines and their interaction with the endogeneous excitant quinolinic acid. Brain Research, 247: 184-187.
34. Peterson, D.W., Collins, J.F. and Bradford, H.F. (1983) The kindled amygdala model of epilepsy: Anticonvulsant action of amino acid antagonists. Brain Research, 275: 169-172.
35. Racine, R. (1972) Modification of seizure activity by electrical stimulation. Ⅱ: Motor seizure. Electroencephalogr. Cli. Neurophysiol., 32: 281-294.
36. Sakai, S. and Wada, J.A. (1987) Reversible suppression of amygdaloid and cortically kindled seizure in Senegalese baboon, Papio papio, by unilateral injection of gabaculline into the substantia innominata. Epilepsia, 28: 618.
37. Sato, T., Mori, N. and Kumashiro, H. (1985) A new model of epileptic seizure utilizing the additive excitatory activity by combining two excitatory amino acids. Folia Psychiatr. Neurol. Jpn., 39: 431-432.
38. Tanaka, A. (1972) Progressive changes of behavioural and electroencephalographic responses to daily amygdaloid stimulation in rabbits. Fukuoka Acta Med., 63:152-164.
39. Thompson, J.L., Holmes, G.L., Taylor, G.W. and Feldman, D.R. (1988) Effects of kynurenic acid on amygdaloid kindling in rats. Epilepsy Res., 2: 302-308.
40. Turski, L., Meldrum, B.S. and Collins, J.F. (1985) Anticonvulsant action of $\beta$-kainic acid in mice. Is $\beta$-kainic acid an N-methyl-D-aspartate antagonist? Brain Research, 336:162-166.

41. Van Gelder, N.M. and Courtois, A. (1972) Close correlation between changing content of specific amino acids in epileptogenic cortex of cats and severity of epilepsy. Brain Research, 40: 447-484.
42. Van Gelder, N.M., Sherwin, A.L. and Rasmussen, T. (1972) Amino acid content of epileptogenic human brain: Focal versus surrounding regions. Brain Research, 40: 385-392.
43. Van Gelder, N.M., Sherwin, A.L., Sacks, C. and Aldermann, F. (1975) Biochemical observations following administration of taurine to patients with epilepsy. Brain Research, 94: 297-306.
44. Van Gelder, N.M., Siatitsas, I., Menini, C. and Gloor, P. (1983) Feline generalized penicillin epilepsy: Changes of glutamic acid and taurine parallel the progressive increase in excitability of the cortex. Epilepsia, 24: 200-213.
45. Wada, J.A. (1980) Amygdaloid and frontal cortical kindling in subhuman primates. In: Limbic Epilepsy and the Dyscontrol Syndrome, M. Girgis and L.G. Kiloh (ed), Elsevier, Amsterdam, pp 133-146.
46. Wada, J.A., Mizoguchi, T. and Komai, S. (1981) Cortical motor activation in amygdaloid kindling: Observation in non-epileptic rhesus monkeys with anterior 2/3 callosal bisection. In: Kindling 2, J.A. Wada (ed), Raven, New York, pp 235-248.
47. Waker, J.E. and Fonnum, F. (1983) Regional cortical glutaminergic projects to the amygdala and thalamus of rats. Brain Research, 267: 371-374.
48. Watkins, J.C. and Evans, R.H. (1981) Excitatory amino acid transmissions. Ann. Rev. Pharmacol., 21: 165-204.

Discussion of Dr. Mori's Presentation

DR. CAIN: I'd like to congratulate you on a nice study; I'm glad somebody did it. To my knowledge it's the only study showing such kindling so far. As a fan of transfer from way back I'm very glad to see the transfer data. I can't remember much of them now, but I liked them when I saw them. So many data there. I did have one question about kynurenic acid. Can you tell me why was it you got a more effective blocking of kindling by kynurenic acid on the one excitatory amino agonist compared to the other. I can't remember which was better at blocking, but it was better at blocking one than the other. And secondly, why did you use a tenth of a molar dose which was so much smaller than the agonist dose you used.

DR. MORI: Kynurenic acid totally blocked the electroclinical response induced by glutamate, and significantly retarded the seizure development induced by aspartate. Kynurenic acid is known to reduce depolarization and excitation elicited by NMDA, kainate and quisqualate in this rank order of potency. NMDA appears to excite aspartate-preferring receptors, and quisqualate appears to excite glutamate-preferring receptors. Our results, however, showed kynurenic acid to have a stronger effect against glutamate than aspartate, strongly suggesting that aspartate is more potent than glutamate in producing epileptiform activity. Indeed, the number of injections required for the development of the first stage 1 manifestation was significantly smaller in aspartate kindling than in glutamate kindling. These results do not conflict with the recent observations that almost all the excitatory amino acid antagonists, which exert a potent anticonvulsant action in many animal models of epilepsy, block excitation caused by aspartate or NMDA. The antiepileptic potency of kynurenic acid has been reported to be comparable to that of 2-APH in several animal models of epilepsy. In this study, intra-amygdaloid injection of 0.1 or 0.2 μ moles of 2-APH was effective in suppressing the electrically-kindled seizure. In addition, the rate of chemical kindling by means of intra-amygdaloid injections of 1.5 μ moles of glutamate or aspartate was similar to that of electrical amygdaloid kindling. For these reasons, we chose 0.1 μ mole of kynurenic acid, which was about one tenth of a molar dose glutamate or aspartate.

DR. UEMURA: I'd like to make some comments since I've been engaged in a study of bicucullin kindling. In my experiment we also found positive transfer between bicucullin amygdaloid kindling and electrical kindling. The animals which are subjected to the bicucullin kindling after the completion of electrical kindling, showed stage 5 seizure on the first injection of bicucullin. Similarly the animals which were subjected to electrical kindling after the completion of bicucullin kindling reached stage 5 at a mean of 2.7 electrical stimulations. So we think that the transfer effect from electrical kindling to bicucullin kindling is stronger than that of the opposite direction. Another interesting thing we found is that in transfer kindling we have never seen the phases of stage 3 and stage 4. That is, all animals reached stage 5 on the first stimulation of transfer kindling or they reached stage 5 by skipping over the phase of stage 3 and stage 4. Did you observe such phenomena?

DR. MORI: Unlike your results, obtained with bicucullin kindling, the electrically-kindled animals did not always respond with stage 5 seizure upon the first injection of glutamate, aspartate, or glutamate plus aspartate mixture. However, the electrically-kindled animals sometimes skipped stages 3 and/or 4 manifestations when subjected to excitatory amino acid kindling. Also, the animals kindled chemically with excitatory amino acid skipped stages 3 and/or 4 seizures when subjected to electrical kindling.

DR. UEMURA: Then I think we can conclude that the generalization mechanism responsible for stage 3 to stage 5 should be the same and established in both types of kindling, but I'm wondering whether the common neuronal pathway responsible for stages 1 and 2 exists or is established after either types of kindling. Do you have any ideas on this?

DR. MORI: The intensity of the electroclinical response produced by chemical agents such as bicucullin and excitatory amino acids, parallels the size of the dosage used. Therefore, like the response to bicucullin injection in your study, if the dose of excitatory amino acids were increased, all the electrically-kindled animals may respond with stage 5 seizure on the first injection. However, the electroclinical response induced by electrical stimulation does not parallel the stimulus intensity. When those animals which were kindled chemically with bicucullin or excitatory amino acids were subjected to electrical kindling, they skipped stages 3 and/or 4 seizures, independent of the stimulus intensity used for electrical kindling. For these reasons, I agree with your assumption.

DR. MODY: It's very interesting that you find that injections of aspartate were more potent in producing kindling than glutamate because from biophysical studies it's more apparent that aspartate may be a selective NMDA agonist as opposed to glutamate. Instead of kynurenic acid, I was wondering if you tried 7-chlorol kynurenic acid which is a more specific antagonist at the glycine site, and if you've ever tried injecting NMDA itself.

DR. MORI: We have tried neither 7-chlorol kynurenic acid nor NMDA.

DR. SATO: I think it's very important to remember that the essential condition with regard to kindling is the appearance of repeated after-discharge. So, whatever you use to induce after-discharge you can kindle the animal, for example, bicucullin, carbachol or glutamate, etc. So I wonder, what is the common mechanism of glutamate kindling? Just for that mechanism to produce after-discharge or the mechanism for inducing trans-synaptic change? I mean, it is important to distinguish the two processes. How to produce after-discharge and how to kindle the animal are two quite different processes. So it is important to make it clear. Your data of interchangeability just shows the mechanism to induce after-discharge. Do you have any comment on this?

DR. MORI: Your view is very important for understanding the mechanisms of the kindling phenomenon. Chemical kindling appears to be a useful tool to study the two mechanisms you mentioned, i.e., how to produce afterdischarge and how to cause the trans-synaptic changes. In electrically kindled animals, an initially ineffective dose of excitatory amino acids became effective in producing sustained spikes. Furthermore, electrical stimulation at the afterdischarge threshold produced more intense and prolonged afterdischarges following completion of chemical kindling. These observations suggest that excitatory amino acids and electrical stimulation share a common mechanism for producing afterdischarge. The rate of electrical kindling was significantly facilitated following completion of excitatory amino acid kindling, and the development of stage 5 seizure was significantly encouraged when the electrically-kindled animals were subjected to electrical kindling. In addition, excitatory amino acid and electrical kindling showed a very similar pattern of seizure development. These results suggest that both excitatory amino acid- and electrically-induced afterdischarges produce the common trans-synaptic changes.

mentioned, i.e., how to produce afterdischarge and how to cause the transsynaptic changes. In electrically kindled animals, an initially ineffective dose of excitatory amino acids became effective in producing sustained spikes. Furthermore, electrical stimulation at the afterdischarge threshold produced more intense and prolonged afterdischarges following completion of chemical kindling. These observations suggest that excitatory amino acids and electrical stimulation share a common mechanism for producing afterdischarge. The rate of electrical kindling was significantly facilitated following completion of excitatory amino acid kindling, and the development of stage 5 seizures was significantly enhanced when the electrically-kindled animals were subjected to chemical kindling. In addition, excitatory amino acid and electrical kindling showed very similar pattern of seizure development. These results suggest that both excitatory amino acid- and electrically-induced afterdischarges produce the common transsynaptic changes.

# INVOLVEMENT OF EXCITATORY AMINO ACIDS IN THE MECHANISMS OF KINDLING

Gérard RONDOUIN, Mireille LERNER-NATOLI, Robert CHICHEPORTICHE and Jean-Marc KAMENKA

CNRS UPR 41 - INSERM U 249, Montpellier, France

There are several lines of evidence that excitatory amino acids (EAA) receptors, which have a rather ubiquitous distribution in the CNS play a role in the genesis of epilepsy. Systemic or intracerebroventricular injections of EAA agonists (glutamate, NMDA, kainate, quisqualate) were all demonstrated to induce epilepsy (22), while their antagonists were potent antiepileptic or anticonvulsant agents in a wide variety of animal models (5,6). Furthermore, epileptic activities were often associated with an increase in the release of the putative endogenous transmitters aspartate and glutamate (7,15,29). There exist at least three different subclasses of EAA receptors (NMDA, quisqualate and kainate) and these multiple receptors may lead different responses to the same transmitter (8). In addition, it appears that ions are able to modulate their activation (18). Among these subclasses, most of the studies focused, in part because of the relative abundance of NMDA-antagonists, on the role of the NMDA-receptor which is mainly involved in the plasticity of the central nervous system and the generation of oscillatory activity (38). Several results indicate that it plays too a crucial role in epilepsy. An application of NMDA into the caudate nucleus of the rat elicited a repetitive bursting response (11). Such a bursting activity can be induced by perfusion of hippocampal slices with a magnesium free medium (1), the presence of this ion being known to modulate the opening of NMDA channels. This epileptiform activity is inhibited by 2-amino-5-phosphono-valerate (APV), a competitive NMDA receptor antagonist (12). APV also blocked epileptiform discharges induced in hippocampal slices by bicuculline (13), indicating the involvement of NMDA receptors even in epileptic models initially related to a blockade of GABAergic transmission. Such effects were partially demonstrated with non competitive antagonists which block the NMDA receptor associated ionic channel (14). To go further insight these mechanisms, we need in vivo models in which we have a progressive development of seizures. This is realized in the kindling model (10). The progressive aggra-

*Kindling 4*
Edited by J. A. Wada
Plenum Press, New York, 1990

vation of seizure severity and of AD duration observed in kindling (26), allows to study the development of epilepsy as well as the mechanisms of generalized seizures. Moreover, from a pharmacological point of view, this model permits to distinguish between an antiepileptic effect of a drug and its anticonvulsant properties. In 1982, phencyclidine (PCP) (which was not yet known as an antagonist of NMDA receptor) was shown to retard the kindling development in rats (2). More recently, APV (3), D,L-2-amino-7-phosphonoheptanoate (APH) (35) and MK-801 (9,19,28) were found to inhibit the kindling rate too. Other clues seemed to indicate a role of NMDA receptors in this phenomenon: the binding of [3H]TCP was decreased (30) in the hippocampus of amygdala kindled rats and an NMDA receptor contribution (normally absent) to electrophysiologic response of dentate gyrus granule cells was demonstrated in kindled animals (20), suggesting a possible modulation of NMDA receptors by the kindling process.

In the experiments reported here, we used thienyl-cyclohexyl-piperidine (TCP), a derivative of PCP with a high affinity for the PCP receptor (36), to approach, both in vivo and in vitro, the mechanisms of kindling related to NMDA receptors and the associated channels. In addition, we determined the effects on kindled seizures of other derivatives of PCP which display different affinities for the TCP sites.

## METHODS

### Effect of TCP on kindling development or kindled seizures

Male Sprague & Dawley rats, weighing 200 to 250g at time of surgery, were used for this study. The animals were anesthetized with equithesin (0.4 ml/100 g). A recording-stimulating electrode, made of two strands of nickel-chrome insulated except at the tip was implanted in the right amygdala, connected to a plug and linked to the skull with an acrylic cement. The rats were divided into groups according to the following protocol: group I rats (n=24) received a daily intraperitoneal injection of saline; group II rats (n=20): TCP 5 mg/kg; group III rats (n=8): TCP 1 mg/kg; group IV rats (n=20): TCP 0.5 mg/kg. The animals were stimulated 15 min after drug injection and their behaviour observed before, during and following the stimulation. After the 15th stimulation, saline was substituted for TCP, except for 9 rats in group II that still received TCP (5 mg/kg) until the 20th stimulation. A fifth group (n = 10) underwent the kindling protocol and was used to compare the effects of a single dose of diazepam (Valium', Roche) (2 mg/kg), TCP, GK13, GK11, GK73, GK108, GK115 (5 mg/kg) or saline (control) on kindled seizures. For each rat, the behavioural stage of kindling was assessed according to the classification of Racine (26) and the afterdischarge (AD) duration, observed on the EEG track, was noted. The AD durations were compared with an analysis of variance and the behavioural stages with a Wilcoxon's rank test. At the end of the experiments, animals of groups III, IV and V were deeply anesthetized with ether oxyde. A 25 µA direct current was applied through the bipolar electrode for 2 min. The brains were removed after an intra-aortic perfusion of an isotonic 4 % formalin and 2 % potassium ferrocyanide buffer. The electrode location was controlled on 40 µm slices cut with a vibratome and stained by 1 % cresyl-violet. Rats of the two first groups underwent binding studies.

TCP binding studies

Group I (24 kindled rats) and group II (20 kindled with TCP) were used in this study. In addition, 25 control rats were handled every day but not stimulated. Three days after the last stimulation, rats of the 3 groups were killed by decapitation, their brain quickly removed and cortice, striata, hippocampi and cerebella dissected at 4 °C and stored in liquid nitrogen until used for binding assays. Homogenates were prepared in 50 mM Tris-HCl pH 7.7 buffer and [3H]TCP binding experiments were performed as previously described (37). Homogenates (0.5-0.9 mg of protein per ml) were incubated for 30 min at 25 °C in a 5 mM Hepes-Tris, pH 7.7 buffer in the presence of concentrations of [3H]TCP ranging from 0.5 nM to 1 µM. The homogenate-bound [3H]TCP was separated from the free one only by rapid filtration on GF/B Whatman glass fibre filters pretreated with 0.05 % polyethyleneimine. Then, the bound radioactivity was counted in 3 ml ACS (Amersham) in 6 ml minivials with a liquid scintillation counter (LKB Rackbeta 1214). Non specific binding was determined in the presence of 100 µM unlabelled TCP and displaceable binding to filters alone was measured in parallel experiments. Binding parameters were determined by a Scatchard plot analysis, using a two sites model program (27). The means ± SEM of the calculated values were compared for significant differences by a Student "t" test.

Hippocampal slices

Male Wistar rats (170-220 g) were decapitated under ether anesthesia, their brains were removed and the hippocampi dissected into chilled (10-15° C), oxygenated artificial cerebrospinal fluid (ACSF) (composition in millimolar : NaCl 124 mM, KCl 2.5 mM, $KH_2PO_4$ 1.25 mM, $MgSO_4$ 2mM, $NaHCO_3$ 26 mM, $CaCl_2$ 2.5 mM, glucose 10 mM). Transverse slices of 400 µm thickness were cut using a Mc Ilwain tissue chopper and transferred into a recording chamber which was continuously perfused at a rate of 1.8 ml/min with oxygenated (95 % $O_2$ - 5% $CO_2$) ACSF. Fourty five minutes later, the slices were then perfused with ACSF, complemented with saccharose 4 mM, progressively warmed until reaching a constant temperature of 35.5 °C. Slices were then allowed to equilibrate in the chamber for at least 60 min before data collection. Electrical activity was recorded by a 1-2 MΩ ACSF filled glass microelectrode placed in the pyramidal cell body layer of CA1. A bipolar platinum-irridium electrode was positioned in the stratum radiatum to stimulate the Schaffer's collateral fibres. Slices were stimulated with monophasic constant voltage current pulses from a lab-made stimulator (9-30 V, 0.1 ms duration). For induction of long term potentiation (LTP), 3 trains of 100 ms duration, 250 Hz frequency, were repeated with an inter-train interval of 2 s, and with a stimulation voltage set to induce half of the maximal population spike (PS) amplitude. The signal was digitalized on line by a computer and stored on disk. Applications of drugs were always preceded by at least a 15 min control period in drug-free ACSF. The amplitude of the PS was measured for different intensities in order to define the input-output (I/O) curve for each slice (PS amplitude vs stimulus voltage). I/O curves were determined before and after tetanus and in drug-free and drug conditions. In some experiments, the slices were perfused with ACSF containing low doses [2-4 μM] of the GABA-A antagonist bicuculline. Slices were also prepared from rats having experienced limbic epilepsy induced by a previous (8 days before the slice experiment) intra-amygdaloid injection of 1 nmole of kainic acid. These slices were studied in normal and low [$Mg^{2+}$] ACSF.

## RESULTS

### Effect of TCP on kindling development

Daily injections of TCP (5 mg/kg), 15 mn prior to the amygdala stimulation, clearly retarded kindling. The progression of both kindling stages and AD duration were delayed. Cessation of TCP treatment resulted in a progressive development of AD and rank of seizures parallel to those of the control group (figure 1).

### Effect of TCP on seizures in kindled rats

TCP (5 mg/kg) elicited a significant decrease ($p < 0.001$) of AD duration compared to controls (from 100-120 s to 70 s), but this effect was clearly lower than that of diazepam (2 mg/kg). None of the tested drug altered the AD threshold. As regards behaviour, TCP treated rats displayed seizures ranging between stage 1 and 3. Similar results were obtained with GK 11, GK73 and GK108 while GK13 and GK115 had no effects (Figure 2). It thus appears that the early mechanisms of kindling were blocked and that the development of kindling was shifted to the right as a function of the TCP treatment. Such an effect was reported by different authors with PCP, ketamine, MK801 (2,9,19,28). This effect was different of what may be observed with a classical anticonvulsant drug which delayed kindling but was unable to block the classic facilitating accumulation effect of repeated stimulations on the further development of kindling (24). While TCP has a high efficacy in blocking kindling development, it has a rather slight effect on AD duration and severity of seizures

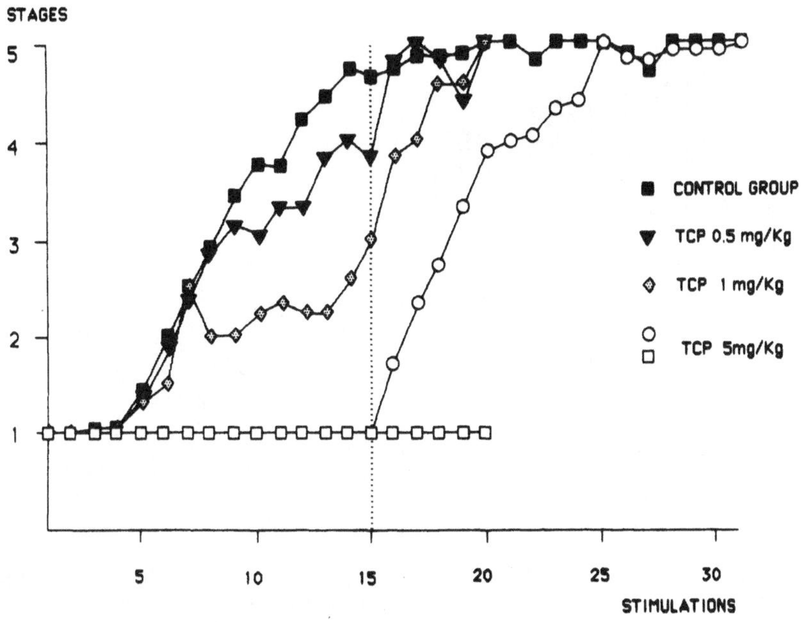


Figure 1. Inhibition of kindling development by daily TCP treatment administered 15 min before the amygdala stimulation.

in previously kindled rats. TCP (5 mg/kg) only reduced the duration of AD and the rank of seizures to stage 3 (moreover, TCP induced ataxia may prevent to correctly discriminate the stage). Other derivatives had effects related to their affinity for TCP sites indicated by their IC50 for [3H]TCP binding. MK801, which as TCP does, binds with the PCP receptor, was recently reported to have anticonvulsant effects in amygdala, hippocampal and cortical kindled animals with different degrees depending of the stimulated structure (9). In addition, it appears that the doses necessary for this anticonvulsant effect were more important than those required to block the development of kindling.

These last results and those reported by Mody and Heinemann (20) suggest that NMDA receptors and associated channels may be modulated by the kindling procedure and support the permanence of the phenomenon. Therefore we investigated [3H]TCP binding in kindled rats.

Kindling and TCP binding

Scatchard plots obtained from saturation curves showed that [$^3$H]TCP bound to two distinct sites. Within all the regions, Kd values of [$^3$H]TCP were in the range of 4.3-7.1 nM for the high affinity binding sites and 340-792 nM for the low affinity binding sites. Neither the Kd nor the Bm of high affinity binding sites were changed by the kindling paradigm or TCP chronic treatment. While no significant modification could be evidenced in the cortex and cerebellum, a slight increase of the number of low affinity sites was observed in the kindled and kindled + TCP rats compared to controls. In

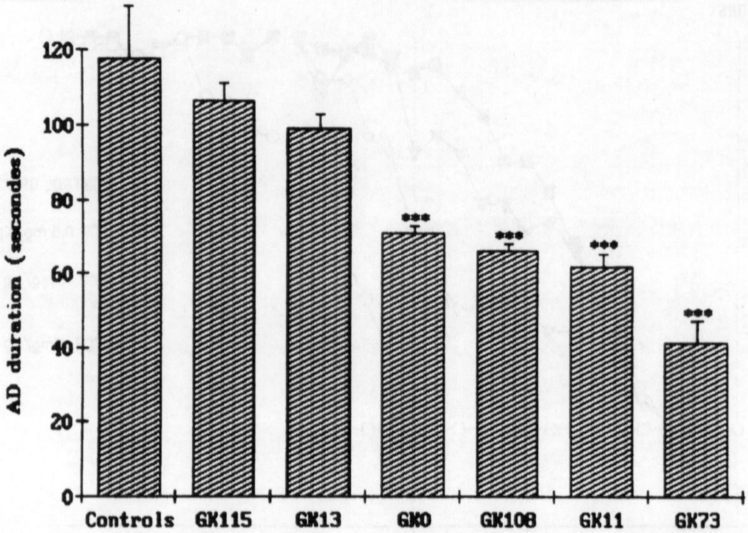


Figure 2. Effect of different derivatives of PCP (5 mg/kg) on AD duration in kindled rats. The results are correlated to the IC50 of each derivative: GK11 (7 μM) < TCP < PCP < GK108 < GK115 (110 μM) < GK13 (8000 μM), measured in Herpes-Tris buffer , pH 7.5, 25° C.

contrast, whereas kindling alone reduced BmL in the striatum, the treatment by TCP increased both KdL and BmL in the hippocampus (Figure 3). These results (i) confirm the existence of two families of [$^3$H]TCP binding sites in the rat brain, (ii) underline the close distribution of the high affinity TCP binding sites (hippocampus > cortex > striatum > cerebellum) to that of NMDA receptors (16), (iii) indicate that kindling did not induce an increase of the number of NMDA receptors as the results of Mody and Heinemann (20) could suggest. Okasaki et al. (21) recently observed no changes in the binding of [$^3$H]glutamate one day after the last seizure and a small but significant reduction of specific binding 28 days after the last stimulation and that may agree with our observations. Our results however are not consistant with those of Sircar et al. (30) who reported a decline of the density of PCP receptors 72 hours after the last class 5 seizure. Chronic TCP treatment did not affect the high affinity [$^3$H]TCP binding sites in the CNS of non stimulated or daily stimulated rats suggesting that the repetitive stimulations do not influence the modulation of NMDA receptors when the associated channels were blocked by TCP. Changes in the low affinity binding sites were only observed in the hippocampus and the striatum of kindled rats. The observation of Okasaki et al. (21) precludes a possible correlation between the increase of low affinity binding sites and an eventual enhancement of NMDA receptors. It seems rather that these sites would be other receptors such as monoamines uptake complexes or ionic channels. Finally, the inverse modifications observed in the striatum of kindled + saline versus kindled + TCP rats are of interest considering the role of the striatum in the generalization of seizures.

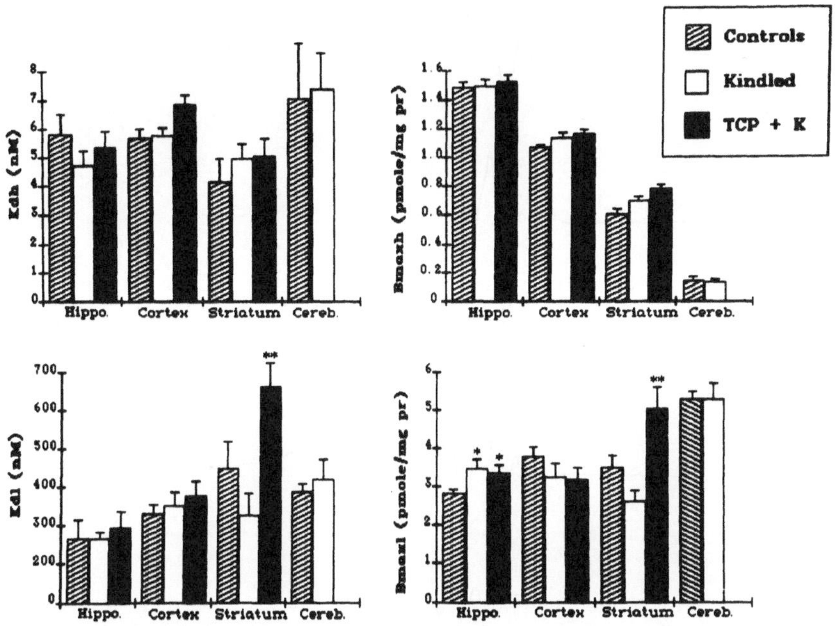


Figure 3. Affinities and number of TCP sites for the high (Kdh and Bmaxh) and low (Kdl and Bmaxl) affinity [$^3$H]TCP binding sites, measured in various structures of control, kindled-saline and kindled-TCP rats.

Kindling is known to be associated with a potentiation of synaptic transmission, at least in the dentate gyrus (17). Therefore we tried to determine in hippocampal slices in vitro, the contribution to epileptogenesis of LTP resulting from tetanic stimulation. In addition, the involvement of NMDA receptors under such conditions was compared with their role in epileptiform activities recorded in slices prepared from epileptic rats.

We first demonstrated (results not shown) that TCP (10-50 µM) prevented the induction of LTP in CA1 by a tetanic stimulation of the Schaffer's collateral fibres and reversed a former induced LTP. The tetanic stimulus not only induced LTP too, when bicuculline (2-4 µM: doses which did not produce epileptiform events per se) was added to the perfusion, but also triggered epileptic responses (Figure 4). We observed the same phenomenon if the tetanus was given before the perfusion of the convulsant agent. These effects were prevented and reversed by the application of TCP (20-40 µM) and APV (10-20 µM) indicating that NMDA receptors were involved in this potentiation of epileptic activities. It can be assumed on one hand that, as already demonstrated, LTP was easily obtained in such disinhibitory conditions and on the other hand that LTP may decrease the threshold of epileptic events. These results indicate that a LTP like mechanism might, at least partially, be implicated in the generation of epileptiform activities in CA1. Such mechanisms associated with primary or secondary impairment of inhibition, could be involved in kindling or in stimulus train induced bursting (STIB) and could account for the progression of AD from the primary focus to secondary foci, the repetitive epileptic

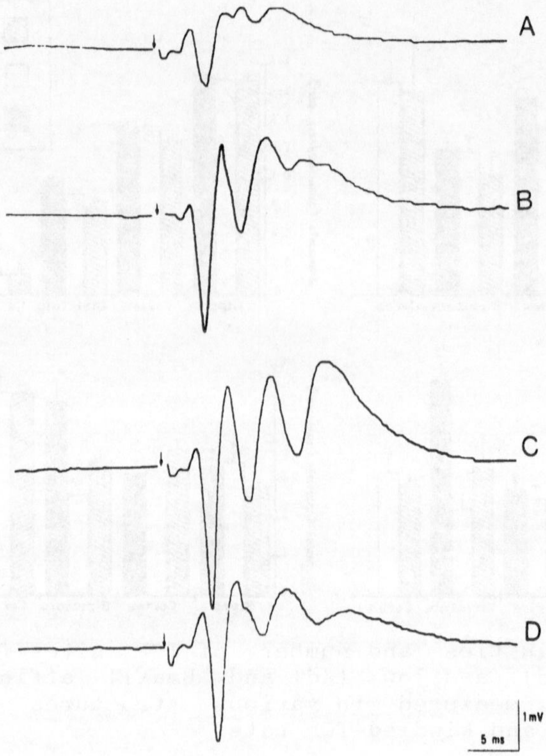


Figure 4. Evidence that a tetanic stimulation potentiates epileptiform activities. Bicuculline (4 μM) was added to ACSF A: CA1 population spike (stimulation 15 V) 45 min after the beginning of the Bic + ACSF perfusion. B: 5 min after tetanus. C: 20 min after tetanus. D: 20 min after TCP application (30 μM - 15 min).

discharges inducing first a LTP-like phenomenon in the target structure and a secondary impairment of inhibition as demonstrated in CA1 by Steltzer et al. (31). Some presumptions of the existence of such a phenomenon could be found in the kindling litterature (32). Moreover, it was demonstrated that repetitive subliminar (i.e. not inducing AD) high frequency stimulations (similar to those inducing LTP) facilitated the further development of kindling.

Animals injected into the amygdala with 1 nmole of KA, consequently displayed in the hippocampus continuous spiking during a few hours occasionnally resulting in mild seizures. Slices prepared from these rats 8 days later, showed an enhanced susceptibility (i) to bicuculline [2 μM] and (ii) to the lowering of $Mg^{2+}$ concentration. Low $Mg^{2+}$ concentrations (0.5-0.1 mM) induced epileptiform evoked responses and spontaneous bursting in CA3 and CA1, similar to the interictal-like activity reported by Wilson et al. (39) during the exposure of hippocampal slices to $Mg^{2+}$ free ACSF. TCP (30 μM) suppressed these activities only in CA1. Coan et al. (4) clearly showed that, in this ionic conditions, a NMDA receptor-mediated component participated to the synaptic transmission.

## CONCLUSION

From these results, there is a good evidence that NMDA receptors are implied in the mechanisms of limbic epilepsy. The blockade of the early steps of kindling by NMDA antagonists may be interpreted as a focal involvement of these receptors. However, these antagonists may also act at distant secondary sites and at key sites crucial to the development of kindling. It was indeed demonstrated that a part of the anticonvulsant action of NMDA antagonists could be explained by their action in the deep prepyriform cortex (25) the substantia nigra (34) and the entopedoncular nucleus (23). The involvement of NMDA receptors may induce or be concomitant of plastic change as early sprouting in gyrus dentatus (33) and long term potentiation of synaptic transmission. It is however evident from clinical and experimental data that such phenomenons are not sufficient to account for epilepsy. They could be followed if the epileptic activity is sustained by an excessive activation of these systems resulting in an enhancement of $Ca^{2+}$ influx and secondary excitotoxic lesions, undesirable phenomenons which may also be prevented by NMDA antagonists.

## REFERENCES

1) Anderson, W.W., Lewis, D.W., Swartzwelder, H.S. and Wilson, W.A., 1986, Magnesium-free medium activates seizure-like events in the rat hippocampal slice, Brain Research, 398: 215-219.
2) Bowyer, J.F., 1982, Phencyclidine inhibition of the rate of development of amygdaloid seizures, Exp. Neurol., 75: 173-183.
3) Cain, D.P., Desborough, K.A. and McKitrick, D.J., 1988, Retardation of amygdala kindling by antagonism of NMDA and muscarinic cholinergic receptors: evidence for the summation of excitatory mechanisms in kindling, Exp. Neurol., 100: 179-187.
4) Coan, E.J., Saywood, W., Collingridge, G.L., 1987, MK-801 blocks NMDA receptor-mediated synaptic transmission and long term potentiation in rat hippocampal slices, Neurosci. Lett., 80: 111-114.
5) Croucher, M.J., Collins, J.F. and Meldrum, B.S., 1982, Anticonvulsant action of excitatory aminoacid antagonists. Science, 216: 899-901.
6) Czuczwar, S.J. and Meldrum, B.S., 1982, Protection against chemically-induced seizures by 2-amino-7-phosphonoheptanoic acid, Eur. J. Pharmacol., 83: 335-338.
7) Dodd, P.R., Bradford, H.F., Abdul-Ghani, A.S., Cox, D.W.G., Coutinhonetto, J., 1980, Release of amino acids from chronic epileptic and sub-epileptic foci in vivo, Brain Research, 193: 505-517.
8) Foster, A.C., and Fagg, G.E., 1984, Acidic amino acid binding sites in mammalian neuronal membranes: their characteristics and relationship to synaptic receptors. Brain Research Rev., 7: 103-164.
9) Gilbert, M.E., 1988, The NMDA-receptor antagonist, MK801, suppresses limbic kindling and kindled seizures, Brain Research, 463: 90-99.

10) Goddard, G.V., McIntyre, D.C. and Leech, C.K., 1969, A permanent change in brain function resulting from daily electrical stimulation, Exp. Neurol., 25: 295-330.
11) Herrling,P.L., Morris, R. and Salt, T.E., 1983, Effects of excitatory amino acids and their antagonists on membrane and action potentials of cat caudate neurons. J. Physiol., 339: 207-222.
12) Herron, C.E., Williamson, R. and Collingridge, D.L., 1985, A selective N-methyl-D-aspartate antagonist depresses epileptiform activity in rat hippocampal slices, Neurosci. Lett., 61: 255-260.
13) Hynes, M.A. and Dingledine, R., 1984, Attenuation of epileptiform burst firing in the rat hippocampal slice by antagonists of NMDA receptors. Soc. Neurosci. Abstr., 10: 229.
14) Kemp, J.A., Foster, A.C. and Wong, E.H.F., 1987, Non competitive antagonists of excitatory aminoacid receptors, Trends in Neurosci., 10: 294-298.
15) Leach, M.J., Marden, C.M., Miller, A.A., O'Donnel, R.A., Weston, S.B., 1985, Changes in cortical amino acids during kindling in rats, Neuropharmacology, 24: 937-940.
16) Maragos, W.F., Penney, J.B. and Young, A.B., 1988, Anatomic correlation of NMDA and [$^3$H]TCP labelled receptors in rat brain, J. Neurosci., 8:493-501.
17) Maru, E. and Goddard, G.V., 1987, Alteration in dentate neural activivties associated with perforant-path kindling. I. Long term potentiation of excitatory synaptic transmission, Exp. Neurol., 96: 19-32.
18) Mayer, M.L. and Westbrook, G.L., 1987, The physiology of excitatory amino acids in the vertebrate central nervous system. Prog. in Neurobiol., 28: 197-276.
19) McNamara, J.O., Russell, R.D., Rigsbee, L. and Bonhaus, D.W., 1988, Anticonvulsant and antiepileptogenic actions of MK-801 in the kindling and electroshock models, Neuropharmacology, 27: 563-568.
20) Mody, I. and Heinemann, U., 1987, NMDA receptors of dentate gyrus granule cells participate in synaptic transmission following kindling, Nature, 326: 701-704.
21) Okasaki, M., McNamara, J.O. and Nadler, V., 1989, N-methyl-D-aspartate receptor autoradiography in rat brain after angular bundle kindling, Brain Research, 482: 359-364.
22) Olney, J.W., 1986, Excitotoxic aminoacids, News Physiol., Sci., 1: 19-23.
23) Patel, S., Millan M.H., Mello, L.M. and Meldrum B.S., 1986, 2-amino-7-phosphonoheptanoic acid (2APH) infusion into entopeduncular nucleus protects against limbic seizures in rats, Neurosci. Lett., 64: 226-230.
24) Peterson, S.L., Albertson, T.E., Stark, L.G., Joy, R.M. and Gordon, L.S., 1981, Cumulative afterdischarge as the principle factor in the acquisition of kindled seizures, Electroencephalogr. Clin. Neurophysiol., 51: 192-200.
25) Piredda, S. and Gale K., 1986, Role of excitatory amino acid transmission in the genesis of seizures elicited from the deep prepiriform cortex, Res. Reports, 377: 205-210.
26) Racine, R.J., 1972, Modification of seizure activity by electrical stimulations. II. Motor seizures, Electroencephalogr. Clin. Neurophysiol., 32: 281-294.
27) Rosenthal, H.E., 1967, Graphic method for the determination and presentation of binding parameters in a complex system, Annal. Biochem., 20: 525-532.

28) Sato, K., Morimoto, K. and Okamoto M., 1988, Anticonvulsant action of a non-competitive antagonist of NMDA receptors (MK-801) in the kindling model of epilepsy, Brain Research, 463: 12-20.
29) Sherwin, A., Robitaille, Y., Quesney, F., Olivier, A., Villemure, J., Leblanc, R., Feindel, W., Andermann, E., Gotman, J., Andermann, F., Ethier, R., Kish, S., 1988, Excitatory amino acids are elevated in human epileptic cerebral cortex, Neurology, 38: 920-923.
30) Sircar, R., Ludwig, N., Zukin, S.R. and Moshe, S.L., 1987, Down-regulation of hippocampal phencyclidine (PCP) receptors following amygdala kindling, Eur. J. Pharmacol., 141:167-168.
31) Stelzer, A., Slater, N.T. and Bruggencate, G.T., 1987, Activation of NMDA receptors blocks GABAergic inhibition in an in vitro model of epilepsy, Nature, 236: 698-701.
32) Sutula, T. and Steward, O., 1987, Facilitation of kindling by prior induction of long-term potentiation in the perforant path, Brain Research, 420: 109-117.
33) Sutula, T., Xiao-Xian, H., Cavazos, J. and Scott, G., 1988, Synaptic reorganization in the hippocampus induced by abnormal functional activity, Science, 239: 1147-1150.
34) Turski, L., Cavalheiro, E.A., Turski, W.A. and Meldrum, B.S., 1986, Excitatory neurotransmission within the substantia nigra pars reticulata regulates threshold for seizures produced by pilocarpine in rats: effects of intranigral 2-amino-7-phosphonoheptanoate and N-methyl-D-Aspartate, Neurocscience, 18: 61-77.
35) Vezzani, A., Hui-Qiu, W., Moneta, E. and Samanin, R., 1988, Role of the N-methyl-D-aspartate-type receptors in the development and maintenance of hippocampal kindling in rats, Neurosci. Lett., 87: 63-68.
36) Vignon, J., Chicheportiche, R., Chicheportiche, M., Kamenka, J.M., Geneste, P. and Lazdunski, M., 1983, [$^3$H]TCP: a new tool with high affinity for the PCP receptor in rat brain, Brain Research, 280: 194-197.
37) Vignon, J., Privat, A., Chaudieu, I., Thierry, A., Kamenka, J.M. and Chicheportiche, R., 1986, [$^3$H] thienyl-phencyclidine ([$^3$H]TCP) binds to two different sites in rat brain. Localization by autoradiographic and biochemical techniques, Brain Research, 378: 133-141.
38) Wallen, P. and Grillner, S., 1987, NMDA receptor-induced, inherent oscillatory activity in neurons active during fictive locomotion in the lamprey, J. Neurosci., 7: 2745-2755.
39) Wilson, W.A., Swartzwelder, H.S., Anderson, W.W., Lewis, D.W., 1988, Seizure activity in vitro : a dual focus model, Epilepsy Research, 2: 289-293.

Discussion of Dr. Rondouin's Presentation

DR. LOSCHER: I have a question regarding the selectivity of the drugs which you are using. We have heard of a lot on the anticonvulsant action of so-called NMDA antagonists such as TCP, PCP and MK801. From drug discrimination studies it is known that, for example, PCP but also MK801 generalize to drugs which stimulate 5HT receptors. For instance, MK801 generalizes in such discrimination studies to compounds which selectively stimulate the 5HT 1A sub-type. So could you comment on the selectivity of TCP.

DR. RONDOUIN: I think that the experiment with different derivatives of PCP, which has a different affinity for the TCP sites, could answer your question because it seems to me that their effect is very well related to this affinity to the TCP receptor. I don't know if they are able to generalize, as you mentioned, to 5HT, because if so the 5HT hypothesis would be completely discarded. But I think that there is good evidence from our, and also from the work of McNamara, and also from others about NMDA receptors, that these mechanisms are clearly related to NMDA receptor. For example, BTCP which is a very powerful inhibitor of monoamines has no effect on kindling development and the kindled seizures that we saw.

DR. CAIN: You concluded that TCP blocked the early stages of kindling, you used the word blocked, and I'm not sure I saw that in the data. In other words, as I recall, when you terminated the drug treatment and gave your first subsequent stimulation, the animals were at about stage 1.7 or so. Then they took about another 7 after-discharges to fully kindle, so that is very consistent with all the other results we've been discussing here, similar nature, and that tells me something slightly different, that there is, to be picky, a small but non-trivial amount of savings that's going on there. I just wonder if you'd like to comment on that.

DR. RONDOUIN: You know that the blockade of the channel by TCP is voltage dependent, so probably repeated depolarization could in part unblock the system. This is very interesting to see and we can see it very well in the slice preparation. Let's look at the slide.

DR. CAIN: Now, what I see in the high dose group is that on the first non-drug stimulation they're at about 1.6 or 1.7, stage 1.6 or 1.7, then after another 6 or 7 AD's they're right up there. Is that not an indication of some savings due to the first 15 drug stimulations?

DR. RONDOUIN: What we see, for example on slice, is that the blockade by TCP is not complete. For example, in low magnesium you can block the increase of the population spike you see in this condition by TCP, you can reverse it. If you put a normal slice in magnesium-free you see first an increase of the population spike, and then if you put TCP block, you reverse it. If you then stimulate the slice with tetanic stimulation, you can see it, you reverse the block. So probably, it is known for ketamin that the blockade of the channel is voltage dependent. This repetitive depolarization of the neurons could indicate a partial unblock of the channels, and that allows for such - if we could see the next slide, we can see that there is a little prohibition of the after-discharge duration during this period.

# KINDLING AND EXCITATORY AMINO ACIDS

Mitsumoto Sato[1)], Kiyoshi Morimoto[2)],
Kazufumi Akiyama[2)] and Motoi Okamoto[2)]

1)Department of Psychiatry, Tohoku University, Sendai
2)Department of Neuropsychiatry, Okayama University
Okayama

## INTRODUCTION

The kindling effect provides an excellent animal model of human epilepsy, which has been defined as a chronic brain disorder characterized by recurrence of seizures due to excessive discharge of cerebral neurons. In clinical respests, a long lasting seizure susceptibility of the brain relates to outcome of epilepsy, while recurrence of seizures, especially epileptic status, results in difficulties in social functioning of the patients with epilepsy. Using kindling model, these two different processes (seizure susceptibility and seizure induction) can be clearly identified by separating the processes of seizure development during kindling sessions and the fully kindled seizures. In addition, it seems noteworthy that the fully kindled convulsive seizure is different from acute convulsions such as pentetrazol- and penicillin-induced convulsions, due to involvement of kindling-induced seizure susceptibility of the brain in induction of seizures.

This report includes our recent evidence to discuss a possible relationship of excitatory amino acid (EAA)-mediated neurotransmission to both the seizure susceptibility and the seizure induction mechanisms.

## EAA AND KINDLED SEIZURE SUSCEPTIBILITY

Kindling is a phenomenon of progressive and lasting increase in seizure susceptibility produced by repeated induction of localized afterdischarges (ADs) in a brain structure. Kindling rates may represent a process of enhancement of seizure susceptibility in originally non-epileptic brain. To examine a relationship of EAA-mediated neurotransmission to the kindling-induced seizure susceptibility, prophylactic effects of N-Methyl-D-aspartate (NMDA) antagonists on kindling were first examined in amygdala (AM) kindling. Secondly, effects of NMDA antagonists on long-term potentiation (LTP) were examined to understand a mechanism of the effects of the NMDA antagonists on kindling. Thirdly, specific binding of NMDA receptors in the AM and hippocampus (HIPP) were determined in the AM kindled rat brains to examine whether a change in NMDA receptors relates to the long-lasting seizure susceptibility. Finally, our recent findings of altered ibotenate-stimulated hydrolysis of polyphosphoinositide (PPI) after kindling were presented to discuss a relationship between EAA receptor-mediated intracellular signal transduction and the seizure susceptibility.

*Kindling 4*
Edited by J. A. Wada
Plenum Press, New York, 1990

Effects of NMDA Antagonists on AM Kindling

(+)-5-methyl-10, 11-dihydro-5H-dibenzo[a,d]cyclohepten-5, 10-imine maleate (MK-801, a non-competitive NMDA antagonist) and 3-(2-carboxy-piperazin-4-yl) propyl-1-phosphonate (CPP, a competitive NMDA antagonist) were used.

Methods. Male Srague Dawley (SD) rats weighing between 250 and 300 g at the time of surgery were used. A tripolar electrorodes of isolated nichrome wire (0.18 mm in diameter) was implanted in the left AM (coodinate, AP: 0, L: 5.0, D: 8.0) under pentobarbital anesthesia. After one week of recovery from electrode implantation, kindling was commenced by AM stimulation. Parameters of stimulus were 60 Hz, sine wave and 1-sec duration at current intensity of 200 μA. On the first day, AM was stimulated without drug treatment to examine AD duration and stages of behavioral seizure manifestations, which was rated by Racine's classification [1]. Depending upon AD duration and seizure stages, rats were divided into 3 groups; control, MK-801-treated and CPP-treated group. From the second day to 4th day (drug sessions), either saline, MK-801 (1 and 2 mg/kg, i.p.) or CPP (5 and 10 mg/kg, i.p.) were administrated at 8:00 a.m.. Two hrs after the injections, six AM stimulations a day were delivered at 2 hr-intervals. After the drug sessions, stimulation was applied once daily without drug treatment until the rats developed to stage-5 convulsions on 5 consecutive days (drug-free sessions). Electrographic monitoring and behavioral observation were made at each AM stimulation.

Results and Discussion. As compared with control, both the prolongation of AD duration and development of seizure stages were significantly suppressed in MK-801-treated group during the drug sessions. The suppression was maintained for 3 to 4 days in the subsequent drug-free sessions (Fig. 1). The suppression of seizure development of the MK-801 treated group in the drug-free sessions was dependent of the dose administered during the drug sessions. While all rats of saline-treated control developed to stage-5 convulsions at 17th stimulation, all rats in MK-801-treated group remainedin stage-1 or 2 at 15th stimulation. Since each kindling stimulus induced ADs during the drug sessions and for subsequent few days, the prophylactic effect of MK-801 on AM kindling was not due to anticonvulsant action to diminish ADs which has been reported to be essential to produce kindling [1]. It is presumed that the prophylactic action of MK-801 may depend upon prevention of trans-synaptic changes resulting from synchronous bombardment of repeated ADs.

Ten mg/kg of CPP also suppressed the development of kindling during the drug sessions (Fig. 1). The suppression was more prominent in behavioral seizure development than the prolongation of AD duration. In the drug-free sessions after the drug sessions, 3 or 4 additional kindling stimuli were required to develop stage-5 seizures. However, duration of ADs initially produced in the drug-free sessions in the CPP-treated group did not differ from that of control. It is presumed that CPP may not prevent a basic mechanism for electrographic seizure development, which may be masked by anticonvulsant action of CPP to shorten duration of ADs during the drug sessions.

These results indicate that NMDA antagonists, especially a non-competitive antagonist MK-801 rather than competitive antagonist CPP, had an action to prevent AM kindling in rats. This finding is consistent with our previous report [2] that systemic administration of 0.25 or 1 mg/kg of MK-801 prior to each daily kindling stimulation prevented AM kindling in rats. The reason why non-competitive NMDA antagonist, rather than competitive antagonist, blocks trans-synaptic changes underlyed kindling

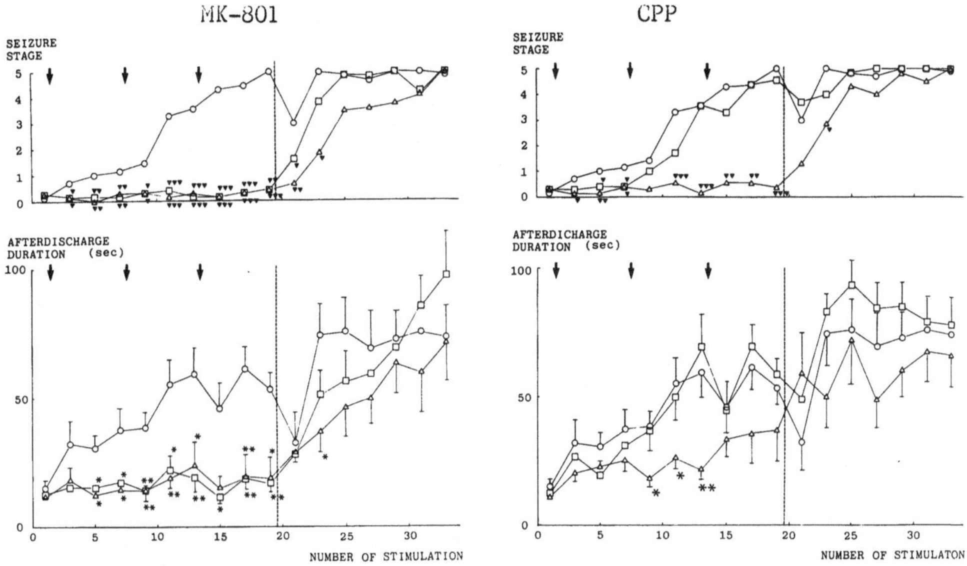


Fig. 1 Effects of MK-801 and CPP on Amygdala Kindling

Open circle : Control (n=7), Open square : MK-801 1mg/kg or CPP 5 mg/kg (n=7), and Open triangle: MK-801 2 mg/kg or CPP 10 mg/kg (n=7). * , ** and *** indicate $p < 0.05$, 0.01 and 0.001, respectively. ( Katayama, K. et al., unpublished data)

is not clear. One possibility is a insufficient dosis of CPP used in this study. The prophylactic action of CPP on kindling should be futher examined using dosis more than 10 mg/kg. As an another possibility, a kindling-induced change in NMDA-associated ion channels rather than NMDA receptors itselves might be critical for induction of kindling, since MK-801 has been suggested to block NMDA receptors via NMDA-associated ion channels [3]. Although, there are many discussion whether LTP shares a common trans-synaptic change in kindling, kindling-induced potentiation similar to LTP has been reported after HIPP [4] and perforant path [5] kindling. Accordingly, we examined the effects of these NMDA antagonists on LTP in next.

## Effects of NMDA Antagonists on LTP

Methods. Male SD rats weighing between 400 and 500 g at the time of electrode implantation were used. Under urethane anesthesia (1.2 g/kg, i.p.), monopolar recording electrodes and stimulating electrodes were implanted into the hilus of the dentate gyrus and the ipsilateral perforant path in one hemisphere, respectively. Test pulses consisted of 250 µsec and 200 - 500 µA at 30 sec-intervals that is enough to produce the population spike. LTP was induced by a set of trains, which consisted of 10 high-frequency (400 Hz, 20 msec) and high-intensity trains (250 µsec) at 1 sec-intervals. After 2 hrs of MK-801 (1 or 2 mg/kg, i.p.) or CPP (5 or 10 mg/kg, i.p.) administration, the LTP-inducing trains were delivered to the perforant path. All evoked potentials were sampled and stored on magnetic disc for data analysis. The slope of rising phase of the EPSP and the height of the population spike (PSH) were measured for each evoked potential. LTP was evaluated in terms of the percent increase in the EPSPs and PSHs between the average of 5 potentials evoked by 250 µsec test stimuli before and after induction of LTP. Rectal temperature was monotored and maintained at 36.5 to 37°C by a heating lamp.

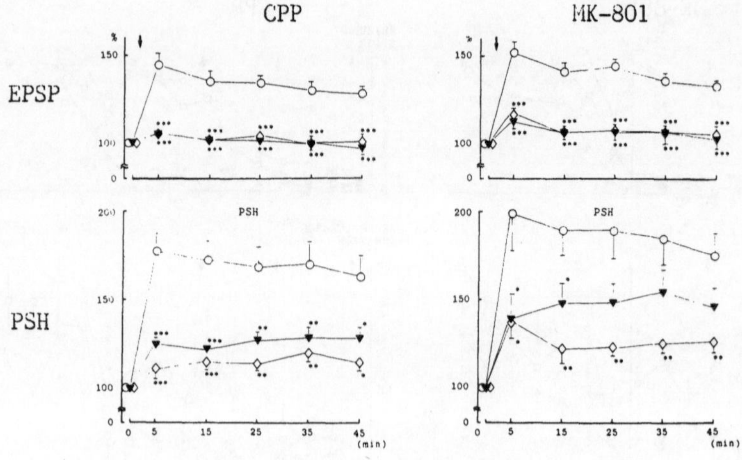


Fig. 2 Effects of CPP and MK-801 on LTP

Open circle, solid triangle and open square represent control, 5 mg/kg of CPP- or 1 mg/kg of MK-801-treated group, and 10 mg /kg of CPP- or 2 mg/kg of MK-801-treated group, respectively. PSH: population spike height. EPSP: slope of rising phase of EPSP. Value represents mean ± S.E.M. of percent change. *, ** and *** indicate $p<0.05$, 0.01 and 0.001 by Student' t-test compared with control. (Katayama, K. et al., unpublished data)

Results and Discussion. Both MK-801 and CPP had a significant action on LTP of the EPSPs, though these antagonists had no effect on evoked potentials before the tetanic stimulation. The early phase as well as later phase of LTP of the EPSPs was almost completely suppressed by these agents. It seems important to note that both NMDA antagonists suppressed LTP of synaptic components (EPSPs) more than cell discharge component (PSHs)(Fig. 2). This finding is partially consistent with ealier report by Coan et al.[6] that MK-801 prevented induction of LTP of the hippocampal CA1 region.

The present evidences that NMDA antagonists suppressed both the kindling seizure development and LTP indicate that activation of NMDA receptors and NMDA-associated ion channels is critical for induction of a long-lasting increase in synaptic transmission in these two phenomena. However, it still remains unclear whether a change in NMDA-mediated neurotransmission participates in induction of kindling-induced potentiation or maintainance of the induced potentiation. For this question, it appears necessary to examine a long-lasting change in NMDA receptors in the fully kindled rat brain.

Specific NMDA Receptor Binding of AM-kindled Rat Brains

[$^3$H]MK-801 and [$^3$H]CPP binding assays were performed in the AM and HIPP of AM kindled rats.

Methods. Male SD rats weighing 300 g at surgery were housed with free access to food and water under a 12-h light : 12-h dark (06.00 -18.00 light) cycle. A tripolar electrode was stereotaxically implanted into the left AM under pentobarbital anesthesia. Sham operated rats were used as controls. The rats were allowed to recover for 10 days before commencing kindling. The rats of kindling group received kindling stimulus once daily in a 1-sec train of 60 Hz sine wave at a current intensity of 200-400 μA until stage-5 of the kindled convulsion appeared on 5 successive days.

For $[^3H]$MK-801 binding, AM kindled rats and matched control were decapitated either 24 hrs or 7 days after the last stage-5 convulsion. For $[^3H]$CPP binding, those were decapitated either 24 hrs or 28 days after the convulsion.

Specific $[^3H]$MK-801 and $[^3H]$CPP binding

The brain tissues were homogenized using a Polytron in 50 vol of appropriate bufferes: 5 mM Tris-HCl buffer pH 7.4 at 25°C for $[^3H]$MK801; 50 mM Tris-acetate buffer pH 7.6 at 25°C for $[^3H]$CPP. The homogenates for both binding assays were centrifuged at 30,000 g for 15 min. Supernatant is discarded, and the resultant pellet was frozen at -70°C for at most one week. On the day of experiments, the frozen pellet was thawed and resuspended in 50 vol of the same buffers. For $[^3H]$MK-801 binding assay, the membrane was sonicated for 30 sec and washed 8 times with the same buffer by repeated centrifugations and resuspensions. For $[^3H]$CPP binding assay, the membranes were incubated with 0,1 % Triton X-100 at 25°C for 15 min and centrifuged at 30,000 g for 15 min. The membrane was washed four times with the same buffer without Triton X-100 by repeated centrifugations and resuspensions. The pellets for $[^3H]$CPP binding were finally suspended in 50 vol. of 50 mM Tris-acetate buffer. Saturation isotherm of specific $[^3H]$MK-801 binding was determined by incubating $[^3H]$MK-801 with the homogenates (40-100 μg protein) in the presence of 100 μM of L-glutamate in 1,0 ml of 5 mM tris-HCl buffer at 25°C for 60 min. Nonspecific $[^3H]$MK-801 binding was determined in the presence of 100 μM of MK-801. The reaction was terminated by rapid filtration through Whatman GF/B filters using a Brandel cell harvester. Saturation isotherm of specific $[^3H]$CPP with the homogenates (40-100 μg protein) in 0.5 ml of 50 mM Tris-acetate buffer at 25°C for 30 min. Nonspecific $[^3H]$CPP binding was determined in the presence of 1 mM L-glutamate. The reaction was terminated by rapid centrifugation. The resultant pellets were washed once with 50 mM Tris-acetate buffer and digested with Scintillamine®. protein was determined according to the method of Lowry et al. using human serum albumine as a standard.

Data Analysis

Pharmacological parameters (the maximal number of binding sites (Bmax) and dissociation constant (Kd) were analysed by the nonlinear least squares analysis on an Apple II microcomputer. The statistical significance of the radioreceptor assay data was judged using one way ANOVA and Student' t-test.

Results and Discussion. Specific $[^3H]$MK-801 binding increased in a dose dependent manner with increaseing concentrations of L-glutamate (data not shown). This seems to be in accordance with the use dependent increase of NMDA receptor function. There was no difference in the Kd or Bmax among the 3 groups (control, 24 hrs and 7 days after the last kindled seizures) in the AM and HIPP (Fig. 3). Neither Bmax nor Kd of specific $[^3H]$CPP binding was significantly different between the kindled and control rats in the AM and HIPP. 28 days (Fig. 4) or 24 hrs (data not shown) after the last kindled seizures.

These results suggest no postictal or interictal change in NMDA receptors of the AM and HIPP in AM kindling model of epilepsy, in which seizure susceptibility based upon kindling-induced potentiation must be maintained. Although NMDA antagonists, especially MK-801, prevented induction of kindling-induced potentiation, the negative findings in specific binding of $[^3H]$ NMDA antagonists in the kindled brain suggest no dorect relationship of change in NMDA receptors to maintain the potentiation. Up to date, no consistent data is available on long-term change in content and reuptake of glutamate and aspartate or glutamate receptor binding assay in the kindled animal brain [7].

Despite a lack of change in NMDA receptors of the kindled brain, intracellular signal transduction after activation of a subset of EAA

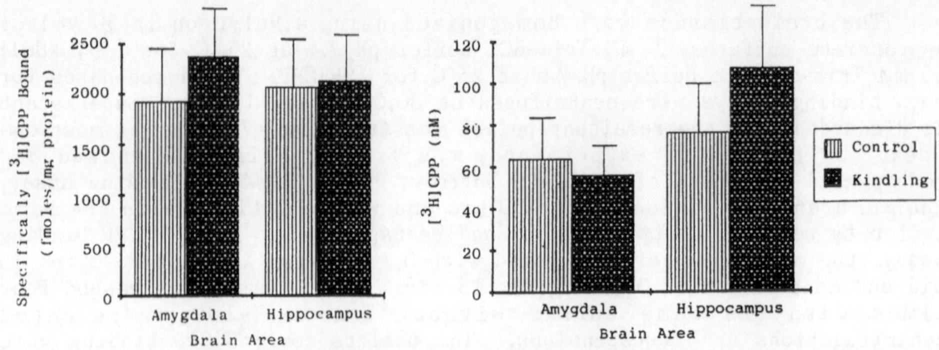


Fig. 3 Specific [$^3$H]MK-801 Binding

Saturation isotherm of specific [$^3$H]MK-801 binding in the amygdala/piriform cortex and hippocampus of amygdala kindled rats. A: Bmax and B: $K_D$.

The kindled rats were decapitated 24 hrs or 7 days after the last seizure. There is no significant difference between the control and kindled rats in Bmax or $K_D$.

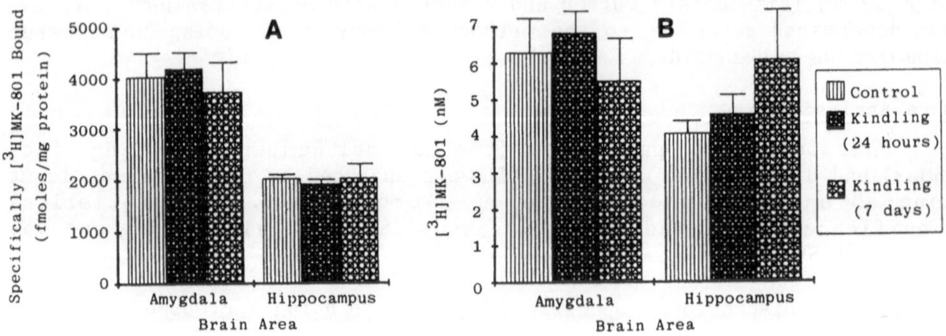


Fig. 4 Specific [$^3$H]CPP Binding

The kindled rats were decapitated 28 days after the last seizure. There is no significant difference between the control and kindled rats in Bmax or $K_D$.

receptors may be involved in some aspects of kindling model of epilepsy. In fact, we have reported a marked increase in ibotenate-stimulated hydrolysis of membrane phospholipid polyphosphoinositol (PPI) in the AM plus pyriform cortex of AM kindled rats [8]. Prior to our study, Iadorola et al. [9] had reported a significant increase of ibotenate-stimulated PPI hydrolysis in the HIPP of AM kindled rats at 24 hrs, but not at one month after the last convulsions. Akiyama et al. [10] and Yamada et al. [11] reported recently that the kindling-induced increase in PPI hydrolysis of the AM lasted for at least 30 days after the last convulsion and that the increase was observed also in the unstimulated AM plus pyriform cortex after AM kindling or in the AM plus pyriform cortex after HIPP kindling. These findings suggest the increased PPI hydrolysis of the brain region independently of a local effect of the electrical stimulation given to the primary site. The mechanisms for the long-term enhancement of ibotenate-stimulated PPI hydrolysis in the AM plus pyriform cortex remain unclear. As Iadorola et al.[9] reported no change in carbamylcholine- and norepinephrine-stimulated PPI hydrolysis, it seems interesting to identify a receptor site whose stimulation with ibotenate increases PPI hydrolysis. Lack of change in specific NMDA receptor binding in the AM after kindling and no activation of PPI hydrolysis by NMDA itself suggest a possible role of non-NMDA receptors in maintainance of kindling-induced potentiation. Akiyama and Sato[7] postulated that APB-sensitive quiaqualate receptors or ibotenate p, quisqualate p type as a candidate non-NMDA receptors, by which voltage dependent block of NMDA receptor-ion channels may be changed. Geula et al.[12] reported a lasting increase in potassium evoked $Ca^{++}$ dependent release of glutamate from hippocampal slices for 30 days after entorhinal kindling. These findings may represent lasting changes in pre- and post-synaptic mechanisms in EAA-mediated neurotransmission, which may relate to maintainance of kindling-induced seizure susceptibility. It is presumed that NMDA receptors are not primarily involved in maintainance of the seizure susceptibility of the kindling model of epilepsy.

## Anticonvulsant Action of NMDA Antagonists

The above described data of prophylactic action of MK-801 and CPP on kindling and of specific binding study using [$^3$H]NMDA antagonists suggested that NMDA-associated ion channels and/or NMDA receptor-mediated changes play a key role in induction, but not in maintainance, of

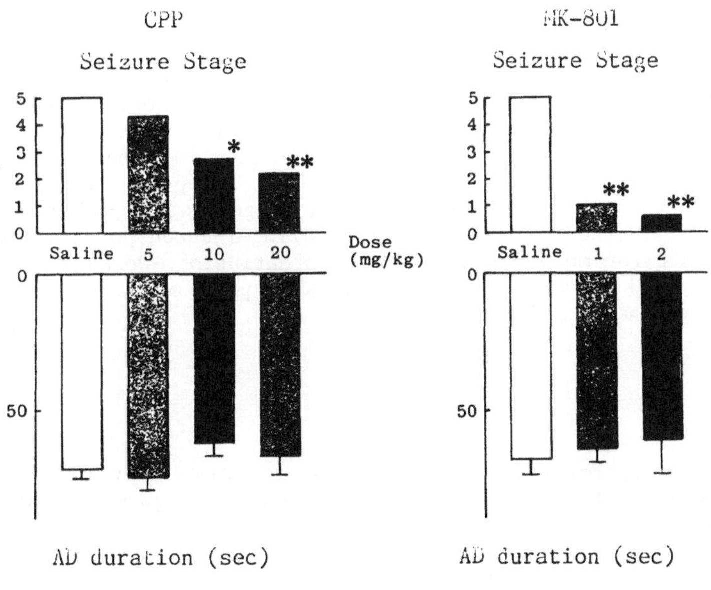


Fig. 5 Effects of CPP and MK-801 on Kindled AM Seizures

Value represent mean ± S.E.M.. *: $p < 0.1$, **: $p < 0.005$ by Mann-Whitney's U test compared with control. AD: afterdischarges.

kindling-induced potentiation. To examine a role of NMDA receptor in seizure induction mechanisms, anticonvulsant action of MK-801 and CPP were examined in AM kindled rats.

Methods. Male SD rats weighing between 250 and 300 g at surgery were used. AM kindling was achieved by once daily AM stimulation at 200 μA intensity with the same stimulus parameters described above. After confirming stage-5 convulsions on 5 consecutive days, stimulus intensity was decreased by 50 μA step to decide minimum intensity to produce stage-5 convulsion. This stimulus intensity was used to evaluate anticonvulsant action of the agents. Different dose of MK-801 (1 and 2 mg/kg), CPP (5, 10 and 20 mg/kg) or same volume of saline was administered intraperitoneally to the rats of MK-801-treated, CPP-treated groups and control, respectively. Anticonvulsant action was evaluated at 2 and 24 hrs after administration of the agents.

Results and Discussion. Neither MK-801 nor CPP had a significant action on the AD duration. Although behavioral seizures were suppressed significantly after more than 1 mg/kg of MK-801 or 10 mg/kg of CPP, the effects was hardly distinguished from non-specific inhibition on motor seizure manifestations due to behavioral toxicity of the drugs administered. Negative action of MK-801 on AD duration in this study is different from the previous report of Sato, K. et al.[2], in which 2 mg/kg of MK-801 markedly shortened the AD duration in previously fully kindled AM rats. The difference between these two results may be due to different stimulus intensity that applied for evaluation of anticonvulsant effects : 200 μA in this study v.s. stimulus intensities below 200 μA to produce kindled AM seizures in the latter study. Accordingly, the present data may not neglect the previously reported action of MK-801 to elevate AD thresholds.

## SUMMARY

NMDA antagonists, especially the non-competitive antagonist MK-801, had an action to prevent the development of AM kindling. This action was more prominent than the anticonvulsant action on kindled AM seizures. The present data that NMDA antagonists blocked LTP of EPSPs rather than PSH are in consistent with the previous reports by others. These results suggest a critical role of NMDA receptors-mediated signal tranduction in induction of both long-term kindling-induced potentiation and LTP. On the other hand, negative results in specific NMDA receptor binding of AM and HIPP were found in the fully kindled rats, suggesting negative role of the NMDA receptors in the maintainance of kindling-induced potentiation. As a critical change for maintainance of the potentiation, we suggested a long-term change in PPI hydrolysis of the AM plus pyriform cortex, which is stimulated by non-NMDA receptors.

## REFERENCES

1. R. J. Racine, Modification of seizure activity by electrical stimulation: I. Afterdischarge threshold. Electroenceph. clin. Neurophysiol., 32: 269-279 (1972)
2. K. Sato, K. Morimoto, and S. Otsuki, Anticonvulsant action of amino acid antagonists against kindled hippocampal seizures. Brain Res., 463: 12-20 (1988)
3. J.A. Kempf, A.C. Foster, and F.H.F. Wong, Non-competitive antagonists of excitatory amino acids receptors. Trends Neurosci., 10: 294-298 (1987)

4. R. J. Racine, F. Newberry, and W. M. Burnham, Post-activation potentiation and the kindling phenomenon. Electroenceph. clin. Neurophysiol., 39: 261-271 (1975)
5. R. M. Douglas, and G.V. Goddard G.V., Long-term potentiation of the perforantpath - granule cell synapse in the rat hippocampus. Brain Res., 86: 205-215 (1975)
6. E. J. Coan, W. Saywood, and G.L. Collingridge, MK-801 blocks NMDA receptor-mediated synaptic transmission and long term potentiation in rat hippocampal slices. Neurosci. Lett., 80: 111-114 (1987)
7. K. Akiyama, and M. Sato, Neurochemistry of kindling (in Jpn). Protein, Nucleic Acid and Enzyme, 34: in press, (1989)
8. K. Akiyama, N. Yamada, and M. Sato, Increase in ibotenate-stimulated phosphatidylinositolhydrolysis in slices of the amygdala/pyriform cortex and hippocampus of rat by amygdala kindling. Exp. Neurol., 98: 499-508 (1987)
9. M. J. Iadolora, F. Nicoletti, J. R. Naranjo, F. Putnam, and E. Costa, Kindling enhances the stimulation of inositol phospholipid hydrolysis elicited by ibotenic acid in rat hippocampal slices. Brain Res., 274: 174-178 (1986)
10. K. Akiyama, N. Yamada, and S. Otsuki, Lasting increase in excitatory amino acid receptor-mediated polyphosphoinositide hydrolysis in the amygdala/pyriform cortex of amygdala kindled rats. Brain Res. 485: 95-101 (1989)
11. N. Yamada, K. Akiyama, and S. Otsuki, Hippocampal kindling enhances excitatory amino acid receptor-mediated polyphosphoinositol hydrolysis in the hippocumpus and amygdala/pyriform cortex. Brain Res., 490: 126-132 (1989)
12. G. Guela, P.A. Jarvie, T.C. Logan, and J.T. Slevin, Long-term enhancement of $K^+$-evoked release of L-glutamate in entorhinal kindled rats. Neurochem., 48: 999-1017 (1988)

Discussion of Dr. Sato's Presentation

DR. LOSCHER: You have shown in one of your first slides that CPP injected during kindling had no effect on the after-discharges but it blocked the development of motor seizures. At the dosage that you used, 10 mg/kg IP, CPP has been shown to reduce muscle tone, so it could be that this effect on motor seizures is just the result of the reduction in the motor tone. For instance, we have shown recently that in different models of motor seizures, CPP is only active at dosages which reduce muscle tone, which is in contrast to conventional anti-epileptic drugs.

DR. SATO: Yes, it might be. But we examined at various dosages. With a less toxic dose we found motor seizure was blocked.

DR. BURNHAM: I'm a little out of my depth here. Does the MK801 bind to the same site as the TCP that's been used? Is there conflict in the long-term binding data here? Could someone explain.

DR. MCNAMARA: I'm just trying to answer his question. When Dr. Sato looked at 1 day and 7 days, he found no difference in MK801 binding or CPP binding at either time. We looked at 1 day and 28 days, we found no difference at 1 day, in parallel to what he found. We found the difference only at 28 days. There is no conflict at all.

DR. SATO: Our data are 1 day after and also 1 week after the last convulsion, but not 4 weeks after.

DR. MCNAMARA: To clarify further, what we're studying in terms of PI hydrolysis is very different as Dr. Sato mentioned. And to further expand, what we were studying is a NMDA receptor mediated event, it is very different pharmacologically and directionally from the ibotenate. Ibotenate alone stimulates PI hydrolysis and the pharmacology of that receptor is not that of the NMDA receptor, it is a novel or a unique form of a receptor. What we were studying is specifically an NMDA receptor mediated event and it involves inhibition of carbachol stimulated PI hydrolysis. So there is no conflict whatsoever in these data. In fact, I think that one thing that's really gratifying is everybody does the MK801 experiments and the CPP experiments, and everybody finds the same answer. You have all of these different drugs, and I realize, Wolfgang, you could say, well, there is something wrong, you know, the MK801, the PCP, etc., it's all 5-HT 1A receptor agonists, and then we've got to have a different explanation for the CPP, we're going to find another explanation for the APP and the APH. But it seems to me that you've got structurally distinct compounds. They do share one common effect, as every drug does, they have multiple effects; but the commonality of the findings with all these different drugs acting at different sites, makes it extraordinarily likely that the mechanism of these drugs is due to NMDA channel activation or blocking.

DR. SATO: May I ask you, do you have any evidence that indicates long-lasting increase in carbachol stimulated PI hydrolysis in the amygdala?

DR. MCNAMARA: We don't know if it's in the amygdala. We didn't look in the amygdala; we just looked in the hippocampus.

DR. SATO: Can you explain how the NMDA inhibits carbachol induced PI hydrolysis, because NMDA doesn't couple to the PI.

DR. MCNAMARA: Correct; we believe that it's an indirect effect of NMDA'S depolarizing NMDA receptive neurons. We speculate that the mechanism is that the hydrogen ion concentration within the NMDA receptive cell is modified as a consequence of the depolarization. The idea is that depolarization alkalinizes the cytosol and thereby inhibits phospolipase C activity.

# EFFECT OF SELECTIVE LESIONS WITHIN THE SUBSTANTIA NIGRA ON THE ANTICONVULSANT EFFECT OF ANTIEPILEPTIC DRUGS IN FULLY-KINDLED RATS

Wolfgang Löscher, Ulrich Wahnschaffe, Dagmar Hönack and Chris Rundfeldt

Department of Pharmacology, Toxicology, and Pharmacy
School of Veterinary Medicine, D-3000 Hannover 71, F.R.G.

## INTRODUCTION

A substantial body of evidence suggests that the substantia nigra (SN), particularly the pars reticulata region, is a critical site involved in seizure propagation (1,2). Autoradiographic studies of 2-deoxyglucose metabolism have shown that activation of the SN is a stable feature of fully generalized convulsions induced by chemoconvulsants (3) or amygdala-kindling (4). Extracellular, single-cell recording studies have shown that SN neurons exhibit a change in firing pattern, to a bursting mode, during amygdala-kindled seizures (5). Bilateral intranigral microinjection of either muscimol, a GABA agonist, or gamma-vinyl GABA (GVG), a GABA-elevating drug, were shown to attenuate or block seizures induced by electroshock (6), amygdaloid kindling (7,8), chemoconvulsants such as pentylenetetrazol and bicuculline (6), flurothyl-induced seizures (7), ethanol withdrawal seizures (8), pilocarpine-induced seizures (9), seizures obtained in genetically epilepsy-prone rats (10) as well as spontaneous generalized non-convulsive seizures in rats (11). Furthermore, bilateral microinjection of GVG into the SN retarded kindling development (8,12). In fully kindled rats, biochemical evidence of impaired GABAergic transmission has been found in the SN (13,14). Because GABA has been shown to inhibit nigral efferents, it is likely that the anticonvulsant effect of intranigral application of muscimol or GVG relates to an inhibition of nigral projections that are permissive or facilitative to seizure propagation (1). In support of this, bilateral destruction of the SN by electrocoagulation or microinjection of the neurotoxin kainic acid attenuated bicuculline-induced clonic and tonic seizures as well as tonic electroshock seizures (15). In amygdala-kindled rats, bilateral destruction of SN by N-methyl-D,L-aspartate (NMDA) not only suppressed motor seizures but also reduced afterdischarges recorded at the amygdala, suggesting that the nigra not only transmits seizure activity from rostral to caudal sites but is actively involved in the generation of limbic seizures at the site of origin (7).

The various microinjection and lesion studies strongly imply that inhibition of some neuronal population within the nigra can terminate the spread of seizure activity to motor centers. Thus, the SN might represent a target of clinically used antiepileptic drugs, especially those which act by potentiating GABAergic neurotransmission (1). Substantiating this possibility, bilateral intranigral microinjection of

*Kindling 4*
Edited by J. A. Wada
Plenum Press, New York, 1990

clonazepam has been shown to produce a marked elevation of generalized seizure threshold in the kindling model (16). Furthermore, systemic injections of the antiepileptic drugs clonazepam, diazepam, valproic acid (VPA), and phenobarbital were shown to inhibit SN neurons in rats, whereas phenytoin and carbamazepine were not effective in this regard (17). For further exploration of whether the SN might constitute a site of antiepileptic drug action, we studied the anticonvulsant effect of diazepam, VPA, phenobarbital, and carbamazepine before and after bilateral destruction of the SN in fully-kindled rats.

## METHODS

### Preparation of Animals

All experiments were carried out in female Wistar rats, weighing 210-230 g. The animals had bipolar electrodes implanted stereotaxically in the right piriform cortex under anesthesia with chloral hydrate (360 mg/kg i.p.). Coordinates for electrode implantation derived from the atlas of Paxinos and Watson (18) were: AP -0.8, L -4.8, V -8.5. For comparison with rats electrically kindled from stimulation of the piriform cortex, a second group of animals received implantation of a bipolar electrode into the right basolateral amygdala (AP -2.2, L -4.8, V -8.5). In all rats, guide cannulae were implanted over the area of the SN bilaterally (AP -4.9, L ± 2.2, V -1.5; for microinjection V was -7.5). Skull screws served as the indifferent reference electrode. The electrode assembly and guide cannulae were attached to the skull by dental acrylic. Since results obtained in amygdala-kindled rats were not different from rats kindled in piriform cortex, only results of the piriform cortex group will be presented in this paper.

### Kindling

After a postoperative period of 2 weeks, constant current stimulations (500 µA, 1 ms, monophasic square-wave pulses, 50/s for 1 s) were delivered to the piriform cortex at intervals of 1 day until at least 10 reproducible stage 5 seizures were elicited. The electrical activity of the piriform cortex was recorded before and after each stimulation. In fully-kindled rats, the following parameters of kindled seizures were measured. Afterdischarge threshold (ADT) was determined by administering a series of stimulations at 1-min intervals beginning at 10 µA and increasing in steps of about 20% of the previous current (19). The threshold was defined as the lowest intensity producing AD. Since all animals exhibited generalized seizures (stage 4-5) at the ADT current, it was not necessary to determine the threshold for generalized seizures (GST) separately. Severity and duration of the seizures and afterdischarge duration (ADD) were recorded both during threshold determinations and at suprathreshold stimulation with 500 µA. Seizures were classified behaviorally according to Racine (20): 1, immobility, eye closure, twitching of vibrissae, sniffing, facial clonus; 2, head nodding associated with more severe facial clonus; 3, clonus of one forelimb; 4, rearing, often accompanied by bilateral forelimb clonus; 5, rearing with loss of balance and falling accompanied by generalized clonic seizures. Seizure duration was the duration of limbic (stage 1-2) and/or motor seizures (stage 3-5); limbic seizure activity sometimes occurring <u>after</u> termination of stage 3-5 seizures was not included in seizure duration. ADD was the total time of spikes in the EEG recorded from the piriform cortex.

### Drug Testing

The effects of diazepam (5 mg/kg i.p.), VPA (200 mg/kg i.p.), carbamazepine (20 mg/kg i.p.) and phenobarbital (30 mg/kg i.p.) were compared before and after destruction of the SN (see below) in fully kindled rats. For anticonvulsant drug testing, all stimulations were carried out with 500 µA. The drugs were injected 15 min (VPA), 30 min (diazepam, carbamazepine) or 60 min (phenobarbital) prior to stimulation. Dosages and pretreatment times were chosen on the basis of previous experiments with the drugs in kindled rats (21). Control readings were determined 2-3 days prior to and after each drug administration. At least 4 days were interposed between 2 drug injections.

## Destruction of the SN

For lesioning of SN, the injection cannulae were inserted through the guide cannulae and 5 min later the neurotoxin ibotenic acid (22) was injected at a dose of 5 µg in 0.5 µl of buffered saline at a rate of 0.1 µl/min. This dose of ibotenic acid was chosen on the basis of dose-effect and evolution experiments in rats showing an almost complete neuron degeneration in the SN with minimum destruction of surrounding regions (see Results) within 2-3 days after injection of this dosage. The injection cannula was withdrawn 5 min after completion of the injection. All injections were made unilaterally; the SN of the other hemisphere was destroyed 3 days after the first lesion. The major behavioural symptom after unilateral injection of ibotenic acid was contralateral circling for several hours.

In order to examine the effect of SN lesions on kindled seizure parameters, the ADT was determined 1 day before the first lesion as well as 1 week after bilateral destruction of the SN. Furthermore, control stimulations with 500 µA were carried out before and after the lesions. Anticonvulsant testing was started 12 days after bilateral destruction of the SN using the protocol described for anticonvulsant evaluation prior to the lesions.

## Histology

After termination of anticonvulsant testing, placement of stimulating electrodes and injection cannula tips and the extent of the SN lesions were examined histologically in all animals. The rats were anesthetized with chloral hydrate (360 mg/kg i.p.) and perfused with a fixative consisting of 4% phosphate buffered formaldehyde (pH 7.3). Two h later, the brains were removed and processed for paraffin embedding. Subsequently, serial sections of the entire brain were cut coronally at 7 µm, and every 20th section was mounted on a glass slide and stained with cresyl violet or hematoxylin eosine. Only animals with correct placement of the kindling electrodes and selective bilateral lesions of the SN were used for evaluation of the kindling experiments described below. Lesions that did not involve near-complete bilateral destruction of SN were not considered SN lesions.

## Statistics

All data are given as means ± S.E. Significance of differences was calculated by the Wilcoxon signed-rank test for paired replicates.

# RESULTS

Placement of kindling electrodes in the piriform cortex in the 8 rats which fulfilled the criteria of our study (see above) is shown in Fig. 1. Kindling development and the behavioral characteristics of kindled seizures in these rats were not different from amygdala-kindling, which is in line with previous studies with rats electrically kindled in the piriform cortex (23). The only significant difference to amygdala-kindled rats was that the ADT was markedly lower in rats kindled in the piriform cortex (see Table 1).

Anticonvulsant activity of the 4 antiepileptic drugs evaluated in piriform cortex-kindled rats prior to SN lesions is shown in Fig. 2. At the dosages used, all drugs significantly reduced severity and duration of the behavioral seizures and duration of afterdischarges recorded from the site of stimulation.

After completion of these experiments, the SN was destroyed bilaterally by means of ibotenic acid. The morphological changes induced by ibotenic acid in the SN are shown in Fig. 3, while reconstructions of the extent of the lesions in representative animals are depicted in Fig. 4. Histological examination of the SN demonstrated an almost complete loss of neurons, which were replaced by glia cells (Fig. 3). With the dosage of ibotenic acid used (5 µg), the neuronal damage was largely restricted to the SN and zona incerta (Fig. 4).

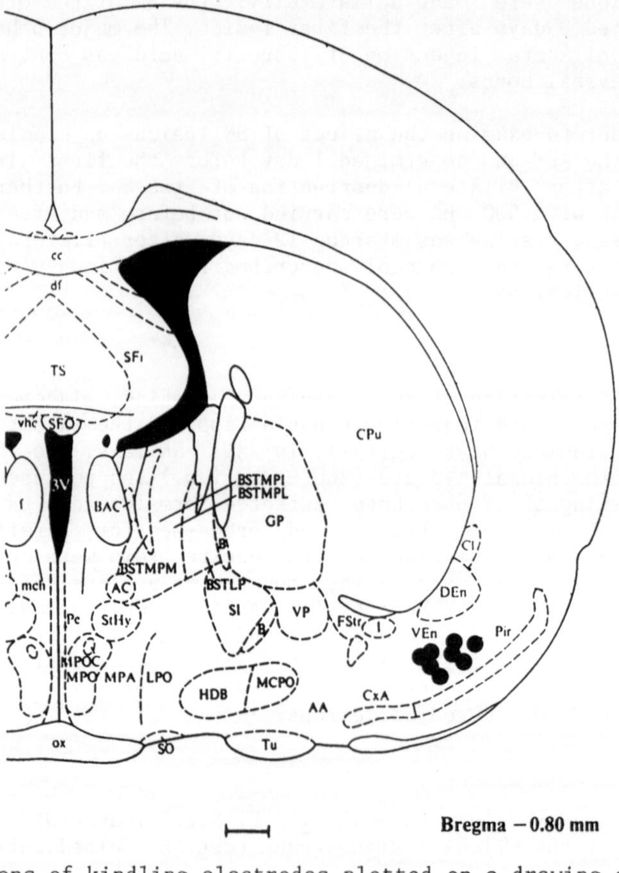


Fig. 1. Locations of kindling electrodes plotted on a drawing of a coronal brain section according to the stereotaxic atlas of Paxinos and Watson (18). The right side of the brain corresponds to the right side of the figure. Each filled circle represents one animal of the group of 8 animals used for final evaluation of kindling data. AA, anterior amygdaloid area; Cl, claustrum; CxA, cortex-amygdala transition zone; DEn, dorsal endopiriform nucleus; Pir, piriform cortex; VEn, ventral endopiriform nucleus. The bar indicates 0.6 mm.

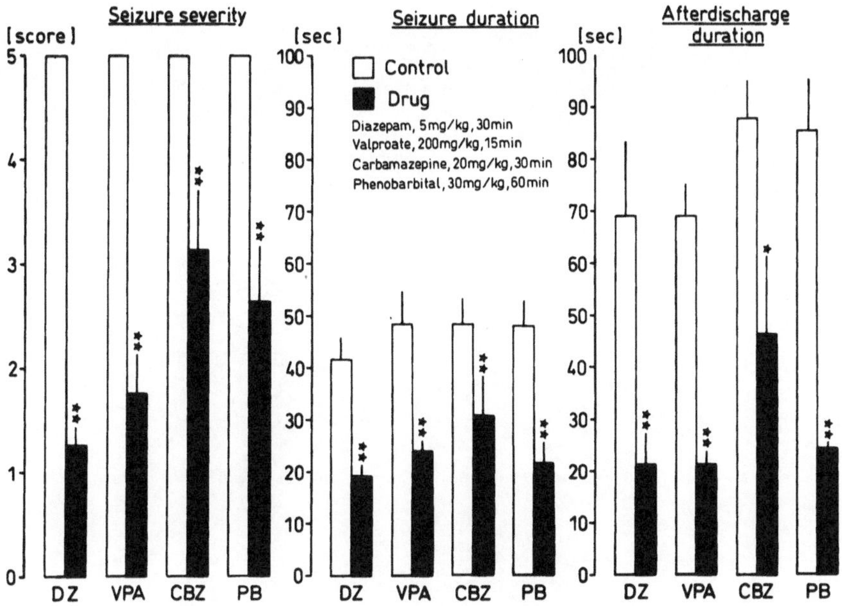


Fig. 2. Anticonvulsant effect of diazepam, VPA, carbamazepine, and phenobarbital in fully kindled rats prior to destruction of the SN. Data are given as means ± S.E. of 8 rats kindled from stimulation of the piriform cortex (see Fig. 1). Control data were determined 2-3 days prior to drug administration. Absence of S.E. indicates that all animals had the same readings. Significance of differences between control and drug data is indicated by asterisks (*$P<0.05$; **$P<0.01$).

Table 1. ADT and seizure parameters at ADT before and after bilateral destruction of SN

| | ADT (µA) | Seizure parameters at ADT | | |
|---|---|---|---|---|
| | | Seizure severity (score) | Seizure duration (sec) | Afterdischarge duration (sec) |
| pre-lesion | 32.0 ±8.5 | 4.9 ±0.12 | 48.5 ±5.5 | 83.8 ±5.0 |
| post-lesion | 33.6 ±6.8 | 5.0 ±0 | 52.9 ±4.8 | 73.8 ±8.5 |

Values are means ± S.E. of 8 fully-kindled rats which were kindled from stimulation of the piriform cortex. The data were determined 1 day before and 1 week after bilateral destruction of SN.

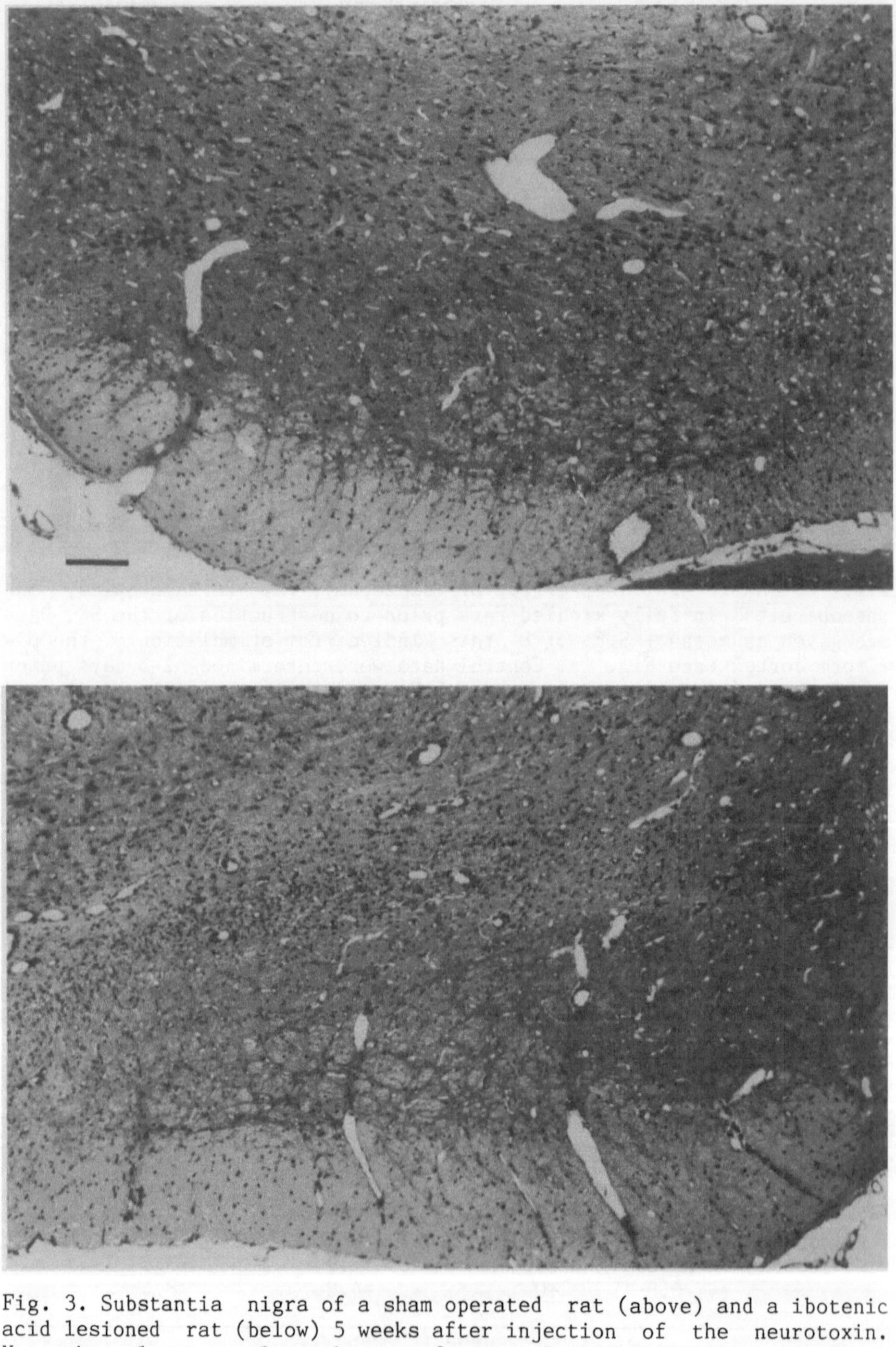

Fig. 3. Substantia nigra of a sham operated rat (above) and a ibotenic acid lesioned rat (below) 5 weeks after injection of the neurotoxin. Note the almost complete absence of neuronal perikarya and the striking glial response in the lesioned SN. Note also the preservation of axonal bundles in the lesioned SN. The bar indicates 0.2 mm.

The almost complete bilateral destruction of SN had no influence on susceptibility of the rats to electric stimulation of the pirifirm cortex or severity and duration of the kindled seizures (Table 1). All lesioned rats exhibited generalized seizures at threshold or suprathreshold stimulation, and there was not even a tendency for alterations in seizure threshold.

The anticonvulsant activity of diazepam, VPA and phenobarbital determined after bilateral SN lesions was not different from that prior to the lesioning (Fig. 5). In contrast, the anticonvulsant effect of carbamazepine was markedly reduced after destruction of the SN (Fig. 5).

## DISCUSSION

The principal findings of this study are: (1) Near-complete, bilateral, selective destruction of the SN does not alter kindled seizure threshold or severity and duration of kindled seizures; (2) except in case of carbamazepine, SN lesions do not alter the anticonvulsant activity of major antiepileptic drugs.

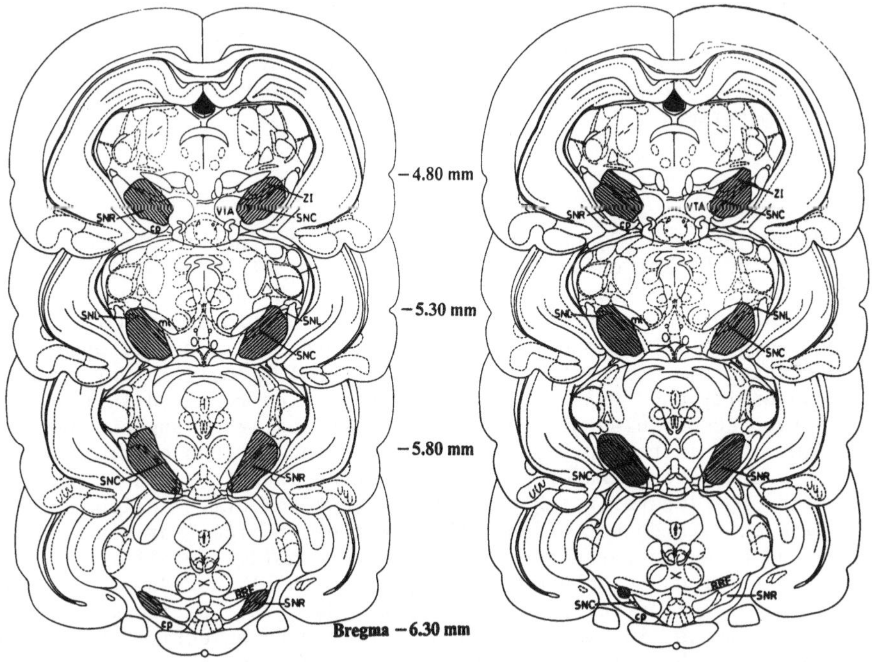


Fig. 4. Schematic reconstruction of ibotenic acid-induced lesions of the SN as plotted on the drawings of coronal brain sections according to the stereotaxic atlas of Paxinos and Watson (18). Shaded areas represent the lesioned area as determined by histological examination (see Fig. 3). The animal with the largest lesions of the group of 8 rats used for final evaluation of the kindling experiments is represented by the drawings on the left side, whereas the drawings on the right side represent the animal with the smallest lesion. Cp, cerebral peduncle; RRF, retrorubral field; SNR, substantia nigra pars reticulata; SNC, substantia nigra pars compacta; SNL, substantia nigra pars lateralis; VTA, ventral tegmental area; ZI, zona incerta.

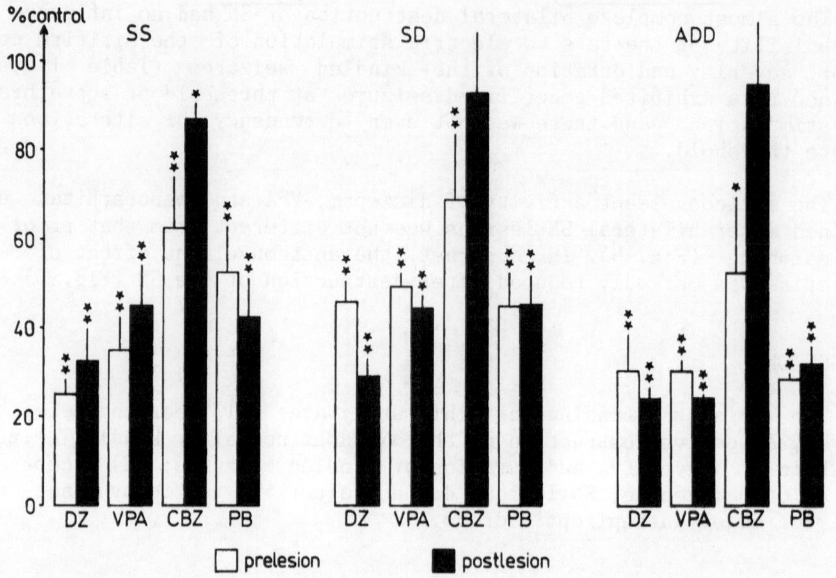


Fig. 5. Anticonvulsant effect of diazepam, VPA, carbamazepine, and phenobarbital before and after bilateral destruction of SN. Seizure severity (SS), seizure duration (SD) and afterdischarge duration (ADD) are shown as percent of individual control values. Significance of differences to individual controls is indicated by asterisks (*$P<0.05$; **$P<0.01$). Data are means ± S.E. of 8 rats kindled from stimulation of the piriform cortex.

McNamara et al. (7) have reported previously that bilateral destruction of the SN markedly reduced the duration of motor seizures and afterdischarges in amygdala-kindled rats stimulated with currents 10% above the generalized seizure threshold. In some lesioned animals, motor seizures were abolished and brief or no afterdischarges were obtained. In these animals, administration of additional stimuli in which both current intensity and train duration were doubled resulted in long afterdischarges accompanied by generalized seizures, thus suggesting that the lesions had increased seizure threshold rather than eliminated the ability to generate seizures. This suggestion is not substantiated by the present experiments, which demonstrate that selective lesions of the SN in kindled rats do not affect the duration of focal or generalized seizures and have no influence on seizure threshold. The difference to the results reported by McNamara's group is certainly not due to the location of the kindling electrode (i.e., amygdala vs. piriform cortex), because experiments in amygdala-kindled rats with selective bilateral lesions of SN by ibotenic acid also failed to demonstrate an effect of the lesion on kindled seizures (Wahnschaffe and Löscher, in preparation). The findings of McNamara et al. (7) thus might relate to the fact that, in contrast to our experiments, the lesions produced in their study by 20 µg of the neurotoxin NMDA were not selective but included several structures other than the S.N., e.g. most parts of the thalamus, which were not damaged in our animals. Interestingly, a subsequent study of McNamara's group (12) showed that more selective lesions of the SN by NMDA or thermocoagulation facilitated rather than inhibited kindling development in rats. Thus, the inhibitory effects of lesions in fully kindled rats reported by McNamara et al. (7) might have been due to destruction of structures other than the SN. In this respect, it should be noted that in the studies reported by Garant and Gale (15), kainate or thermocoagulative lesions of SN were also not selective but included damage to the cerebral peduncles and/or medial lemniscus, which were not lesioned in our experiments. With respect to lesions induced by excitotoxins, it has been demonstrated recently for other brain regions that ibotenate, which is thought to act by stimulation of NMDA receptors, generally produces much more direct and uniform lesions than kainate or NMDA itself (22), thus suggesting that the choice of excitotoxin might influence data obtained from lesion studies.

The present data on anticonvulsant testing before and after SN lesions demonstrate that the SN is not a crucial anatomic site at which antiepileptic drugs, such as benzodiazepines, VPA and phenobarbital, act to suppress seizures. Since benzodiazepines, VPA and barbiturates are thought to enhance GABAergic transmission (24), the data seem to refute the proposal that the SN is the anatomical site which is responsible for exerting anticonvulsant effects in response to drug-induced augmentation of GABA transmission (1,2,6). With regard to studies on the effect of intranigral microinjection of GABAergic drugs, such as muscimol, GVG or benzodiazepines, on generalized seizures described in the introduction, one should note that microinjection of these drugs into SN pars reticulata has been shown to reduce muscle tone (25,26) so that the effect of intranigral injection of such drugs on generalized motor seizures might be secondary due, at least in part, to their muscle relaxant action.

The only drug for which an altered efficacy was found after SN destruction was carbamazepine. This was unexpected since experiments of Waszczak et al. (17) had indicated that carbamazepine, 1.25 - 40 mg/kg, did not alter neuronal firing of nondopaminergic neurons of the substantia nigra pars reticulata in awake rats. It could thus be that differences exist in the effects of carbamazepine on SN neurons in kindled and non-kindled rats. Furthermore, the fact that the SN lesions included both pars reticulata and pars compacta might be important for the loss of anticonvulsant activity of carbamazepine in the lesioned animals.

In conclusion, these results do not support a pivotal role of the SN in seizure propagation or suppression in the kindling model. Indeed, as already suggested by the experiments of Waszczak et al. (17), the SN is certainly not a unique or uniformly important site of anticonvulsant drug action, but the present experiments with carbamazepine indicate that this structure might be involved in anticonvulsant action of certain antiepileptic drugs, at least in the seizure model selected for these studies.

ACKNOWLEDGEMENTS

The study was supported by grants from the Deutsche Forschungsgemeinschaft (Lo 274/3-1). We thank Dr. L. Turski (Schering, Berlin, F.R.G.) for helpful discussions and advice. We thank Mrs. C. Bartling for technical assistance.

REFERENCES

1. Gale, K., Role of substantia nigra in GABA-mediated anticonvulsant actions, in: Basic Mechanisms of The Epilepsies. Molecular and Cellular Approaches, A.V. Delgado-Escueta, A.A. Ward, D.M. Woodbury and R.J. Porter, eds., p. 343, Raven Press, New York (1986).

2. Gale, K., Progression and generalization of seizure discharge: anatomical and neurochemical substrates, Epilepsia 29 (Suppl. 2):S15 (1988).
3. Ben Ari, Y., E. Tremblay, D. Riche, G. Ghilini, and R. Naquet, Electrographic, clinical and pathological alterations following systemic administration of kainic acid, bicuculline and pentylenetetrazol: metabolic mapping using the deoxyglucose method with special reference to the pathology of epilepsy, Neuroscience 6:1361 (1981).
4. Engel, J. Jr., L. Wolfson, and L. Brown, Anatomical correlates of electrical and behavioral events related to amygdala kindling, Ann. Neurol. 3:538 (1978).
5. Bonhaus, D.W., J.R. Walters, and J.O. McNamara, Activation of substantia nigra neurons: role in the propagation of seizures in kindled rats, J. Neurosci. 6:3024 (1986).
6. Iadarola, M.J., and K. Gale, Substantia nigra: site of anticonvulsant activity mediated by gamma-aminobutyric acid, Science 218:1237 (1982).
7. McNamara, J.O., M.T. Galloway, L.C. Rigsbee, and C. Shin, Evidence implicating substantia nigra in regulation of kindled seizure threshold, J. Neurosci. 4:2410 (1984).
8. Löscher, W., S.J. Czuczwar, R. Jäckel, and M. Schwarz, Effect of microinjections of $\gamma$-vinyl GABA or isoniazid into substantia nigra on the development of amygdala kindling in rats, Exp. Neurol. 95:622 (1987).
9. Turski, L., E.A. Cavalheiro, M. Schwarz, W.A. Turski, L.E.A. De Moraes, Z.A. Bortolotto, T. Klockgether, and K.-H. Sontag, Susceptibility to seizures produced by pilocarpine in rats after microinjection of isoniazid or $\gamma$-vinyl GABA into the substantia nigra, Brain Res. 370:294 (1986).
10. Millan, M.H., B.S. Meldrum, C.A. Boersma, and C.L. Faingold, Excitant amino acids and audiogenic seizures in the genetically epilepsy-prone rat. II. Efferent seizure propagating pathway, Exp. Neurol. 99:687 (1988).
11. Depaulis, A., M. Vergnes, C. Marescaux, B. Lannes, and J.-M. Warter, Evidence that activation of GABA receptors in the substantia nigra suppresses spontaneous spike-and-wave discharges in the rat, Brain Res. 448:20 (1988).
12. Shin, C., J.M. Silver, D.W. Bonhaus, and J.O. McNamara, The role of substantia nigra in the development of kindling: pharmacologic and lesion studies, Brain Res. 412:311 (1987).
13. Löscher, W., and W.S. Schwark, Evidence for impaired GABAergic activity in the substantia nigra of amygdaloid kindled rats, Brain Res. 339:146 (1985).
14. Löscher, W., and W.S. Schwark, Further evidence for abnormal GABAergic circuits in amygdala-kindled rats, Brain Res. 420:385 (1987).
15. Garant, D.S., and K. Gale, Lesions of substantia nigra protect against experimentally-induced seizures, Brain Res. 273:156 (1983).
16. King, P.H., C. Shin, H.H. Mansbach, L.S. Chen, and J.O. McNamara, Microinjection of a benzodiazepine into substantia nigra elevates kindled seizure threshold, Brain Res. 423:261 (1987).
17. Waszczak, B.L., E.K. Lee, and J.R. Walters, Effects of anticonvulsant drugs on substantia nigra pars reticulata neurons, J. Exp. Pharmacol. Ther. 239:606 (1986).
18. Paxinos, G., and C. Watson, The Rat Brain in Stereotaxic Coordinates, 2nd ed., Academic Press, Sydney (1986).
19. Freeman, F.G., and M.F. Jarvis, The effect of interstimulation interval on the assessment and stability of kindled seizure thresholds, Brain Res. Bull. 7:629 (1981).

20. Racine, R.J., Modification of seizure activity by electrical stimulation. II. Motor seizure, Electrencephalogr. Clin. Neurophysiol. 32:281 (1972).
21. Löscher, W., R. Jäckel, and S.J. Czuczwar, Is amygdala kindling in rats a model for drug-resistant partial epilepsy? Exp. Neurol. 93:211 (1986).
22. Coyle, J.T., Excitotoxins, in: Psychopharmacology: The Third Generation of Progress, H.Y. Meltzer, ed., p. 333, Raven Press, New York (1987).
23. Cain, D.P., M.E. Corcoran, K.A. Desborough, and D.J. McKitrick, Kindling in deep prepyriform cortex of the rat, Exp. Neurol. 100:203 (1988)
24. Macdonald, R.L., and M.J. McLean, Anticonvulsant drugs: mechanisms of action, in: Basic Mechanisms of the Epilepsies. Molecular and Cellular Approaches, A.V. Delgado-Escueta, A.A. Ward, D.M. Woodbury and J.R. Porter, eds., p. 713, Raven Press, New York (1986).
25. Schwarz, M., W. Löscher, L. Turski, and K.-H. Sontag, Disturbed GABAergic transmission in mutant Han-Wistar rats: further evidence for basal ganglia dysfunction, Brain Res. 347:258 (1985).
26. Turski, L., M. Schwarz, T. Klockgether, and K.-H. Sontag, Substantia nigra: a site of action of muscle relaxant drugs, Neurosci. Abstr. 11:1162 (1985).

Discussion of Dr. Loscher's Presentation

DR. MCNAMARA: Just to comment first of all, obviously, Wolfgang, the evidence implicating the role of the substantia nigra in the propagation of kindled seizures was based not only on the basis of lesions but also on the basis of muscimol injections, gamma vinyl GABA injections, and benzodiazepine injections, so it is pharmacological as well as lesion. There is a discrepancy, as you point out, between your lesion results and our lesion results. We found no effect of the lesion on kindling development, and that is a very different experiment obviously from examining the effects of nigra lesions on seizures in a previously kindled animals. I wonder whether apart from the anatomic distribution of the lesions, however, whether other differences in the experimental approach might account for some of the discrepancies that we have found. So what I'm trying to understand is, when you did these lesions with the ibotenic the question is, you kindled those animals first, and then you figured out what their thresholds were, and then you lesioned the animals, and how long after did you stimulate them, and how did you determine the thresholds?

DR. LOSCHER: How long after the lesion, you mean?

DR. MCNAMARA: I'm just trying to figure out how you did the experiment because I suspect that's the cause of the discrepancy.

DR. LOSCHER: I don't think so, because what we did, as I stated, we determined the ADT starting from 10 microamps and then increasing by 20% steps up to the first after-discharge threshold several times before and after. So the last ADT determination before the lesion was 1 day before. But in the fully kindled animals the ADT was constant before the lesion. Then after the lesion, the first ADT determination was after 1 week. And then we repeated this determination after 4 or 5 weeks; again, absolutely the same ADT, which corresponded to the values determined prior to the lesions.

DR. MCNAMARA: And the way you determined the ADT is how?

DR. LOSCHER: By starting with 10 microamps, and then we increased in 20% steps; so the next step would be 12 microamps, at one minute intervals.

DR. MCNAMARA: Well, that's a clear difference from the way we did it. We did it at 3 days rather than 1 week after the lesion, and we figured out exactly what their thresholds were before by a different method from yours. We were doing one stimulus a day and did it the way Mac Burnham does it. OK? Then we stimulated them at a single shock 3 days later at 20% over the threshold. It was clear, we could clearly overcome the effect of the lesion by further increasing the stimulus intensity and furthermore, if we stimulated them at that suprathreshold stimulus at which we could see inhibitory effects, if we did that repeatedly, we would rediscover the kindling phenomenon. So I think it's obviously not an all or none effect, and I suspect, it could be a function of the anatomic distribution of the lesions, but the other point is that it also could be differences in the experimental paradigm.

DR. LOSCHER: Just one comment to the microinjection studies. One problem, a similar problem as with the NMDA antagonists, is that if you inject GABA mimetic drugs such as muscimol or gamma vinyl GABA into the nigra, we also did this, one should be very careful and one should note that intranigral injections of such drugs decreases muscle tone. Torsky has a paper in press showing that all drugs which exert effects on motor seizures after bilateral intranigral injections, exert a marked effect on muscle tone. It's very difficult to separate a selective anticonvulsant effect from this effect on the muscle tone, which of course could retard motor seizures or reduce the intensity of motor seizures, and so on. So, you are on the safe side, of course, if you look on after-discharge durations, something like that, but in terms of generalized motor seizures, most of the studies were done without EEGs, one has to be very careful. For example, with electroshock seizures, if you reduce the muscle tone, you will see no tonic extension and this has nothing to do with selective anti-convulsant effects.

DR. MCNAMARA: Yes, I agree completely, and that's why our EEG recordings in those kindling studies, I think, were very important in demonstrating that what Karen had found was not - I don't think what she found was simply due to muscle relaxation. But I agree completely. That's a very interesting series of observations. Why didn't you look at phenytoin?

DR. LOSCHER: It was just a matter of time, you know. We needed about one year for all these lesion studies and so we decided 4 drugs are enough for the first step.

DR. MOSCHE: I think that is a very nice study. I have two comments. One possibility is that with your lesion you may be destroying norepinephrine pathways nearby, and although we do say that this may not have an effect on the development of kindled seizures, I would like to know what is the catecholamine content of those brains in your animals. See, it is easy to say that there is no norepinephrine component if there is no facilitation. But if there is a worsening of the seizures or if they are still the same, the possibility that you may be destroying fibers en passant from the BTA, it may play a role. The other comment that I would like to make is that the substantia nigra seems to have an effect beyond the effect on muscle tone, because we have age-related differences on the effects of the nigra and we have worsening of the seizures in the mature animal which will not account for changes in the muscle tone.

DR. CORCORAN: It wasn't clear to me whether your lesions destroyed the pars compacta as well as the pars reticulata?

DR. LOSCHER: It's not possible with ibotenic acid or such neurotoxin to selectively destroy only one part of the nigra.

DR. CORCORAN: So you were getting the whole?

DR. LOSCHER: But we did also some studies with incomplete destruction in which we, for instance, had a 2/3 destruction of the reticulata and only one side of the compacta. We had a lot of animals with incomplete destructions, but in all these animals there was no effect on our parameters.

DR. CORCORAN: My further question is, in the animals that had extensive lesions, were they Parkinsonian rats, were they aphagic and adipsic, and did you have to tube feed them?

DR. LOSCHER: No. We did it so that we first injected ibotenic in one substantia nigra and 3 days after that injected ibotenic acid in the second one, because the animals show circling for several hours after the injection, and if you do it bilaterally at the same time some of the animals will die. So the way we did it, with 3 days in between the 2 lesions, none of our animals died, and it was not necessary to feed them. The circling was for about 3-4 hours, and after that they recovered, and 3-4-5 days after the bilateral lesions the animals had absolutely no changes in behaviour. And we observed them for several weeks, so there was nothing to see in terms of feeding problem or anything like that.

DR. FERNANDEZ-GUARDIOLA: I think it is very important the timing after the lesion. We have several very classic examples of the changes after one lesion, for example in the reticular formation. If you make that bilateral lesion the animal is comatose, loses consciousness. If you wait long enough, for example, for one month or a month-and-a-half, then the animal recovers. And all the assumptions on the result of the lesion change with time. I think this is very important and proves that for that time after the lesion there is no effect, but I think more experiments need to be done at times after the lesions.

DR. LOSCHER: Right. What we've done which I did not mention, we also studied the dynamics after the lesion with one drug - diazepam. We injected diazepam 2 times with 4 weeks in between after the lesion. The first injection was about 2 weeks, the second injection was after about 6 weeks, and again, there was absolutely no difference. So the anticonvulsant activity of the diazepam was exactly the same 2 weeks and 6 weeks after the lesion, and so was the kindled seizure activity of the control stimulations. So at least in this time period, 6 weeks after the lesion, there was nothing which would point to a dynamic development of something.

# PHARMACOLOGICAL AND TOXICOLOGICAL EVALUATION OF POTENTIAL ANTIEPILEPTIC DRUGS IN THE KINDLING MODEL OF EPILEPSY

Larry G. Stark[1], Steven L. Peterson[2] and Timothy E. Albertson[1]

*Departments of Pharmacology[1], School of Medicine, University of California, Davis, CA and Medical Pharmacology/Toxicology[2] College of Medicine, Texas A & M University, College Sta., TX*

## INTRODUCTION

The history of the pharmacology of antiepileptic drugs is now over 50 years old and much of the pioneering work can be traced to the discovery and development of phenytoin. The history and details of that particular research have recently been summarized and reviewed [7]. The use of pentylenetetrazol to produce chemoshock seizures and the subsequent discovery of trimethadione brought the science of pharmacology into the domain of antiepileptic drug testing in an even more permanent way. Toman, Swinyard, Goodman, Brown, Richards and Everett published a series of papers that further developed and quantified maximal electroshock and pentylenetetrazol models of epilepsy [31,30,29,28,5] that remain key components in the preclinical testing and evaluation of antiepileptic drugs (AEDs). Among this pioneering work one finds constant emphasis on not only refinement of methods useful for detection of antiepileptic drug efficacy, but on estimation of drug selectivity as a very important consideration.

Selectivity of drug effects is often the goal of investigators in pharmacology, regardless of the particular class of drugs they may be studying. The ability of a drug to produce a desired pharmacological effect without producing undesired side effects or toxicity was partially achieved with the discovery of phenytoin. Certainly it had fewer sedative side effects than phenobarbital, even though it turned out to have its own characteristic set of side effects and toxicities. The general principles of pharmacology associated with risk versus benefit were applied and the concept of a "protective index", or ratio of toxic dose to effective dose, came into common use.

Toman [29] also emphasized the importance of using an array of preclinical animal tests to evaluate AEDs. Indeed, his advice is still taken seriously. Kindling has become a valuable new addition to the array of available models in which to evaluate critically the selectivity of AEDs. The work described below has been completed keeping these early aspects of the preclinical development of antiepileptic drugs well in mind. We will illustrate some of these points applied to the kindling model as we describe endpoints used for efficacy and toxicity in drug studies done both during acquisition and following completion of kindling in adult and juvenile rats.

*Kindling 4*
Edited by J. A. Wada
Plenum Press, New York, 1990

## ENDPOINTS USED TO DETERMINE DRUG EFFICACY IN KINDLING STUDIES

### Experimental design of acquisition studies

Kindling experiments, especially those involving prophylactic drug treatment throughout the acquisition phase, are very labor intensive. In addition to the usual decisions about the experimental design regarding electrode placement, stimulation parameters and number of stimulations per day, one must make choices concerning pharmacological aspects of the design in a somewhat empirical fashion. Very often one has very incomplete information about newly synthesized drugs with potential antiepileptic effects. How soluble is it and in what solvent or combination of solvents? How stable is it in solution in the solvent chosen? An experimental group must generally be assigned to solvent treatment as a control for the drug study. How much time should be allotted between pretreatment and the kindling stimulation to permit adequate absorption of the drug via a given route of injection? And perhaps most importantly, what should the dose or doses be for chronic treatment during acquisition? This decision may follow some preliminary dose range-finding in animals that have not been surgically implanted. Once these decisions have been made, a number of endpoints are available for quantification of the efficacy of a drug treatment.

Daily observations and measurements on each animal regarding a rank score of seizure severity, afterdischarge duration and perhaps seizure threshold values would ordinarily complete an experimental session. When followed to completion, the data collected permit an investigator to compare the control and treated groups in several ways. For all the treatment groups it is possible to determine the number of stimulations required to elicit the first seizure of each type including the number required for the first +5 generalized seizure. One may ascertain the efficacy of the test substance with respect to its ability to suppress the focal seizure, afterdischarge duration, or some behavioral component of the seizure such as forelimb clonus. When a range of doses is used it is possible to compute an ED50 value corresponding to any one of these endpoints.

### Studies in fully-kindled rats

Once an acquisition study has been completed, the permanence of the kindling phenomenon makes it possible to continue study of other aspects of the pharmacology of the drug. One may wish to determine the time course of an ED50 dose, for example. Alternatively, one might study the effect of the same range of doses used previously on the kindled threshold in fully kindled rats. The basis for determination of ED50 values might again rest on efficacy for reducing the focal seizure, the spread of the afterdischarge, or on the ability of the drug to alter some behavioral component of the elicited seizures.

## ENDPOINTS USED TO DETERMINE DRUG TOXICITY IN KINDLING STUDIES

A number of different endpoints have been utilized to quantify side effects and toxicity associated with test substances in kindling experiments. These have ranged from no mention of the method used, if any, to simply the general appearance of the animal, presence or absence of ataxia, presence or absence of righting reflexes and whether or not placing reflexes remain intact after treatment. Subjective observations regarding sedation may be mentioned. More recently, experimenters have been reporting rotorod performance for animals during either acquisition trials or in fully kindled animals in an attempt to further quantify motor neurological deficits have also been reported [27]. The neurological tests used in the Saffan and glycine experiments were as follows: 1) Rotorod. The animal was placed on a 6 cm diameter rod that rotated at 8 rpm. Inability to remain on the rod for one min using any of 3 separate trials was considered a failure; 2) Position sense. The right hindlimb was lowered over the edge of a table and the inability to correct the position of the limb within 5 sec was considered a failure; 3) Gait and Stance. The animal was placed on a table surface so that the gait could be observed and any circling or staggering gait, abnormal limb or torso posture, tremor, hyperactivity, somnolence, or catalepsy was considered a failure. Any of these endpoints may be used in conjunction with studies done using multiple doses to calculate a TD50 (toxic dose for 50% of the subjects).

## EVALUATION OF SELECTIVITY BASED ON TOXICITY AND EFFICACY TESTING

The selectivity of the test agent, expressed as the ratio of the dose producing side effects or toxicity to the dose producing anticonvulsant effects, may be calculated using the TD50 and ED50 values determined according to a given set of endpoints. Since many combinations of endpoints for toxicity and efficacy exist, it follows that there is no one unique value for a calculated "protective index".

## SOME STUDIES OF DRUGS DURING ACQUISITION OF THE KINDLING RESPONSE AND IN FULLY-KINDLED ADULT RATS

### I. Studies on LY-201116

LY-201116, [4-amino-N-(2,6-dimethylphenyl)benzamide], classified as a 4-aminobenzamide has been reported to be an effective anticonvulsant in the maximal electroshock seizure test in mice [4,13,3]. The compound has virtually no activity at non-toxic doses against other seizure models such as subcutaneous pentylenetetrazol, bicuculline, picrotoxin or strychnine [22]. Several reports have noted the similarity in profile between this new class of anticonvulsant and phenytoin [13,22]. It has been suggested that the compound or its derivatives may be useful for the treatment of generalized tonic clonic and complex partial seizures in humans [22]. We have examined the effects of LY-201116 in the amygdaloid kindling model, both during acquisition of kindled seizures and in fully kindled rats.

A. Acquisition study. Male Sprague-Dawley rats weighing approximately 300g were the subjects. They were housed individually during the experiment and had access to food and water throughout the day. They were anesthetized and implanted stereotaxically with bipolar electrodes for stimulation and recording in the amygdala (coordinates: 1.0 mm posterior to the bregma, 4.75 mm lateral, 7.5 mm ventral from the surface) and with other cortical electrodes for recording the EEG. Details of the surgical techniques have been published previously [1]. Kindling trials were conducted once per day, 30 min after the i.p. injection of either the solvent (DMSO, 0.25 ml/Kg) or the drug. Stimulus parameters: 1 sec train of 60Hz biphasic square waves, each 1 msec in duration and 400 $\mu$A in amplitude. Seizure severity was measured using both afterdischarge duration and a behavioral ranking scale. The afterdischarge duration (AD) following the stimulus was equated with the period during which 3 or more spikes, at least twice the amplitude of the prestimulus amplitude, occurred at a frequency of 1 per sec or faster in the amygdala and/or cortex. Assessment of behavioral seizure severity was determined using a ranking scale such that a score of (0) was assigned for no behavioral response; (1) indicated facial clonus or eye closure; (2) indicated 1 plus head nodding or head and neck clonus; (3) indicated 2 plus forelimb clonus; (4)

rearing on the hindlimbs; and (5) indicated 4 plus repeated rearing and/or loss of postural balance.

The dose of LY-201116 chosen for the acquisition study was 7.5 mg/Kg. This dose was estimated from range-finding studies to be below that which would be clearly neurotoxic, but perhaps effective. Previous work had been done using the oral route with longer time intervals between dosing and testing. Subjects were treated for 14 consecutive days and then all injections were stopped. Kindling acquisition trials continued for an additional 10 days following cessation of treatment. Rotorod performance was judged by a single trial on the rod 30 min after pretreatment. The rod rotated at 15 rpm and the total time the rat remained on the rod, or a criterion performance of 30 sec, was recorded. Group performances on parametric measures (days to kindle, seconds of AD, rotorod performance) were compared using a one-way ANOVA. Where significant differences were found, comparisons between control and treated groups were done using either Student's t-test or Dunnett's Test. Mann-Whitney U-tests were done to compare rank scores for the two groups.

## Results

The results of the acquisition study are shown in Fig. 1, 2 and 3. The 7.5 mg/Kg dose did not significantly alter the number of stimulations required to produce the first stage 5 seizure (control mean = 12.2 + 1.9 vs drug mean = 14.3 + 1.7). The one-way ANOVA test did not reveal any treatment effect on afterdischarge duration, graphically shown in Fig. 1. Although the rank scores for the treated group tended to be lower, there were significant differences only on day 6, day 10 and day 14 (Fig. 1).

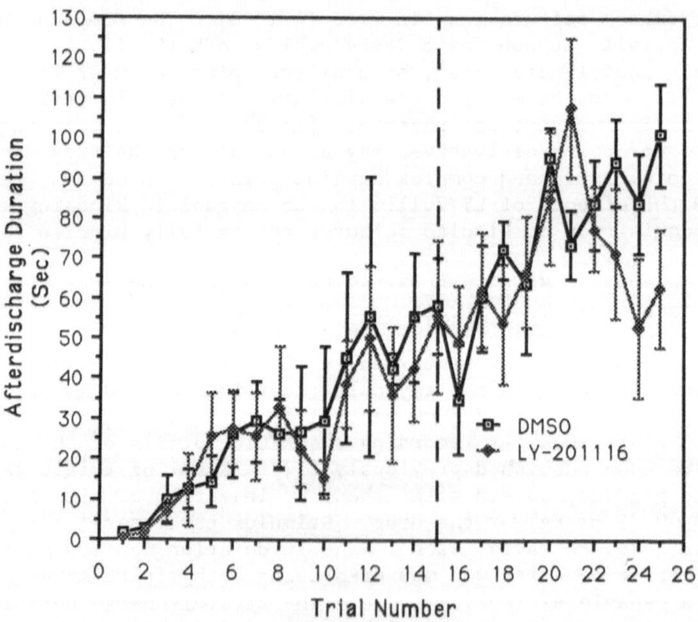


**Figure 1.** Effects of LY-20116 on afterdischarge duration during kindling acquisition trials. Data to the left of the vertical dashed line was obtained following daily treatment with drug or vehicle; data to the right of the line was obtained without treatment. Vertical bars signify 1 S.E.M.

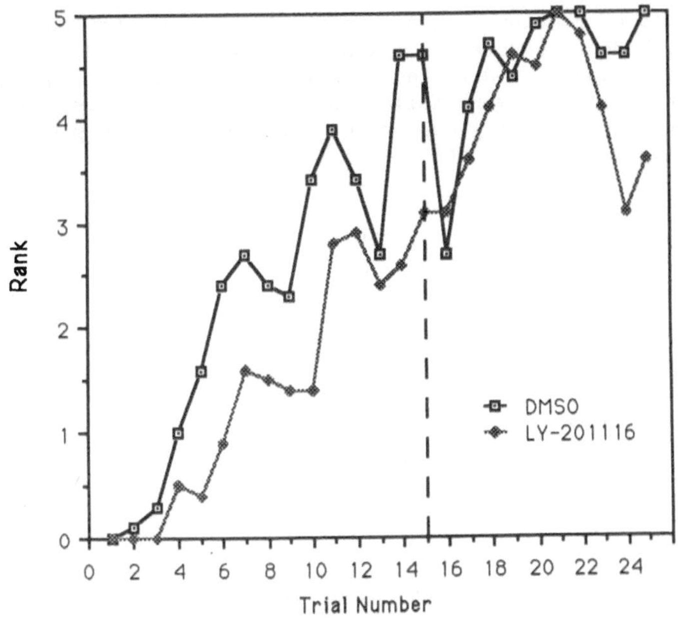


**Figure 2.** Effects of LY-201116 on rank scores during kindling acquisition trials. See Fig. 1 legend for details.

The one-way ANOVA test did reveal a significant treatment effect on rotorod performance, graphically shown in Fig. 3. Further evaluation showed that significant differences occurred only on day 4 and day 11.

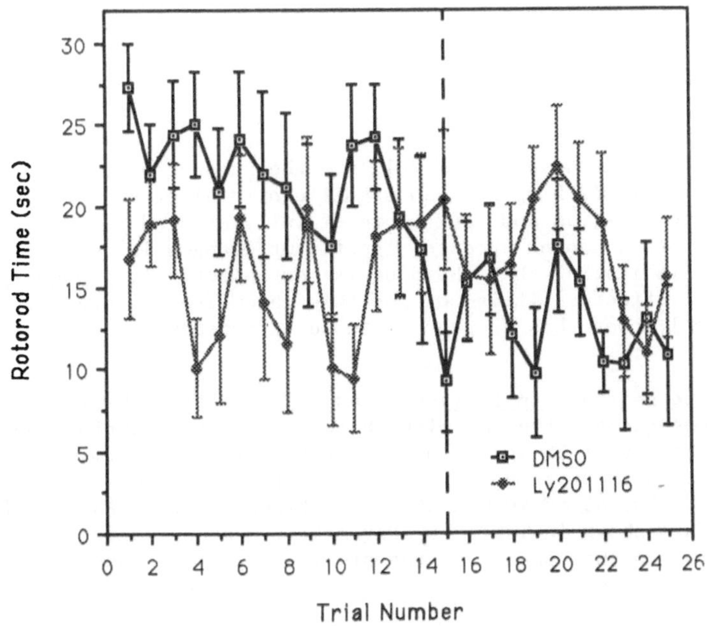


**Figure 3.** Effects of LY-201116 on rotorod performance during kindling acquisition trials. See Fig. 1 for details.

The results of continuation of the kindling trials beyond day 14 reveal that the afterdischarge duration for both groups continued to increase (Fig. 1) and the apparent difference in rank scores was diminished (Fig. 1) indicating that all rats were fully kindled. It is interesting that the performance of the control group on the rotorod tended to progressively decrease during the experiment.

B. <u>Dose-response study in fully-kindled rats</u>. Once all the kindled rats in the acquisition study had received 10 additional trials without any treatment, other doses of LY-201116 were evaluated and compared to solvent treatment for their effects on the fully kindled seizures. Parametric measures and rank scores were compared to the previous trial results as percentage of control response. The results of pretreatment with DMSO, 7.5 mg/Kg, 11.25 mg/Kg or 15.0 mg/Kg 30 min prior to the kindling stimulation are shown in Table 1.

**TABLE 1**

Dose-Response Study of LY0201116

| Treatment | %ControlRR | %ControlAD | Rank score |
|---|---|---|---|
| DMSO | 189.7 ± 51 | 147.7 ± 45 | N.S. |
| 7.50 | 83.6 ± 14* | 83.3 ± 19 | N.S. |
| 11.25 | 67.4 ± 19** | 29.5 ± 15** | P ≤.001 |
| 15.00 | 20.3 ± 11** | 3.8 ± 2 ** | P ≤.001 |
| F(3,39) | 6.27 | 6.26 | P ≤.001 |

* = P ≤0.05
** = P ≤0.01

It is clear that doses of LY-201116 higher than that used for the acquisition trials (7.5 mg/Kg) can significantly alter both the afterdischarge duration and rank score of fully kindled rats. It is also apparent that these doses produce significant changes in rotorod performance, indicating that the drug is not selectively producing antiepileptic effects. It is not possible calculate a protective index greater than 1.0 for this drug under these circumstances.

C. <u>Time-response study in fully-kindled rats</u>. All of the fully kindled rats were subsequently treated with the effective (but performance impairing) dose of 11.25 mg/Kg on 3 separate days to determine the time-response relationship of the antiepileptic effects noted earlier. Rats were tested at 30, 60 or 90 min after treatment and the results of those experiments are shown in Table 2. Parametric measures and rank scores were again compared to the results on the previous trial and expressed as a percentage of control response.

It is apparent that the antiepileptic effect of LY-201116 following administration of 11.25 mg/Kg on afterdischarge duration lasts no longer than a maximum of 90 minutes, although the behavioral severity of the seizures appears to return somewhat earlier (Table 2).

**TABLE 2**

Time-Response Study of LY-201116

| Time | %ControlRR | %ControlAD | Rank score |
|---|---|---|---|
| 30 | 68.8 ± 18 | 32.8 ± 14 | 2.1 ± .43 |
| 60 | 74.4 ± 11 | 59.4 ± 10 | 3.6 ± .52* |
| 90 | 122.9 ± 30 | 90.2 ± 10** | 4.4 ± .35** |
| F(2,28) | 1.62, N.S. | 5.6, P ≤.001 | |

* = P ≤0.05, 30 min vs 60 min
** = P ≤0.01, 30 min vs 90 min

D. Effects of LY-201116 on kindled seizure threshold. Further inspection of the results of the dose-response study suggested that this drug might influence seizure threshold since animals seemed to react in an all-or-none fashion to the dosages. Studies on seizure threshold were done to compare the effects of the solvent with 15 mg/Kg LY-201116 30 minutes after treatment. Thresholds were determined using the same stimulus parameters described above, but the amplitude of the stimulus was only 50$\mu$A during the first stimulation. Stimulus intensity was increased by 50$\mu$A every minute until a sustained afterdischarge lasting more than 5 sec was obtained, or until a maximum of 500$\mu$A was reached. (See results Table 3.)

**TABLE 3**

Effects of LY-201116 on Kindled Seizure Threshold

| Treatment | DMSO | | LY-201116 (15 mg/Kg) | | | | |
|---|---|---|---|---|---|---|---|
| Thr(uA) | AD (sec) | Rank | Thr(uA) | %Thr | AD(sec) | %AD | Rank |
| 284.4 ±26.9 | 102.7 ±9.9 | 4.7 | 440.6** ±22.5 | 174.8** ±18.4 | 32.3** ±10.9 | 32.6** ±9.8 | 1.9** |

N = 16
** = P ≤0.01 drug compared to DMSO

Thus, from these results it appears that LY-201116 raises the threshold for induction of the focal seizure as well as decreasing the spread of the seizure away from the focus.

Summary of LY-201116 Results

LY-201116, in contrast to its apparent selective effects when given orally in the maximal electroshock seizure test in mice, had no selective effects against fully kindled amygdaloid seizures in rats. Toxicity, as measured by rotorod performance, was apparent at all doses tested and persisted for 60-90 minutes following treatment with 11.25 mg/Kg intraperitoneally. The drug raised seizure threshold and decreased spread of the afterdischarge. The brainstem or spinal action of the drug responsible for the hindlimb ataxia and flaccidity may, of course, contribute to the loss of the tonic hindlimb extension component on the MES test, the usual indicator of an "anticonvulsant effect" in that model. The results obtained here also raise some doubts with respect to its potential use in human complex partial epilepsy if kindling is a valid preclinical indicator of clinical efficacy.

Parli et al. recently reported [16] that LY-201116 is rapidly metabolized by N-acetylation in several species, including the rat. Analogs of LY-201116 have been successfully synthesized which retain the anticonvulsant activity while altering the metabolic pathway and it would be of interest to determine whether this new analog would have greater selectivity in the kindling model.

II. Studies on an anesthetic steroid (Saffan)

Recent investigations have demonstrated that some steroids exert anesthetic properties by a GABAergic mechanism of action [11]. Because of this agonist activity, it was postulated that some steroids might also possess anticonvulsant activity. Saffan (Glaxo, Inc.) is the trade name for a veterinary anesthetic product that is a combination of the synthetic steroid alphaxalone and alphadalone. The active compound is thought to be alphaxalone with alphadalone being included only to enhance the solubility of alphaxalone. The purpose of this study was to investigate the anticonvulsant properties of Saffan.

The general methods used were identical to those described previously for LY-201116. An alternative method was used to assess neurological deficits [27]. Rats were tested for drug-induced neurological deficits during the 5 min period preceding each seizure test. Saffan was administered intramuscularly to the hindlimb, and a range of doses were tested (1.25 mg/kg to 10 mg/kg).

The following test schedule was used to evaluate the effect of the drug on kindled amygdaloid seizures. On day 1 the rat received stimulation without drug. On day 2 the animal received either the drug or vehicle, and then the kindling stimulus. On day 3 the animals were not handled. Days 4 and 5 were repeats of days 1 and 2, respectfully, except that the rats received a different dose of drug. If an animal did not respond with a maximal seizure response (rank 5) on days 1 or 4, it was not used on the following day for drug treatment. The duration of the AD on the drug day was expressed as a percentage of the control response.

Nonlinear probit analysis was employed to determine the TD50 and confidence interval on the quantal neurological deficit response. A nonlinear regression analysis was performed on the AD data. The ED50 was also determined from the nonlinear regression analysis.

The results on the Saffan experiment are shown in Fig. 5. It reduced the cortical AD and behavioral seizure response associated with kindled amygdaloid seizures, but only at doses that induced neurological deficits (Fig. 5). Pretreatment 1 hr prior to the stimulation reduced the cortical AD with a calculated ED50 of 3.2 mg/kg and TD50 of 3.9 mg/kg. The protective index for Saffan against the cortical AD was thus 1.2. Saffan also reduced the severity of the convulsions, but again, only at doses associated with neurological deficits.

The steroids alphaxalone and alphadolone as combined in Saffan have efficacy against kindled amygdaloid seizures, but they lack selectivity (P.I. = 1.2) with respect to their anticonvulsant properties. The anticonvulsant profile of Saffan resembles that of pentobarbital [21] in that it is effective only at doses that induce neurological deficits. Perhaps the steroids and pentobarbital have similar mechanisms of action at the GABA receptor-chloride ionophore complex inducing both anticonvulsant and anesthetic effects [11]. Enhancement of the GABA response has been correlated with anticonvulsant activity while increases in chloride conductance are associated with anesthetic properties of barbiturates [23]. The lack of selectivity demonstrated here suggests that presumed GABA-mediated anticonvulsant effects of the steroids occur only at doses which also produce neurological deficits.

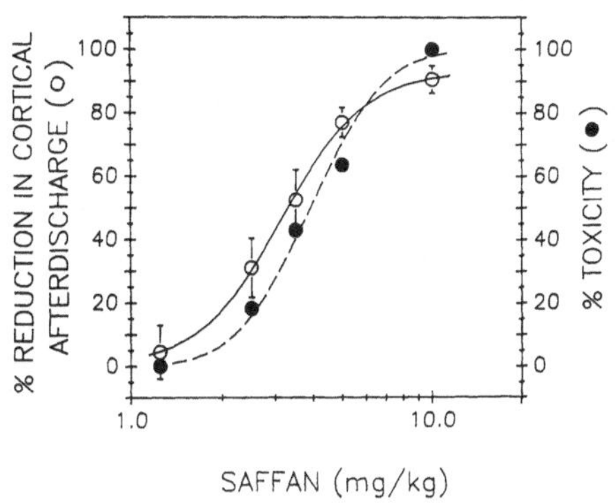


**Figure 4.** Effect of Saffan on seizure response and neuro-logical deficit in kindled amygdaloid seizures. Saffan was administered i.m. 30 min prior to the seizure test. Each animal was tested for neurological deficit during the 5 min period preceding the seizure test and the percent with neurological deficit is presented as % toxicity. The results of the kindled seizures represent the averages 7-14 rats at each dose.

## III. Therapeutic adjuncts to antiepileptic drug effects in kindled seizures

Glycine, systemically administered, has been shown to possess anticonvulsant activity and to potentiate clinically effective anticonvulsants [25,32]. When tested in standardized models of epilepsy in rats, doses of glycine with no anticonvulsant activity of their own have been shown to potentiate the effect of other anticonvulsants in maximal electroshock [20], subcutaneous pentylenetetrazol [19], and kindled amygdaloid seizures [18]. These findings indicate that glycine might be useful as an adjunct to anticonvulsant therapy, allowing a decrease in the dose of a standard anticonvulsant without losing efficacy. The purpose of the present study was to further define the dose-response and time-response relationships of glycine in rats with kindled amygdaloid seizures.

The general methods of surgery, kindling and handling of the animals were the same as described above. Glycine was administered orally as a 3M solution in 0.9% saline. A near-saturation concentration of glycine was employed so that the largest doses could be tested in volumes as small as possible; nevertheless, doses greater than 40 mM/kg could not be tested due to the limited gastric capacity of the rats. Glycine was administered 1, 2, 4 or 8 hours prior to the seizure test. All animals (N=14) received both dosage levels at all four pretreatment intervals. The schedule for testing the animals twice per week was the same as described above for the Saffan experiment.

All statistical comparisons were done using multivariate analysis of variance for repeated measures. The Scheffe multiple comparison procedure was used for tests of individual pairs of means.

The results of the glycine experiment are shown in Figure 5. It reduced the duration of the cortical AD elicited by the kindling stimulus in a dose-dependent manner ($P < .025$) and the time-response effects were also statistically significant ($P < 0.002$). The Scheffe multiple comparison procedure established a significant decrease in the cortical afterdischarge when 40 mM/kg glycine was administered 1 or 2 hrs prior to seizure testing, compared to control (no drug) or the corresponding 30 mM/kg glycine dose. No other significant differences in cortical afterdischarge were found.

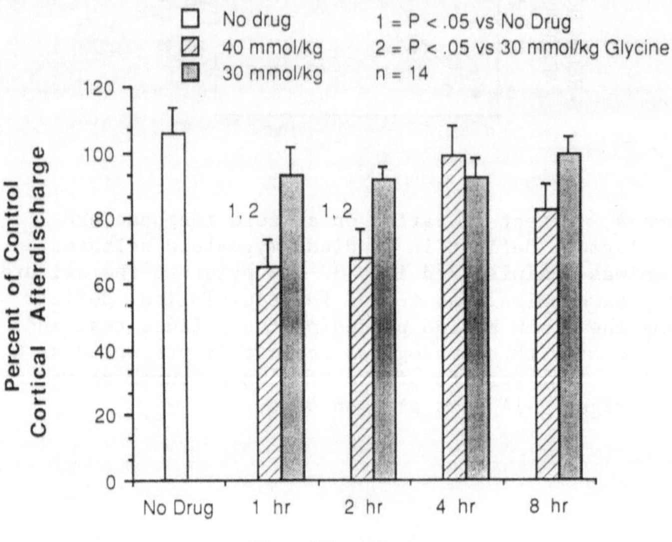


**Figure 5.** Effect of glycine on cortical afterdischarge duration in kindled amygdaloid seizures. Glycine was administered at the times indicated prior to the seizure test. No Drug indicates a seizure control test in which neither glycine nor the vehicle was administered prior to the seizure test. Each column represents the average response of the same 14 rats. The 40 mmol/kg glycine dose administered 1 and 2 hours prior to the seizure test significantly reduced the duration of the corticol afterdischarge duration as compared to the control seizure (No Drug) as well as the 30 mmol/kg glycine dose administered at the corresponding times of administration.

Glycine was found to have anticonvulsant activity in rats exhibiting kindled amygdaloid seizures. This extends the findings of previous work demonstrating that a dose of glycine, which by itself is inactive, potentiates the anticonvulsant activity of phenobarbital and diazepam in kindled amygdaloid seizures [18]. The present study demonstrates that glycine alone can reduce the duration of cortical afterdischarge without inducing neurological deficits. Although a possible interaction with GABA receptors has been proposed as the mechanism by which glycine inhibits kindled seizures [18], there is other evidence that glycine potentiates the activity of the excitatory amino acids NMDA and glutamate [12]. On the basis of *in vitro* experiments it has been proposed that excess glycine overactivates NMDA receptors and therefore exerts proconvulsant activity, exacerbating seizure disorders [6]. Glutamate has been implicated in amygdaloid kindling since increased concentrations of it have been found in ventricular cerebral spinal fluid during rank 5 seizures [17] and because NMDA antagonists inhibit kindled seizures [15]. Because of the apparent role of glutamate in kindling, it was expected that glycine might influence the kindled seizure response. However, the *in vivo* studies reported here suggest that glycine is anticonvulsant and, from other studies, also potentiates the activity of clinically effective anticonvulsants. Although the apparent discrepancy between the *in vitro* and the *in vivo* experiments requires further investigation, these findings support the importance of employing suitable as well as multiple models of epilepsy to investigate potential mechanisms of anticonvulsant drug action.

## A STUDY OF DIAZEPAM DURING ACQUISITION OF THE KINDLED RESPONSE AND IN FULLY-KINDLED JUVENILE RATS

A number of investigators have developed methods for study of neonate and juvenile rats during acquisition of the kindled response [8,26,9] and have compared the results in these immature nervous systems to those obtained in adults. Several reports have suggested that benzodiazepine agonists, including diazepam, produce seizure-like behavioral responses in immature rats [14,2]. We recently completed a drug study using diazepam to determine whether or not the drug would influence kindling acquisition or the fully kindled response since these reports suggested a possible proconvulsant action for the drug in these immature rats. The efficacy of anticonvulsant drugs in the immature, kindled rat brain has not been widely explored.

In the first part of the study, weanling rats (23-25 days old) were implanted chronically with amygdaloid electrodes. They were treated with DMSO, 0.5 mg/Kg or 1.0 mg/Kg diazepam once daily 30 min before the first of two kindling stimulations spaced 240 minutes apart. After 5 days the treatments were stopped , but the stimulations continued for an additional 10 days. The number of stimulations required to elicit the first stage 5 seizure was recorded along with the average length of the afterdischarge and the seizure rank for each trial. (See results, Table 4, Fig. 6.)

**TABLE 4**

Effects of Diazepam on the Acquisition of Amygdaloid Kindled Seizures

| Dose (mg/Kg) | N | Stimulations to First Rank "5" Seizures |
|---|---|---|
| Control | 16 | 15.9 ± 1.2 |
| 0.5 | 9 | 19.8 ± 1.0* |
| 1.0 | 5 | 22.0 ± 0.5** |

* = P ≤0.05 compared to Control
** = P ≤0.01 compared to Control

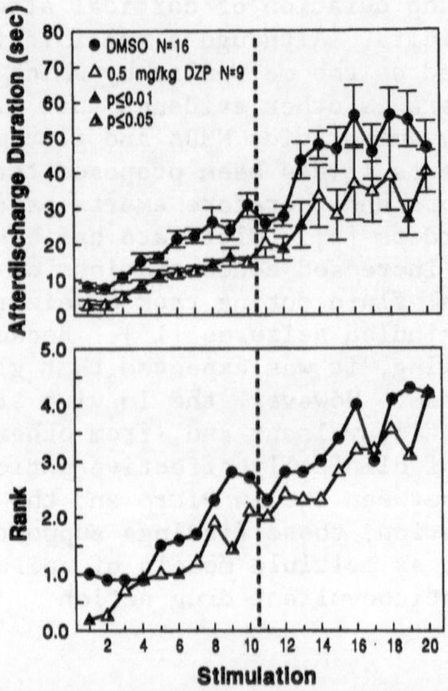


**Figure 6.** The effect of diazepam (DZP) pretreatment on kindling acquisition. Mean afterdischarge duration (± SEM) and seizure ranks are presented for daily 0.5 mg/Kg i.p. diazepam or DMSO control pretreatments. The dotted line represents when daily drug pretreatment was terminated. Weanling rats were stimulated twice daily 30 minutes and 270 minutes after single daily drug dosing for the first days and 240 minutes apart for the last 5 days.

Five days of treatment with either dose of diazepam significantly increased the number of stimulations required to reach the first stage 5 seizure compared to the control group. No evidence of spontaneous seizure activity was seen after daily doses of diazepam.

In the second part of the study, weanling rats were stimulated twice daily 240 minutes apart for a total of 12 days until stable stage 5 seizures predominated. Prior to the next stimulation they were treated with 0.5ml/Kg DMSO i.p., or diazepam (0.25, 0.5, 1.0, 2.0, or 4.0 mg/Kg i.p. in DMSO). Thirty minutes later animals were stimulated, the elicited afterdischarge recorded and seizure ranks were assigned. Twenty-four hours later all animals were stimulated without pretreatment to verify that the seizure response had returned to baseline values. Forty-eight hours later animals were again pretreated with another dose of diazepam and stimulated 30 minutes later. (See results, Table 5.)

Diazepam significantly reduced both seizure rank and afterdischarge duration of fully kindled seizures in these juvenile rats. The highest dose produced obvious behavioral sedation. The ED50 for suppression of afterdischarge was 0.5 mg/Kg (0.2 and 1.4 mg/Kg = 95% confidence intervals).

These two experiments demonstrate that diazepam has anticonvulsant properties in weanling and juvenile rats that include slowing the acquisition of the kindled response and reducing the severity and length of fully kindled seizures. These effects are similar to, and perhaps more exaggerated than, the effects of the drug in kindled adult rats [1,10,24].

**TABLE 5**

Effects of Diazepam on Fully Kindled Amygdaloid Seizures

| Dose(mg/Kg) | N | AD(Sec) | AD% | R |
|---|---|---|---|---|
| Control | 31 | 61.2 ± 4.8 | ---- ---- | 4.4 |
| 0.25 | 4 | 34.8 ± 7.2 | 72.8 ± 13.2 | 4.0 |
| 0.50 | 7 | 27.4 ± 9.2* | 45.9 ± 15.6* | 1.6* |
| 1.0 | 7 | 6.6 ± 2.5** | 17.6 ± 5.9** | 0.4 |
| 2.0 | 7 | 19.5 ± 4.7** | 33.9 ± 8.5** | 1.6* |
| 4.02 | 6 | 7.3 ± 1.3** | 12.7 ± 2.4** | 0.0* |

1 = ± of control trial AD for same animals
2 = 6/6 animals sedated with righting reflex intact prior stimulation
* = P ≤0.05 compared to Control
** = P ≤0.01 compared to Control

There was no EEG or behavioral evidence of spontaneous seizure activity in contrast to earlier reports in immature rats [14,2]. This preliminary study suggests that kindling in immature rats may again join chemoshock and electroshock methods as a useful tool for the study of the potential efficacy of AEDs.

## SUMMARY COMMENTS

The kindling model of epilepsy remains a fascinating and productive tool for investigations of various aspects of this pathophysiologic state. Compared to some of the older models such as MES and chemoshock, the generalization phenomena associated with kindling occur at a much slower rate and provide us with an extended window of opportunity in which to test various hypotheses as well as drugs. We have mentioned that the endpoints with which one may evaluate drugs may vary and we have stressed the need to do both anticonvulsant testing and toxicity testing at the whole animal level.

## REFERENCES


1. Albertson, T.E., Peterson, S.L., and Stark, L.G. (1980): Anticonvulsant drugs and their antagonism of kindled amygdaloid seizures in rats. Neuropharmacology 19:643-652.
2. Barr, G.A. and Lithgow, T. (1983): Effect of age on benzodiazepine induced behavioral convulsions in rats. Nature 302:431-432.
3. Clark, C.R. (1987): Comparative anticonvulsant activity and neurotoxicity of 4-amino-N-(2,6-dimenthylphenyl)benzamide and prototype antiepileptic drugs in mice and rats. Epilepsia 29(2):198-203.
4. Clark, C.R., Sansom, R.T., Lin, C-M, and Norris, G.N. (1985): Anticonvulsant activity of some 4-aminobenzanilides. J. Med. Chem. 28:1259-1262.
5. Everett, G.M. and Richards, R.K. (1944): Comparative anticonvulsant action of 3,5,5-trimethyloxazolidine-2,4-dione (tridione), dilantin and phenobarbital. J. Pharm. & Exp. Ther. 81:402-407.
6. Foster, A.C. and Kemp, J.A. (1989): Glycine maintains excitement. Nature 338:377-378.

7. Freidlander, W.J. (1986): Putnam and Meritt and the discovery of dilantin. Epilepsia 27:(Suppl. 3).
8. Gilbert, M.E. and Cain, D.P. (1980): Electrode implantation in infant rats for kindling and chronic brain recording. Dev. Brain Res. 1:553-555.
9. Gilbert, M.E. and Cain, D.P. (1982): A developmental study of kindling in the rat. Dev. Brain Res. 2:321-328.
10. Harris, Q.L., Lewis, S.J., Young, N.A., Vajda, F.J. and Jarrott, B. (1988): Relationship between the dose-response effects of diazepam and clobazam on electroencephalographic parameters and on kindled amygdaloid seizure activity in rats. Clin. Exp. Pharmacol. Physiol. 15:753-764.
11. Harrison, N.L., Majewska, M.D., Harrington, J.W. and Barker, J.L., (1987): Structure-activity relationships for steroid interaction with gamma-aminobutryic $acid_A$ receptor complex. J. Pharmacol. Exp. Ther. 241:346-353.
12. Johnson, J.W. and Ascher, P. (1987): Glycine potentiates the NMDA response in cultured mouse brain neurons. Nature 325:529-531.
13. Leander, J.D., Lawson, R.R., and Robertson, D.W. (1988): Anticonvulsant effects of a novel aminobenzamide (LY2011116) in mice. Neuropharmacology 27(6):623-628.
14. Nutt, D.J. and Little, H. (1986): Benzodiazepine-receptor mediated convulsions in infant rats: Effects of beta-carbolines. Pharmacol. Biochem. Behav. 24:841-844.
15. Okazaki, M.M., McNamara, J.O. and Nadler, J.V. (1989): N-methyl-D-aspartate receptor autoradiography in rat brain after angular bundle kindling. Brain Research 428:359-364.
16. Parli, C.J., Potts, B.D., Kovach, P.M., Robertson, D.W. (1987): Metabolic fate of the anticonvulsant LY201116 in rats, mice, dogs, and monkeys. Pharmacologist 29:176.
17. Peterson, D.W., Collins, J.F. and Bradford, H.F. (1983): The kindled amygdala model of epilepsy: Anticonvulsant action of amino acid antagonists. Brain Research 275:169-172.
18. Peterson, S.L., (1986): Glycine potentiates the anticonvulsant action of diazepam an phenobarbital in kindled amygdaloid seizures of rats. Neuropharmacology 25(12):159-1363.
19. Peterson, S.L. (1989): Glycine potentiates diazepam and sodium divalproate in subcutaneous pentylenetetrazol seizures. Soc. Neurosci. Abstr., in press.
20. Peterson, S.L., Trzeciakowski, J.T., Boehnke, L.E. and Riegel, R.N. (1988): Glycine potentiates phenobarbital, carbamazepine and MK-801 in maximal electroshock seizures. Soc. Neurosci. Abstr. 14:865.
21. Raines, A., Blake, G.J., Richardson, B., and Gilbert, M.B. (1979): Differential selectivity of several barbiturates on experimental seizures and neurotoxicity in the mouse. Epilepsia 20:105-113.
22. Robertson, D.W., Lawson, R.R., Rathbun, R.C., and Leander, J.D. (1988): Pharmacology of LY201409, a potent benzamide anticonvulsant. Epilepsia 29(6):760-769.
23. Schulz, D.W. and MacDonald, R.L. (1981): Barbiturate enhancement of GABA-mediated inhibition and activation of chloride ion conductance: Correlation with anticonvulsant and anesthetic actions. Brain Res. 209:177-188.
24. Schwark, W.S. and Loscher, W. (1985): Comparison of the anticonvulsant effects of two novel GABA uptake inhibitors and diazepam in amygdaloid kindled rats. Naunyn-Schmiedeberg's Arch. Pharmacol. 329:367-371.
25. Seiler, N. and Sarhan, S. (1984): Synergistic anticonvulsant effects of a GABA agonist and glycine. Gen. Pharmac. 15:367-369.
26. Stark, L.G., Albertson, T.E., Joy, R.M., He, P. and Streisand, J. (1986): The acquisition of a kindled response in developing rats using 24-h intertrial intervals. Dev. Brain Res. 24:291-294.

27. Swinyard, E.A. and Woodhead, J.H. (1982): Experimental detection, quantification, and evaluation of anticonvulsants. In: Antiepileptic Drugs, (Woodbury, D.N., Ed.), pp.111-126. Raven Press, New York.
28. Swinyard, E.A., Brown, W.C., and Goodman, L.S. (1952): Comparative assays of antiepileptic drugs in mice and rats. J. Pharmacol. Exp. Ther. 106:319-330.
29. Toman, J.E.P. (1964): Animal techniques for evaluating anticonvulsants. In: Animal and Clinical Pharmacologic Techniques in Drug Evaluation, (Nodine, J.H. and Siegler, P.E., Eds.), Chapter 46, pp. 348-352, Year Book Medical Publishers, Chicago.
30. Toman, J.E.P., Goodman, L.S. (1948): Anticonvulsants. Physiological Reviews 28(1):409-432.
31. Toman, J.E.P., Swinyard, E.A., and Goodman, L.S. (1946): Properties of maximal seizures and their alteration by anticonvulsant drugs and other agents. J. Neurophysiol. 9:231-239.
32. Toth, E. and Lajtha, A. (1984): Glycine potentiates the action of some anticonvulsant drugs in some seizure models. Neurochem. Res., 12:1711-1718.

ACKNOWLEDGEMENT

Supported in part by N.I.N.C.S. #24566.

# CONTINGENT TOLERANCE TO THE ANTICONVULSANT EFFECTS OF DRUGS ON KINDLED CONVULSIONS

John P.J. Pinel, C. Kwon Kim, and Michael J. Mana

Department of Psychology
University of British Columbia
Vancouver, B.C., Canada

At the Kindling 3 Conference in 1985, we reported that the development of tolerance to ethanol's anticonvulsant effect on kindled convulsions elicited in rats by amygdalar stimulation is greatly influenced by the temporal relation between the administration of ethanol and the convulsive stimulation (18). We reported that substantial tolerance developed to ethanol's anticonvulsant effect if the kindled rats were stimulated following each of five bidaily (once every 48 hr) intraperitoneal ethanol injections, but not if they were stimulated before each injection (see also 16). The fact that both groups of rats received the same exposure to ethanol and the same number of stimulations suggested that the critical factor in the development of the tolerance was the administration of convulsive stimulation durinq the periods of ethanol exposure. On the basis of this and similar observations (e.g, 13;17;20), we have proposed a theory of tolerance that emphasizes the idea that functional drug tolerance is a reaction to the expression of a drug's effect rather than to the mere presence of the drug in the system (19). Although a drug's effects are often an inevitable consequence of drug exposure, in some instances they may not be fully expressed unless the nervous system (or other target tissue) is involved in a specific pattern of neural activity during periods of drug exposure (cf., 4). We believe that anticonvulsant drug effects can belong to this latter category; in our experiments, the anticonvulsant effects of drugs that are administered on an intermittent basis cannot be fully expressed unless convulsive activity occurs during the periods of drug exposure. In such situations, the drug-effect theory of tolerance predicts that functional tolerance will not fully develop unless the subject is repeatedly stimulated so that it experiences the drug effect (the anticonvulsant effect in this case); drug exposure is not sufficient for maximal tolerance development. Drug tolerance that is contingent upon the occurrence of a particular pattern of neural or behavioral activity during periods of drug exposure is referred to as contingent tolerance (see also 5).

Although our early studies of contingent tolerance focused on the development and dissipation of tolerance to ethanol's anticonvulsant effect, we have recently completed a series of experiments that focused on the role that convulsive stimulation can play in the development of tolerance and cross-tolerance to the anticonvulsant effects of several antiepileptic drugs that are commonly prescribed in clinical practice. This chapter summarizes these recent experiments.

*Kindling 4*
Edited by J. A. Wada
Plenum Press, New York, 1990

## GENERAL METHODS

The procedure that we used to study the development of tolerance to the anticonvulsant effects of various clinical antiepileptics involved five phases.

Phase I: Kindling. First, each rat received a total of 45 stimulations (400 microamps, 60 Hz, 1 sec) through a single bipolar electrode implanted in the left basolateral amygdala. These kindling stimulations were administered three times per day, five times per week, over a 3-week period, with at least 2 hours between successive stimulations.

Phase II: Baseline. Beginning 48 hours after the last stimulation of the kindling phase, each rat received a total of four baseline stimulations. These stimulations were administered once every 48 hours; this bidaily stimulation schedule was maintained for the duration of the experiment. The duration of forelimb clonus elicited by each stimulation was our primary measure of motor seizure severity. We also recorded both motor seizure class according to Pinel and Rovner's (21) modification of Racine's (22) scale and the latency to the initiation of forelimb clonus elicited by each stimulation; however, because these measures were generally less sensitive than, more variable than, and highly correlated with the duration-of-forelimb-clonus measure, we do not report them here. The consistency of the forelimb clonus duration measure is impressive; in study after study, we have found the mean forelimb clonus duration on the last baseline trial to be between 40 and 45 seconds, with the forelimb clonus duration for individual subjects almost always within 10 seconds of this value.

Phase III: Drug-Screen Trial. A drug-screen trial occurred 48 hours after the last baseline stimulation; the test dose of the drug under investigation was administered to each rat 1 hour before a convulsive stimulation in order to establish a baseline against which to assess the subsequent development of tolerance. Drug doses were selected on the basis of pilot studies; doses were selected that were the lowest doses able to eliminate all forelimb clonus on the drug-screen trial.

Phase IV: Treatment Phase. The treatment phase of each experiment began 48 hours after the drug-screen trial. During the treatment phase, the bidaily stimulation schedule established during the baseline phase was maintained, and the drug of interest (at the same dose administered on the drug-screen trial) or an equal volume of its vehicle were administered to each subject either 1 hour before or 1 hour after each stimulation.

Phase V: Tolerance Test. Following the last trial of the treatment phase, all subjects received an injection of the drug at the same dose that was administered on the drug-screen trial, and then 1 hour later it received a convulsive stimulation so that the degree of tolerance to the drug's anticonvulsant effects could be assessed.

Phase VI: Cross Tolerance Phase. In our studies of cross-tolerance, the tolerance test was followed by a series of cross-tolerance trials. During these trials, the bidaily stimulation schedule was continued, and 1 hour before each stimulation, each subject received a test dose of a second anticonvulsant drug so that the degree of cross-tolerance to its effects could be assessed.

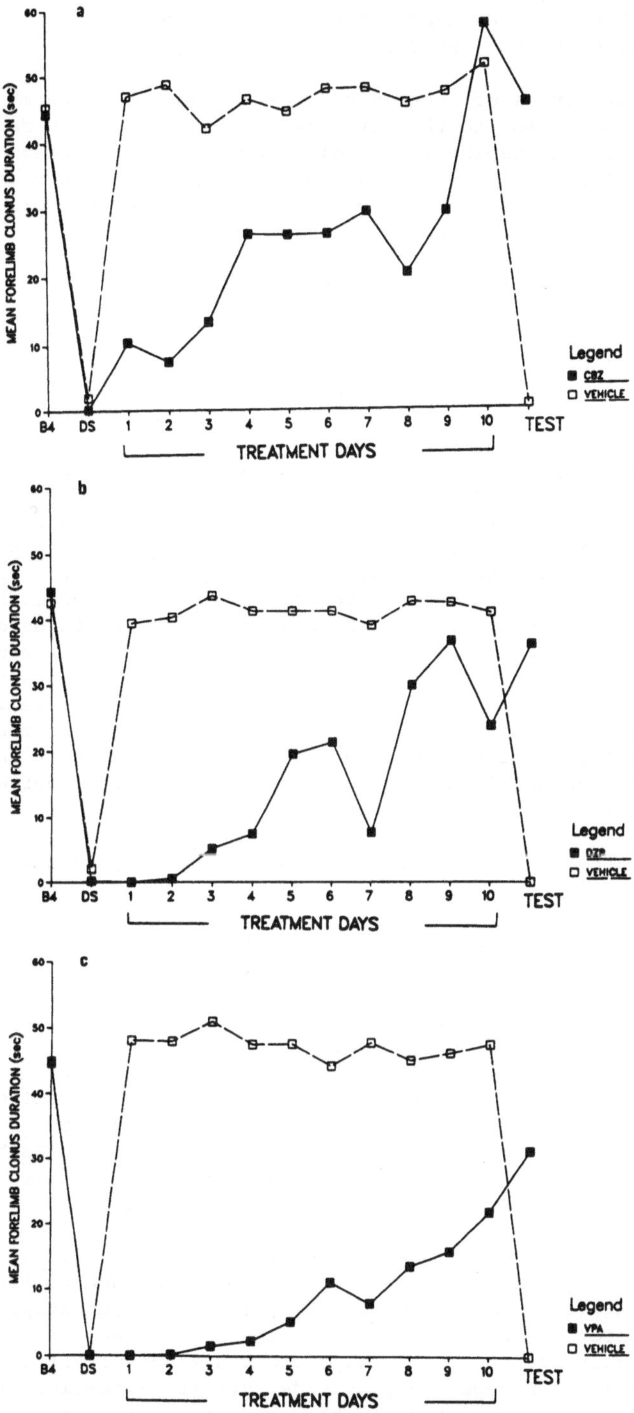


Figure 1. Tolerance to the anticonvulsant effect of carbamazepine (Panel a), diazepam (Panel b), and sodium valproate (Panel c) on kindled convulsions elicited by amygdalar stimulation 1 hour after injection.

TOLERANCE TO THE ANTICONVULSANT EFFECTS OF CARBAMAZEPINE, DIAZEPAM, AND SODIUM VALPROATE

Our first series of experiments was designed to determine whether tolerance would develop to the anticonvulsant effects of bidaily injections of carbamazepine (CBZ), diazepam (DZP), and sodium valproate (VPA) on convulsions elicited by amygdalar stimulation in rats. These drugs were chosen because each had been shown to exert a reliable anticonvulsant effect on kindled seizures (e.g., 1;2;3), and because each of these antiepileptic drugs belongs to a different drug family with a different putative mechanism of action.

Each of the three drugs was administered intraperitoneally in a 1% Tween 80/isosaline vehicle at a volume of 4 ml/kg. The diazepam was injected in solution; both sodium valproate and carbamazepine were injected as suspensions. There were six groups of subjects: a CBZ-drug group (n=12) and a CBZ-control group (n=8); a DZP-drug group (n=12) and a DZP-control group (n=8); and a VPA-drug group (n=12) and a VPA-control group (n=8). On the drug-screen trial, the rats in the two CBZ groups received carbamazepine (70 mg/kg, IP); those in the two DZP groups received diazepam (2 mg/kg, IP); and those in the two VPA groups received sodium valproate (250 mg/kg, IP). One hour later, the rats in all six groups received a convulsive stimulation. During each of the ten subsequent treatment trials, the same doses of the three drugs were administered to the rats in the CBZ-drug, DZP-drug and VPA-drug groups 1 hour before each bidaily stimulation. The rats in the respective control groups each received an appropriate volume of the vehicle, also 1 hour before each bidaily convulsive stimulation. On the test trial, the rats in all six groups received the same drug that they had received on the drug-screen trial, at the same dose, so that the degree of tolerance development could be assessed.

As illustrated by the three panels of Figure 1, each of the drugs (CBZ, Panel a; DZP, Panel b; and VPA, Panel c) initially exerted a marked anticonvulsant effect on the kindled convulsions; there was a total blockade of forelimb clonus in almost every rat on the drug-screen trial. However, over the ensuing treatment trials, the rats in each of the drug groups developed a substantial degree of tolerance. Accordingly, the duration of the forelimb clonus displayed on the tolerance test trial was substantially greater than that displayed on the drug-screen trial for the rats in all three drug groups. In contrast, the rats in the control groups displayed no significant increase in forelimb clonus duration after the same period. Accordingly, the forelimb clonus displayed by the rats in each of the three drug groups on the test trial was significantly longer than displayed by the rats in the respective control groups.

These results confirm and extend previous observations in two ways. First, they confirm earlier reports of tolerance to the anticonvulsant effects of diazepam (e.g., 12) and carbamazepine (e.g., 7), and they provide what we believe is the first clear experimental evidence of tolerance to sodium valproate's anticonvulsant effects (however, see 15). Second, they provide further validation of the kindling model as a useful tool in the study of tolerance to anticonvulsant drug effects. The anticonvulsant effects of all three drugs were large, reliable, and easily quantified; there was no seizure-related subject attrition; and the development of tolerance could be traced in individual subjects from trial to trial.

It is not clear whether a metabolic or a functional change underlies the tolerance to the anticonvulsant effects of carbamazepine, diazepam, and sodium valproate reported in the present experiment. Acceleration of metabolism by the induction of hepatic enzymes has been reported for carbamazepine (e.g., 7;23), raising the possibility that metabolic tolerance may underlie its loss of efficacy. However, accelerated metabolism does not appear to play a significant role in the development of tolerance to the anticonvulsant effects of either sodium valproate (cf., 11) or diazepam (cf., 8), suggesting that some sort of functional change must underlie the decrease in their effectiveness at suppressing kindled convulsions following repeated bidaily administration.

The magnitude of the tolerance that developed to the anticonvulsant effects of diazepam and valproate on kindled convulsions in the present experiments is much greater than that reported from other laboratories. In comparison to the present results, Löscher and Schwark (12) observed much less tolerance to the anticonvulsant effects of diazepam on kindled convulsions, and Young, Lewis, Harris, Jarrott, and Vadja (24) reported no tolerance whatsoever to the anticonvulsant effects of sodium valproate on kindled convulsions. There are several plausible explanations for the differences between the present results and those of Löscher and Schwark (12) and Young et al. (24). For example, Löscher & Schwark (12) used a much higher dose of diazepam (5 mg/kg) than was used in the present experiments; and both Löscher and Schwark (12) and Young et al. (24) used a different kindling procedure. However, from our perspective one of the most interesting explanations for these differences has to do with the relation between the periods of drug exposure and the administration of convulsive stimulation during the treatment phases of each of the respective experiments. As mentioned earlier, we have found that the development of tolerance to ethanol's anticonvulsant effect can be critically dependent upon our subjects' receiving convulsive stimulation during periods of drug exposure. In the present experiments, the rats in the drug treatment groups were stimulated 1 hour after they received the drug on each of the tolerance-development trials; thus, all of the rats in the three drug groups repeatedly experienced the drug's anticonvulsant effect. In contrast, this condition was present on only half of the treatment trials in the studies reported by Löscher and Schwark (12) or Young et al. (24). The possibility that the relation between drug exposure and convulsive stimulation plays an important role in the development of tolerance to the anticonvulsant effects of carbamazepine, diazepam and sodium valproate was explored in our next series of experiments.

## CONTINGENT TOLERANCE TO THE ANTICONVULSANT EFFECTS OF CARBAMAZEPINE, DIAZEPAM, AND SODIUM VALPROATE

Most of our work on contingent tolerance to the effects of anticonvulsant drugs has employed a before-and-after design (10). During the treatment phase of these experiments, one group of kindled rats received the drug <u>before</u> the convulsive stimulation on each treatment trial so that the stimulation occurred while the animal was under the influence of the drug. The rats in a second group did not receive the drug until <u>after</u> they were stimulated, so that convulsive stimulation never occurred during periods of drug exposure. Thus, the kindled rats in one group repeatedly experienced the anticonvulsant effects of the drug while those in the other group, which received the same number of drug injections and stimulations, did not. On the test trial, the rats in both groups received the drug before they were stimulated so that their tolerance to its anticonvulsant effect could be assessed. In this type of design, any evidence of greater tolerance in the <u>drug-before-stimulation group</u> than in the <u>drug-after-stimulation</u>

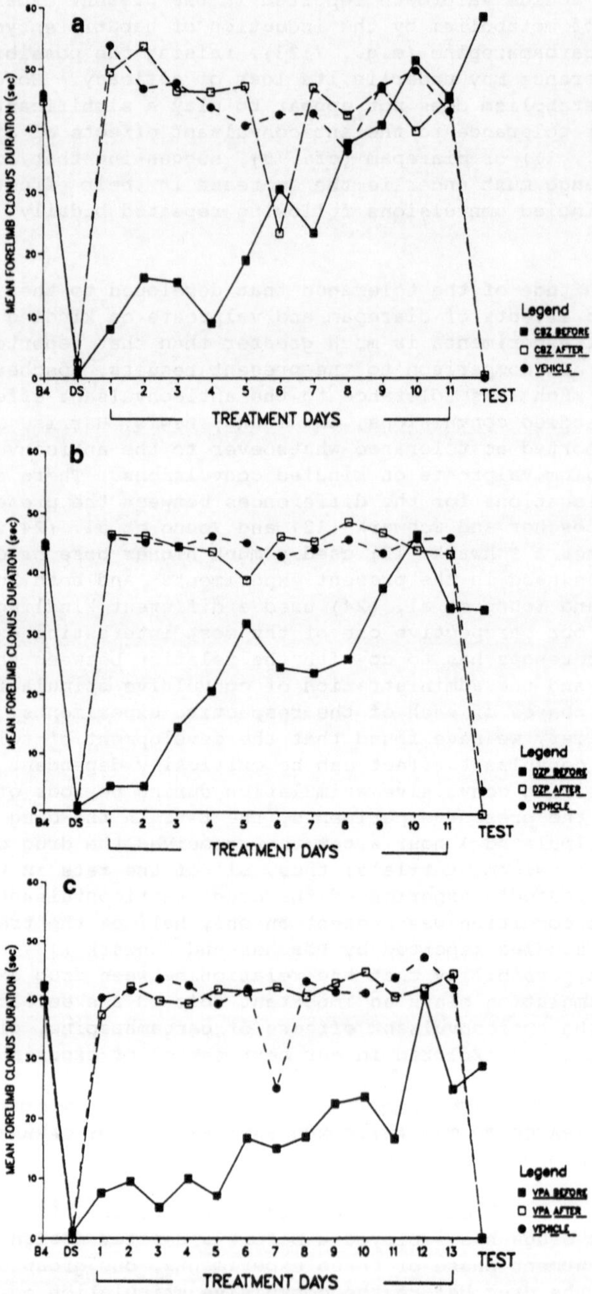


Figure 2. Contingent tolerance to the anticonvulsant effect of carbamazepine (Panel a), diazepam (Panel b), and sodium valproate (Panel c) on kindled convulsions elicited by amygdalar stimulation. Kindled rats receiving a drug injection 1 hour before convulsive stimulation on the treatment days became tolerant; those receiving a vehicle injection, or a drug injection 1 hour after convulsive stimulation, on the treatment days did not become tolerant.

group is attributable to the temporal relation between convulsive stimulation and drug exposure because the two groups do not differ in any other way.

The five-phase experimental protocol followed in our studies of contingent tolerance was identical to that described previously in this chapter: there was a kindling phase, a baseline phase, a drug-screen trial, a treatment phase (which incorporated the before-and-after design), and the test trial. As in our earlier work, diazepam (2 mg/kg) and carbamazepine (70 mg/kg) were administered intraperitoneally in a 1% Tween 80/isosaline vehicle, at a volume of 4 ml/kg. However, sodium valproate was administered by gavage in a 1% Tween 80/isosaline vehicle, at a volume of 4 ml/kg. This change in the drug-delivery protocol was made because we found that the sodium valproate caused less distress when administered by gavage than by IP injection.

During the treatment phase, the rats in the three drug-before-stimulation groups (CBZ-before-stimulation, n=11; DZP-before-stimulation, n=11; and VPA-before-stimulation, n=10) received their respective drugs 1 hour before each convulsive stimulation. The rats in the three drug-after-stimulation groups (CBZ-after-stimulation, n=10; DZP-after-stimulation, n=12; and VPA-after-stimulation, n=10) received their respective drugs 1 hour after each convulsive stimulation. The rats in three control groups (CBZ-control, n=9; DZP-control, n=8; and VPA-control, n=9) received the appropriate volume of vehicle either 1 hr before or 1 hr after convulsive stimulation. Because there were no significant differences between the vehicle-before-stimulation and vehicle-after-stimulation conditions, the scores of before and after vehicle subjects were combined in each of the three drug-control conditions.

It is readily apparent in Figure 2 that the test dose of each of the drugs initially exerted a powerful anticonvulsant effect; as in our previous experiments, the duration of forelimb clonus was zero for almost every subject receiving carbamazepine (Panel a), diazepam (Panel b), or sodium valproate (Panel c) on the drug-screen trial. It is also readily apparent that the rats in each of the drug-before-stimulation groups developed considerable tolerance to the anticonvulsant effects of their respective drugs over the course of the 10 treatment trials; almost every drug-before-stimulation rat demonstrated a substantial amount of forelimb clonus on the test trial. In contrast, the rats in the three drug-after-stimulation groups and in the three vehicle control groups displayed no more forelimb clonus on the test trial than they had on the drug-screen test.

These results challenge the assumption that drug exposure is sufficient for the development of tolerance to a drug's effects. Although the rats in all three of the drug-before-stimulation groups displayed a substantial amount of tolerance, the rats in the three drug-after-stimulation groups displayed no tolerance whatsoever, despite the fact that they were exposed to the same regimen of drug injections. The key factor in the development of tolerance to the anticonvulsant effects of bidaily injections of carbamazepine, diazepam, and sodium valproate in the present experiments was the occurrence of convulsive stimulation during the periods of drug exposure. This observation provides strong support for a drug-effect interpretation of the development of tolerance to the anticonvulsant effects of these drugs.

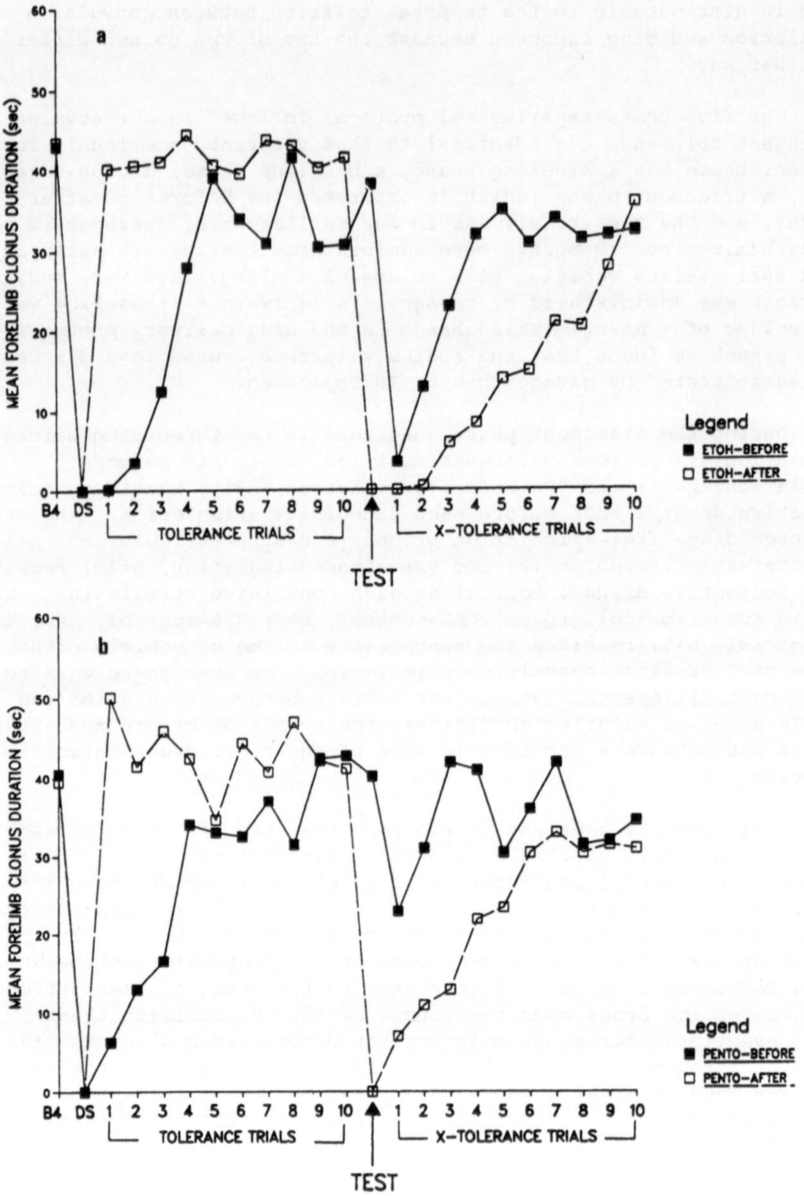


Figure 3. Panel a: Contingent transfer of tolerance to anticonvulsant effects from ethanol to pentobarbital. Kindled rats from the ethanol-before-stimulation group subsequently developed tolerance to pentobarbital's anticonvulsant effects significantly faster than rats from the ethanol-after-stimulation group.

Panel b: Contingent transfer of tolerance to anticonvulsant drug effects from pentobarbital to ethanol. Kindled rats from the pentobarbital-before-stimulation group subsequently displayed more tolerance to the anticonvulsant effect of ethanol than did rats from the pentobarbital-after-stimulation group.

## CONTINGENT CROSS-TOLERANCE TO ANTICONVULSANT DRUG EFFECTS: PENTOBARBITAL AND ETHANOL

Can the administration of convulsive stimulation during periods of exposure to one anticonvulsant drug effect influence the transfer of tolerance to another drug's anticonvulsant effects? To answer this question, we (17) used a before-and-after design to assess the transfer of tolerance between the anticonvulsant effects of ethanol and pentobarbital.

After the baseline phase, the subjects were assigned to one of four groups: an ethanol-before-stimulation group (n=14), an ethanol-after-stimulation group (n=14), a pentobarbital-before-stimulation group (n=13), and a pentobarbital-after-stimulation group (n=12). Both drugs were administered intraperitoneally in isotonic saline, at a volume of 7.5 ml/kg; ethanol was administered at a dose of 1.5 g/kg and pentobarbital at a dose of 15 mg/kg. As in our earlier experiments, the drug-screen trial was followed by a treatment phase. During the 10 bidaily treatment trials, each rat received the same drug as on the drug-screen trial, either 1 hour before or 1 hour after each convulsive stimulation. After the treatment phase, all of the subjects received a test injection of the same drug that they had received during the drug-screen and treatment phases, so that the development of tolerance to its anticonvulsant effect could be assessed. Following this test trial, the cross-tolerance phase began. During this phase, each subject received a series of 10 bidaily stimulations; 1 hour before each of these 10 stimulations, the rats that had received ethanol during the treatment phase received pentobarbital, whereas the rats that had received pentobarbital received ethanol.

There were three noteworthy findings from this experiment, which are illustrated in the two panels of Figure 3. First, as can be clearly seen in the first half of each panel, the development of tolerance to the anticonvulsant effects of both ethanol (Panel a) and pentobarbital (Panel b) were dependent upon the administration of convulsive stimulation during periods of drug exposure. The rats in both the ethanol-before-stimulation group and the pentobarbital-before-stimulation group displayed significantly more forelimb clonus on the test trial than on the drug-screen trial. In contrast, the rats from the ethanol-after-stimulation and pentobarbital-after-stimulation groups developed no tolerance whatsoever. Second, as can be seen in the second half of each panel, there was a significant degree of cross-tolerance between the anticonvulsant effects of ethanol and pentobarbital in both directions. And third, the manifestation of cross-tolerance was influenced by the relation between convulsive stimulation and drug exposure during the treatment phase of each experiment; rats from both drug-before-stimulation groups displayed cross-tolerance, whereas there was no evidence of cross-tolerance in either of the drug-after-stimulation groups. The cross-tolerance was evident on the first cross-tolerance trial when the pentobarbital-before-stimulation rats were switched to ethanol (Panel b). In contrast, the cross-tolerance to pentobarbital's anticonvulsant effects in the rats from the ethanol-before-stimulation group manifested itself as an acceleration in the appearance of tolerance to pentobarbital's anticonvulsant effects relative to the rate of tolerance acquisition displayed by the rats from the ethanol-after-stimulation treatment condition (Panel a).

## DISCUSSION

The results of the present experiments clearly support the idea that convulsive stimulation can play an important role in the development of tolerance to anticonvulsant drug effects. We found that kindled rats that receive bidaily injections of ethanol, carbamazepine, diazepam, sodium valproate, or pentobarbital rapidly develop tolerance to the anticonvulsant effects of these drugs only if they receive convulsive stimulation during the periods of drug exposure. Furthermore, we found that the administration of convulsive stimulation during the periods of bidaily exposure is also important for the transfer of tolerance to the anticonvulsant effects of other drugs. These findings together provide strong support for the drug-effect theory of tolerance.

The drug-effect theory of tolerance does <u>not</u> imply that convulsive stimulation is a critical factor in the development of all instances of tolerance to anticonvulsant drug effects. Tolerance to anticonvulsant drug effects has been shown to develop to many drugs in the absence of contingent convulsive stimulation--including the drugs employed in the current studies (cf., 6). We believe that one explanation for the results of these studies is the fact that they often involve a schedule of chronic drug administration; it has been emphasized by several authors, that this type of administration schedule is most likely to produce a state of tolerance (e.g., 6;9). Moreover, in our own laboratory, we have demonstrated the development of tolerance to the anticonvulsant effects of both ethanol and diazepam in the absence of convulsive stimulation when these drugs are administered on a chronic basis (14; Mana & Pinel, in preparation). In contrast, in most of our work on contingent tolerance to the anticonvulsant effects of drugs we have employed a bidaily schedule of drug administration; tolerance does not seem to develop to the anticonvulsant effects of periodic drug exposure in the absence of concomitant convulsive activity.

Why should the schedule of drug administration have such a potent effect on the importance of convulsive stimulation to the development of tolerance to the anticonvulsant effects of drugs? From the perspective of a drug-effect theory of functional tolerance, one explanation is that intermittent exposure to anticonvulsant drugs does not disrupt the normal activity of the nervous system enough to elicit the neural adaptations that are responsible for functional tolerance to the drug's effect. The increase in CNS activity elicited by convulsive stimulation during periods of intermittent exposure to anticonvulsant drugs may potentiate the effects of these drugs on the nervous system to the extent that the neural adaptations that are responsible for functional tolerance will manifest themselves. From this perspective, convulsive stimulation during periods of drug exposure may play a less important--or even a redundant--role in the development of tolerance when an anticonvulsant drug is administered on a chronic basis because chronic, but not intermittent, exposure to anticonvulsant drugs disrupts the basal activity of the CNS enough that the changes underlying the development of tolerance are elicited. Thus, chronic but not intermittent exposure to an anticonvulsant drug will result in the development of tolerance to the drug's anticonvulsant effects even if the recipient of the drug does not experience convulsions during the periods of drug exposure--not because of simple exposure to the drug, but because the effects of the drug during chronic exposure are disruptive enough to elicit an adaptive response from the nervous system in the absence of any convulsive activity.

Supported by a Canadian Medical Research Council grant to JPJP, and by a Canadian Medical Research Council postgraduate scholarship to MJM.

## REFERENCES


1. Albertson, T.E., Peterson, S.L. and Stark, L.G. (1980). Anticonvulsant drugs and their antagonism of kindled amygdaloid seizures in rats. Neuropharmacology, 19:643-652.
2. Albright, P.S. and Burnham, W.M. (1980). Development of a new pharmacological seizure model: Effects of anticonvulsants on cortical- and amygdala-kindled seizures in the rat. Epilepsia, 21:245-248.
3. Ashton, D. and Waquier, A. (1979). Behavioral analysis of the effects of 15 anticonvulsants in the amygdaloid-kindled rat. Psychopharmacology, 65:7-13.
4. Balster, R.L. (1985). Behavioral studies of tolerance and dependence. In L.S. Seiden and R.L. Balster (Eds.), Behavioral pharmacology: The current status (pp. 403-418). New York: Alan R. Liss.
5. Carlton, P.L. and Wolgin, D.L. (1971). Contingent tolerance to the anorexigenic effects of amphetamine. Physiology and Behavior, 7:221-223.
6. Frey, H.H. (1987). Tolerance to antiepileptic drug effects. Experimental evidence and clinical significance. Polish Journal of Pharmacology and Pharmacy, 39:495-504.
7. Frey, H.H. and Löscher, W. (1980). Pharmacokinetics of carbamazepine in the dog. Arch Int Pharmacodyn Ther, 243:180-190.
8. Frey, H.H., Phillipin, H.P. and Scherk., R. (1986). Development of tolerance to the anticonvulsant effect of benzodiazepenes in dogs. In H.H. Frey, W. Fröscher, W.P. Koella and H. Neinardi (Eds.), Tolerance to beneficial and adverse effects of antiepileptic drugs (pp. 71-79). New York: Raven Press.
9. Kalant, H., LeBlanc, A.E. and Gibbins, R.J. (1971). Tolerance to, and dependence on, some non-opiate psychotropic drugs. Pharm Rev. 23:135-191.
10. Kumar, R. and Stolerman, J.P. (1977). Experimental and clinical aspects of drug dependence. In L.L. Iversen and S.D. Iversen (Eds.), Principles of behavioral pharmacology, Vol. 7 (pp. 321-367). New York: Plenum Press.
11. Löscher, W. (1985). Valproic acid. In H.H. Frey and D. Janz (Eds.), Antiepileptic drugs (pp. 507-536). New York: Springer-Verlag.
12. Löscher, W. and Schwarck, W.S. (1985). Development of tolerance to the anticonvulsant effect of diazepam in amygdala-kindled rats. Experimental Neurology, 90:373-384.
13. Mana, J.J. and Pinel, J.P.J. (1987). Response contingency and the dissipation of ethanol tolerance. Alcohol and Alcoholism Research, Suppl. 1:413-416.
14. Mana, M.J., Pinel, J.P.J. and Lê, A.D. (1988). Tolerance to ethanol's anticonvulsant effect in the absence of a response contingency. Proceedings of the Fourth Congress of the International Society for Biomedical Research on Alcohol, A 45.
15. Paule, M.G. and Killam, E.K. (1986). Behavioral toxicity of chronic ethosuximide and sodium valproate in the epileptic baboon, Papio papio. Journal of Pharmacology and Expermental Therapeutics, 238:32-38.

16. Pinel, J.P.J., Colborne, B., Sigalet, J.P. & Renfrey, G. (1983). Learned tolerance to the anticonvulsant effects of alcohol in rats. Pharmacology Biochemistry & Behavior, 18 (Suppl. 1): 507-510.
17. Pinel, J.P.J., Kim, C.K., Paul, D.J. and Mana, M.J. (1989). Contingent tolerance and cross-tolerance to anticonvulsant drug effects. Psychobiology, 17:165-170.
18. Pinel, J.P.J. and Mana, M.J. (1986). Kindled seizures and drug tolerance. In J. Wada (Ed.), Kindling 3 (pp. 393-408). New York: Raven Press.
19. Pinel, J.P.J., Mana, M.J. and Kim, C.K. (in press). Effect-dependent tolerance to ethanol's anticonvulsant effect on kindled seizures. In R.J. Porter, R.H. Mattson, J.A. Cramer and I. Diamond (eds.), Alcohol and seizures: Basic mechanisms and clinical implications. Philadelphia: F.A. Davis.
20. Pinel, J.P.J., Mana, M.J. and Renfrey, G. (1985). Contingent tolerance to the anticonvulsant effect of ethanol. Alcohol, 2:495-499.
21. Pinel, J.P.J. and Rovner, L. (1978). Electrode placement and kindling-induced experimental epilepsy. Experimental Neurology, 58:335-346.
22. Racine, R. (1972). Modification of seizure activity by electrical stimulation. II. Motor seizure. Electroencephalography and Clinical Neurophysiology, 32:281-294.
23. Schmutz, M. (1985). Carbamazepine. In H.H. Frey and D. Janz (eds.), Antiepileptic drugs (pp. 479-506). New York: Springer-Verlag.
24. Young, N.A., Lewis, S.J., Harris, Q.L.G., Jarrott, B. and Vajda, F.J.E. (1987). The development of tolerance to the anticonvulsant effects of clonazepam, but not sodium valproate, in the amygdaloid kindled rat. Neuropharmacology, 26:1611-1614.

Discussion of Dr. Pinel's Presentation

DR. LOSCHER: As you know I belong to the people who do such studies with chronic administration of antiepileptic drugs in fully kindled rats. We found some things that are a bit different from your data. We compared the effect of 10 different anti-convulsant drugs administered on a chronic basis in fully kindled rats and the drugs were administered depending on their kinetic difference. Some drugs had to be administered 3 times daily, others only once, and all of these 10 drugs were injected at a dosage which displayed about the same anticonvulsant potency. In our hands these drugs were injected as a .. potent dosage. The difference to your study is that we did not stimulate the kindled animals every day, but 3 times a week for up to 6 weeks. And so if your hypothesis is right, one would suspect that with all these drugs there is development of tolerance because the kindling paradigm was the same, the frequency of the stimulation was the same, the anticonvulsant effect of the first doses was the same, but what came out was that there were several drugs with absolutely no tolerance, or a very slowly developing tolerance. For instance, with valproic acid we did studies up to 6 weeks with 3 times daily administration and found an increase in activity on kindled seizure rather than a decrease. With benzodiazepine there were marked differences in terms of development of tolerance; for example with clobazam we had tolerance, complete loss of activity within several days, with clonazepam it took 2 weeks or even more. So, it was not so easy as shown by your experiments. My second comment, if your hypothesis is right, then I would suspect that the same would be true in epileptic patients, but again, there are antiepileptic drugs with no evidence of tolerance in epileptic patients. For example, with valproate you have no escape from its antiepileptic efficacy, whereas with benzodiazepines you have an escape phenomenon depending on the type of the seizure which you are studying and the patients. So how could you explain that by your theory?

DR. PINEL: Well, we've done the work with the drugs that we've described here, and we just can't say whether these kinds of ideas will hold for all anticonvulsant drugs. The other thing is that I think one of the reasons why we went to this bi-daily schedule in the first place is that it gets around the confounding effects of drug accumulation. I think that when you get into more frequent injections, that starts to be a problem that can offset tolerance development. I guess another point would be, I think people in clinical practice would frequently not see tolerance develop, because they tend to start with low doses and gradually increase those doses over time, and it's just not possible in that kind of situation to see whether tolerance is developing. The third point, those of you who have attended these conferences before know that I'm always ranting and raving about the difference between elicited and spontaneous seizures, and this is certainly a factor here. In a patient there's no telling how many electrographic events are going on, and we really don't know what the critical effect is. I'm sure it's not the expression of the behaviour; it's certainly something that's going on in the nervous system. So I just don't know what to say about the clinical situation. I think in a sense your work is set up to kind of mimic the clinical condition; and I think that's a very legitimate kind of approach. We sort of started from the other end, that is, from a more theoretical mechanistic end, and we really attempted not to do many of the things that are present in the clinical situation because we had a specific idea about mechanisms and we sort of did this work for other reasons. But we are now branching out in trying to do some work more akin to yours and trying to make some comparisons between these two research approaches.

DR. SUSAN WEISS: We've done work that's similar to your work with carbamazepine looking at contingent tolerance, and we were bothered also by the fact that tolerance is not reported in the clinical literature and yet we saw it so profoundly in this kind of situation. One of the things that we considered is that in the clinical circumstance, or in the circumstances that you describe, you're giving the drug in both the non-contingent fashion and in a contingent fashion and in other kinds of other learning paradigms that slows down the type of learning that you get, in this case the learning is tolerance development in your situation. In ours we're always pairing the drug with the kindling stimulation which actually facilitates tolerance development. We've done a study in which we've given animals non-contingent and contingent drugs, so we had animals that were chronically being treated with carbamezapine in addition to treating them with carbamezapine before each kindling stimulation, and tolerance development was slower in those animals than in animals that had just gotten a control diet in carbamazepine only presented in a contingent fashion right before each seizure. And we've done some controls to show that it's not a matter of an increased dosage with the chronic and intermittent carbamazepine compared to the control diet, and intermittent carbamezepine. So I think it may be the way in which you're giving the drug in relation to the effect that's important enough.

DR. LOSHER: Just to comment on this. There is at least one seizure type in humans which is a model for what you tried to say; these are very frequent spike wave discharges in patients. In such patients which have over the day very frequent spike wave discharges and the EEG can study the effect of even one single administration. If you do that with benzodiazepine several groups have shown that there's even acute tolerance. So even the second dosage will have reduced activity compared to the first one, whereas for instance with valporate again, there's an increased action after the second or third dosage. So I guess it's more complicated than your series.

DR. PINEL: Maybe I could just describe one pilot study that we just finished that has some important clinical implications. We thought that perhaps this pattern that clinicians use, which makes good sense and I'm not criticizing it, I do it the same way which is start with a little dose and if that doesn't work try something a little more and so on. Well, we thought, gee, if our ideas are right, if allowing animals to experience or patients to experience convulsions and only partially blocking them, maybe this is going to contribute to tolerance development. So Kwon Kim just finished a study of pentobarbital tolerance in which we tried several regimens; one in which we used a very large dose, 50 mg/km IP, and we had another ascending regimen of doses in which we started off at 10 mg/kg and gradually moved them up, and then at the end of I think probably 3 weeks of bi-daily injections we compared the tolerance in those two groups and the animals that were on the ascending condition were extremely tolerant, those that had received the high dose all along were not very tolerant at all. I don't want to mislead people; you look at our studies and it's like 0 and 100, and that's because we've gone to a lot of work to develop this paradigm to study this particular effect. I'm really not saying it's going to be like that in other people's paradigms. But I think this is a nice paradigm for studying that effect, if you're interested in it.

develop this paradigm to study this particular effect. I'm really not saying it's going to be like that in other people's paradigms. But I think this is a nice paradigm for studying that effect, if you're interested in it.

CAN KINDLING-INDUCED SLEEP PATHOLOGY BE CORRECTED BY PHENOBARBITAL?*

Toshio Hiyoshi

National Epilepsy Center, Shizuoka Higashi Hospital
Shizuoka, Japan, 420

Juhn A. Wada

Neuroscience and Neurology, University of British Columbia
Vancouver, B.C., Canada, V6T 1W5

## INTRODUCTION

Kindled seizure susceptibility, as indicated by the incidence of interictal discharge (IID) and the development of kindled convulsion with an 'all-or-none' response to threshold stimulation, is known to persist in amygdaloid (AM) kindled cats for more than 12-months without applying stimulation.[1] Many antiepileptic drugs such as Phenobarbital (PB), Phenytoin, Carbamazepine, Valproic Acid and Primidone have been shown to exert either a prophylactic effect on kindling or an anticonvulsive effect on established kindled seizure.[2-8] However, the question of whether medication with these drugs can reverse the once established kindled seizure susceptibility remains unanswered.

Recently, a growing body of evidence strongly suggests that kindled seizure susceptibility also manifests itself in a sleep disorder. Although reports vary considerably on slow wave sleep (SWS), they consistently indicate that the percentage of time spent in rapid-eye-movement sleep (REMS) decreased progressively in conjunction with seizure development during AM and hippocampal (HIPP) kindling. This reduction appeared during the early stage of kindling and became maximal at the completion of kindling.[9-11] The significance of REMS suppression in kindled animals was emphasized by the finding that this suppression persists for at least a one-month period without applying kindling stimulation.[10,12] These findings strongly suggest that REMS suppression following seizures relates not to the induced kindled convulsion but to a neuronal change underlying acquired seizure susceptibility following kindling.[10]

The REMS period in cats has been shown to be seizure resistant.[9,13-16] Artificial deprivation of REMS was reported to facilitate AM kindling[15] and decrease the threshold of kindled generalized convulsion.[17] Therefore, it is possible to hypothesize that the suppression of REMS, resulting from repeated electrical stimulation and induced seizures, consequently partici-

*Supported by grants from Medical Research Council of Canada
to JAW

*Kindling 4*
Edited by J. A. Wada
Plenum Press, New York, 1990

pates in the enhancement of seizure susceptibility on the next stimulation; thus, kindling develops and kindled seizure susceptibility persists. However, significant suppression of REMS was also observed during administration of PB and Phenytoin in AM-kindled cats,[18] and this suppression with Phenytoin administration was found in intact cats to be lacking a rebound increase.[19]

These findings led us to perform a series of experiment in cats extending the observation period to 11 months to determine whether the persistence of acquired seizure susceptibility following AM-kindling was causally related to the kindling-induced alteration of sleep organization, and whether chronic administration of PB could reverse the once-established kindled seizure susceptibility, as well as the kindling-induced change in sleep organization.[20-22] PB was chosen because of its longer plasma half-life and wider therapeutic range in human epileptic patients.

## I. EFFECT OF AM KINDLING ON THE SLEEP-WAKING PATTERN

### A. OBSERVATION ON 6-HR RECORDING

#### Materials and Methods

Seven adult male cats weighing 3.6 - 5.1 kg were surgically prepared under pentobarbital anesthesia (32 mg/kg, i.v.) for sleep monitoring, according to the manual by Ursin & Sterman,[23] and for AM-kindling as described in our previous reports.[24,25] Animals were kept in the colony under a 12-hr light/dark cycle (light on at 8 a.m. and off at 8 p.m.) and on an ad libitum feeding schedule throughout the experiment.

Two weeks following the surgery, the animals underwent baseline sleep recording. For this purpose, they were placed in a recording chamber 6 hrs a day, starting at 10 a.m., for 4 consecutive days and polygraphic recordings were performed on the third and fourth day. In 2 cats (#894 & #895) the recordings were extended to a period of 22.5 hrs on the fourth day. This 4-day schedule was repeated for subsequent sleep monitoring in this study. The animals' behaviour was monitored on TV and the EEG-machine was set in a separate room so that a sound attenuated condition could be maintained. Six-hr polygraphs were recorded with 10 mm/sec paper speed. The state of vigilance was scored for each successive 30 sec epoch for waking (W), SWS and REMS according to Ursin and Sterman.[23]

After the completion of baseline sleep recording, left AM kindling began in a manner well established in our laboratory. Upon completion of kindling and establishment of the generalized seizure triggering threshold (GST), sleep organization was examined in a manner identical of that used for pre-kindling baseline recording except that the kindled seizure was induced about 30 min before each recording. Following a rest interval of 3-4 weeks, sleep organization was again examined without prior convulsion.

#### Results

Animals spent a mean of 35, 48 and 17 percent time in W, SWS and REMS, respectively, during the baseline 6-hr recording (Fig. 1). When sleep organization was examined over a 22.5-hr period in 2 cats, the percentages of time spent in SWS and REMS decreased to 37 and 13, respectively. Thus, the timing of our 6-hr recording period (10 a.m. - 4 p.m.) appeared to be highly relevant to the study of sleep.

Upon completion of AM kindling, the percentage of time spent in REMS significantly decreased, although the animals were spending significantly more time in sleep. Thus, a significant suppression of the REMS/total sleep time (TST) ratio occurred. In some animals, the REMS percentage remained unchanged or even increased; however, the increase of SWS was so marked as to suppress this ratio. These findings of sleep organization change during the immediate postictal 6-hr period continued during the extended 22.5-hr period except that the decrease in W became less profound in the latter period. However, when sleep recording was performed with a rest interval of 3-4 weeks following the last kindled convulsion, the animals' sleep pattern returned completely to the pre-kindling baseline level (Fig. 1).

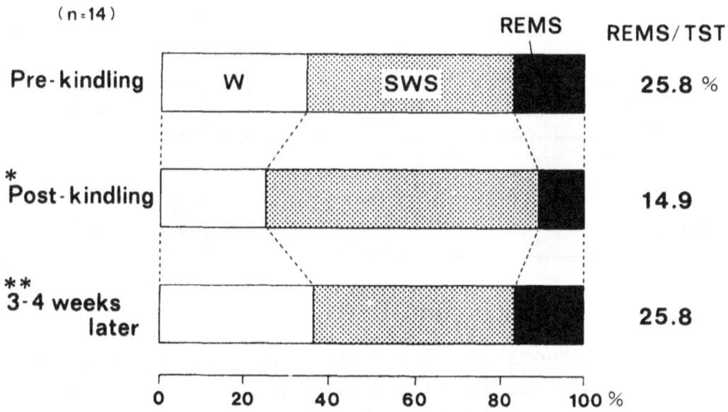


Fig. 1. Mean percents of 6-hr recording time spent in different stages of alertness and those of total sleep time (TST) in REMS in 7 cats.
*Recordings started 30 min after kindled convulsion.
**Recordings without prior kindled convulsion.

## B. OBSERVATION ON 22-HR RECORDING

Since our findings are completely different from the previous observations, which indicated a persistent nature to the AM-kindling induced sleep organization change,[10,12] we performed a separate series of experiments employing daily 22-hr polygraphic recordings. At this time, special attention was paid to the length of time the kindled-seizure induced sleep organization change persisted and how it is manifested in the circadian modulation of sleep-waking pattern.

### Materials and Methods

Four adult male cats weighing 4.1 - 6.3 kg were used. The general procedure was the same as in the 6-hr recording study. For sleep recording, the animals were placed in a recording chamber 22 hrs daily, starting at 10 a.m.. Food, water and saw dust were prepared in the chamber and were replenished every morning. A 2-day adaptation period was allowed before each series of sleep recording. Baseline recording was performed consecutively for 5 days. The first series of post-kindling recordings was performed consecutively for 8 days in which kindled seizure was induced 30 minutes before each recording on the first, 2nd, 3rd and 8th recording days. Following a 2-month rest interval, the recording was again performed for a consecutive 5 days without prior seizure. Then the primary site AM was restimulated and the recording was repeated for 4 days with daily induction of kindled seizures.

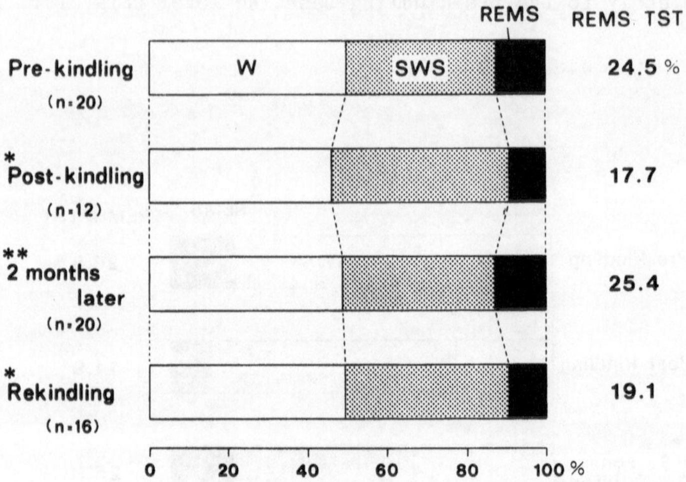


Fig. 2. Mean percents of 22-hr recording time spent in different stages of alertness and those of total sleep time in REMS in 4 cats.
*Recordings started 30 min after kindled convulsion.
**Recordings without prior kindled convulsion.

<u>Results</u>

The animals spent a mean of 49.6, 38.0 and 12.4 percent time in W, SWS and REMS, respectively, during the baseline 22-hr recording (Fig. 2). These figures were comparable to those reported in the literature for 23- or 24-hr period recording.[10,26-29]

When each of these three states were analyzed on an hourly basis, it became evident that there existed a circadian fluctuation in their more general occurrence in a day as has been reported.[26] Thus, in our stable environment, the animals were most active before sunrise and sunset and slept more often around noon and midnight. In addition to this circadian rhythm in the occurrence of W, SWS and REMS, the hourly REMS/TST ratio also showed a circadian fluctuation (Fig. 3).

At the completion of AM kindling, the percentage of time spent in REMS significantly decreased while that of SWS increased, resulting in a significant reduction of the REMS/TST ratio. The percentage of time spent in W slightly decreased but this was not statistically significant (Fig. 2). During the 22-hr period, the suppression of REMS was most profound during the immediate post-ictal hours and it continued without a rebound increase.

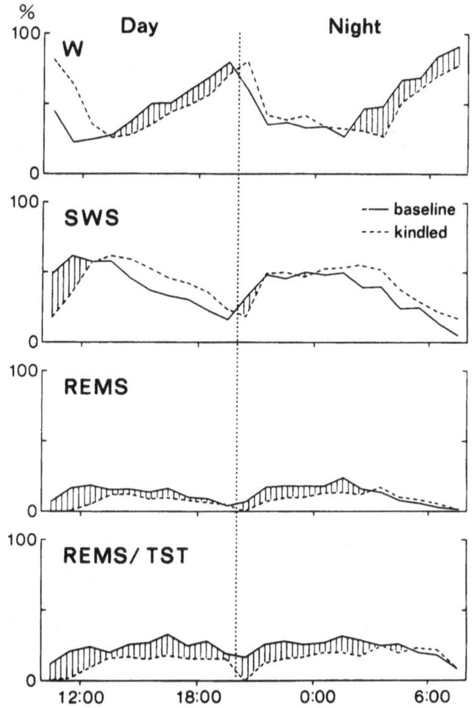


Fig. 3. Hourly percents of time spent in different stages of alertness and those of total sleep time spent in REMS on the pre-kindling baseline and post-kindling recordings. In the latter, kindled convulsion was induced 30 min before the recording. Each hourly value represents a mean of 20 and 12 recordings in 4 cats performed on 5 and 3 consecutive days before and after kindling, respectively. Shadowed area indicates the suppression of each value following kindled convulsion.

SWS was also suppressed during the immediate post-ictal hours. However, this was exceeded by the suppression of REMS as indicated by the decreased REMS/TST ratio, which was consistently observed throughout the 22-hr period. Moreover, suppression of SWS was followed by a rebound increase which completely compensated for the loss of both SWS and REMS (Fig. 3). These changes were transient. Without preceding kindled convulsion, decreased REMS was gradually restored and returned to the pre-kindling baseline level almost completely on the 4th day. Subsequent induction of kindled seizure again reversed this recovery (Fig. 4). Following a 2-month rest interval, the following parameters showed exactly the same values as those of the pre-kindling baseline: (i) the percentage of time spent in W, SWS and REMS,

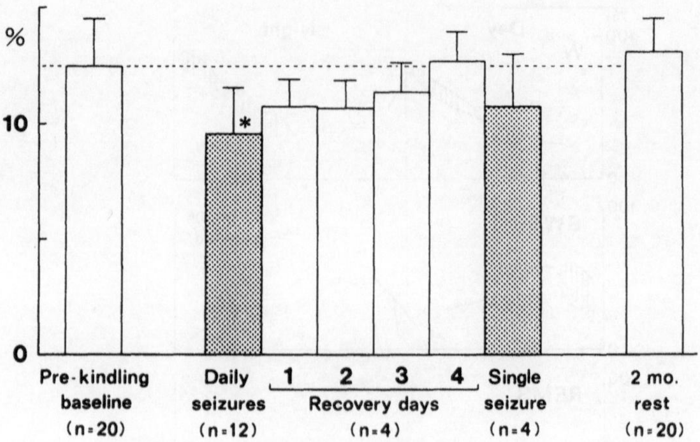


Fig. 4. Recovery of REMS suppression following daily induction of kindled seizures and reappearance of REMS suppression upon single induction of seizure. Values are shown as a mean percents of 22-hr recording time in 4 cats. Shadowed bars indicate that the recordings are performed 30 min following seizures. *$P < 0.001$ compared with pre-kindling baseline.

(ii) the total number and the mean duration of these episodes, (iii) the latencies for the onset of SWS and REMS, and (iv) the circadian sleep-waking pattern and the REM cycle. Subsequently, all animals responded to the previously established GST stimulation with a fully kindled convulsion, followed by an abrupt suppression of REMS (Fig. 2).

## II. PB EFFECT ON THE SLEEP-WAKING PATTERN AND KINDLED SEIZURE SUSCEPTIBILITY

### Materials and Methods

Upon completion of post-kindling sleep recording, 4 out of 7 cats used in the 6-hr sleep study underwent chronic PB medication. PB was given either subcutaneously (20 mg/kg every 3 days) or periorally (6 - 8 mg/kg/day) for a period of up to 9 months without AM stimulation. Plasma PB concentrations were monitored monthly and were confirmed to be maintained above 15 - 20 μg/ml. The PB dose was reduced to one half for a 1-month period before dis-continuation. The remaining 3 animals were used as nonmedication controls. A preliminary study indicated that the PB withdrawal effect on the GST and on the electro-clinical manifestation of AM-kindled seizure is negligible when a 3-week interval followed abrupt discontinuation of the drug. Based on this information and with confirmation of plasma PB washout, the primary site AM was re-stimulated 1 month after PB discontinuation. Six-hr polygraphic recordings were repeated during PB medication and just before AM rekindling following PB washout.

### Results

Sleep-Waking Pattern. During the administration of PB, the percentage of time spent in REMS significantly decreased while that of SWS remained unchanged. The decrease in REMS was compensated for by the increase in W, although this increase was not statistically significant (Fig. 5).

However, when the recordings were extended to a 22.5-hr period in 1 of 3 cats, these changes completely disappeared (Fig. 6). Although the information on 22.5-hr recording was limited to 1 animal, the discrepancy between the observations in the 6- and 22.5-hr periods appeared to be due to a modification of the circadian sleep-waking pattern under PB administration.

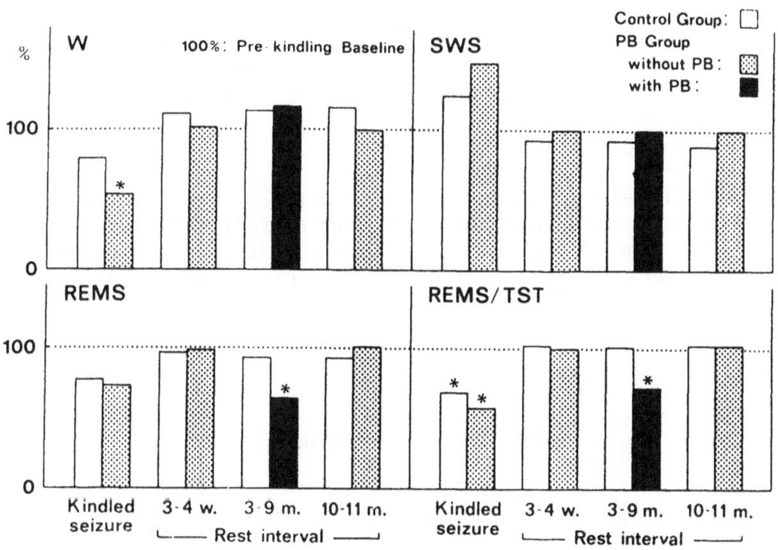


Fig. 5. Chronological profiles of sleep-waking pattern following kindling studied by means of 6-hr recording in PB medication and non-medication control animals. Each bar represents a mean of 6 recordings in 3 cats. *$P < 0.05$ compared with baseline.

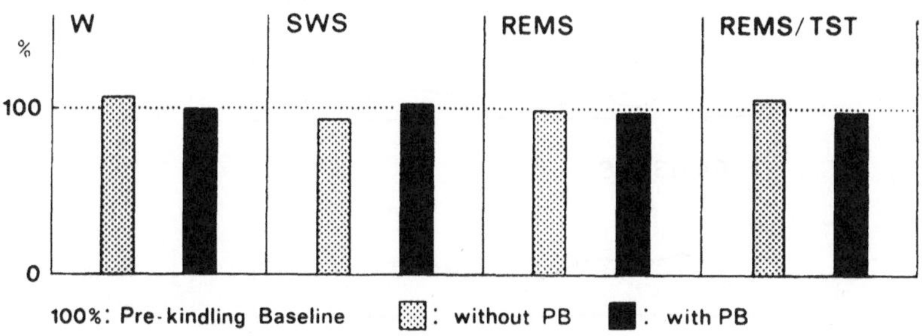


Fig. 6. Effect of chronic PB administration on sleep-waking pattern in kindled cat (no. 894) examined by means of 22.5-hr recording. Each bar represents a mean of 2 recordings.

Thus, the circadian fluctuation of W, SWS, REMS and the REMS/TST ratio diminished and this particular animal slept almost equally at any time period during a day (Fig. 7).

These changes returned completely to the premedication/prekindling level upon PB washout. The control animals showed a stable sleep-waking pattern identical to that of the prekindling baseline throughout the 11-month observation period (Fig. 5).

Kindled Seizure Susceptibility. During the up-to-11-month observation period, the interictal discharges (IIDs) were recorded bilaterally in the AM and were mostly independent of each other. In general, their frequency was highest during the immediate postictal recording, and gradually decreased in both the control and medicated animals during the subsequent period without AM stimulation. However, the IIDs never disappeared except from the contralateral AM in 1 medicated cat. The persistent nature of IIDs was also indicated by the fact that there were no changes in their morphology, although a considerable decrease in their amplitude were present in both the medicated and nonmedicated animals.

Upon the first re-stimulation of the primary site AM at the previously established GST, 2 animals responded with fully kindled seizure. Another 2 animals failed to respond with electro-clinical seizure manifestations; however, once the AD was induced by increasing the stimulus intensity from 100 uA to 200 uA, a Stage 6 seizure was evoked.

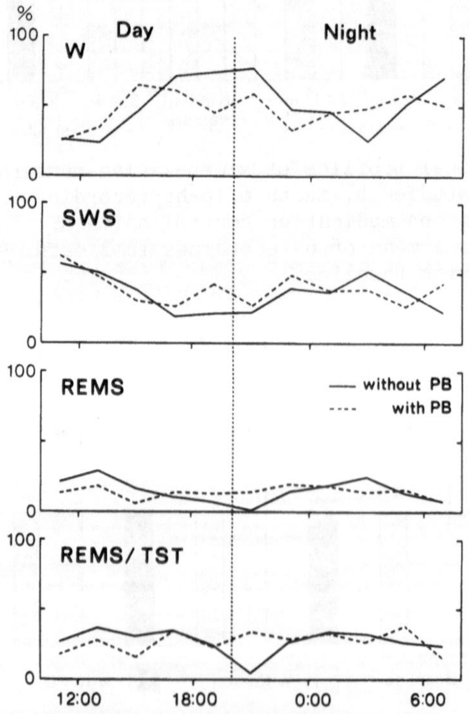


Fig. 7. Circadian sleep-waking pattern of kindled cat no. 894 examined with or without administration of PB. Each curve represents a mean of 2 recordings.

## DISCUSSION

### Effect of AM Kindling on Sleep-Waking Pattern

The findings of both the 6- and 22-hr recording study clearly indicate that the percentage of time spent in REMS significantly decreased, while that of SWS increased at the completion of kindling. Hourly examination of the sleep-waking pattern indicated that SWS also decreased in the immediate post-ictal hours; however, the decrease of REMS exceeded that of SWS, resulting in the suppression of the REMS/TST ratio which was consistently observed during the 22-hour period. In addition, the suppression of SWS was followed by an increase which completely compensated for the loss of both SWS and REMS, while the suppression of REMS was observed throughout the 22-hour period without a rebound increase. These findings strongly suggest that the fundamental sleep organization change induced by kindled convulsion is a selective suppression of REMS but not SWS. These changes returned completely to the pre-kindling baseline level when kindled seizure was no longer evoked. Upon AM restimulation with the previously established GST, the animals responded with fully kindled convulsion, and the induced seizures again reversed the recovery of the sleep-waking pattern. Therefore, we must conclude that sleep organization changes following AM kindling, particularly those characterized by the suppression of REMS, are completely independent of the acquisition of kindled seizure susceptibility. Rather, they are the direct correlates of the preceding generalized convulsion as has been documented by Cohen and Dement upon electroconvulsive shock.[30]

It is well known that mechanical interruption of REMS is followed by a compensatory increase of REMS if the subjects are allowed uninterrupted sleep.[31,32] This finding, along with the fact that the subject sometimes went directly from wakefulness into REMS without passing through the SWS, has been regarded as evidence of a physiological need for REMS.[33] In this study, however, suppression of REMS induced by kindled convulsion was most profound during the immediate post-ictal hours and it gradually returned to the baseline level without rebound increase. Therefore, this state of REMS suppression is completely different from the state under artificial deprivation. From the viewpoint of homeostatic regulation of REMS, it is possible that kindled convulsion transiently fulfills a physiological need for REMS.

### Effect of Chronic PB Administration on Sleep-Waking Pattern

It has been well documented that administration of barbiturate suppresses REMS in man[34,35] and in intact cats.[36] The single administration of PB in kindled cats is also reported to suppress REMS.[18] The findings of our 6-hr recording confirmed this report. However, the findings of our 22.5-hr recording suggest that a tolerance develops to the suppression of REMS by PB, as reported on the administration of another barbiturate,[34] and yet, a modification of the circadian sleep-waking pattern was observed during chronic administration of PB. Since a limitation exists in the number of animals investigated, further study is needed to verify our findings. Nevertheless, PB modification of sleep organization returned completely to the pre-medication/pre-kindling levels upon PB washout and the control animals showed a stable sleep-waking pattern identical to the pre-kindling baseline throughout the 11-month observation period. These findings further strengthen the transient nature of sleep organization change following AM-kindled convulsion.

### Effect of Chronic PB Administration on Kindled Seizure Susceptibility

The elevation of GST in 2 of 4 animals cannot be contributed to the effect of PB medication since this was also observed in our previous study upon re-stimulation following a 12-month rest interval.[1] It has been reported in cats that the administration of PB suppresses AM kindled seizures at plasma levels of about 15 - 20 μg/ml.[2,3,7,18] In this study, PB levels were maintained above this value without producing overtly apparent toxic manifestations. Therefore, we must conclude that long-term PB administration at therapeutic doses cannot reverse the fundamental processes underlying AM kindled seizure susceptibility as reflected in the incidence of IID and the development of generalized convulsion with an 'all-or-none' response to the stimulation.

## CONCLUSION

It is concluded that (1) sleep organization change following AM kindling is characterized by the suppression of REMS and a compensatory increase of SWS, but this change is independent to the acquisition of kindled seizure susceptibility; rather, this is a direct correlate of convulsion which may transiently fulfill the physiological need for REMS, and (2) subsequent chronic PB administration at therapeutic doses does not affect sleep organization and the once-established AM kindled seizure susceptibility as reflected in the incidence of IID and the development of kindled generalized convulsion with an 'all-or-none' response to threshold stimulation.

These findings strengthen the persistent nature of AM-kindled seizure susceptibility which appeared to be entirely independent of kindled seizure-induced sleep organization change.

## REFERENCES


1. J. A. Wada, M. Sato, and M. E. Corcoran, Persistent seizure susceptibility and recurrent spontaneous seizures in kindled cats. Epilepsia, 15: 465-478 (1974).
2. J. A. Wada, T. Osawa, M. Sato, A. Wake, M. E. Corcoran, and A. S. Troupin, Acute anticonvulsant effects of Diphenylhydantoin, Phenobarbital and Carbamazepine: a combined electro-clinical and serum level study in amygdaloid kindled cats and baboons. Epilepsia, 17: 77-88 (1976).
3. J. A. Wada, M. Sato, A. Wake, J. R. Green, and A. S. Troupin, Prophylactic effects of Phenytoin, Phenobarbital, and Carbamazepine examined in kindling cat preparations. Arch. Neurol., 33:426-434 (1976).
4. J. A. Wada, Pharmacological prophylaxis in the kindling model of epilepsy. Arch. Neurol., 34:389-395 (1977).
5. J. A. Wada, Kindling, antiepileptic drugs, seizure susceptibility and a warning. In: "Epilepsy updated: causes and treatment," P. Robb, ed., Miami, Symposia Specialists Inc. (1980).
6. V. Leviel and R. Naquet, A study of Valproic Acid on the kindling effect. Epilepsia 18: 229-233 (1977).
7. N. Kakegawa, An experimental study on the modes of appearance and disappearance of suppressive effect of antiepileptic drugs on kindled seizure. Psychiat. Neurol. Jpn., 88: 81-98 (in Japanese) (1986).
8. T. Hiyoshi, S. Suzuki, K. Yagi, and M. Seino, On the anticonvulsive effect of Primidone: an experimental study through overkindling of cat. Jpn. J. Psychiat. Neurol., 40: 505-506 (1986).
9. T. Tanaka, and R. Naquet, Kindling effect and sleep organization in cats. Electroencephalogr. Clin. Neurophysiol., 39: 449-454 (1975).

10. G. Rondouin, M. Baldy-Moulinier, and P. Passouant, The influence of hippocampal kindling on sleep organization in cats. Effects of alphamethylparatyrosine. Brain Res., 181: 413-424 (1980).
11. M. N. Shouse and M. B. Sterman, Sleep and kindling: II. Effects of generalized seizure induction. Exp. Neurol. 71: 563-580 (1981).
12. M. N. Shouse and M. B. Sterman, "Kindling" a sleep disorder: degree of sleep pathology predicts kindled seizure susceptibility in cats. Brain Res., 271: 196-200 (1983).
13. H. Cohen, J. Thomas, and W. C. Dement, Sleep stages, REM deprivation and electroconvulsive threshold in the cat. Brain Res., 19: 313-317 (1970).
14. M. Sato and T. Nakashima, Kindling: Secondary epileptogenesis, sleep and catecholamines. Can. J. Neurol. Sc., 2: 439-446 (1975).
15. R. Kawahara and J. A. Wada, Effect of REM sleep deprivation on amygdaloid kindling in cats. Jpn. J. EEG/EMG., 11: 176-184 (in Japanese) (1983).
16. M. N. Shouse, State disorders and state-dependent seizures in amygdala-kindled cats. Exp. Neurol., 92: 601-608 (1986).
17. M. N. Shouse and M. B. Sterman, Acute sleep deprivation reduces amygdala-kindled seizure threshold in cats. Exp. Neurol., 78: 716-727 (1982).
18. R. Kawahara, I. Okubo, T. Tanaka, H. Takeshita, and T. Inomaru, The effect of three anticonvulsants on REM sleep and generalized seizure in amygdaloid-kindled cats. Jpn. J. EEG/EMG., 15: 273-281 (in Japanese) (1987).
19. H. B. Cohen, R. F. II. Duncan, and W. C. Dement, The effect of diphenylhydantoin on sleep in the cat. Electroencephalogr. Clin. Neurophysiol., 24: 401-408 (1968).
20. T. Hiyoshi, N. Mori, and J. A. Wada, Feline amygdaloid kindling and sleep. Electroencephalogr. Clin. Neurophysiol., 73: 254-259 (1989).
21. T. Hiyoshi and J. A. Wada, Feline amygdaloid kindling and sleep-waking pattern: observation on daily 22-hour polygraphic recording. Epilepsia, (in press).
22. T. Hiyoshi and J. A. Wada, Nine-month phenobarbital administration failed to reverse amygdaloid kindled seizure susceptibility in cats (submitted).
23. R. Ursin and M. B. Sterman, "A manual for recording and scoring of sleep and waking stages in the adult cat." Brain Information Service/Brain Research Institute, University of California, Los Angeles (1981).
24. J. A. Wada and M. Sato, Generalized convulsive seizures induced by daily electrical stimulation of the amygdala in cats. Neurology, 24: 565-574 (1974).
25. T. Hiyoshi and J. A. Wada, Midline thalamic lesion and feline amygdaloid kindling. I. Effect of lesion placement prior to kindling. Electroencephalogr. Clin. Neurophysiol., 70: 325-338 (1988).
26. M. B. Sterman, T. Knauss, D. Lehmann, and C. D. Clemente, Circadian sleep and waking patterns in the laboratory cat. Electroencephalogr. Clin. Neurophysiol., 19: 509-517 (1965).
27. R. Ursin, The two stages of slow wave sleep in the cat and their relation to REM sleep. Brain Res., 11: 347-356 (1968).
28. R. Ursin, The effects of 5-hydroxytryptophan and L-tryptophan on wakefulness and sleep patterns in the cat. Brain Res., 106: 105-115 (1976).
29. E. A. Lucas and M. B. Sterman, The polycyclic sleep-wake cycle in the cat: Effects produced by sensorimotor rhythm conditioning. Exp. Neurol., 42: 347-368 (1974).
30. H. B. Cohen and W. C. Dement, Sleep: suppression of rapid eye movement phase in the cat after electroconvulsive shock. Science, 154: 396-398 (1966).

31. W. Dement, The effect of dream deprivation. Science, 131: 1705-1707 (1960).
32. W. Dement, P. Henry, H. Cohen and J. Ferguson, Studies on the effect of REM deprivation in humans and in animals. Res. Publ. Ass. Nerv. Ment. Dis., 45: 456-468 (1967).
33. W. C. Stern and P. J. Morgane, Theoretical view of REM sleep function: Maintenance of catecholamine systems in the central nervous system. Behavioral Biology, 11: 1-32 (1974).
34. I. Oswald and R. G. Priest, Five weeks to escape the sleeping-pill habit. Brit. Med. J., 2: 1093-1095 (1965).
35. E. Hartmann, The effect of four drugs on sleep patterns in man. Psychopharmacologia (Berl.), 12: 346-353 (1968).
36. D. J. Hinman and M. Okamoto, Sleep patterns in cats during chronic low-dose barbiturate treatment and withdrawal. Sleep, 7: 69-76 (1984).

Discussion of Dr. Hiyoshi's Presentation

DR. SHOUSE: Very nice presentation. We, of course, observed a long-term reduction in total sleep time 1 month after kindling. But we did find, later on, that there was no rebound as some others have seen and that is abnormal. The absence of rebound is clearly abnormal. But I agree with you that is a dissociated phenomenon and is not related to seizure events. I'm glad that you clarified that point. Thank you.

DR. CALVO: I have a methodological question. We also have studied the effect of the development of kindling over sleep organization in longitudinal studies of 23 hours daily recordings and I have one question. The habituation of the animals to the recording cage; do you habituate the animals long enough, because we did not find any changes in REM sleep or in sleep organization of these animals. We habituate the animals for 15 days and then we start the kindling process and for this longitudinal purpose. For the control situation we woke the cats instead of giving the kindling and we didn't find any effect on REM sleep. That means it is the same thing to wake cats or stimulate them to produce the seizure. We observed that there is a shift in the REM sleep distribution and if it is possible I will show this tomorrow in some of our slides. The second comment that I have, is that it seems that phenobarbital administration is worse than the kindling effect over REM sleep. What do you think about this?

DR. HIYOSHI: For the first question, we habituated the cats to the recording chamber for 2 days before every series of recordings. The animals readily adapted to the chamber for eating and sleeping, and they showed a stable sleep-waking pattern during the baseline recording. I think adaptation was not a problem according to our experiences. And for your second question, I have another slide. We administered phenobarbital to cats used for the 6 hour recording study and we found REM sleep suppression under phenobarbital. However, in one cat, when the recordings were extended from 6 hours to 22.5 hours on the last day of every series of recordings, we could not find any difference in the REM sleep percentages between the recordings performed with and without phenobarbital. When these recordings were analyzed on a 2-hourly basis, the circadian fluctuation became evident in the recordings without phenobarbital. However, under phenobarbital, the circadian modulation disappeared as shown in this dashed line. So, during our recording time from 10:00 a.m. to 4:00 p.m., REM sleep decreased. But during the subsequent period, REM sleep increased. With data from only 1 cat it is not so conclusive, but at least we can say that REM sleep suppression following kindled seizures and under phenobarbital is different.

# BASIC MECHANISMS UNDERLYING SEIZURE-PRONE AND SEIZURE-RESISTANT SLEEP AND AWAKENING STATES IN FELINE KINDLED AND PENICILLIN EPILEPSY

M.N. Shouse[1], A. King[1], J. Langer[1], K. Wellesley[1], T. Vreeken[1], K. King[1], J. Siegel[2], and R. Szymusiak[3]

Sleep Disturbance[1], Neurobiology[2], and Neurophysiology[3]
Research, VA Medical Center, Sepulveda, CA

Anatomy and Cell Biology[1], Psychiatry[2], and Psychology[3]
UCLA School of Medicine, Los Angeles, CA

Epilepsy is a chronic neurological disorder which is manifested at some times and masked at others. Sleep-waking state physiology is one of the most well documented factors affecting the clinical expression or suppression of human epilepsy[1,2]. Specifically, non-rapid-eye-movement (NREM) sleep and the gradual process of awakening from NREM sleep are the most vulnerable periods for seizures, especially convulsions. Moreover, the type of epilepsy is an important consideration in the timing of convulsions. Temporal lobe epilepsy with secondary generalized convulsions is the most common pure sleep epilepsy, with convulsions occurring in NREM or the transition from NREM to rapid-eye-movement (REM) sleep in nearly 60% of the patients[2]. In contrast, over 90% of patients with primary generalized, "petit mal" epilepsy display convulsions exclusively after awakening[2]. Finally, type of epilepsy is not a factor in the suppression of seizures during REM sleep. REM sleep is the most anti-epileptic state in the sleep-wake cycle for all generalized electrographic (EEG) and clinical seizures [1].

Our laboratory is identifying the brain mechanisms for these seizure-prone and seizure-resistant states in humans. The initial objective was to establish animal models of human generalized epilepsies with sleep or awakening convulsions.

## FELINE MODELS OF SLEEP AND AWAKENING EPILEPSY

The amygdala kindling model[3] of secondary generalized, temporal lobe epilepsy was used to study sleep-activated seizures. The systemic penicillin model of "petit mal" epilepsy[4] was used to study convulsions after awakening. These experimental models resemble human counterparts in the timing of seizures during the sleep-wake cycle, as detailed below.

### Amygdala kindling model of sleep epilepsy

Amygdala kindling is a viable model of "sleep epilepsy," based upon the timing of evoked and spontaneous seizures in the sleep-wake cycle.

*Kindling 4*
Edited by J. A. Wada
Plenum Press, New York, 1990

Conclusions are derived from studies in amygdala kindled kittens (n=9) and adult cats (n=12). Table 1 summarizes the characteristics of kindling development in cats at different ages because subcortical kindling has never been described in kittens[5].

The ontogeny of basolateral amygdala kindling in kittens resembles adults in the progression of seizure stages and kindling rates[3,6], but there are two main differences. First, initial and immediate post-kindling afterdischarge (AD) thresholds are much higher in 2.5 to 4-month-old kittens than in older kittens and adults. Second, the five youngest kittens (2.5 to 5 months) developed spontaneous seizures within three months of the first kindled convulsion. None of the older kittens ($\geq$5.5 months) or adults ($\geq$ 1 year) had spontaneous seizures.

One factor in the development of spontaneous seizures might be reduced capacity to inhibit seizure discharge in young animals. The three youngest kittens rarely had the normal post-ictal refractory period after elicited convulsions; rather, they relapsed into generalized seizures (stage 3 to 6) shortly after an evoked generalized tonic-clonic convulsion (GTC). Spontaneous seizures occurred one hour to several days after failure of post-ictal depression, mostly during slow-wave-sleep (SWS), the feline equivalent of human NREM sleep.

Kittens could have convulsions in the waking state, but the majority of waking seizures we observed were nonconvulsive. One kitten had unusual, nonconvulsive seizures which we have never seen before in a kindled cat. These events are called "catnip" seizures because the kitten purred and wagged its tail continuously. The kitten often assumed a stereotyped posture with dorsoflexion of head and hips, punctuated by jacknife-like jerks. Clinical signs accompanied bilateral amygdala spiking and spike-wave activity in thalamus and cortex. Symptomatology is consistent with a complex-partial seizure, although some aspects are reminiscent of West syndrome[7]. Electroclinical events could persist 1.5 hours during waking;

Table 1. Kindling development in kittens and adult cats.

| AGE at Initial AD | Initial AD Threshold (mA) | Kindling Rate (ADs to GTC) | Post-Kindling AD Threshold (mA) | Spontaneous Seizures |
|---|---|---|---|---|
| 2.5-4.0 months (n=3) | $15.3 \pm 0.6^{*}$ | $21.0 \pm 17.3$ | $3.7 \pm 1.5^{*}$ | 74% GTCs in SWS; "catnip" seizures in waking |
| 5 months (n=2) | $1.3 \pm 1.1$ | $23.5 \pm 7.8$ | $0.7 \pm 0.2$ | 100% GTCs in SWS |
| 5.5-6.5 months (n=4) | $1.2 \pm 4.0$ | $24.3 \pm 6.7$ | $1.3 \pm 0.3$ | NONE |
| >1.0 year (n=12) | $1.1 \pm 0.7$ | $22.0 \pm 8.7$ | $0.9 \pm 0.5$ | NONE |

* p <.05 from adult cats (>1 year)

EEG seizures continued in SWS and could be accompanied by a convulsion; all EEG and clinical seizures were suppressed during REM sleep.

The timing of spontaneous convulsions during the sleep-wake cycle in young kittens (Table 1) corresponds to the threshold data in older kittens and adult cats. Table 2 shows that stable post-kindling thresholds were higher in kittens than adult cats, but state-dependent seizure patterns in evoked seizure susceptibility were the same, regardless of age. SWS, particularly the transition into REM sleep, is the most seizure prone state for both spontaneous (Table 1, Figure 1) and elicited (Table 2) convulsions. REM sleep is the least vulnerable state for spontaneous and elicited seizures.

Table 2. Timing of elicited convulsions, expressed as GTC thresholds, during the sleep-wake cycle in kindled kittens and adult cats. Threshold is the <u>inverse</u> of seizure susceptibility. Susceptibility is highest during SWS, especially the REM transition, and lowest during stable REM sleep.

| | GTC Thresholds(mA) | | | |
|---|---|---|---|---|
| AGE at Initial AD | Alert Waking | SWS | SWS to REM transition | REM sleep |
| 5 to 6.5 month old kittens (n=6) | $.9 \pm .4^{+}$ | $.8 \pm .4^{*+}$ | $.7 \pm .4^{*+}$ | $1.0 \pm .5^{*+}$ |
| Adult cats > 1 year (n=12) | $.6 \pm .3$ | $.4 \pm .3$ | $.2 \pm .2$ | $0.7 \pm .3$ |

* $p<.05$ from alert waking baseline; + $p<.05$ from adult cats.

The temporal distribution of kindled seizures in kittens and adult cats resembles the clinical literature for human temporal lobe epilepsy, where partial and generalized seizures can occur during waking, convulsions occur predominantly in NREM sleep, and all generalized seizures are rare in REM sleep[1,2]. We have kindled a spontaneous sleep epilepsy in kittens ($\leq$ 5.5 months) and shown that the older animals, although not developing spontaneous epilepsy, are still most vulnerable to elicited seizures during SWS and the transition into REM sleep.

<u>Kindled sleep epilepsy vs. penicillin-induced awakening epilepsy</u>

Figure 1 shows the timing of generalized EEG and motor seizures during the sleep-wake cycle after amygdala kindling and during systemic penicillin epilepsy. The patterns of spontaneous seizures are clearly different. Whereas kindled kittens have SWS and REM transition seizures, penicillin epilepsy is most pronounced after awakening from SWS[5]. 59% of kindled convulsions occurred in the REM sleep transition. Over 65% of generalized myoclonic seizures and tonic-clonic convulsions (GTCs) during penicillin epilepsy occurred during the extended drowsy period after awakening. Systemic penicillin epilepsy thus mimics its human counterpart, myoclonic petit mal epilepsy, in which 96% of myoclonus and convulsions occur exclusively after awakening[2]. Finally, both epilepsy models are resistant to generalized EEG and motor seizures during REM sleep.

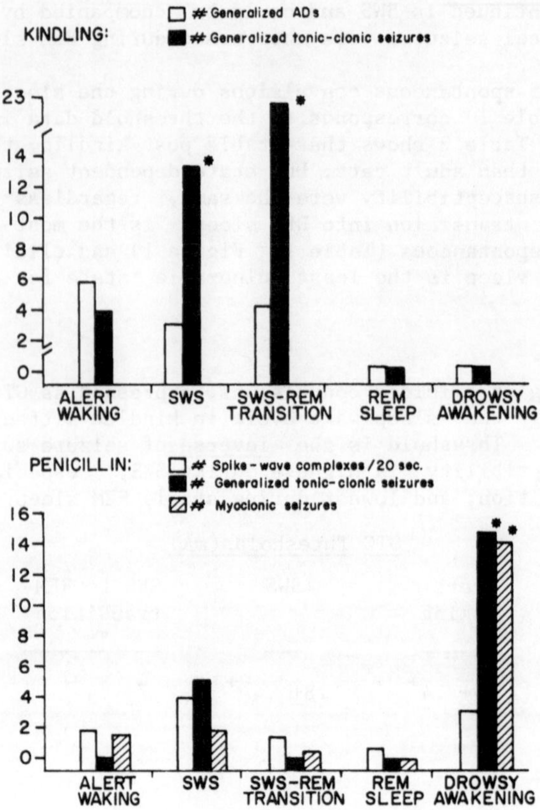


Fig. 1. Timing of spontaneous seizures during the sleep-wake cycle in nine kindled kittens (top) and 12 cats with systemic penicillin epilepsy (300,000-400,000 IU/kg; bottom).

## Anatomical substrates of seizure prone and seizure resistant states

Initial studies on the basic mechanisms that might explain these sleep vs. arousal activated seizure patterns used evoked potential methodology to study thalamocortical excitability in the two models. Many sleep-waking and sensorimotor functions are integrated at the thalamocortical level, making it a likely candidate as a final common pathway in the timing of convulsions in the sleep-wake cycle. We focused on the somatomotor pathway, in part because both models present motor seizures. Generalized motor seizures, especially convulsions, are more entrained to specific sleep and awakening states than any other focal or generalized seizure manifestation[1]. Equally important, motor cortex and its thalamic relay nucleus, ventralis lateralis (VL) are modulated quite differently by hypothalamic and reticular inputs involved in sleep and arousal, and this modulation shows up well with evoked population amplitude measures[8,9]. Accordingly, it has been possible to distinguish sleep vs. arousal seizure mechanisms at this level of the neuraxis using evoked response amplitude measures[10].

Evoked response studies (Figure 2) suggest that thalamocortical relays of the somatomotor system become hyper-responsive (hyperexcited) throughout the sleep-wake cycle during the development of feline generalized epilepsy. However, it is important to note: 1) Thalamic cells are most hyperexcited during seizure-prone sleep states in kindled cats (A) and could combine with high or variable motor cortex excitability to trigger convulsions during SWS and the REM transition. 2) Motor cortex is most hyperexcited during seizure-prone states in penicillin epilepsy (B). Peak cortical response amplitudes correspond to frequent spike-wave activity during drowsiness and SWS and to frequent spontaneous convulsions during drowsy awakening; and 3) Thalamocortical cells are least hyperexcited during the seizure resistant state of REM sleep in both epilepsy models, seen as a minimal increase from baseline in thalamic response amplitudes for kindled epilepsy (A) and in cortical response amplitude for penicillin epilepsy (B). These findings suggested that the thalamocortical relay might be important in motor seizure generalization and might also provide a site for sleep-waking state modulation of motor seizures in both epilepsy models.

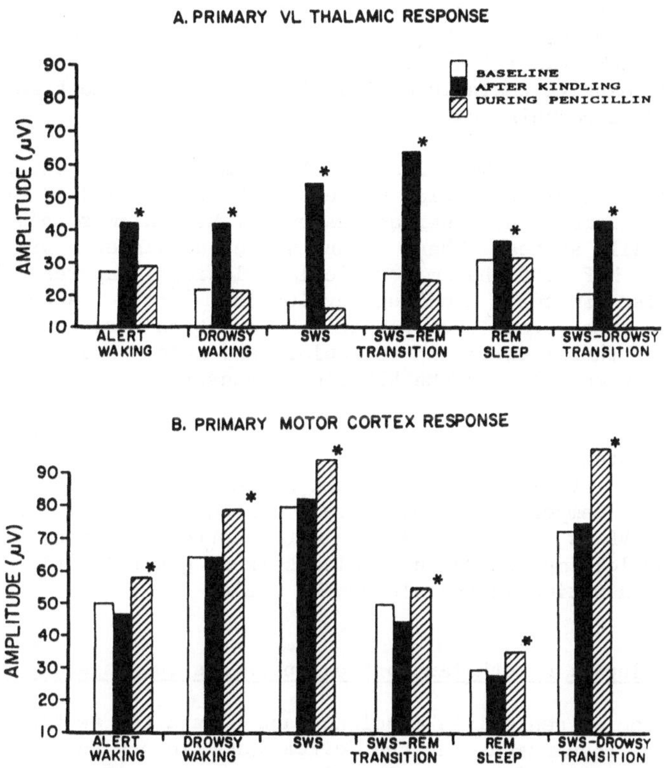


Fig. 2. Mean amplitude of primary evoked population responses (uV) in ventral lateral (VL) thalamus (A) and motor cortex (B) during baseline (n=8), after kindling (n=4) and during penicillin epilepsy (n=4). VL responses were elicited by stimulation of dentate nucleus. Motor cortex responses were elicited by stimulation of VL. Recurrent, low intensity stimulation of either dentate or VL was provided every 2.5 seconds. Evoked responses were sampled in each sleep or waking state (n=16 to 32 responses averaged per state in VL or motor cortex). Amplitude is an index of excitability. Kindling increased thalamic excitability. Penicillin increased cortex excitability. * = $p < .05$ from baseline.

Follow-up studies used spontaneous and evoked extracellular unit and lesion methodology to examine the hypothalamic and reticular pathways thought to regulate thalamocortical excitability patterns as well as sleep and awakening states. Preliminary findings suggest testable anatomical models for seizure prone and seizure resistant states in kindled and penicillin epilepsy, as follows.

<u>Kindling: Reticulothalamocortical pathways in sleep-activated epilepsy</u>

The timing of kindled seizures in the REM sleep transition, coupled with our previous finding that kindled cats show abnormal behavioral arousal at this time, suggest brain stem participation[11]. It is well known that the ultradian REM sleep transition and REM sleep cycle are controlled by the brain stem, as they persist only in the brain stem after midcollicular transection[12]. The specific mechanism for the REM sleep transition is obscure. However, a likely candidate for propagation of abnormal excitation at this time is the midbrain reticular formation and its rostral continuation, the reticular nucleus of thalamus. The thalamic reticular nucleus exerts a tonic excitatory influence on VL thalamus during shifts from states of EEG synchronization to EEG desynchronization, such as the REM transition[8,9]. Thus, ascending brain stem reticular influences increase VL thalamic excitability during the most seizure prone state for kindled convulsions.

Figure 3a shows that NMDA lesions of the midbrain reticular formation (MRF) and the thalamic reticular nucleus (TRN) block kindled GTCs during the REM sleep transition, whereas lesions of VL thalamus block kindled seizures in all states. These results are consistent with previous findings that MRF and thalamic reticular lesions suppress behavioral arousal and also amygdala kindling development[13,14]. Accordingly, we propose that the brain stem reticular formation propagates abnormal excitation to thalamocortical reticular and motor relays to trigger convulsion preferentially in the REM sleep transition.

The fact that VL lesions block kindled GTCs throughout the sleep-wake cycle is consistent with evidence that VL is chronically hyperexcited in kindled cats[10]. VL lesions also block seizure generalization in other feline models of temporal lobe epilepsy[15] as well as intractable seizures in humans[16]. VL is a major relay to motor cortex for many extrapyramidal motor nuclei implicated in kindled seizure generalization and could elicit GTCs at any time in the sleep-wake cycle.

<u>Penicillin epilepsy: Hypothalamocortical pathways in awakening convulsions</u>

The timing of penicillin myoclonus and GTCs after awakening suggested forebrain regulation, as the "circadian" sleep-wake cycle persists only in the forebrain after midcollicular transection[12]. The sleep-waking cycle has been further localized to the antagonistic divisions of the hypothalamus, where the preoptic basal forebrain induces SWS and the posterior hypothalamus induces awakening[17]. Recent evidence suggests that cells in the posterior lateral hypothalamus discharge in relation to EEG and behavioral arousal and might elicit spontaneous awakening by direct projection to neocortex[18,19].

Preliminary lesion and extracellular unit studies suggest that this hypothalamocortical pathway could also mediate penicillin seizures after awakening. Figure 3b shows that lesions of the posterior hypothalamus (PH) block penicillin myoclonus and GTCs on awakening (Fig 3b), whereas lesions of the reticular and thalamic relays implicated in kindled seizures did not.

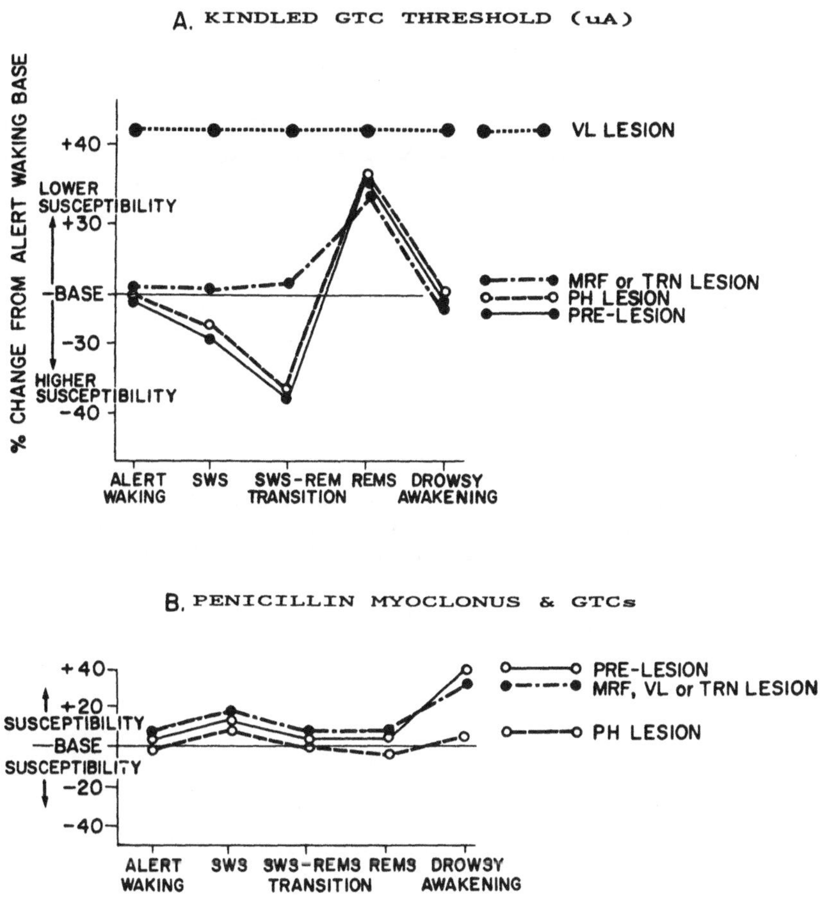


Fig. 3. Motor seizure susceptibility before (n=24) and after NMDA lesions in one of four sites (n= 6 each): ventral lateral (VL) thalamus, the midbrain reticular formation (MRF), thalamic reticular nucleus (TRN) or posterior hypothalamus (PH). A: Kindling (n=12). Pre-lesion GTC thresholds are lowest (susceptibility is highest) in SWS and the REM transition. MRF and TRN lesions eliminated vulnerability only at these times. VL lesions protected against GTCs in all states. PH lesions had no effect. B. Penicillin epilepsy (n=12). Before lesions, myoclonus and convulsions peak during drowsiness after awakening. PH lesions eliminated vulnerability at this time, whereas the other lesions did not.

Figure 4 illustrates spontaneous discharge rates of a posterior lateral hypothalamic (PLH) neuron during and after awakening from sleep before and after a subconvulsive dosage of penicillin. PLH discharge rates normally increase during and after awakening (top), and awakening discharge is enhanced by penicillin (bottom). Figure 5 shows effects of penicillin on the spontaneous and evoked orthodromic response of a PLH neuron. Compared to pre-penicillin baseline (A), penicillin increased spontaneous discharge rates, enhanced evoked excitation and reduced the duration of post excitatory discharge suppression (B). Thus, penicillin could increase posterior hypothalamic cell activity by direct excitation and/or reduced inhibition.

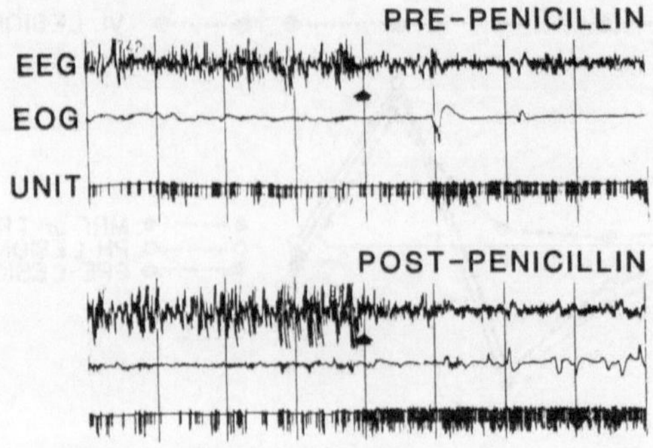


Fig 4. Spontaneous unit activity in a posterior lateral hypothalamic (PLH) neuron during SWS and at awakening (arrow) before and after a subconvulsive dose of penicillin (200,000 IU/kg).

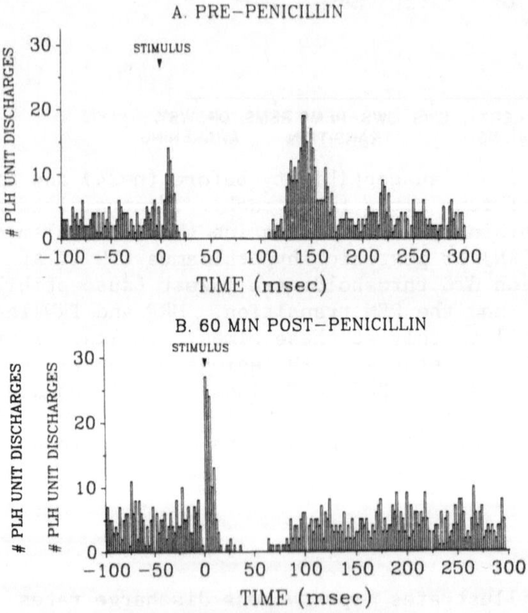


Fig. 5. Evoked orthodromic response in a PLH neuron, induced by stimulation of external capsule (100 pulses, .8 mA), before and after penicillin. Stimulus onset at time 0. Note increased excitation and reduced discharge suppression in B.

Collectively, the findings suggest that awakening convulsions might be triggered by abnormal discharges from the posterior hypothalamus, a region thought to underlie awakening from sleep in humans and animals. Discharges from this region are propagated to motor cortex, which is already hyper-excited in penicillin epilepsy. The interaction of hypothalamus and motor cortex could elicit penicillin GTCs preferentially after awakening.

REM sleep suppression of seizures in kindled and penicillin epilepsy

REM sleep is the most antiepileptic state in the sleep-wake cycle for human generalized epilepsy, yet the neural mechanism is unknown[1]. As in humans, REM sleep in cats retards the spread of EEG seizure discharges and has even more potent anticonvulsant effects[6,20]. The fact that generalized EEG seizures can occur without clinical motor accompaniment in REM sleep suggested differential mechanisms for EEG vs. motor seizure suppression. We have dissociated these mechanisms in both penicillin [20] and kindled epilepsy. Thalamocortical EEG desynchronization protects against the spread of EEG seizures, whereas lower motor neuron inhibition of REM sleep blocks clinical motor accompaniment.

Figure 6 demonstrates differential modulation of seizure manifestations by REM sleep components. The top tracing shows penicillin spike-wave and myoclonic seizures during SWS and REM sleep before manipulation of REM sleep components. Spike-wave activity, the prominent EEG seizure manifestation of penicillin epilepsy, is common in SWS but rare during normal REM sleep. Epileptic myoclonus can accompany spike-wave activity in SWS but does not occur during normal REM sleep, which is characterized by profound lower motor neuron inhibition (note EMG silence during REM sleep with spike-wave paroxysm).

The middle tracing shows effects of a selective syndrome of REM sleep without thalamocortical EEG desynchronization, created by systemic administration of the anticholinergic agent, atropine[20]. During atropine and penicillin trials, REM sleep has a SWS-like EEG, and spike-wave incidence during REM sleep is identical to SWS; however, no clinical accompaniment occurs, evidenced by continued silence in the EMG.

Opposite effects occur after a syndrome of REM sleep without lower motor neuron inhibition, called REM sleep without atonia. As indicated above, "sleep paralysis" (skeletal muscle atonia) occurs during normal REM sleep and is induced by descending pathways from the medial-lateral pontine tegmentum to the spinal cord[12]. Lesions of the pontine "atonia" center disinhibit muscle tone in REM sleep[22,23] (note activity in bottom right EMG channel), but all other REM sleep components remain intact. The animal is also mobile during REM sleep without atonia and appears to be "acting its dreams out." The bottom tracing of Figure 6 shows the effects of the pontine lesion on penicillin epilepsy. During REM sleep without atonia, penicillin spike-wave activity is as uncommon as in normal REM sleep, but when it does occur, it is associated with myoclonus. The loss of lower motor neuron inhibition after medial-lateral pontine lesions appears to release epileptic myoclonus in the penicillin epilepsy model.

The same differential effect of these REM sleep syndromes was obtained in kindled cats, as shown in Figure 7. The top tracing shows a kindling trial during normal REM sleep. Normal REM sleep retards EEG discharge generalization from amygdala and further delays clinical motor accompaniment. The middle tracing shows a kindling trial during atropine-induced REM sleep without thalamocortical EEG desynchronization. Atropine facilitates EEG discharge generalization but does not affect clinical motor accompaniment, evidenced by continued atonia (EMG silence) during

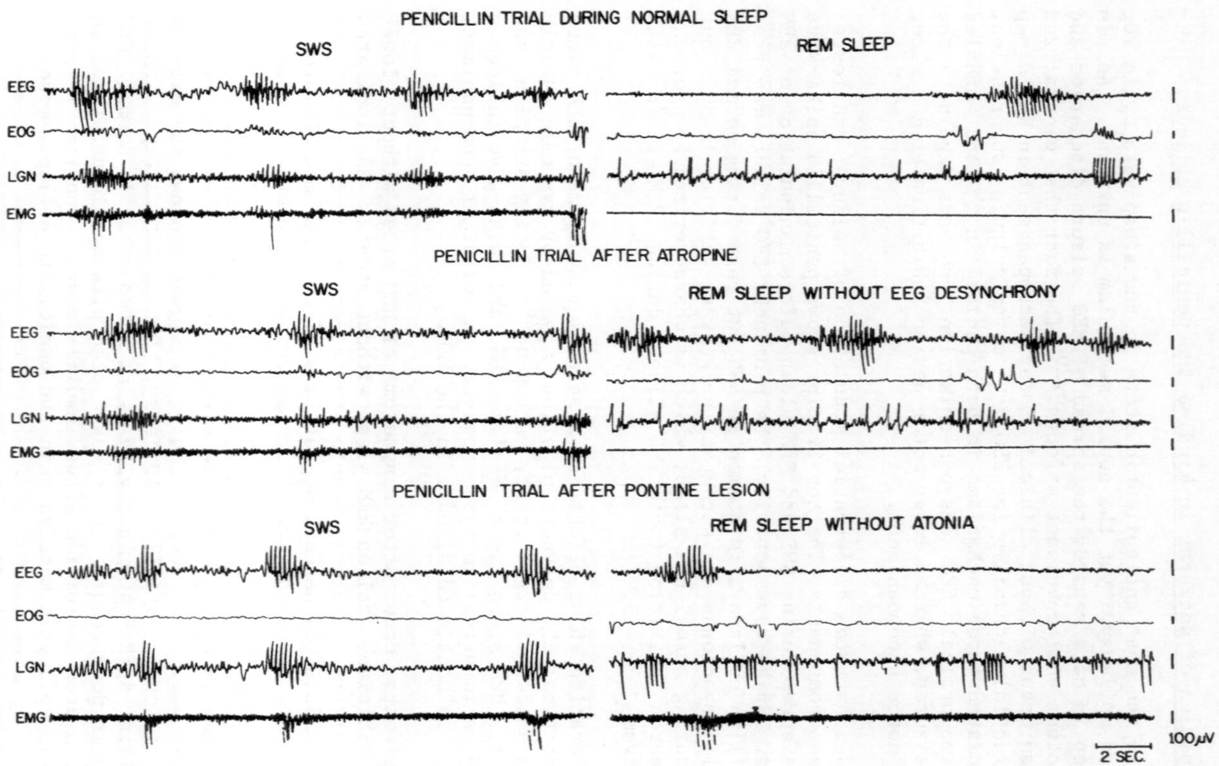


Fig. 6. Systemic penicillin epilepsy during SWS and REM sleep before and after dissociation of REM sleep components. Spike-wave paroxysms are visible in the EEG tracing, and myoclonic seizures were associated with discharges in the EMG tracing in this cat. TOP: Normal REM sleep suppresses EEG and motor seizures. MIDDLE: Atropine creates a SWS-like EEG in REM sleep and increases spike-wave activity during REM. BOTTOM: Pontine lesions induce REM sleep without atonia and release myoclonus in REM. From [20], with permission.

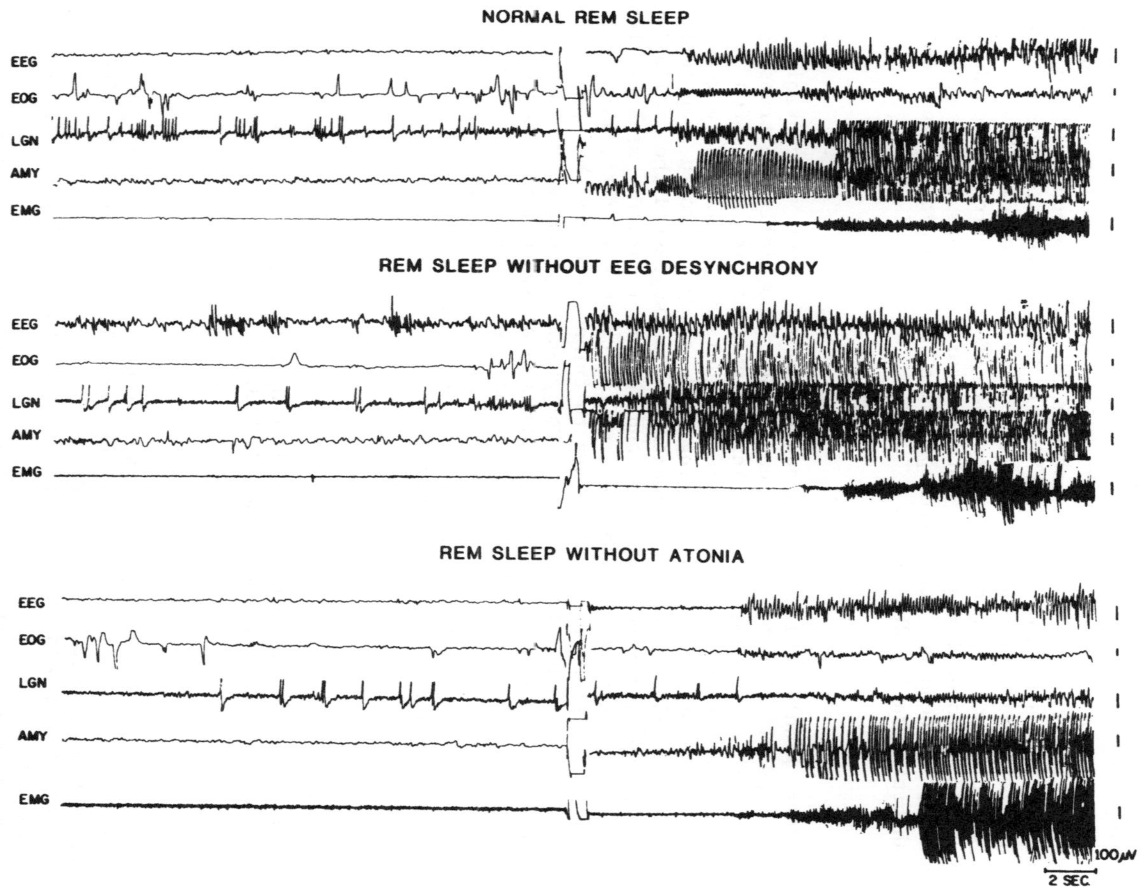


Fig. 7. Kindling trials before and after REM sleep syndromes. See text.

the first 10 seconds of generalized AD. This result is consistent with preliminary findings that atropine facilitates EEG discharge generalization during the development of kindling in cats [24]. The bottom tracing shows the opposite effect during REM sleep without atonia. AD generalization is delayed as in normal REM sleep, but when the EEG seizure spreads from amygdala, there is immediate clinical motor accompaniment.

Figures 6 and 7 suggest that thalamocortical EEG synchrony abolishes protection against EEG but not clinical motor seizures; in contrast, loss of lower motor neuron inhibition during REM sleep abolishes protection against motor but not EEG seizures. These effects were obtained in 26 cats exposed to amygdala kindling (n=6), systemic penicillin epilepsy (n=10) or electroconvulsive shock (n=10) [20]. Because REM sleep components have been localized to the brain stem, it seems likely that protection against the spread of EEG seizure discharges is mediated by the ascending brain stem pathways which induce intense thalamocortical EEG desynchronization during REM sleep[25,26]. Protection against motor seizures appears to be regulated by separate, descending brain stem pathways which mediate lower motor neuron inhibition during REM sleep.

SUMMARY

Amygdala kindled kittens and adult cats are most susceptible to spontaneous and elicited convulsions during SWS, especially the transition from SWS to REM sleep. Systemic penicillin epilepsy is associated with convulsions after awakening. Both epilepsy models are resistant to generalized EEG and motor seizures during REM sleep. Evoked population response, extracellular unit and lesion studies suggested the following anatomical pathways for seizure prone and seizure resistant states. For sleep epilepsy, abnormal excitation from the brain stem reticular formation may activate thalamocortical relays to trigger convulsions preferentially in the REM sleep transition. For awakening epilepsy, abnormal excitation from the posterior hypothalamus may activate motor cortex to elicit convulsions on awakening. Finally, we dissociated the factors for REM sleep suppression of EEG vs. motor seizures. Thalamocortical EEG desynchronization during REM discourages the spread of EEG seizures; in contrast, motor inhibition during REM sleep blocks only clinical motor accompaniment.

ACKNOWLEDGMENTS

This research was supported by the Veterans Administration and by PHS grants NS25629, NS14610, MH43811 and MH42903.

REFERENCES

1. M.N. Shouse, Seizures and epilepsy during sleep, in: "Principles and Practice of Sleep Medicine," M.H. Kryger, T. Roth and W.C. Dement, eds., Saunders, Philadelphia, 1989.
2. D. Janz, The grand mal epilepsies and the sleeping-waking cycle, Epilepsia 3: 69 (1962).
3. J.A. Wada and S. Sato, Generalized convulsive seizures induced by daily stimulation of the amygdala in cat: correlative electroencephalographic and behavioral features, Neurology 24: 565 (1974).
4. P. Gloor, Generalized epilepsy with spike-wave discharge: A reinterpretation of its electrographic and clinical manifestations, Epilepsia 20:571 (1977).
5. S.L. Moshe, E.F. Sperber and B.J. Albala, Kindling as a model of epilepsy in developing animals, in: "Kindling and Synaptic Plasticity:

The legacy of Graham Goddard," F. Morrell, ed., Dirkhauser, Boston, in press.
6. M.N. Shouse, Differences between two feline epilepsy models in sleep and waking state disorders, state dependency of seizures and seizure susceptibility: Amygdala kindling interferes with systemic penicillin epilepsy, Epilepsia 28:399 (1987).
7. W.J. West, On a peculiar form of infantile convulsion, Lancet 1: 724 (1841).
8. M. Steriade, Ascending control of motor cortex responsiveness, Electroenceph. Clin. Neurophysiol. 26: 25 (1969).
9. M. Steriade, Ascending control of thalamic and cortical responsiveness, Int. Rev. Neurobiol. 12:87 (1970).
10. M. Shouse, Thalamocortical mechanisms of state-dependent seizures during amygdala kindling and systemic penicillin epilepsy in cats, Brain Res. 425:198 (1987).
11. M. Shouse, State disorders and state dependent seizures in amygdala kindled cats, Exp. Neurol. 92: 601 (1986).
12. J. M. Siegel, Brainstem mechanisms generating REM sleep, in: "Principles and Practice of Sleep Medicine," M.H. Kryger, T. Roth and W.C. Dement, eds., Saunders, Philadelphia, 1989.
13. J.A Wada and M. Sato. Effects of unilateral lesions in the midbrain reticular formation on kindled amygdaloid convulsions in cats. Epilepsia 16: 693 (1975).
14. M.N. Shouse and M.B. Sterman. Sleep pathology in experimental epilepsy, in: "Sleep and Epilepsy," M.B. Sterman, M.N. Shouse and P. Passouant, eds., Academic Press, New York, 1982.
15. M.N. Shouse and M.B. Sterman, Changes in seizure susceptibility, sleep time and sleep spindles following thalamic and cerebellar lesions, Electroenceph. Clin. Neurophysiol. 46: 1 (1979).
16. S. Mullan, G. Valiati, J. Karasick and M. Mailias, Thalamic lesions for the control of epilepsy, Arch. Neurol. 16: 277 (1967).
17. W.R Hess, "Diencephalon-Autonomic and Extrapyramidal Functions," Grune and Stratton, New York, 1954. Consciousness Symposium" J.F. Freshaye, ed., Blackwell, Oxford, 1954.
18. R. S. Szymusiak, T. Iriye and D. J. McGinty, Sleep-waking discharge of neurons in the posterior lateral hypothalamic area of cats. Brain Res. Bull. In Press, 1989.
19. C.B. Saper, Organization of cerebral cortical afferent systems in the rat. II. Hypothalamocortical projections, J. Comp. Neurol. 237: 21, (1985).
20. M.N. Shouse, J.M. Siegel, M.F. Wu, R.S. Szymusiak and A.R. Morrison. Mechanisms of seizure suppression during rapid-eye-movement (REM) sleep in cats. Brain Res., In Press, 1989.
21. C.H. Vanderwolf and T.E. Robinson, Reticulocortical activity and behavior: A critique of the arousal theory and a new synthesis (with commentaries), Behav. Brain Sci. 4: 459 (1981).
22. M. Jouvet and F. Delorme, Locus coeruleus et sommeil paradoxique, C.R. Soc. Biol. 159: 4790 (1988).
23. J.C. Hendricks, A. R. Morrison and G. L. Mann, Different behaviors during paradoxical sleep without atonia depend on pontine lesions site. Brain Res. 239: 81 (1982).
24. M. E. Corcoran, J.A Wada, A Wake, and H Urstad. Research note: Failure of atropine to retard amygdala kindling. Exp. Neurobiol. 51: 271 (1976).
25. B.E Jones and H. H. Webster. Neurotoxic lesions of the dorsolateral pontomesencephalic tegmentum-Cholinergic cell area in the cat. I. Effect upon the cholinergic innervation of the brain, Brain Res., 458: 285 (1989)
26. S.R. Vincent and P. B. Reiner, The immunohistochemical localization of choline acetyltransferase in the cat brain, Brain Res. Bull. 18: 371 (1987).

Discussion of Dr. Shouse's Presentation

DR. MOSHE: Thank you for keeping us awake for a sleep study. You know in babies when they have severe seizures, the EEG does not show much of REM sleep. There is a syndrome called 'infantile spasms' where REM sleep almost totally disappears and if you treat it successfully, when the REM sleep period increases, the seizures may be suppressed. I have some questions concerning the type of seizures that you saw in the kittens. Were they different during wakefulness from sleep?

DR. SHOUSE: I'm glad you asked that question, because we're moving right into a possible model of West syndrome and in certain ways might have picked that up, I don't know. We got some very strange non-convulsive seizures in these kittens and the younger the kitten the more strange the seizures. The one in particular that I think Nick is talking about, we're calling them right now catnip seizures because the kitten wagged its tail continuously and purrs. You never see this in amygdala kindling or straight motor seizures, so we didn't expect this. Purring is not abnormal but it is abnormal when it goes on for 1.5-3 hours continuously and wagging. Moreover, in this period they adopt this stereotyped posture where the head is thrown back, the tail sometimes comes up, and they're just sitting there basically just like this, and every now and again they jack-knife; and that's a West syndrome-like phenomenon. Fifty paediatric neurologists have asked me, do you have hypsarrythmia? And the answer is, I don't know.

DR. MOSHE: I won't ask you that. I will ask you if the EEG changes from wakefulness to sleep? A wake seizure versus a sleep seizure, if the EEG pattern changes. I don't care if they don't have hypsarrythmia.

DR. SHOUSE: They do have an EEG pattern which is spike wave and also they have bilateral amygdala spiking, and yes, this does change. No, it doesn't change in slow-wave sleep, they keep the same pattern; it's not clear whether the seizure is quite the same. They can certainly have convulsions and maybe not. But in REM it does disappear.

DR. MOSHE: You may try and do the phenomenon of antagonism that Jim started; you may be able to get ... because hypsarrythmia is a multifocal syndrome, so if you have two foci you may be getting more seizures of the West type. The other question is, you know the studies of Ferrendelli about the mammilothalamic pathways involved in 3/second spike and wave discharges? The penicillin model that you propose is almost another model for generalized epilepsy of the petit-mal type. How could you relate his studies with yours?

DR. SHOUSE: I'm sorry, you've got to be more specific, what pathway are you referring to?

DR. MOSHE: The mammilothalamic activation that occurs in rats that have a model of absence seizures.

DR. SHOUSE: I have not studied the mammilothalamic tract.

DR. MOSHE: You don't think that it is anything related to the posterior hypothalamus?

DR. SHOUSE: Well, it might be; they're certainly, as you very well know, adjacent to each other.

DR. MOSHE: OK. So you don't believe the theory of the mammilothalamic pathways.

DR. SHOUSE: No, I don't disbelieve it; we haven't looked at that in our penicillin epilepsy model. But MacIntyre suggested that I should look at that, and I think that we will.

# KINDLING, ANXIETY AND LIMBIC EPILEPSY: HUMAN AND ANIMAL PERSPECTIVES

Robert Adamec

Departments of Psychology and Basic Medical Science
Memorial University
St. John's, Newfoundland, Canada

## INTRODUCTION

Recent clinical data in humans suggests that two types psychopathology associated with human epilepsy are anxiety and depression[1,2]. There is little agreement, however, regarding which type of epilepsy creates the greatest risk for interictal psychopathological complications[2]. Epileptics experience many life disturbances (such as illness, family and employment problems[3], which are suspected precipitants of anxiety and depressive disorders[4,5]. Since there is no clear association between type of seizure disorder and risk for anxiety and depression, it is possible that the interictal psychopathology is a response to the stresses of being an epileptic.

More recent data, however, favour the view that when the seizure disorder involves the limbic system, conditions are created which predispose such epileptics to anxiety and depression. A recently completed study of 114 epileptics, and 143 normal and chronic illness controls found that those patients who experience auras indicative of limbic system activation report psychiatrically significant problems resembling anxiety and depression[6]. Since frequency and intensity of auras predict psychopathology[6], it is likely that limbic activation by seizures must be repeated for interictal mood disturbance to occur.

A number of studies in animals suggest that repeated epileptic activation of the limbic system, including limbic kindling, disturb behavior interictally[7,8]. None of the behaviors reportedly changed by limbic seizures, however, are clearly analogous to human anxiety or depression. Recent work in my laboratory, however, suggests that increases in defensive behavior of felines produced by partial limbic kindling model some aspects of anxiety associated with human limbic epilepsy.

The remainder of this chapter will be devoted to reviewing the effects of partial limbic kindling and the anxiogenic compound N-methyl-beta-carboline-3-carboxamide (FG-7142) on cat defense, and approach-attack behavior. In addition the effects on limbic physiology will be reviewed. It will be contended that increases in defensive behavior produced by partial limbic kindling and FG-7142 involve parallel mechanisms. Thus the study of the effects of one should help to explain the effects of the other. In addition, the data suggest that studying the effects of partial limbic kindling on cat defense is a viable model of the impact of human limbic seizure disorders on human anxiety.

*Kindling 4*
Edited by J. A. Wada
Plenum Press, New York, 1990

## Partial Limbic Kindling, Anxiety and Behavior

In this section the effects of partial kindling on defense and approach-attack will be reviewed, and new findings described. Then the effects of FG-7142 on the same behaviors will be discussed and compared to the effects of kindling on behavior.

Partial Kindling. Repeatedly evoking afterdicharges (AD's), but not motor seizures, (partial kindling), in the feline limbic system produces a number of lasting interictal behavioral changes which are not dependent on seizures or interictal spiking for their maintenance. These changes include: increases in defensive response to species characteristic threats, such as rats, mice, and recordings of conspecific threat howls[9,10]; increased response to provocation by a human[11]; and increased response to electrical evocation of defense by stimulation of the ventromedial hypothalamus (VMH)[11,12]. AD's may be evoked from the amygdala (AM)[9,12] or from the ventral hippocampus (VHP) by electrical or chemical stimulation of the VHP directly, or indirectly by stimulation of the perforant path (PP)[10,11].

Behavioral changes do reverse[9,11,12], though in my studies[7,9,10] they are usually very long lasting (over 100 days after the last AD). The cats are not only more defensive, but they approach and attack rats less. This change has been interpreted to mean that approach-attack is diminished as a secondary consequence of the increase in defensiveness[10].

A recent replication of the effects of partial kindling of the PP-VHP pathway on behavior suggests that the suppression of approach-attack is independent of the increases of defensiveness. Thus seizures potentiate some response suppression system which is not part of the defensive substrate. Data from 14 cats which support this view follow.

Defense in response to prey was measured as the average time spent away from a rat or mouse following a withdrawal of the head or body from the prey (see 10 for details of testing environment and procedures and behavioral measures). Several approach-attack measures were also taken. These are latencies and frequencies of hard (HPS) and soft paw striking (SPS), and using the paw to pin the prey (single paw pin or SPP). Latency and duration of biting was also measured. The target of each of these attacks was also noted. Attacks on the front, middle and rear of the prey were scored separately. Latency to approach and time spent near the prey was measured, as well as latency to sniff and time sniffing the prey. Finally, activity was measured as number of 1 foot square segments of the testing floor crossed. Response of the prey to the approach-attack of the cat was scored as frequency of active, passive or escape defensive responses.

Fourteen cats were tested for their behavioral responses to prey 14 days before repeatedly triggering AD's in the VHP by PP stimulation. Two to six tests were done, and the responses averaged, since the tests did not differ. On any day, prey were presented in the order - rat followed by mouse. Cats were exposed to prey for either 5 or 10 min, with 5 min separating rat and mouse presentations. Eight days before repeated VHP AD's, threshold for VHP AD (afterdischarge threshold or ADT) was determined (see 10 for method). Response to prey was reassessed on the following day. Then 12 VHP AD's were triggered by PP stimulation (see 10 for method) in the morning and late afternoon over six days (6 hours separated the two stimulations). No further AD's were elicited. Response to prey was then monitored on days 1, 6, 13, 21 and 40 following the twelfth AD.

Responses to mice did not change. Two types of behavioral change were noted in response to rats. The first involved behavioral changes appearing on the first day after repeated AD. These behaviors equalled baseline following ADT determination. Then they either increased or decreased on day 1 after repeated AD, and remained changed over the 40 day retest interval. Those increasing were: withdrawal, latencies to bite, HPS and SPP (all main Day effects, $F[5,65] \geq 2.36$, $p<.05$, all $t[65]$, $p<.05$). Those decreasing were frequency to HPS front and middle, and SPP rear targets (all Day x Target, $F[10,130] \geq 1.93$, $p<.05$, all $t[130]$, $p<.05$). These findings are consistent with those previously reported[10]. Moreover, since withdrawal and attack suppression occur simultaneously, suppression of attack may be due to the increase in withdrawal. To test this, variation in withdrawal from rats over pre and post kindling tests was covaried out of the attack measures just listed. If withdrawal were determining the attack suppression, one would expect that covarying out the influence of withdrawal should eliminate the change in attack. This did not happen. The patterns of change over days remained the same. These findings indicate that there is a suppression of attack following repeated AD which is independent of the increase in withdrawal.

The second type of change appeared after ADT determination and persisted over the 40 days retest interval. Measures increasing were latency to SPS all targets (Day effect, $F[5,65]=4.37$, $p<.01$, all $t[65]$, $p<.05$). Measures decreasing were frequency to SPP middle targets, and frequency of prey self defense; and time spent sniffing and near the rat (Day effects, all $F[5,65] \geq 2.44$, $p<.05$ and Day x Target $F[10,130]=2.20$, $p<.03$). This second pattern of change indicates that suppression of some components of approach-attack are produced by very few AD's (two to four are elicited in determining ADT). Moreover, this suppression is independent of increases in withdrawal, since withdrawal does not increase until repeated AD are elicited. Finally, the prey are responsive to diminished attack occurring after ADT. Prey self defense diminishes on the day after ADT and remains diminished thereafter. Since prey defense is diminished after repeated AD, the increase in withdrawal from rats cannot be due to the behavior of the rat, rather it represents an enhanced defensive sensitivity to a diminishing threat.

Taken together the data indicate that repeated AD's increase defensive sensitivity to threat posed by large prey (rats). At the same time repeated AD's suppress approach-attack on rats. These data replicate previous findings[10]. The suppression of approach-attack, however, appears to be independent of increases in withdrawal. This finding suggests that the seizures increase the interictal action of different substrates. One substrate is that which controls defensive response to threat. The second is an attack suppressing mechanism. The increase in withdrawal and the suppression of approach-attack are not general, however, since withdrawal from, and attack on mice are unchanged. Thus the neural changes underlying the increase in defense and the suppression of approach-attack must be in those parts of their neural substrates where complex perceptual stimuli interface with the response controlling systems. Finally, as has been found previously, all of these behavioral changes persist interictally, and are not dependent on further seizures, or interictal spikes for their maintenance.

Feline Defense and Feline Anxiety. It has been suggested that species characteristic defensive behaviors model some aspects of human anxiety[13]. In view of the anxiety proneness of limbic epileptics, it is of interest to determine if the changes in defensive behavior produced by partial limbic kindling model seizure induced increases in anxiety. One way to begin to determine this is to show that behavioral changes produced by compounds which diminish and provoke anxiety in humans, also diminish or provoke defensive behavior in the cat. Moreover, the changes in defense and approach-attack produced by anxiogenic compounds should resemble closely those produced by partial limbic kindling. In this section, data will be reviewed which indicate that anxiogenic compounds administered to cats mimic the effects of repeated AD's on defense and approach-attack behavior.

Feline defensive behavior satisfies most of the criteria for an animal model of anxiety[6,13,14,15,16,17]. Nevertheless, it has yet to be demonstrated that compounds which are anxiogenic in humans increase feline defense. One way to test this is to examine the effects of FG-7142 (FG) on feline defense. FG is a beta carboline inverse agonist of the benzodiazepine receptor (BZR). FG increases behaviors indicative of anxiety in animal models of anxiety[18], precipitates intense anxiety in humans[19], and reduces the action of GABA in opening the chloride ionophore component of the BZR[13,20]. The behavioral and physiological actions are due to binding of FG to the BZR, to which it binds with high affinity[21].

If cat defensive behavior models human anxiety, one would expect that FG would increase defensive behavior. Furthermore, if the behavioral action of FG involves the BZR, then its action should be blocked by RO-15-1788 (RO), a high affinity BZR antagonist[22]. In initial studies, 11 cats were given two injections of FG intrathorasically (ith) (10 mg/kg) spaced 48 hours apart[23]. The second injection of FG was preceded (5 or 10 min) by an injection of RO (10 m/kg, ith). Prior to any FG injections, RO and vehicle were administered alone on separate days. Responses to rats and mice were tested 10 and 20 min after the injections. Vehicle or RO alone had no effect on withdrawal or approach-attack as compared to a no injection baseline. FG alone greatly enhanced withdrawal to rats and mice, and suppressed approach-attack. When RO preceded the FG injection, the effects of FG were blocked completely, the behavior returning to baseline. Thus FG enhances defensive behavior in the cat via an action on the BZR.

Of greater interest were the after effects of injection of FG. These effects were at first detected in 9 of the 11 cats. Lasting increases in defensive behavior and suppression of approach-attack on rats, but not on mice, were observed to persist for over 70 days[23]. This finding was replicated as follows. Ten additional cats were given a single injection of FG (10 mg/kg, ith). Five of them were exposed to rats 10 min after the injection, and five were returned to their cages without exposure to prey. Then response to rats were monitored from 6 to 74 days after the injections. All ten cats showed the same pattern of increase of defensive response to rats. The increase was as large as that seen in the cats given two injections of FG. On the other hand, there was no lasting suppression of approach-attack in the single injection cats. These data indicate that FG can change the substrates of defense and attack suppression lastingly and independently.

Many of the lasting effects of FG on behavior involve the BZR. An injection of RO (10 mg/kg, ith), administered 27 days after the last injection of FG, reversibly reduced the lasting increase in withdrawal[23]. This reversal was incomplete, since RO returned behaviors to levels between those seen in the pre and post FG tests. These findings have been interpreted to mean that there is a change in the preferred conformation of the BZR to an inverse agonist conformation, which RO temporarily changes to a neutral state[23]. Lasting suppression of approach-attack also proved reversible with RO. However, not all measures of approach-attack were reversed by RO. Latencies to approach, SPP, bite, HPS and frequency to HPS rats were not returned to baseline by RO, suggesting that some of the lasting suppression of approach-attack produced by FG is independent of the BZR[23].

It is of interest to compare the lasting effects of FG on behavior to those produced by partial limbic kindling. The lasting change in withdrawal from rats produced by FG or partial kindling are the same in magnitude (Figure 1; no Group X Day, Day effect, $F[4,124]=6.10$, $p<.001$; mean contrasts, Days -4,-14 are equal and less than the remainder, which do not differ, all $t[124]$, $p<.001$). Preliminary data also indicate an involvement of the BZR in lasting increases in defense produced by repeated AD. RO (10 mg/kg, ith) reversibly decreased withdrawal from rats (in two cats) which had been increased by partial kindling of the VHP (mean withdrawal: 19.1, 106.6, 32.7, 116.6 sec for baseline, Pre RO [following kindling and one day prior to RO], RO and Post RO [two days after RO], respectively).

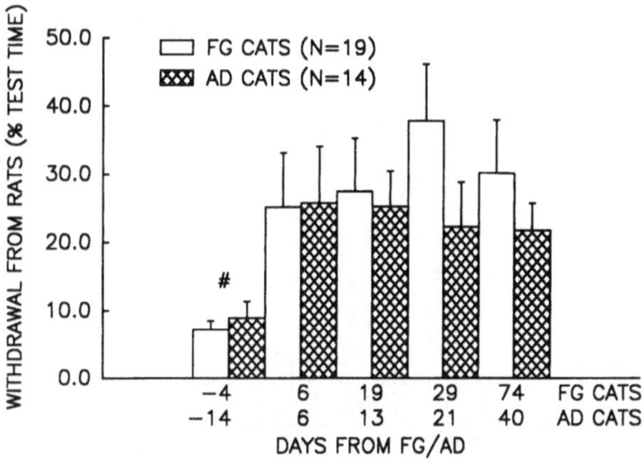


Figure 1. Means ± SEM of withdrawal from rats as a percentage of test time for cats exposed to FG (FG cats) and cats exposed to VHP kindling are plotted versus days from FG or kindling. The "#" indicates those means for both groups differ from all others.

Moreover, the pattern of changes in approach-attack behavior are nearly identical. First, the cats given two shots of FG (FG cats) and cats experiencing repeated VHP AD (AD cats) show similar suppression of approach-attack on rats. With the exception of latency and duration sniffing, all other measures of approach-attack are lastingly suppressed in both groups of cats. FG cats, however show increased latencies to sniff whereas AD cats do not. Conversely, AD cats spend less time sniffing rats, whereas FG cats do not[23]. Another similarity is the fact that neither FG nor AD cats show lasting changes in their behavior toward mice.

Finally FG and AD cats show a generalized change in defensive response to threat, in that there is a lasting change in their defensive response to conspecific threat howls. Response to threat howls was assessed as follows. Cats were exposed to four tests of response to threat. The first two tests were given on consecutive days before FG or AD. The last two tests were given on days 17 and 18 after FG or AD. Each test consisted of four test periods of Pre howl (60 sec), Howl (255 sec), Fight (25 sec) and Post (60 sec). During the howl and fight periods, tape recorded sounds of an aggressive male cat threatening another (Howl), and sounds of a fight between two male cats (Fight) were played through loudspeakers in the test chamber. The aftereffects of this experience were assessed both by the Post test period and by the second day of testing.

A composite measure of defensive response to howls was constructed by averaging the duration of three possible responses to threat. These were seeking shelter under a dark canopy enclosure in the testing chamber, assuming a crouched position with the tail wrapped around the body, a position considered to be a fear motivated self-warming response[24], and freezing by becoming totally immobile for more than one second at a time. A composite was used because individual cats may adopt one or all of the above in response to this type of threat. Duration spent in defense was measured separately in each test period and expressed as a ratio of the test period duration. Analysis of variance was used to assess Group (FG vs AD cats), Retest (Pre and Post FG or AD), Test Day (Test 1 and 2 in a given set of two tests) and Test Period (Pre, Howl, Fight and Post) effects. There were no Group effects, but there was a Retest X Test Day X Test Period Interaction ($F[3,96]=3.447$, $p<.03$, Figure 2). First, within test periods, the

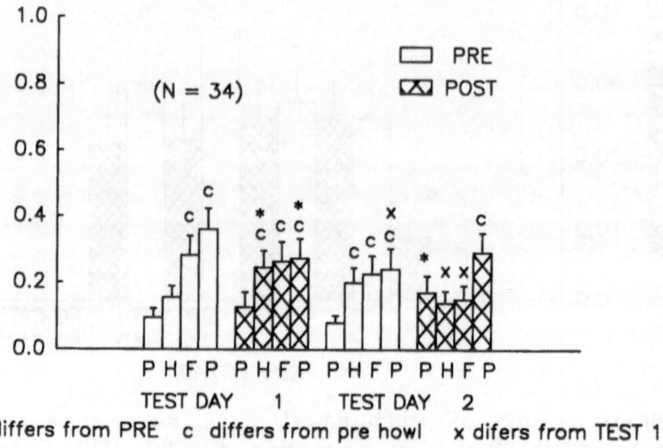


Figure 2. Duration in defense in response to howls as a ratio of test period duration (mean ± SEM) are plotted versus test period (P - Pre howl, H - Howl, F - Fight, P-Post), and consecutive tests on Test Days 1 and 2. Data from Pre and Post FG/AD are plotted separately. Data are combined for FG and AD cats, who do not differ.

threat vocalizations produced an increase in defense over the Pre howl levels, which persisted into the Post Fight test period (Figure 2). Thus the cats were clearly responding to the threat itself, and not the room. Second, the effect of FG or AD (FG/AD) was to sensitize the cats to threat howls on the first test following FG/AD. Cats spent more time in defense in the Howl period on Test Day 1 Post FG/AD than they did in the corresponding Howl period Pre FG/AD (Figure 2). This was not a simple retest effect for two reasons. First, the response to howls did not change from the first to second test days in the Pre FG/AD testing. Second, it has been found repeatedly, and recently replicated with 6 control cats, that response to howls with repeated testing over a retest of interval of 18 days does not change[23]. Furthermore, the cats appear more frightened in the room on the second Test Day following FG/AD. Time in defense is greater in the Pre period of Test Day 2 post FG/AD compared to the Pre period of Test Day 2 prior to FG/AD (Figure 2). There is also a suppression of defense in the howl and fight periods on Test Day 2 post FG/AD relative to Test Day 2 pre FG/AD. This suppression does not represent an habituation, since cats appeared to be agitated, seeking means of escape from the room. This increase in agitation resulted in less crouching, freezing and hiding under the canopy (all $t[96]$, $p<.05$).

Taken together, the data indicate that FG or repeated AD have the same effects on defensive response to conspecific threat, sensitizing the cats to those threats. Lasting increases in response to conspecific threat have been observed before to accompany increased defensive response to prey[10]. Thus, with very few exceptions, the lasting behavioral changes produced by FG or AD are identical. These similarities are not a function of differences in baseline behavior between the two groups of cats. FG and AD cats do not differ with respect to baseline measures of any of the approach-attack or defensive behaviors. The similarities between the groups is also not due to the production of limbic seizures. In sufficient doses, FG can produce limbic seizures and increase limbic glucose utilization in rats[25]. Moreover, repeated doses of FG produce a form of pharmacological kindling which is long lasting[26,27]. Nevertheless, Ongini[28] has shown that 10 mg/kg of FG does not produce limbic seizures in cat dorsal hippocampus or parietal cortex. Moreover, seven cats (3 of which were used in the behavioral studies just described) were implanted with electrodes in the amygdala, VHP and VMH and given 10 mg/kg (ith) of FG. The drug did not produce any seizure activity. On the other hand, FG increased multiple unit activity in the amygdala, VHP (area CA3) of 3/3 cats and it increased neural activity in 2/3 cats in the VMH (measured relative to vehicle control injections). It decreased neural activity in the entorhinal cortex, however, suggesting that the changes in the VHP were intrinsic and not projected via the perforant path.

At this point, it is of interest to compare FG and AD cats with respect to physiological changes which accompany behavioral changes. The lasting changes in interictal physiology following repeated AD will be discussed first, followed by the available data on the effects of FG on limbic excitability.

<u>The Effects of Limbic AD and High Frequency Stimulation on Amygdala-VMH Transmission.</u> The Effects of Limbic AD and High Frequency Stimulation on Amygdala-VMH Transmission. It has been reported that partial kindling of the VHP lastingly increases the size of the neural excitatory VMH potential evoked by amygdala stimulation[29]. These physiological changes persist interictally, and last as long as the behavioral changes. This change has been replicated in the present studies. Nine of the 14 cats described above had electrodes properly placed in the amygdala and VMH. The size of the VMH potential was measured as peak height (PH) from the most negative point of the first component to a tangent line drawn from the beginning of the negative component to its end (see 29 for method). Potentials were evoked with single pulses, 1/10sec, of constant intensity (145% of threshold) over the course of the experiment. Withdrawal from rats and VMH PH remained unchanged from baseline (Pre) and following ADT determination. After partial VHP kindling, however, withdrawal and VMH PH increased and remained elevated from 1 through 40 days post kindling (Days 1, 21 and 40 post kindling, Day effect for withdrawal and VMH PH, all $F[4,32] \geq 2.893$, $p<.04$, mean contrasts all $t[32]$, $p<.02$). In addition, it has also been observed in two cats that fewer than 12 VHP AD's (5-6 AD's) can result in both a lasting (2 week) suppression of the VMH PH and a reduction in withdrawal from rats (means as a percentage of pre kindling baseline: withdrawal: 100.0, 56.4, 80.9; VMH PH: 100.0, 75.2, 89.1 over Pre, 1 day post kindling and 2 weeks post kindling).

Together, these findings suggest that the increase in VMH PH is at least necessary for there to be an increase in withdrawal from rats. These data fit other literature which indicate that the amygdala is a necessary component in the flow of complex sensory information into the medial hypothalamic substrate of defense. For example, withdrawal from prey, as measured here, is disrupted by lesions which interrupt the ventral amydalofugal pathway (VAF)[30]. The VAF mediates the VMH potential and carries excitatory activity from the amygdala to the VMH in cats[31]. Furthermore, lesions of the feline amygdala eliminate the facilitating effect of visual threat on defense elicited by electrical stimulation of the VMH[32].

On the other hand, other data suggest that a decrease in VMH PH is not necessary for a decrease in withdrawal from rats. Since the increase in VMH PH resembles LTP, an attempt was made to induce LTP in the AM-VMH pathway using brief high frequency trains. The purpose of these experiments was to induce both potentiation of VMH PH, and an increase withdrawal from rats[33]. The amygdala was stimulated bilaterally with brief trains of 300 and 400 Hz containing 8-15 pulses, with a train rate of 1/5 sec and 1/10 sec. All of the trains produced the same short term after effects on the VMH potential[33]. There was a short post tetanic potentiation (PTP) which appeared after a delay of 100 sec and returned to baseline by 150 sec, with a declining trend from 150 to 300 sec after the trains. After PTP faded, a long term depression (LTD) of the VMH PH appeared. There may have been a depression at the outset, which could have masked the potentiation at 50 sec post train. Following the first 300 Hz train, LTD lasted at least until forty five minutes following the first 300 Hz train. LTD of VMH PH appeared in the left hemisphere only, however, and was associated with a significant reduction in withdrawal from rats[33]. Withdrawal from rats, and VMH PH, recovered to baseline levels when measured 24 hours after the train, and remained stable on retest 8 days later. These findings fit the view that variation in AM-VMH transmission causes similar variation in withdrawal from rats. Nevertheless, when a 400 Hz train was used, there was a lasting depression of withdrawal from rats which persisted at least 10 days after the train, with no change in the VMH PH of either hemisphere[33]. These data suggest that depression of withdrawal from rats may be accomplished independently of alterations of excitability of the AM-VMH pathway. It is likely that other outputs of the amygdala to the VHP were potentiated, since it has been shown that potentiation of this output is associated with decreased withdrawal from rats[34].

A third experiment with trains supports the view that depression of withdrawal and depression of VMH PH can be independent. Cats (N=4) were given a set of 300 Hz trains delivered over an interval from 10 min (left amygdala stimulation) and 20 min (right amygdala stimulation) after an injection of FG-7142 (10 mg/kg ith). Though it has been found that a single injection of FG will lastingly increase withdrawal from rats, in this case, the high frequency trains blocked the expected increase in withdrawal. Instead there was a decrease in withdrawal which persisted for 2 days after the FG and trains. Then behavior returned to pre drug baselines (withdrawal was measured 2 days prior to FG, and 2, 9 and 24 days after FG, ($t[3]=2.833$, $p<.04$). There was a decrease in the VMH PH of the left hemisphere only. PH fell below baseline on days 2 and 24 after FG (PH was expressed as a percentage of pre drug baseline, all $t[3]\geq 4.076$, $p<.01$). These findings are of interest from two perspectives. First they indicate the importance of the amygdala in the lasting suppression of withdrawal produced by FG. Second, they provide another example of the disassociation of the decreases in the VMH PH and decreases in withdrawal from rats.

Taken together, the data indicate that increases in transmission from the AM to the VMH are at least necessary for increases in withdrawal. Decreases in withdrawal, however, may be accomplished without changing excitability in the AM-VMH pathway. When decreases in withdrawal do accompany decreases in AM-VMH transmission, a change in the left hemisphere is sufficient. This finding may suggest a lateralization of emotional processing in the cat, but the data are not strong enough to draw firm conclusions.

The Effects of Limbic AD on Amygdala-BNST Transmission. Recently the monosynaptic projection from the amygdala to the bed nucleus of the stria terminalis (BNST) in the cat has been found to be excitatory[35]. One can record a compound evoked potential with initial negative and positive after components. The negative component is associated with excitation of single cells. The negative component has a minimal onset latency of 5.8 msec. This potential was investigated in 3 cats, who also had electrodes in the VMH. Stimulation of the AM evoked potentials in both VMH and BNST in these cats. Three additional cats were studied who had electrodes in the VMH and electrodes aimed at the BNST. The BNST electrodes were found to lie in the lateral preoptic area (LPOA) from which a small negative potential could be evoked with AM stimulation. Peak height of the BNST and LPOA potential was measured as described above.

All six cats showed changes in their withdrawal from rats and VMH PH like those reported for the 14 and 9 cats respectively described above. That is VHP ADT determination had no effect on behavior, but partial VHP kindling lastingly increased both withdrawal and VMH PH (Days effects all $F[3,12]\geq 6.85$, $p<.01$, all $t[12]$, $P<.01$, measures taken 14 days before kindling, 7 days before kindling, 1 day after ADT determination, 1 and 21 days after partial kindling). There was no change in the LPOA potentials. There was a lasting increase in the BNST potential, but this increase appeared after VHP ADT determination, before the partial kindling and the increase in withdrawal from rats (Area (BNST vs LPOA) X Days effect, $F[3,12]=4.34$, $p<.03$, all $t[12]$, $p<.05$). Thus other outputs of the amygdala are very sensitive to very few AD's which spread from the VHP to the amygdala.

Potentiation of the projection to the BNST is not sufficient to increase withdrawal from rats. On the other hand, increase in transmission over this pathway may participate in the suppression of approach-attack seen in some measures following VHP ADT (SPS, SPP, sniffing and time near rats).

The Effects of Limbic AD on Hippocampal Recurrent Inhibition. Partial kindling of the cat VHP produces long lasting increases in recurrent inhibition in the dentate area of the VHP[29]. Increased inhibition parallels increases in defensiveness. This has been interpreted to mean that there is a reduction in the excitability of the VHP which functions normally to facilitate attack and suppress defense in the cat[29,36]. These findings have been extended to area CA3 of the VHP. Partial kindling of the PP-VHP lastingly increases recurrent inhibition, measured over intervals from 20 msec to 1 sec after an input to area CA3[37]. Effects last at least two weeks. These findings closely resemble those reported by others in rats[38]. They are also consistent with the findings in the dentate in the cat. Thus not only is the gateway to the trisynaptic circuit less excitable, so is the second stage and a major source of output of the VHP.

The story is different when CA1 is examined. Partial kindling of the VHP reduces recurrent inhibition observed at 20 msec after a pulse to the PP. Inhibition at later intervals (out to 150 msec) is unaffected[37]. Similar findings have been reported in rats[39]. Thus one part of the trisynaptic circuit is made more excitable.

Together, these findings suggest that if increases in VHP inhibition contribute to behavioral change, it is likely that the outputs over the fornix are more important than the CA1 projections to the subiculum. It is possible that the increased excitability in area CA1 contributes to increased defensiveness. On the other hand, one would expect that CA1 response would be secondarily reduced because of the reduction in CA3 excitability.

Summary of the Effects of Limbic AD on Interictal Limbic Physiology. It is apparent from the above that a variety of interictal physiological changes accompany increases in defensiveness following partial limbic kindling. It is not clear which of these changes are necessary and sufficient to produce lasting increases in defensiveness. Many could be epiphenomenal to behavioral change, resulting from the spread of seizure discharges and the severe activation of the seizure event. The available data do suggest that an increase in transmission of excitatory activity from the amygdala to the VMH is at least necessary for an increase in withdrawal. The suppression in approach-attack, however, may be independent of such changes. In this case both the increase in transmission from the amygdala to the BNST as well as the disabling of the VHP by increased recurrent inhibition may mediate the behavioral changes.

Effects of FG-7142 on Limbic Physiology. Since the lasting changes in behavior produced by both limbic kindling and FG-7142 are nearly identical, examining the physiological changes produced by FG should help to uncover those physiological changes which are essential to behavioral change. The behavioral changes produced by FG are not due to seizure activity, so the physiological changes will not be epiphenomenal to seizure activation.

Only preliminary work has been done on this question, and it has concentrated on excitability of the AM-VMH pathway. The effects of FG-7142 on VMH PH were assessed in the left hemisphere only in one cat, and bilaterally in 4 cats. VMH PH was measured after vehicle (ith injection equivolume to FG injection) on one day, and after an injection of FG-7142 (10 mg/kg ith) two days later. Ten samples at 1/5 sec were taken at 10 min (left hemisphere) and 20 min (right hemisphere) post injection and averaged. VMH PH was determined from computer averages of the evoked potentials. There was a 156% increase in the size of the VMH potential evoked by AM stimulation in the left hemisphere only (increase differs from 100% $t[4]=3.99$, $p<.02$, means $\pm$ SEM left and right VMH respectively: $156.00 \pm 14.04$, $99.03 \pm 25.56$).

The laterality of this effect is consistent with the laterality of the effects of high frequency stimulation. There may be some neurochemical peculiarity of the left amygdala which accounts for these hemisphere specific effect. On the other hand, there may be an order effect, since the right hemisphere was stimulated later after drug administration than the left. It is not likely that drug effects had waned, since behavioral effects of FG have been observed 20 min after an injection[23]. More work is required before one can accept a unilateral drug action.

As yet, no studies have been done to determine if the effects on VMH PH are lasting, though these are planned. Moreover, it has yet to be determined if the effects of FG in the VMH are BZR dependent. It will also be of interest to investigate the effects of FG on VHP excitability. As described above, FG increases population cell firing in CA3 while reducing firing in the entorhinal cortex. This could reflect either a directly induced increase in CA3 cell excitability, or an indirectly induced increase in CA3 cell firing due to a lifting of local inhibition. The latter is likely because FG-7142 has been shown to produce a BZR dependent decrease in recurrent inhibition in area CA1 in hippocampal slices[40]. If FG has such an effect, it is opposite in direction to what would be expected from the behavioral data, unless the acute effects of FG on hippocampal inhibition differ from its lasting aftereffects. Further work is needed and is planned.

## CONCLUSIONS

The following conclusions seem warranted. The aftereffects of partial limbic kindling and FG-7142 on behavior may provide a model of the effects of limbic epilepsy on anxiety disorder in humans. This model should be useful in uncovering those changes in limbic physiology which are necessary and sufficient for increased anxiety. The data gathered so far suggest that increased transmission of excitatory activity from amygdala to VMH is at least necessary for increases in defensiveness. Potentiation of other outputs of the amygdala, though, contribute to behavioral changes, such as suppression of defense and approach-attack. The molecular mechanism of some of these changes appears to involve the BZR, in that the specific BZR antagonist, RO-15-1788 blocks the increases in defensiveness produced by FG, and may reverse the seizure induced increases in defensiveness. Continued study of these phenomena seems warranted, and promises progress toward understanding of the molecular basis of at least one form of psychopathology associated with human epilepsy.

## REFERENCES


1. M. R. Trimble and M. M. Perez, Quantification of psychopathology in adult patients with epilepsy in: "Epilepsy and behavior '79," B. M. Kulig, H. Meinardi, G. Stores, ed. Swets and Zeitlinger, Lisse (1980).
2. B. P. Hermann, and S. Whitman, Behavioral and personality correlates of epilepsy: a review, methodological critique, and conceptual model, Psych. Bull. 95:451-497 (1984).
3. R. J. Mittan and G. E. Locke, The other half of epilepsy: psychosocial problems, Urban Health. (Jan-Feb):38-39 (1982).
4. E. S. Paykel, Recent life events in the development of depressive disorder, in: "The Psychobiology of depressive disorders: Implications for the effects of stress," E. S. Paykel ed., Academic Press, New York, (1979).
5. T. Takeuchi, T. Takahashi, H. Kotsuki, S. Aizawa, S. Maruyama and K. Kodama, Life events related to the inception of anxiety neurosis, Jap. J. Psychiat. Neurol. 40(2):137-142, (1986).
6. R. Adamec, Kindling, anxiety and personality, in: "The Clinical Relevance of Kindling," M. R. Trimble and T. G. Bolwig, eds., John Wiley and Sons, Chichester, (1989) (in press).
7. R. Adamec and C. Stark-Adamec, Limbic kindling and animal behavior - Implications for human psychopathology associated with complex partial seizures, Biol. Psychiat. 18(2):269-293 (1983).
8. R. Adamec, Does kindling model anything clinically relevant, Biol. Psychiat. (accepted for publication) (1990).
9. R. Adamec, Normal and abnormal limbic system mechanisms of emotive biasing, in: "Limbic Mechanisms," K. E. Livingston and O. Hornykiewcz, eds., ), Plenum Press, New York, (1978).
10. R. Adamec and C. Stark-Adamec, Partial kindling and emotional bias in the cat: Lasting after effects of partial kindling of the ventral hippocampus. I. Behavioral Changes, Behav. Neur. Biol. 38:205-22 (1983).
11. N. Griffin, N., J. Engel and R. Bandler, Ictal and enduring inter-ictal disturbances in emotional behavior in an animal model of temporal lobe epilepsy, Brain Res. 400:360-364 (1987).
12. A. Siegel, Anatomical and functional differentiation within the amygdala, in: Modulation of Sensorimotor Activity During

Alterations in Behavioral States.," R. Bandler, ed., Allan R. Liss, New York (1984).

13. P. Skolnick, P. Ninan, T. Insel, J. Crawley and S. Paul, A novel chemically induced animal model of human anxiety Psychopath. 17:25-36 (1984).
14. H. Maeda, Effects of psychotropic drugs upon the hypothalamic rage responses in cats Fol. Psychiat. Neurol. Jap. 30:539-546 (1976).
15. L. Wolgin, and S. Servidio Disinhibition of predatory attack in kittens by oxazepam, Soc. Neurosci. Abs. 5:2282-0 (1979).
16. S. L. Stoddard, V. K. Bergdall, D. W. Towsend and B. E. Levin, Plasma catecholamine associate with hypothalamically-elicited defense behavior, Physiol. Behav. 36:867-873 (1986).
17. S. L. Stoddard, V. K. Bergdall, P. S. Conn and B. E. Levin, Increases in plasma catecholamines during naturally elicited defensive behavior in the cat, J. Auton. Nerv. Sys., 19: 189-197 (1987).
18. M-H. Thiebot, P. Soubrie and D. Sanger, Anxiogenic properties of beta-CCE and FG 7142: a review of promises and pitfalls Psychopharm. 94:452-463 (1988).
19. R. Dorrow, R. Horowski, F. Paschelke, M. Amin and C. Braestrup, Severe anxiety induced by FG-7142, a carboline ligand for benzodiazepine receptors, Lancet 2:98-99 (1983).
20. J. N. Crawley, P. T. Ninan, D. Pickar, G. P. Chrousos, M. Linnoila, P. Skolnick and S. M. Paul, Neuropharmacological antagonism of the carboline-induced "anxiety" response in rhesus monkeys, J. Neurosci. 5:477-485 (1985).
21. C. Braestrup, R. Schmiechen, G. Neef, M. Nielsen and E. N. Peterson, Interaction of convulsive ligands with benzodiazepine receptors, Science 216:1241-1242 (1982).
22. H. Mohler and J. G. Richards, Agonist and antagonist of benzodiazepine receptor interactions in vitro, Nature. 294:763-765 (1981).
23. R. Adamec, FG-7142 and "Anxiety" in the cat: Acute and Lasting after effects. Europ. J. Pharmacol., (accepted for for publication and under revision) (1990).
24. R. J. Andrew, The information potentially available in mammal displays, in: "Non-verbal communication," R. A. Hinde, ed., Cambridge University Press, London (1972).
25. A. Ableitner and A. Herz Changes in local cerebral glucose utilization induced by the beta-carbolines FG 7142 and DMCM reveal brain structures involved in the control of anxiety and seizure activity. J. Neurosci. 7:1047-1055 (1987).
26. H. J. Little, D. J. Nutt and S. C. Taylor, Kindling and withdrawal changes at the benzodiazepine receptor, J. Psychopharmacol. 1:35-46 (1987).
27. M. G. Corda, O. Giorgi, F. Gatta, and G. Biggio,Long-lasting proconflict effect induced by chronic administration of the beta-carboline derivative FG 7142. Neurosci. Lett. 62:237-240 (1985).
28. é. Ongini, Behavioral and EEG effects of benzodiazepines and their antagonists in the cat. in: "Benzodiazepine recognition site ligands: Biochemistry and pharmacology," G. Biggio and E. Costa Costa, eds., Raven Press, New York (1983).
29. R. Adamec and C. Stark-Adamec, Partial kindling and emotional bias in the cat: Lasting after effects of partial kindling of the ventral hippocampus. II. Physiological changes, Behav. Neur. Biol. 38:223-239 (1983).
30. S. M. Pellis, D. P. O'Brien, V. C. Pellis, P. Teitelbaum, D. L. Wolgin and S. Kennedy, Escalation of feline predation

along a gradient from avoidance through "play" to killing. Behav. Neurosci. 102(5):760-777 (1988).

31. J. T. Murphy, The role of the amygdala in controlling hypothalamic output. in: "The neurobiology of the amygdala," B. E. Eleftheriou, ed., Plenum Press, New York (1972).
32. H. Maeda and K. Hirata, Two-stage amygdaloid lesions and hypothalamic range: A method useful for detecting functional localization, Physiol. Behav. 21:529-530 (1978).
33. R. Adamec, R. The role of the amygdala and ventromedial hypothalamus in partial kindling induced increases in defensiveness in the cat, Agg. Behav. (1989) (Accepted for publication).
34. R. E. Adamec and C. Stark-Adamec, Partial kindling and behavioral change-Some rules governing behavioral outcome of repeated limbic seizures, in: "Kindling 3", J.A. Wada, ed., Raven Press, New York (1986).
35. R. Adamec, The relationship of the amygdala and the bed nucleus of the stria terminalis of the cat: An evoked potential and single cell study. Behav. Neur. Biol. (in press) (1989).
36. R. Adamec and C. Stark-Adamec, Limbic Control of Aggression, Neuro-Psychopharm. Biol. Psychiat. 7:505-512 (1983).
37. R. Adamec, The effects of partial kindling on inhibition in the cat ventral hippocampus, (in preparation) (1989).
38. L. P. Tuff, R. J. Racine and R. Adamec, The effects of kindling on GABA-mediated inhibition in the dentate gyrus of the rat. I: Paired-pulse depression, Brain Res., 277:79-90 (1983).
39. J. Kapur, J. L. Stringer and E. W. Lothman, Evidence that repetitive seizures in the hippocampus cause a lasting reduction of GABAergic inhibition J. Neurophysiol. 61:417-426 (1989).
40. P. G. Gluchankov, V. S. Vorobyov and V. G. Skrebitsky, Influence of carboline derivative FG7142 on the inhibition in hippocampal sections. Bull. Exp. Biol. Med. Moscow 12:724 (1985).

# PHARMACOLOGICAL DISSOCIATION BETWEEN THE MECHANISMS OF KINDLING AND LONG-TERM POTENTIATION BY APV AND URETHANE ANESTHESIA

Donald P. Cain, Francis Boon and Eric L. Hargreaves

Department of Psychology
University of Western Ontario
London, Ontario, Canada N6A 5C2

## INTRODUCTION

Kindling and long term potentiation (LTP) are among the most widely studied models of neural plasticity. In many respects they are strikingly similar. In addition to the fact that they both model CNS plasticity, the method of inducing them and the neural responses that result are very similar. For example, both are usually induced by the localized application of brief, high-frequency trains of electrical pulses through implanted electrodes, and both result in a lasting increase in the neural response to a constant stimulus (2,10). In addition, recent research has suggested that the two models may share aspects of a common underlying neural mechanism, and this has led to the suggestion that LTP might constitute the cellular mechanism of kindling (1,7,17).

This is an attractive idea. If one kept in mind the fact that both LTP and kindling involve an increase in the electrophysiological response to a constant stimulus, one might view kindling as an exaggerated form of LTP, and this idea has been put forward (1). Thus, LTP might be a relatively weak manifestation of the neural changes involved in this form of plasticity, and kindling might be a relatively strong manifestation of the same changes, to which has been added an increase in the general excitability of the cell that reveals itself as a tendency to fire in bursts. According to this view, the epileptiform properties of kindled tissue might be accounted for in terms of both the strong increases in synaptic efficacy resulting from the exaggerated LTP-like changes that occur, plus the tendency to fire in bursts, which can result in seizure activity and convulsions. The fact that the induction of LTP does not normally result in kindling can be explained by the fact that normal LTP-inducing currents are purposely kept below the threshold current necessary to evoke afterdischarge (AD), a critical requirement for kindling (14).

On the other hand, there is considerable evidence that LTP and kindling are not completely similar and probably do not share a common underlying mechanism. Some of this evidence indicates that LTP and kindling do not always mutually facilitate one another, are not induced at comparable speed, do not persist for comparable durations, and are not similarly affected by pharmacological and other experimental treatments (see refs. 4 and 16 for reviews). However, to date there has been no direct test of the question whether a treatment that interferes with the development of one model also interferes similarly with the development of the other model in the same preparation. The following series of experiments was designed to provide such a direct test using urethane anesthesia, and in different animals, the n-methyl-d-aspartate (NMDA) receptor blocker DL-2-amino-5-phosphonivaleric acid (APV).

*Kindling 4*
Edited by J. A. Wada
Plenum Press, New York, 1990

## EFFECTS OF URETHANE ANESTHESIA ON KINDLING AND LTP

Urethane anesthesia does not block LTP. In fact, many LTP studies, including the pioneering LTP work of Bliss and Lomo (2), were conducted while the subjects were anesthetized with urethane. In an effort to find an anesthetic suitable for use in acute kindling studies, we studied the effect of urethane on the development of amygdala kindled seizures and on the expression of previously kindled seizures (6). This study showed that an anesthetic dose of urethane (1.5 g/kg) almost completely eliminated evoked AD and completely eliminated convulsive behavior in both situations. Subanesthetic doses of urethane (0.25 and 0.5 g/kg) strongly attenuated the expression of previously kindled seizures. These findings suggested that urethane might provide a means of dissociating between the mechanisms of LTP and kindling.

Our first step was to select a preparation that we could use for both kindling and LTP. It also had to be a preparation that allowed the measurement of any kindling-induced potentiation effects that occurred. We adopted the common LTP preparation in the rat in which a stimulating electrode is chronically implanted in the perforant path at the level of the angular bundle, and a recording electrode is implanted in the hilus of the dentate gyrus. This allowed us to evoke large potentials with relatively small intensity pulses of current, and to measure the properties of the potentials using a microcomputer-based averaging and analysis system.

We allowed 1 to 2 weeks to elapse after surgery, and then began monitoring the response to single test pulses applied to the perforant path. When the responses appeared to be stable over a number of days we obtained an input/-output (I/O) curve that related the range of stimulus intensities to the range of responses for both the maximum slope of the e.p.s.p. and the population spike amplitude. Averages of the responses to 10 biphasic stimulation pulses, each 0.1 msec in duration, were taken. All of the potentials were obtained while the rats were immobile, as movement has been shown to affect the properties of potentials evoked in this circuit (3). The interval between the delivery of individual pulses was a minimum of 10 sec and usually not more than 30 sec. An I/O curve was always taken before experimental manipulations, and also 24 hrs later for the APV group (see below). Additional measures were taken at other times, and for this purpose the averaged response to 10 pulses at an intensity 80% of the maximum value used in determining the I/O curve was obtained.

In the first study, 10 rats were prepared as described above, and an I/O curve was taken. Next, the AD threshold was determined by applying a standard kindling 1 sec train of biphasic pulses, each 1.0 msec in duration, at 60 Hz. The electroencephalogram (EEG) was recorded throughout this procedure from the dentate hilus lead. The stimulation intensity was initially set at 20 mA, and subsequently raised in small steps until an AD appeared in the EEG. All rats exhibited ADs. Any potentiation of the evoked potential measures that occurred was then measured periodically during the next few weeks until it had decayed to within 10% of initial values.

At this point 10 potentials were evoked by the 80% intensity stimulus and averaged (PRE URE, Figs. 1 and 2), and urethane (ethyl carbamate, 1.5 g/kg i.p.) was administered. After complete anesthetization had occurred, 10 potentials were again averaged (POST URE, Figs. 1 and 2), following which LTP stimulation was applied. This consisted of 20 high-frequency trains, each 20 msec long, made up of 8 biphasic 0.1 msec pulses at 400 Hz. The trains were applied while the animals were immobile at intervals of approximately 20-30 sec. LTP effects were then measured immediately, and at 0.5, 1 and 2 hrs (POST LTP, Figs. 1 and 2). After the 2 hr measure, we attempted to evoke an AD using kindling stimulation delivered at the previously determined AD threshold. If this failed to evoke an AD, the kindling stimulation intensity was raised gradually until a maximum of 1600 $\mu$A (base-to-peak) had been reached.

The results of this study are summarized in Fig. 1, where e.p.s.p. maximum slope is shown, and in Fig. 2, where population spike amplitude is shown as a percent change from the value determined just before administration of urethane. Administration of urethane did not block increases in either measure due to the LTP stimulation, and both measures taken immediately after the LTP trains (0 hr) were significantly higher than those taken just before administration of urethane (PRE URE). The e.p.s.p. measures returned to near-baseline values thereafter, but the population spike measures remained elevated throughout. All attempts to evoke AD after the 2 hr measure were unsuccessful, even when stimulation intensities of up to 1600 $\mu$A were used.

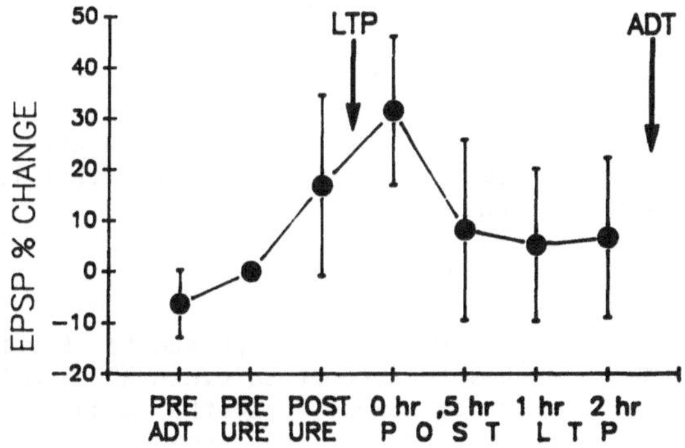


Fig. 1. E.P.S.P. maximum slope percent change relative to the measure taken just before administration of urethane (PRE URE). The measure taken immediately after the LTP trains were applied is significantly greater than that taken before urethane. Afterdischarge could not be evoked immediately after the 2 hr measure was taken.

These results confirm those reported in many previous studies in demonstrating a significant LTP effect in subjects anesthetized with urethane. However, urethane completely blocked our ability to evoke AD, even when very high stimulation currents were used. This was true even though all of the electrode placements were confirmed to support AD at the beginning of the study, and even though we attempted to evoke AD at a time when the population spike was still significantly potentiated. If LTP is the cellular mechanism of kindling, we would predict that the LTP effects should contribute to the occurrence of AD, yet AD could not be evoked at this time.

A tentative conclusion from this study might be that urethane anesthesia blocks kindling through the complete suppression of AD, but does not block LTP. Urethane thus seems to provide a pharmacological dissociation between the two models.

This argument has a flaw. It is true that urethane blocked kindling in both this study and our previous one (6). However, it did so by suppressing AD. One could imagine that if one could evoke AD there might be evidence of kindling-induced potentiation of the evoked potential, and if the AD were evoked repeatedly, there might be evidence of increases in the strength and duration of the AD, just as there is in normal kindling. This result would be strong evidence against the idea that urethane provides a pharmacological dissociation between the two models.

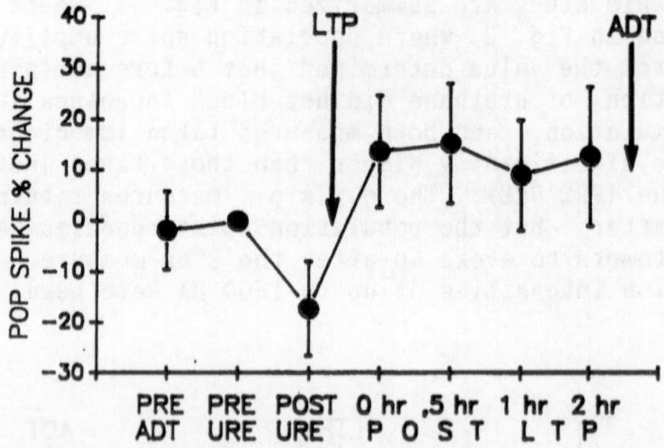


Fig. 2. Population spike amplitude percent change relative to the measure taken just before administration of urethane (PRE URE). The measures taken after the LTP trains are significantly greater than that taken before urethane.

In order to approach this objection, we repeated the first experiment but made two changes. The first was to attempt to evoke AD repeatedly, at hourly intervals, beginning as soon as complete anesthesia had been achieved with urethane (no LTP stimulation was given). ADs evoked at hourly intervals in normal rats show clear kindling effects reflected in an increase in the duration of AD (6,15). Thus, any tendency for the repeated ADs to kindle the rats should manifest itself as an increase in AD duration, even though the urethane anesthesia prevented the display of motor seizures. The second change was to adjust the kindling stimulation so as to be certain to evoke AD after each hourly stimulation. This was done by increasing the intensity or duration, or both, of the stimulation.

In this study 6 rats were prepared as described above with perforant path stimulating electrodes and dentate hilus recording electrodes. A baseline I/O curve was again determined and the rats received urethane in the same dose and manner as in the first study. A second baseline was then taken, after which we attempted to evoke AD. The only successful method for this was to increase the duration of the stimulation to 4 or 8 sec using stimulation intensities of 1800-6400 µA (base-to-peak). These are very high durations and intensities, and are much higher than those normally required for perforant path kindling.

The results are summarized in Fig. 3, where e.p.s.p. maximum slope is shown, and in Fig. 4, where population spike amplitude is shown as a percent change from the value determined initially. All rats displayed AD at each of three stimulations at hourly intervals. Urethane blocked the increase in AD duration that would normally have occurred, as is indicated by means of 10.4, 11.7 and 11.5 sec, respectively. Urethane also blocked the increases in the evoked potential measures that normally would have occurred as a result of repeatedly evoking AD (see below). One hr after the third AD the measures continued to show no evidence of potentiation effects.

These results seem to effectively counter the argument that urethane merely blocks AD but might not necessarily block AD- (and kindling-) induced potentiation effects, and kindling itself (as manifested in increases in the duration of AD). This study indicates that even if AD is evoked repeatedly, urethane completely blocks kindling. Since urethane does not block LTP, we can conclude that urethane provides a pharmacological dissociation between LTP and kindling.

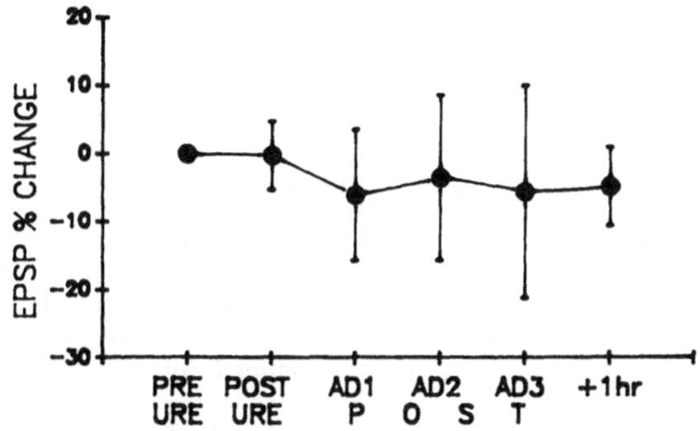


Fig. 3. E.P.S.P. maximum slope percent change relative to the measure taken just before administration of urethane (PRE URE). Elicitation of three ADs did not lead to an increase in this measure.

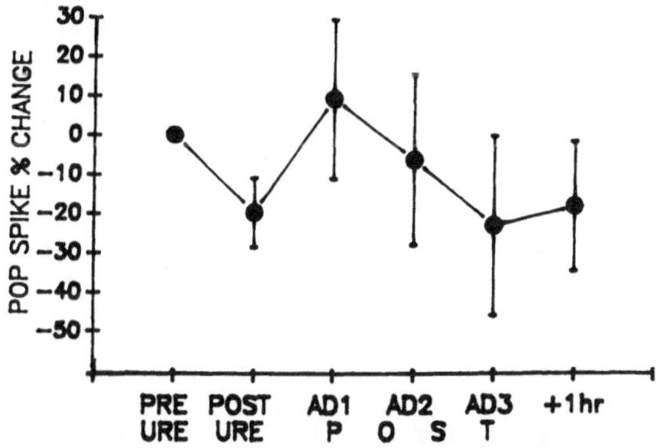


Fig. 4. Population spike amplitude percent change relative to the measure taken just before administration of urethane (PRE URE). Elicitation of three ADs did not lead to an increase in this measure.

## EFFECTS OF APV ON KINDLING AND LTP

APV is a specific blocker of the NMDA receptor, and it completely blocks LTP (11,13). However, we and others have found that APV retards but does not completely block amygdala kindling (5,9). What would be the effect of APV on LTP and kindlng-induced potentiation in the perforant path-dentate hilus circuit? We used 11 rats prepared as previously described, with the exception that they carried chemitrodes rather than electrodes implanted into the hilus of the dentate. The chemitrodes were constructed by attaching an electrode to the side of a 23-ga stainless steel guide cannula. The electrode tip ended 1.0 mm below the tip of the guide cannula, and the 30-ga injection cannula, which could be inserted through the guide cannula, protruded 1.0 mm beyond the tip of the guide cannula. This allowed us to make injections of APV directly into the hilus and to record from the same area.

An initial baseline I/O curve was obtained. The next day, 7.5 µg APV in 0.5 µl saline was injected into the hilus slowly using a Sage syringe pump. A second baseline was taken 15 min later, after which LTP stimulation was administered as described above. Evoked potential measures taken immediately, 1 hr, and 24 hrs after the LTP stimulation showed that the APV completely blocked the LTP that would normally have been produced by the stimulation (see Figs. 5 and 6).

This result confirms the previously reported blocking effect of APV on LTP, but it does not prove that APV blocks an essential component of the mechanism of LTP. One could imagine that the APV simply elevated the threshold for the induction of LTP, but did not block the actual mechanism. If one could overcome the elevation in the threshold, one might see LTP effects.

We attempted to do this by again injecting APV into the hilus of the same rats some days later and applying LTP stimulation, but this time at an intensity three times the intensity used previously. This value was chosen because it far exceeds the value that is typically necessary for LTP effects in the hilus.

The results were identical to those before. The APV completely blocked LTP effects in both the e.p.s.p. maximum slope and the population spike amplitude (see Figs. 5 and 6).

We next asked whether AD evoked in the same rats after administration of the same dose of APV would result in kindling and potentiation of the evoked potential.

We again injected APV into the hilus of the same rats some days later. Fifteen min later we determined the AD threshold in the usual way by applying a 1-sec train of kindling pulses at 60 Hz to the perforant path electrode. One hour later evoked potential measures were again taken, after which another injection of APV was made. In some rats the amount of APV injected was the same as was injected initially; in other rats the amount was half the initial dose, and constituted a 'maintenance' dose. The results did not differ between these subgroups, and the data were collapsed across subgroups. Fifteen min later AD was again evoked, and 1 hr later evoked potential measures were again taken. This process was repeated once more for both subgroups, so that a total of three injections of APV were made and three ADs were evoked, followed at 1-hr intervals by measures of the evoked potential.

The results (Figs. 5 and 6) show that the APV did not block either AD or AD-induced potentiation of the e.p.s.p. maximum slope or the population spike amplitude, and both of these measures were still elevated 24 hrs and 1 wk later (see below). The mean AD durations over the three ADs were 35.5, 62.0 and 64.5 sec, which is indicative of the progression that would normally occur during the first three repetitions of the stimulation at hourly intervals.

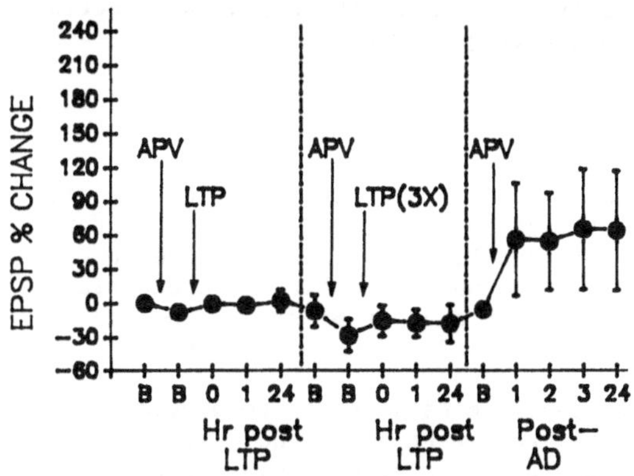


Fig. 5. E.P.S.P. maximum slope percent change relative to the measure taken just before administration of APV (first panel, first 'B'). There was no effect of LTP trains at normal intensity (first panel, 'Hr post LTP'), or 3 times normal intensity (second panel, 'Hr post LTP). Subsequent elicitation of AD elevated this measure (third panel, 'Post-AD), and subsequent LTP trains in the undrugged rat resulted in further potentiation effects (fourth panel).

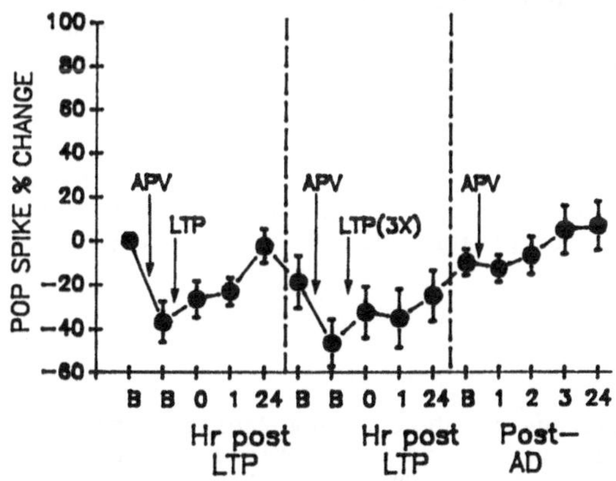


Fig. 6. Population spike amplitude percent change relative to the measure taken just before administration of APV (first panel, first 'B'). There was no effect of LTP trains at normal intensity (first panel, 'Hr post LTP'), or 3 times normal intensity (second panel, 'Hr post LTP'). Subsequent elicitation of AD elevated this measure (third panel, 'Post-AD'), and subsequent LTP trains in the undrugged rat resulted in further potentiation effects (fourth panel).

One wk later a baseline was again taken, which indicated that the evoked potential measures were still elevated. At this point LTP stimulation was applied as described above. Further increases in the measures were obtained 1 and 24 hrs later, indicating that the recording arrangements and brain circuitry were not adversely affected by the repeated injection of APV and that the brain circuitry could undergo further LTP.

## DISCUSSION

The results of this study provide a dissociation between LTP and kindling that is the obverse of that obtained with urethane. Urethane did not block LTP, but it completely blocked kindling and AD-induced potentiation. Here, APV completely blocked LTP even when stimulation intensities three times the normal level were used, but it did not block AD, or the increase in AD duration due to repeated AD, or AD-induced potentiation.

Thus, taken as a whole these data appear to provide a double dissociation between mechanisms underlying LTP and kindling, and argue strongly against the view that LTP constitutes the cellular mechanism of kindling (1,7,17).

This conclusion does not mean that LTP effects do not contribute to normal kindling development. The increase in synaptic efficacy that occurs in response to either normal LTP stimulation or to AD elicited during kindling probably contributes to the progressive development of the seizures. Evidence for this is the finding that prior LTP facilitates subsequent kindling in the same circuit (18). However, even repeated strong LTP in the absence of AD cannot kindle seizures (8,18). On the other hand, there is also evidence that inhibitory effects can be potentiated, and it seems likely that this can work against kindling development (12,19). Thus, in normal kindling a variety of complex pro- and antiseizure influences that result from potentiation might be operating.

Perhaps the most general conclusion to be drawn from this and much other work indicating that LTP and kindling have different underlying mechanisms (4,16) is that the two models should be studied in their own right for the contribution that each can make to a comprehensive understanding of plastic change in the brain. LTP is perhaps the model of choice for the study of plastic change that might underlie learning, and kindling is perhaps the model of choice for the study of epileptiform phenomena and brain change related to epilepsy. We should probably not be surprised to find that there are multiple forms and mechanisms for plastic change in a structure as complex as the mammalian brain.

## ACKNOWLEDGEMENTS

We would like to acknowledge our debt to Ron Racine, who suggested the original experiment in this series, and who provided helpful comments on the results. Supported by a grant from NSERC (Canada).

## REFERENCES

1. Baudry, M., 1986, Long-term potentiation and kindling: Similar biochemical mechanisms?, in: "Advances in Neurology", Vol. 44, A.V. Delgado-Escueta et al., ed., Raven Press, New York.
2. Bliss, T.P. and Lomo, T., 1973, Long-lasting potentiation of synaptic transmission in the dentate area of the anesthetized rabbit following stimulation of the perforant path. J Physiol, 232: 331.
3. Buzsaki, G., Grastyan, E., Czopf, J., Kellenyi, L. and Prohaska, O., 1981, Changes in neuronal transmission in the rat hippocampus during behavior. Brain Res, 1981, 225: 235.

4. Cain, D.P., 1989, Long-term potentiation and kindling: How similar are the mechanisms? Trends Neurosci, 12: 6.
5. Cain, D.P., Desborough, K.A. and McKitrick, D.J., 1988, Retardation of amygdala kindling by antagonism of NMD-aspartate and muscarinic cholinergic receptors: Evidence for the summation of excitatory mechanisms in kindling. Exper Neurol, 100: 179.
6. Cain, D.P., Raithby, A. and Corcoran, M.E., 1989, Urethane anesthesia blocks the development and expression of kindled seizures. Life Sci, 44: 1201.
7. Collingridge, G.L. and Bliss, T.P., 1987, NMDA receptors - Their role in long-term potentiation. Trends Neurosci, 10: 288.
8. de Jong, M. and Racine, R.J., 1985, The effects of repeated induction of long-term potentiation in the dentate gyrus. Brain Res, 328: 181.
9. Gilbert, M., 1988, The NMDA-receptor antagonist, MK-801, suppresses limbic kindling and kindled seizures. Brain Res, 463: 90.
10. Goddard, G.V., McIntyre, D.C. and Leech, C.K., 1969, A permanent change in brain function resulting from daily electrical stimulation. Exper Neurol, 25: 295.
11. Harris, E.W., Ganong, A.H. and Cotman, C.W., 1984, Long-term potentiation in the hippocampus involves activation of N-methyl-D-aspartate receptors. Brain Res, 323: 132.
12. Maru, E. and Goddard, G., 1987, Alteration in dentate neuronal activities associated with perforant path kindling. III. Enhancement of synaptic inhibition. Exper Neurol, 96: 46.
13. Morris, R.G.M., Anderson, E., Lynch, G.S. and Baudry, M., 1986, Selective impairment of learning and blockade of long-term potentiation by an N-methyl-D-aspartate receptor antagonist, AP5. Nature, 319: 774.
14. Racine, R.J., 1972, Modification of seizure activity by electrical stimulation: II. Motor seizure. Electroencephalogr Clin Neurophysiol, 32: 281.
15. Racine, R.J., Burnham, W.M., Gartner, J.G. and Levitan, D., 1973, Rates of motor seizure development in rats subjected to electrical brain stimulation: Strain and interstimulation interval effects. Electroencephalogr Clin Neurophysiol, 35: 553.
16. Racine, R.J., Burnham, W.M., Gilbert, M.E. and Kairiss, E.W., 1986, Kindling mechanisms: I. Electrophysiological studies, in: "Kindling 3" J.A. Wada, ed., Raven Press, New York.
17. Slater, N.T., Stelzer, A. and Galvan, M., 1985, Kindling-like stimulus patterns induce epileptiform discharges in the guinea pig in vitro hippocampus. Neurosci Lett, 60: 25.
18. Sutula, T. and Steward, O., 1987, Facilitation of kindling by prior induction of long-term potentiation in the perforant path. Brain Res, 420: 109.
19. Tuff, L.P., Racine, R.J. and Adamec, R., 1983, The effects of kindling on GABA-mediated inhibition in the dentate gyrus of the rat. I. Paired-pulse depression. Brain Res, 277: 79.

Discussion of Dr. Cain's Presentation

DR. GILBERT: As Peter mentioned, we have been doing similar studies in chronically prepared animals by taking animals out and monitoring after-discharge induced potentiation over the course of kindling development with the NMDA blocker MK801. What we have found is, although perforant path kindling was in fact substantially delayed in animals treated with MK801, they did eventually kindle as we had also found in the amygdala. But when we monitored the after-discharge induced potentiation we found that it was quite effective in blocking both potentiation of the EPSP and the population spike over the course of kindling, which looked initially like evidence that facilitation induced by an LTP-like mechanism may be contributing to the kindling process and the spread of the seizure. However, if you took those animals out and then looked at them after they had fully kindled, you found that there was no evidence of potentiation of either component of the evoked potential even though the animals were displaying generalized seizures.

DR. CAIN: So kindling in the absence of potentiation effects and kindling as reflected by increases in the AD duration and behaviour and so on. Interesting.

DR. MODY: I would like to ask you a question about the last graph. If you could back up a couple of slides. What is exactly that you call LTP on the right-hand side there? What is the baseline for that LTP?

DR. CAIN: This is a measure taken about a week on average after this treatment, after this day. This is a potentiated baseline, if that's what you're getting at, you're right. This is the persistence of the potentiation, I think, due to these after-discharges here, and again here. So we're adding more potentiation.

DR. MODY: That's what I wonder about. Do you have more potentiation there? Are those points significantly different from the baseline or the pre-LTP stimulus baseline?

DR. CAIN: I have not done a statistic on this. They may not be statistically higher. Phenomenally, looking at them they are above, you may be right, I don't know.

DR. MODY: Is that LTP then or not?

DR. CAIN: I think it is.

DR. MODY: I don't think so. I don't know, I mean if they're not statistically significant from the pre-tetanus baseline, I doubt that it is LTP.

DR. CAIN: I didn't say they were not statistically significant, I said I have not done a statistic on these data yet. There's more analysis to be done, for example when we take these data further what I want to do is to provide more data presented by the IO curves. I believe, I cannot say that it will be, but I believe for example that these points will probably end up being statistically above that point. May be that point will be statistically above that point, I don't know.

DR. MODY: But can you call it LTP unless you do those statistics?

DR. CAIN: Well, many people have in the literature. I guess you take your pick.

DR. MODY: Anyhow, the other question I had was about the fact of urethane before you induced after-discharges and after you've induced discharges. It seems to me that urethane actually dissociated the EPSP on the spike before after-discharges were evoked. If that's correct, what do you attribute that to?

DR. CAIN: I don't know. In a previous talk where some of these data were given the question was raised, what does urethane do? I don't have an adequate answer.

DR. MODY: Well, in the previous slide you see that EPSP actually goes up and the population spike goes down, and that doesn't happen after the after-discharges.

DR. CAIN: I have no answer to that question.

DR. MCNAMARA: I'd like to ask Mary a question, and then Peter a question. Did you try to induce LTP in those animals after you gave them MK801 and were kindling them, and then you're finished with the MK801 and they're kindled, is that how you did the experiment? And then you said that there was no evidence of LTP in that pathway, isn't that what you said?

DR. GILBERT: We monitored it during kindling and there was no evidence of LTP after 1, 4, up to 13 after-discharges.

DR. MCNAMARA: I see, but the question is, were they kindling during that time?

DR. GILBERT: Yes. It was delayed, but they were kindling. And when they were fully kindled there was still no evidence of LTP.

DR. MCNAMARA: But the question I have is, then did you take those animals and try to induce LTP in them the way, for example, that Ron did a few years ago.

DR. GILBERT: We let them decay. They did decay, the response decayed after kindling in a couple of animals and then we induced LTP and got LTP.

DR. MCNAMARA: OK. Fine, that's all.

DR. BURNHAM: Peter, your data opens up some complex problems. You've told us that in some pathways like this one, in other words the perforant pathway, LTP occurs, and that would be associated with changes, some of them lasting probably up to a month or more. And kindling occurs and it's a different process, and it's associated with changes. Now, when we find changes in a pathway like this, such as Dr. Geinisman has found, or other people have found, how can we tell whether they're associated with LTP or whether they're associated with kindling?

DR. CAIN: Well, that's a good question. Actually he and I had a discussion at lunch over just that. And maybe I'll try and recall what I said. Maybe he can recall what I said, I can't now. One way to do that, I've said actually in print that something like the following ought to be done. Number 1, it would be nice for people to study these two models in the same circuit; this point has come up at this conference before. Well, I've done it, and it ought to be done more because our comparisons can be more direct. So one thing to do would be to take different groups of animals and induce the two models in them using similar electrode implants in the perforant paths, say. You could then look for your anatomical changes in kindled animals, in animals that were fully kindled and animals that were partially kindled, and in different groups in animals that had different retention times. So that if kindling is a permanent phenomenon you ought to see any kind of anatomical change persist as long as kindling does. Now, for the LTP side, what you could do is put electrodes in the same circuit, you could induce LTP and the beauty of following LTP in a chronic preparation is that you can confirm just before you kill the animal and do your anatomy that LTP was there. And you can then in other animals let them have more survival time and follow any decay that would occur in the LTP, then do your anatomy and you should predict that any anatomical changes that were of consequence would be gone. I don't know. Maybe that wouldn't happen, but it's an approach.

DR. BURNHAM: I know from your writings that growth-like changes occur with LTP. Do those disappear as LTP disappears?

DR. CAIN: Well, there are much better writings than mine because you could go to the people who did the studies. I was just citing them. To my knowledge nobody has followed them for very long. I may not remember the work very well off the top here, but it seems to me that LTP typically is not followed for very long and in the particular studies that you refer to I don't think the effect was followed long enough for decay to likely have occurred. It would be nice to see a study like that.

DR. MCNAMARA: I have no axe to grind one way or the other in this argument, OK? I don't care how it turns out, OK? But the thing is, though, it seems to me that all the arguments you have presented, the hypothesis that LTP is necessary but not sufficient for kindling development remains possible. But furthermore, implicit in the argument that I'm advancing is that in order to kindle an animal one must induce abnormal neuronal excitability in multiple sites within the forebrain, not just one, but multiple. And you're just looking at one. Furthermore, when you induce an after-discharge by stimulating the perforant path and you're recording LTP in one place there, in fact you're antidromically activating the entorhinal neurons which are activating the pyriform cortex neurons, and you're getting LTP there. OK? I'm saying hypothetically it could happen.

DR. CAIN: I agree with that. You can make that argument very well. What I would claim for these studies is that they made the obvious tests, the obvious simple-minded tests that occur to me. But sure, I think may be you can talk about more complicated situations, and may be save the theory. I meant to say, but forgot to say during the talk, that I think that LTP probably does contribute to most instances, if not all instances of kindling. But I disagree that it's mandatory. I don't think that it's crucial. I don't think it explains kindling I don't think it has a great deal of overlap in the mechanism of kindling. I think what it might do is aid the early dissemination of after-discharge activity very well. That's probably an important thing. But not a necessary thing.

DR. MCNAMARA: Yes, I'd argue that it's obviously not sufficient, and Ron demonstrated that years ago. But it seems to me that you still haven't excluded the possibility that it's necessary. These experiments have not excluded the possibility that it's necessary.

DR. CAIN: Can you think of an experiment that would answer that?

DR. MCNAMARA: No. You know I don't know how to do the experiment because it would require doing input output curves of all of these different pathways, and you might say, OK, Mac, you've got a ridiculous hypothesis because it's not testable.

DR. CAIN: That's a good answer to your question.

DR. MCNAMARA: That wasn't kind.

DR. GILBERT: I think I disagree with what you're saying, because the data that I just told you about would argue against what you said in that if you blocked LTP in one excitatory synaptic area, and you've given a systemic injection of a drug that you know ought to block it, and any NMDA, there is no reason to believe that it shouldn't block it in other NMDA related synaptic areas, you'd have to then implicate an NMDA related LTP in order to say that it's not necessary for kindling to occur.

DR. MCNAMARA: And you know that mossy fiber activation of C2 pyramidal cells are non-NMDA dependent LTP, so your argument can be countered.

DR. GILBERT: Then, let me say, an NMDA-related LTP.

DR. MCNAMARA: I would buy that. That's what I think your data support.

AMYGDALA VERSUS LOCAL ANESTHETIC KINDLING:

DIFFERENTIAL ANATOMY, PHARMACOLOGY, AND CLINICAL IMPLICATIONS

Robert M. Post, Susan R.B. Weiss, Mike Clark,
Takashi Nakajima and Agu Pert

Biological Psychiatry Branch, NIMH
Bldg. 10, Room 3N212
9000 Rockville Pike
Bethesda, Maryland 20892

## INTRODUCTION

### Electrical versus Pharmacological Kindling

Electrical kindling, first described by Goddard and associates (14), was elicited by repeated intermittent electrical stimulation of a variety of brain areas culminating in the production of major motor seizures to a previously subconvulsant stimulation. This process is accompanied by growth and spread of afterdischarges and a succession of behavioral seizure stages progressing to the onset of major motor seizures (14,38). With sufficient repetition of the kindled seizures, a stage of spontaneity occurs in which exogenous electrophysiological stimulation is no longer required in order for the animal to experience a seizure (20-22,43).

Tatum and Seevers (42) and Downs and Eddy (10) had originally observed that rats and dogs administered repeated, intermittent doses of cocaine would eventually experience seizures following a previously subconvulsant dose. We replicated these observations in rats and monkeys and were impressed that the time-course, behavioral and electrophysiological evolution, and long-lasting consequences for neural excitability observed with the drug-induced seizure process paralleled that of electrophysiological kindling. We, thus, labeled this process pharmacological kindling (25,30-32). Subsequently, a variety of substances administered parenterally and intracerebroventricularly (i.c.v.) or intracerebrally (i.c.) have been demonstrated to produce a kindling-like seizure evolution, in some instances showing cross-sensitization to amygdala kindling (4). Ehlers et al (12) and Weiss et al (47) have documented that i.c.v. administration of corticotropin-releasing factor (CRF) produced a late onset of seizures resembling those observed with amygdala kindling behaviorally and electrographically, but in this instance, repeated daily administration results in tolerance to the occurrence of seizures rather than sensitization (47). Thus, there is some question as to whether this represents a typical kindling paradigm; nonetheless, CRF seizures produce cross-sensitization to amygdala-kindled seizures (47). A variety of substances, including cocaine and lidocaine, appear to follow a more classical kindling-like time course.

The current manuscript focuses on one specific type of pharmacological kindling; i.e., that associated with local anesthetic administration.

*Kindling 4*
Edited by J. A. Wada
Plenum Press, New York, 1990

Electrical kindling of the amygdala and local anesthetic kindling achieved with lidocaine and cocaine are compared and contrasted (Table 1). While the emphasis is on the differential efficacy of pharmacological interventions in these two types of kindling, anatomical, physiological, and behavioral data are also briefly considered. A striking double dissociation is observed in that the same anticonvulsant, carbamazepine, which is potent in blocking completed amygdala-kindled seizures but not their development, is highly effective in blocking the development of local-anesthetic seizures but not their completed variety. These data suggest that although there are parallels in the phenomenology and time course of electrical and pharmacological kindling, and even cross-sensitization between the two, there are biochemical and, perhaps, anatomical differences. Moreover, these data, in addition to those of Pinel (22), highlight the principle that different stages in kindling evolution (i.e., development, completed, and spontaneous) are clearly differentially amenable to pharmacological intervention.

## RESULTS AND DISCUSSION

### Local-Anesthetic Induced Pharmacological Kindling

Cocaine is a complex drug with dual pharmacological properties. It is both a psychomotor stimulant and a local anesthetic. Its psychomotor stimulant components appear capable of inducing behavioral sensitization similar to that achieved by the psychomotor stimulant amphetamine or the direct dopamine agonist apomorphine (28). Its seizure-inducing properties, on the contrary, are likely related to its local-anesthetic effects. This assertion is largely based on indirect data that other local anesthetics such as procaine or lidocaine, that are essentially devoid of effects on catecholamine reuptake and the associated decreases in firing rates of monoamine neurons (24), are also associated with kindling-like seizure evolution.

Animals administered lidocaine (65 mg/kg i.p., once daily, five times/week) generally do not experience seizures following the first few injections but demonstrate a progressive emergence of seizures in some 40-50% of animals by the end of three weeks (29,32,35). These seizures are generally well tolerated and animals experiencing many weeks of lidocaine seizures have been observed to show spontaneous seizures (Post & Contel, unpublished observations). Cocaine is equally potent as a local anesthetic to lidocaine on a mg/mg basis. However, cocaine (65 mg/kg i.p.) produces seizures in approximately 50% of animals on day one and, usually, the animals succumb to cocaine-related seizures following the first or second convulsion. Elsewhere, we have discussed the possibility that in man convulsive and subconvulsive behavioral consequences of cocaine administration, such as panic attacks, may proceed through parallel stages of kindling-like progression (36,37).

### Glucose-utilization in amygdala- and local anesthetic-kindling

Thus, as summarized in Table 1, repeated intermittent administration of the local anesthetics appears to produce a seizure progression that is temporally and behaviorally similar to that observed with electrical kindling of the amygdala. In addition, there is transfer between lidocaine-induced kindled seizures and the amygdala-kindled variety. That is, animals repeatedly experiencing lidocaine-induced seizures demonstrate a significantly more rapid onset of amygdala-kindled seizures than saline-treated controls (29). Previous studies of regional glucose utilization in restrained animals treated with deoxyglucose i.v. indicated prominent amygdala and hippocampal glucose utilization during lidocaine seizures in some animals although, in others, these areas were not increased and there appeared to be increases in perirhinal cortex instead (29). These differ-

Table 1. Comparison of amygdala and local anesthetic kindling

| | Amygdala Kindling | Local Anesthetic Kindling |
|---|---|---|
| Intermittent subconvulsant stimulus used | + | + |
| Seizures induced upon repetition | + | + |
| Afterdischarge required | + | ? |
| Long-lasting change in excitability | + | + |
| Seizures include clonic movements of head, trunk, forepaws with rearing and falling | + | + |
| Spontaneity eventually observed | + | + |
| Diazepam blocks development and completed | + | + |
| Carbamazepine blocks development | 0 | + |
| Carbamazepine blocks completed seizures | + | 0 |

ences may be related to whether the deoxyglucose is reflecting predominately ictal versus inter-ictal events, although this remains to be adequately delineated.

Recent data (Agu Pert et al, unpublished observations) suggest that cocaine-related seizures (in unrestrained animals kindled with repeated daily doses of 40 mg/kg i.p.) are not associated with a pattern of marked amygdala or dorsal hippocampal increases in glucose utilization like that seen with lidocaine. Rather, compared with acute (40 mg/kg x 1 day) or chronic cocaine (40 mg/day x 15 days without seizures), those experiencing a cocaine-kindled seizure at this dose show increased metabolic activity in the nucleus accumbens, ventral pallidum, olfactory tubercle, anterior olfactory nucleus, subthalamic nucleus, substantia nigra, dorso- and ventro-medial nucleus of the hypothalamus, ventral hippocampus, and several brain stem reticular nuclei. In amygdala-kindled animals, Engel et al (13) and Caldecott-Hazard & Engel (5) had reported increases in glucose utilization in substantia nigra, hippocampus, neocortex, anterior and middle thalamus and globus pallidus in rats showing stage 5 seizures. In stage 1 and stage 2 seizures, ipsilateral amygdala and piriform cortex were invariably involved and hippocampus was involved bilaterally in most, but not all animals. Thus, there would appear to be some areas of convergence in the picture of regional glucose utilization in amygdala-kindled compared with local-anesthetic-kindled seizures (especially hippocampal involvement with lidocaine and substantia nigra with cocaine). Some notable differences are also apparent (i.e., the lack of involvement of amygdala and dorsal hippocampus with cocaine, and dramatic increases in nucleus accumbens, ventral pallidum, and other ventral forebrain structures).

C-fos induction in amygdala- and local-anesthetic-kindled seizures

Recently, another technique has been utilized for mapping potential neuronal structures activated during increases in synaptic activity. *In situ* hybridization of mRNA for the proto-oncogene c-fos has proven to be an interesting marker for processes activated by synaptic transmission, calcium influx, or second messenger changes such as those induced by dibutyryl-cyclic AMP or phorbol esters (which activate protein kinase-C). In contrast to glucose utilization, which is thought to largely involve terminal areas of the neuron, c-fos induction reflects processes at the level of the cell body and nucleus (8,11,18). Nakajima and colleagues (19) and Daval et al (9) have recently used this technique to map different seizure types. Preliminary evidence indicates that different types of seizures activate different neuronal pathways. Particularly in the mouse, electroconvulsive seizures have been demonstrated to increase c-fos induction in a wide variety of limbic and cortical sites, including amygdala and hippocampus (most prominently dentate gyrus), piriform cortex, septum, ventromedial nucleus of the hypothalamus, several brainstem areas, and the granular layer of the cerebellum (9). In contrast, caffeine-induced seizures produce marked increases in c-fos induction in the striatum and olfactory tubercle with little evidence of induction in other areas of brain (19). These changes in c-fos induction may be associated with adenosine $A_2$ receptors as they are blocked by the $A_1$ and $A_2$ agonist NECA but not by preferential $A_1$ agonist CHA. Moreover, they are not inhibited by carbamazepine, which has recently been shown to act selectively at $A_1$ but not $A_2$ receptors (6).

Amygdala-kindled afterdischarges produce increases in c-fos induction in piriform, occipital, and entorhinal cortices prior to activation of the hippocampus on the stimulated side, especially the dentate gyrus (fig. 1, M. Clark, unpublished observation). While one afterdischarge associated with a stage-1 seizure may not produce marked increases in c-fos induction in the dentate gyrus, it appears that either a greater number of stimulations and/or longer afterdischarge durations may be sufficient to induce c-fos in this area. Induction is unilateral first, and then bilateral. While generalized (stage 3, 4, and 5) seizures are usually associated with bilateral c-fos induction in the dentate gyrus, some seizures are without effect in this area (while the piriform, occipital, and entorhinal cortices are again affected), and the precise relationships of kindling evolution to c-fos induction remains to be delineated. The early involvement of piriform, occipital, and entorhinal cortices suggests the possible importance of these structures in the initial stages of afterdischarge spread and kindling evolution. The later, first unilateral then bilateral, induction of c-fos in the hippocampus (especially the dentate gyrus) suggests the possible involvement of this structure in processes leading to seizure generalization. In this regard, the metabolic map achieved with in situ hybridization of c-fos mRNA appears partially convergent with that achieved by deoxyglucose (13); i.e., the early involvement of piriform cortex and later involvement of hippocampus.

In addition to being an alternative marker of neuronal activity and a novel mapping technique, c-fos may also reflect processes importantly involved in mechanisms underlying the long-term changes in neural excitability and plasticity which are involved in the kindling process. This possibility has been suggested by a variety of workers (8,11,18), as c-fos induction has been thought to play a role in processes of neural differentiation, synaptic plasticity, and, possibly, learning and memory. In the lymphocyte, Crabtree (7) has indicated that c-fos is one of several immediate oncogenes induced (15-30 min.) in the activation and commitment of lymphocyte development and is followed by a succession of cascading processes of early or intermediate time frames (30 min. to 48 hours) as well as late oncogenes and related factors induced with a greater lag time (2 to 14 days). Our data (fig. 1) suggest a similar spatio-temporal cascade may occur with kindling evolution with c-fos being only one of many factors associated with changes in synaptic activity and function and, ultimately, even in structure (41).

The pattern of c-fos activation achieved with the local anesthetics lidocaine and cocaine shows areas of common involvement as well as some differences from that associated with electrical kindling of the amygdala.

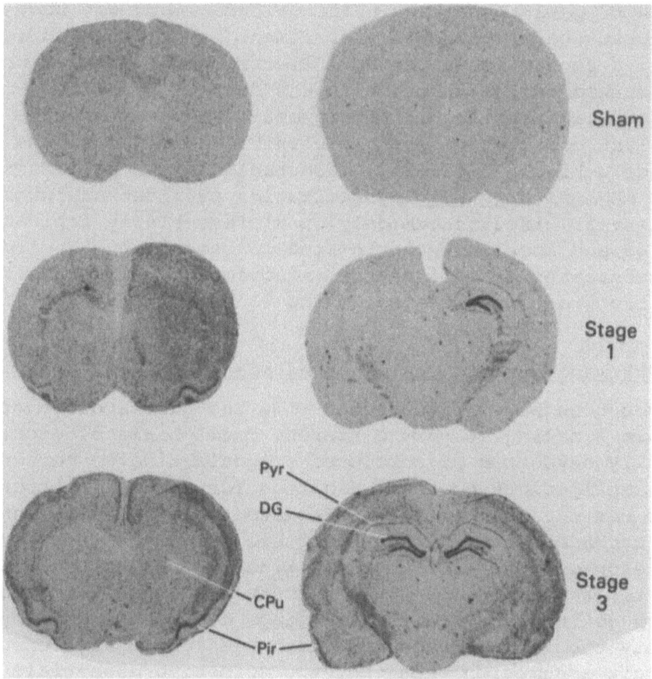


Fig. 1. Expression of c-fos mRNA in amygdala-kindled rats. *In situ* hybridization was performed with Hind III - Bam H1 4.8 kb fragment of mouse c-fos DNA labeled with $^{35}S$ (spec. act. 5 X $10^5$ dpm/ng; Lofstrand Labs, Gaithersburg, MD. Coronal sections are at the level of the striatum (left column) and the hippocampus (right column). Sections in the top row are from a rat (with an electrode implanted) that received the same handling as treated animals but received no electrical stimulation (sham). The middle left section (at the level of the striatum) is from a rat that was stimulated one day and had a long duration of afterdischarge (Stage 1, left); note unilateral induction in piriform cortex. The middle right section (at the level of the hippocampus) is from a rat that was stimulated 2 days with a long duration of afterdischarge (Stage 1, right); note unilateral hippocampal induction of c-fos, especially in the dentate gyrus. Both sections in the bottom row are from a rat that was stimulated several days with long duration of afterdischarge (Stage 3); piriform and hippocampal changes have become bilateral. Abbreviations: Pyr, pyramidal cells of hippocampus; DG, dentate gyrus; CPu, caudate putamen; Pir, piriform cortex)

In the rat and mouse, lidocaine-kindled seizures prominently activate the dentate gyrus of hippocampus (fig. 2), again potentially implicating this structure in either the seizure itself or the process of kindling evolution. In addition, local anesthetics increase c-fos in striatum in the mouse; i.e., similar to caffeine, and this combined pattern is distinctly different from that seen in amygdala-kindled seizures. CRF (100 μg administered i.c.v.) increased c-fos induction markedly in piriform cortex unassociated with a seizure, but the dentate gyrus is again involved when a CRF-induced seizure is produced (M. Clark, unpublished observations). The common activation of the dentate gyrus by a variety of seizure types as well as its activation by repeated and/or prolonged afterdischarges associated with stage-1 amygdala-kindled seizures, suggest that it may be an important substrate of neuronal recruitment in the kindled seizure process. However, a caveat is warranted. Electroconvulsive seizures, as well, produce marked c-fos induction in the dentate gyrus and, rather than being associated with kindling-like progression, appear capable of producing anticonvulsant effects and inhibit both the development and completed phases of amygdala kindling (33,34). Nonetheless, the preliminary "mapping" of c-fos mRNA with in situ hybridization provides an interesting anatomy of potential circuits commonly or differentially activated by amygdala-kindled and local-anesthetic-kindled processes. The role of these neural substrates as well as the induction of c-fos itself in critical phases of the kindling process remains to be further elucidated.

## Differential Pharmacology of Local Anesthetics vs. Amygdala Kindling

While it has been widely recognized that the pharmacology of development of amygdala kindling is very different from pharmacological inhibition of the fully developed or completed amygdala-kindled seizure (1,38), the potential implications of these findings for clinical use of anticonvulsant agents has not been widely appreciated. In the rat, anticonvulsants such as carbamazepine and phenytoin are clearly ineffective in inhibiting the development phase of amygdala kindling, in contrast to ethosuximide, valproate, acetazolamide, phenobarbital, clonazepam, clobazam, and diazepam, which do inhibit this initial stage of seizure development (1,39,44). Yet, the former drugs are often considered anticonvulsants with the implicit assumption that they are effective essentially irrespective of phase of development. This is highlighted in a series of clinical studies that have used phenytoin as an agent to treat post-traumatic epilepsy (52,53 ) with only mixed results. Borromei et al (3), who used phenobarbital in combination with phenytoin, reported effectiveness. To the extent that amygdala kindling is a useful animal model for the development of post-traumatic epilepsy, the data would suggest that lack of efficacy may be related to choice of drugs and that phenytoin may not be the best candidate for blocking the development of an epileptogenic process.

In an elegant study, Pinel (22) has revealed that the late or spontaneous phase of kindled seizures appears to have a differential pharmacology from earlier phases. For example, he found that diazepam, which was highly efficacious in inhibiting the development and completed phases, was not effective in the late spontaneous stage. Conversely, phenytoin, which is largely ineffective or only weakly effective in the developing and completed stages of kindling, was highly effective in suppressing spontaneous seizures. This remarkable dissociation clearly illustrates the changing pharmacological sensitivity and responsivity of amygdala kindling as a function of seizure stage.

Our work with carbamazepine reveals another principle -- that this differential pharmacology as a function of seizure stages can differ across different kindling paradigms. That is, carbamazepine is ineffective in the developing stage of amygdala-kindled seizures, irrespective of dose and mode of drug administration (39,44), while it is highly effective if not the **most** effective anticonvulsant in inhibiting the completed kindled seizure (2). These data are illustrated on the top of figure 3. In marked contrast, carbamazepine is highly effective in blocking the development of lidocaine- and cocaine-kindled seizures, but is ineffective in blocking either completed lidocaine-kindled seizures or high-dose, acute cocaine seizures (fig. 3, middle and bottom). Thus, in another type of double dissociation, the same anticonvulsant is effective in different stages of kindling, depending on kindling type (amygdala versus local anesthetic).

Work with carbamazepine also illustrates another phenomenon; that is, that pharmacoresponsivity appears to differ according to pattern of drug administration. While chronic oral carbamazepine administered in the diet is highly effective in blocking the development of both lidocaine- and cocaine-kindled seizures, repeated intermittent administration of carbamazepine i.p. (in a variety of doses) is completely ineffective in blocking the development of local anesthetic kindling (50). These data are illustrated in fig. 3 (middle and bottom left). When high doses of carbamazepine are administered i.p. (50 mg/kg instead of 15 mg/kg) and given one hour rather than 15 minutes prior to local anesthetic administration, it is not only ineffective, but actually exacerbates the seizure-inducing effects and associated lethality of cocaine. Thus, preliminary data would suggest that chronic rather than repeated intermittent carbamazepine is required to block the development of lidocaine- and cocaine-kindled seizures. They further suggest that the pattern of drug administration and, therefore, chronic rather than intermittent occupancy of appropriate receptor systems and the attendant differential adaptive processes induced, are critical to determining pharmacological responsivity, at least for local-anesthetic kindling development.

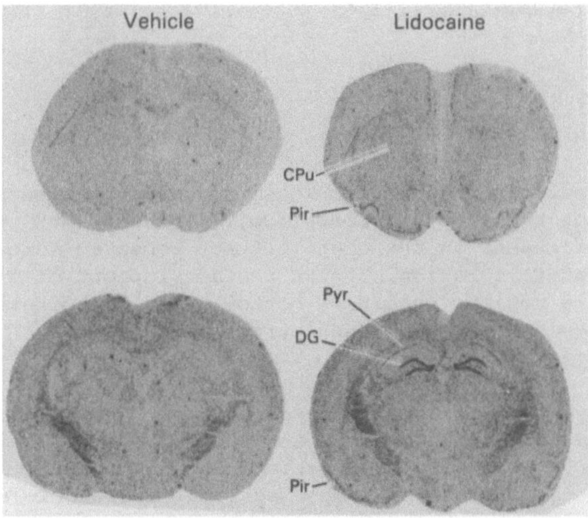


Fig. 2. Expression of c-fos mRNA in the rat in response to a single convulsive dose of lidocaine (right column) or vehicle (left column). Abbreviations: refer to legend of Fig. 1.

Taken together, these data suggest that different biochemical and/or anatomical substrates are involved in different phases of amygdala and lidocaine kindling and that chronic vs. acute intermittent drug administration differentially impacts on these and related neurochemical systems. These data are of considerable importance in conceptualizing kindling as evolving stages in a process rather than as a static event or lesion. They are also important in considering what mechanisms of an anticonvulsant such as carbamazepine underlie differential effects on different seizure types, as a function of both type and stage (26,27). It is likely

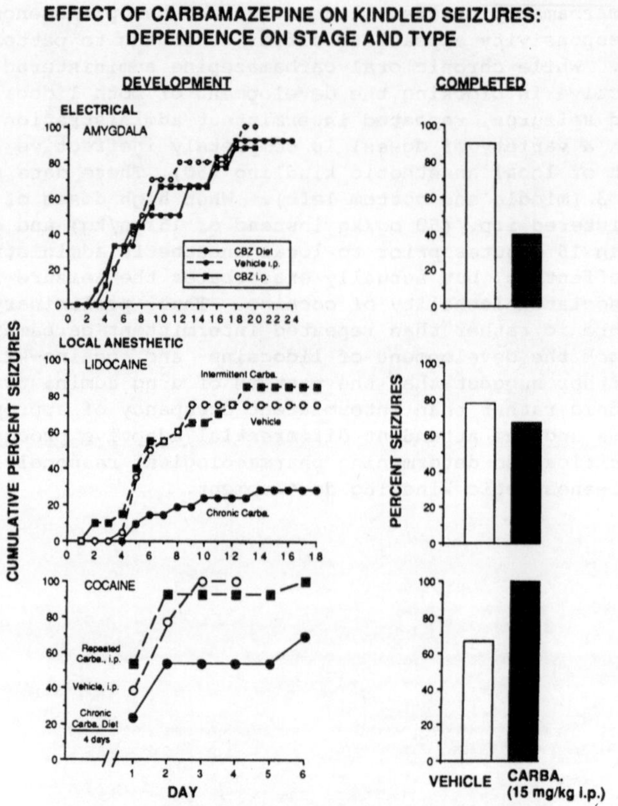


Fig. 3. Carbamazepine blocks completed amygdala-kindled seizures, but not their development (top). In contrast, chronic carbamazepine in the diet (but not repeated intermittent i.p. administration, blocks the development of lidocaine- (middle row) or cocaine- (bottom) kindled seizures, but not completed or high-dose seizures.

that peripheral-type benzodiazepine receptors and alpha-2 adrenergic receptors are involved in carbamazepine's action on completed amygdala-kindled seizures (48,49), but these systems are not sufficient to block amygdala kindling development or completed local anesthetic seizures. The relevance of these systems to carbamazepine's actions on the development of cocaine-kindled seizures, as well as the possible role of common interactions of carbamazepine and local anesthetics at batrachotoxin-sensitive sodium channels (type II) (17,40,51), remains to be explored.

To the extent that the findings that chronic carbamazepine inhibits the development of cocaine kindling in the rat can be extrapolated to other species including man, the current findings emphasize a number of important cautionary notes. They may also be important in relationship to clinical trials of carbamazepine for blocking cocaine-related toxicities and craving. Acute doses of carbamazepine may be ineffective. Controlled clinical trials of chronic carbamazepine would appear indicated in cocaine addicts, based on our preclinical data indicating that many of the physiological consequences of repeated cocaine administration in the kindling paradigm would be blocked with this method (50). In addition, Halikas et al (15) have reported preliminary data that chronic carbamazepine may have assisted chronic cocaine addicts in remaining drug free in comparison with other groups who were not able to comply with chronic carbamazepine administration. Those on chronic treatment also reported a decrease in craving. Clearly, double-blind, randomized clinical trials remain to be conducted in order to confirm these preliminary observations and disentangle possible confounding variables, such as increased motivation for compliance and maintaining a clinically prescribed regimen (i.e., taking chronic carbamazepine), from the effect of this drug on cocaine-induced craving, which was reported subjectively decreased. Moreover, in one subject maintained on carbamazepine, marked cocaine dose escalation was attempted in order to achieve a better "high". This patient had repeatedly experienced cocaine-related seizures, but while on chronic carbamazepine, seizures were not observed (Halikas, personal communication, 1988), again suggesting the possibility that the observation of protection against cocaine-related seizures by carbamazepine in the rat may be pertinent to the human situation. Whether this patient or others who have already reached the stage of "completed" seizure development would eventually enter a phase that was refractory to carbamazepine, as in the preclinical model, remains to be investigated.

Contingent Tolerance to anticonvulsants in amygdala- and local anesthetic-kindled seizures

Repeated administration of carbamazepine (15 mg/kg, i.p.) prior to each once-daily amygdala-kindled stimulation over a relatively brief period of time (3-7 days) results in tolerance to the anticonvulsant effects of carbamazepine (45,46). This phenomenon is labeled contingent tolerance as it does not develop if animals are given carbamazepine following each amygdala-kindled seizure, and is, in fact, reversed by a period of seven or five days of carbamazepine administration following the kindled seizure or by seven days of kindling without any drug administration at all. Tolerance to carbamazepine is not reversed by merely waiting for a period of seven days, or, in a second replication study, 23 days without either kindled stimulation or drug administration. Moreover, administering carbamazepine alone without amygdala-kindled seizures is also insufficient to reverse the contingent tolerance. These data suggest that it is the occurrence of amygdala-kindled seizures in the absence of drug (either no drug or drug administered after the seizure) that is the critical variable in reversing the phenomenon of contingent tolerance. This suggests that a specific unpairing of drug administration and seizure state is required in order to break this pattern of drug-induced refractoriness.

Weiss and Post (46) have explored a series of variables in attempts to discern factors critical to slowing the development of tolerance. They demonstrated that prior exposure to carbamazepine in a non-contingent fashion (i.e., carbamazepine administered after kindled seizures) does not lead to more rapid development of tolerance compared with naive animals, further supporting the view that pharmacokinetic variables are not involved in this phenomenon. Using high doses of carbamazepine (25 mg/kg i.p. rather than 15 mg/kg) not only did not slow development of tolerance but led to its nonsignificantly faster appearance. Alternating carbamazepine with diazepam on a daily basis also did not slow tolerance development, although animals showed lack of cross-tolerance to the central-type benzodiazepine anticonvulsant diazepam. In contrast, they did show cross-tolerance to an anticonvulsant presumably working through the peripheral-type benzodiazepine receptor in brain, PK-11195 (40 mg/kg i.p.). These data support previous observations linking the acute anticonvulsant effects of carbamazepine to peripheral-type benzodiazepine receptor mechanisms (48,49). Chronic oral administration of carbamazepine in the diet did slow the development of tolerance to the once-daily injections of carbamazepine i.p., suggesting that the chronic presence of drug, rather than its precisely paired administration immediately prior to kindled seizures, may be helpful in decreasing the rate of tolerance development. This would also more closely parallel the clinical situation when patients are administered anticonvulsants chronically and not just contingently prior to a presumed time of seizure vulnerability.

In patients showing initial responsivity to an anticonvulsant but then developing emergence of seizures, the current preclinical analysis would suggest the possible utility of a period of time off medication or rotation to another anticonvulsant with a different mechanism of action in order to regain seizure control. Peter Engel (personal communication, 1987) has observed cases of treatment-refractory epileptic patients withdrawn from their medications during the evaluation for epilepsy surgery, who again responded when they were replaced on their same medications. These observations would be consistent with our analysis that a period of seizures in the absence of medication (which would have occurred during the medication-free evaluation) may be sufficient to renew efficacy (at least transiently) in cases of contingent tolerance.

Our data with carbamazepine closely parallel those originally reported for alcohol and diazepam by Pinel and Mana (23) and Mana et al (16). The observation of contingent tolerance with a variety of anticonvulsant drugs, including carbamazepine which is not clinically known for a high incidence of tolerance development, suggests the generality of these observations. Weiss and Post (45,46) have also reported a novel phenomenon of contingent inefficacy in which prior treatment with carbamazepine in the developing stage of amygdala kindling (when it is ineffective) renders animals unresponsive to subsequent carbamazepine administration during the completed stage of kindling (when animals are ordinarily highly responsive).

We have also explored the development of contingent tolerance to the anticonvulsant effects of diazepam in the model of pharmacological kindling with lidocaine. As carbamazepine is ineffective on completed lidocaine seizures, the development of tolerance to this agent could obviously not be studied. Repeated administration of diazepam (0.3 mg/kg i.p. on a once-daily basis), eventually resulted in the reemergence of lidocaine-kindled seizures (anticonvulsant tolerance). Parallel to the findings observed with alcohol, diazepam, and carbamazepine for amygdala-kindled seizures, tolerance was reversed by a period of administering diazepam after the lidocaine-kindled seizures. Again, these data suggest the generality of the contingent tolerance phenomena observed across a series of different anticonvulsant classes in several different types of kindled seizures with differing pharmaco-responsivity.

## CONCLUSIONS

While there is considerable parallelism in the phenomenology and physiology of the evolution of amygdala-kindled and lidocaine-kindled seizures, different biochemical pathways and neural substrates appear to be involved. This is most clearly delineated by the double dissociation in responsivity to the anticonvulsant effects of carbamazepine. Amygdala-kindled seizures in the rat do not respond to carbamazepine in the early development phase but do respond in the completed phase. The converse is true for lidocaine- and cocaine-kindled seizures, where the early development phase is inhibited by chronic carbamazepine, but this drug is without effect in completed phases or on acute high-dose seizures. These differences are further emphasized in mapping strategies where differences in glucose utilization and in induction of c-fos mRNA measured by <u>in situ</u> hybridization are clearly evident. Commonalities in the process as well as distinctions in pharmaco-responsivity may have important clinical implications. The data suggest that different anticonvulsants may be differentially effective in different seizure stages as well as different seizure subtypes. Contingent tolerance may be observed in both amygdala- and lidocaine-kindled seizures, suggesting the importance of temporal contingencies in the development of drug nonresponsiveness and its reversal. An understanding of these principles may yield novel approaches to anticonvulsant therapy as well as to interventions that may ultimately be useful in blocking cocaine-related seizures and their sequelae.

References


1. Albertson, T.E., Joy, R.M. and Stark, L.G., 1984, Carbamazepine. A pharmacological study in the kindling model of epilepsy. Neuropharmacology 23:1117.
2. Albright, P.S. and Burnham, W.M., 1980, Development of a new pharmacological seizure model: effects of anticonvulsants on cortical- and amygdala-kindled seizures in the rat. Epilepsia 21:681.
3. Borromei, A., Caramelli, R., Cipriani, G., Giancola, L.C., Guerra, L., and Lozito, A., 1987, Neurotraumatology and post-traumatic epilepsy. Prevention, treatment and long-term follow-up. Barbexaclone plus phenobarbital (maliasin) versus diphenylhydantoin, phenobarbital, primidone, carbamazepine. Minerva Med. 78:1687.
4. Cain, D.P., 1986, The transfer phenomenon in kindling, in "Kindling 3", J.A. Wada, ed. Raven Press, New York.
5 Caldecott-Hazard, S. and Engel, J. Jr., 1987, Limbic postictal events: anatomical substrates and opioid receptor involvement. Prog. Neuro-Psychopharmacol. Biol. Psychiatry 11:389.
6. Clark, M. and Post, R.M., 1989, Carbamazepine but not caffeine is highly selective for adenosine A1 binding sites. Eur. J. Pharmacol., in press.
7. Crabtree, G.R., 1989, Contingent genetic regulatory events in T lymphocyte activation. Science 243:355.
8. Curran, T. and Franza, B.R. Jr., 1988, Fos and Jun: the AP-1 connection. Cell 55:395.
9. Daval, J.-L., Nakajima, T., Gleiter, C.H., Post, R.M. and Marangos, P.J., 1989, Mouse brain c-fos mRNA distribution following a single electroconvulsive shock. J. Neurochem., 52:1954.
10. Downs, A.W. and Eddy, N.B., 1932, The effect of repeated doses of cocaine on the rat. J. Pharmacol. Exp. Ther., 46:199.
11. Dragunow, M. and Robertson, H.A., 1987, Kindling stimulation induces c-fos protein(s) in granule cells of the rat dentate gyrus. Nature 329:441.
12. Ehlers, C.L., Henriksen, S.J., Wang, M., Rivier, J., Vale, W. and Bloom, F.E., 1983, Corticotropin releasing factor produces increases in brain excitability and convulsive seizures in rats. Brain Res., 278:332.
13. Engel, J., Wolfson, L. and Brown, L., 1978, Anatomical correlates of electrical and behavioral events related to amygdala kindling. Ann. Neurol. 3:538.
14. Goddard, G.V., McIntyre, D.C. and Leech, C.K., 1969, A permanent change in brain function resulting from daily electrical stimulation. Exp. Neurol., 25:295.
15. Halikas, J., Kemp, K., Kuhn, K., Carlson, G. and Crea, F., 1989, Carbamazepine for cocaine addiction? [Letter] Lancet 1:623.
16. Mana, M.J., Pinel, J.P.J. and Kim, C.K., 1986, Contingent tolerance to diazepam's anticonvulsant effect on amygdaloid kindled seizures in the rat. Soc. for Neurosci. Abstracts 12:1564.
17. McLean, M.J. and Macdonald, R.L., 1986, Carbamazepine and 10,11-epoxycarbamazepine produce use- and voltage-dependent limitation of rapidly firing action potentials of mouse central neurons in cell culture. J. Pharmacol. Exp. Ther. 238:727.

18. Morgan, J.I., Cohen, D.R., Hempstead, J.L. and Curran, T., 1987, Mapping patterns of c-fos expression in the central nervous system after seizure. Science 237:192.
19. Nakajima, T., Daval, J.-L., Gleiter, C.H., Post, R.M. and Marangos, P.J., 1989, Adenosine modulation of caffeine-induced c-fos mRNA expression in mouse brain. Brain Res., in press.
20. Pinel, J.P.J. and Rovner, L.I., 1978, Experimental epileptogenesis: kind ling-induced epilepsy in rats. Exp. Neurol. 58:190.
21. Pinel, J.P.J., 1981, Kindling-induced experimental epilepsy in rats: cortical stimulation. Exp. Neurol. 72:559.
22. Pinel, J.P.J., 1983, Effects of diazepam and diphenylhydantoin on elicited and spontaneous seizures in kindled rats: a double dissociation. Pharmacol. Biochem. Behav. 18:61.
23. Pinel, J.P.J. and Mana, M.J., 1986, Kindled seizures and drug tolerance, in "Kindling 3," J.A. Wada, ed., Raven Press, New York.
24. Pitts, D.K. and Marwah, J., 1986, Electrophysiological effects of cocaine on central monoaminergic neurons. Eur. J. Pharmacol., 131:95.
25. Post, R.M., 1977, Progressive changes in behavior and seizures following chronic cocaine administration: relationship of kindling and psychosis, in "Advances in Behavioral Biology, Vol. 21, Cocaine and Other Stimulants," E.H. Ellinwood and M.M. Kilbey, eds., Plenum Press, New York.
26. Post, R.M., 1987, Mechanisms of action of carbamazepine and related anticonvulsants in affective illness, in "Psychopharmacology: A Generation of Progress," H. Meltzer and W.E. Bunney, Jr., eds. Raven Press, New York.
27. Post, R.M., 1988, Time course of clinical effects of carbamazepine: implications for mechanisms of action, J. Clin. Psychiatry 49:35.
28. Post, R.M. and Contel, N.R., 1983, Human and animal studies of cocaine: implications for development of behavioral pathology, in "Stimulants: Neurochemical, Behavioral, and Clinical Perspective," I. Creese, ed. Raven Press, New York.
29. Post, R.M., Kennedy, C., Shinohara, M., Squillace, K., Miyaoka, M., Suda, S., Ingvar, D.H. and Sokoloff, L., 1984, Metabolic and behavioral consequences of lidocaine-kindled seizures. Brain Res. 324:295.
30. Post, R.M. and Kopanda, R.T., 1976, Cocaine, kindling, and psychosis. Am. J. Psychiatry 133:627.
31. Post, R.M., Kopanda, R.T. and Black, K.E., 1976, Progressive effects of cocaine on behavior and central amine metabolism in rhesus monkeys: relationship to kindling and psychosis. Biol. Psychiatry 11:403.
32. Post, R.M., Kopanda, R.T. and Lee, A., 1975, Progressive behavioral changes during chronic lidocaine administration: relationship to kindling. Life Sci., 17:943.
33. Post, R.M., Putnam, F., Contel, N.R. and Goldman, B., 1984, Electroconvulsive seizures inhibit amygdala kindling: implications for mechanisms of action in affective illness. Epilepsia 25:234.
34. Post, R.M., Putnam, F., Uhde, T.W. and Weiss, S.R.B., 1986, ECT as an anticonvulsant: implications for its mechanism of action in affective illness. Ann. N.Y. Acad. Sci., 462:376.
35. Post, R.M., Rubinow, D.R. and Ballenger, J.C., 1984, Conditioning, sensitization, and kindling: implications for the course of affective illness, in "Neurobiology of Mood Disorders," R.M. Post and J.C. Ballenger, eds., Williams & Wilkins, Baltimore.
36. Post, R.M. and Weiss, S.R.B., 1988, Sensitization and kindling: implications for the evolution of psychiatric symptomatology, in "Sensitization of the Nervous System," P.W. Kalivas and C.D. Barnes, eds., Telford Press, Caldwell NJ.
37. Post, R.M., Weiss, S.R.B., Pert, A. and Uhde, T.W., 1987, Chronic cocaine administration: sensitization and kindling effects, in "Cocaine: Clinical and Biobehavioral Aspects," A. Raskin and S. Fisher, Eds., Oxford University Press, New York.

38. Racine, R., 1978, Kindling: the first decade. Neurosurgery 3:234.
39. Schmutz, M., Klebs, K. and Baltzer, V., 1988, Inhibition or enhancement of kindling evolution by antiepileptics. J. Neural Transm., 72:245.
40. Reith, M., Kim, S. and Lajtha, A., 1986, Structural requirements for cocaine congeners to interact with [$^3$H]batrachotoxinin A 20-alpha-benzoate binding sites in sodium channels in mouse brain synaptosomes. J. Biolog. Chem., 261:7300.
41. Sutula, T., Xiao-Xian, H., Cavazos, J. and Scott, G., 1988, Synaptic reorganization in the hippocampus induced by abnormal functional activity. Science 239:1147.
42. Tatum, A.L. and Seevers, M.H., 1929, Experimental cocaine addiction. J. Pharmacol. Exp. Ther. 36:401.
43. Wada, J.A., Sato, M. and Corcoran, M.E., 1974, Persistent seizure susceptibility and recurrent spontaneous seizures in kindled cats. Epilepsia 15:465.
44. Weiss, S.R.B. and Post, R.M., 1987, Carbamazepine and carbamazepine-10,11-epoxide inhibit amygdala kindled seizures in the rat but do not block their development. Clin. Neuropharmacol., 10:272.
45. Weiss, S.R.B. and Post, R.M., 1989, Development and reversal of conditioned inefficacy and tolerance to the anticonvulsant effects of carbamazepine. Submitted to Epilepsia, June 1989.
46. Weiss, S.R.B. and Post, R.M., 1989, Contingent tolerance to the anticonvulsant effects of carbamazepine, in "Carbamazepine: A Bridge Between Epilepsy and Psychiatry. Proceedings of the June 10, 1989 Conference in Milan, Italy", in press.
47. Weiss, S.R.B., Post, R.M., Gold, P.W., Chrousos, G., Sullivan, T.L., Walker, D. and Pert, A., 1986, CRF-induced seizures and behavior: interaction with amygdala kindling. Brain Res., 372:345.
48. Weiss, S.R.B., Post, R.M., Marangos, P.J. and Patel, J., 1986, "Peripheral-type" benzodiazepines: behavioral effects and relationship to the anticonvulsant effects of carbamazepine, in "Kindling 3," J. Wada, ed. Raven Press, New York.
49. Weiss, S.R.B., Post, R.M., Patel, J. and Marangos, P.J., 1985, Differential mediation of the anticonvulsant effects of carbamazepine and diazepam. Life Sci., 36:2413.
50. Weiss, S.R.B., Post, R.M., Szele, F., Woodward, R. and Nierenberg, J., 1989, Chronic carbamazepine inhibits the development of local-anesthetic seizures kindled by cocaine and lidocaine. Brain Res., in press.
51. Willow, M. and Catterall, W.A., 1982, Inhibition of binding of [$^3$H]batrachotoxinin A 20-alpha-benzoate to sodium channels by the anticonvulsant drugs diphenylhydantoin and carbamazepine. Mol. Pharmacol., 22:627.
52. Wohns, R.N. and Wyler, A.R., 1979, Prophylactic phenytoin in severe head injuries. J. Neurosurg. 51:507.
53. Young, B., Rapp, R.P., Norton, J.A., Haack, D. and Walsh, J.W., 1983, Failure of prophylactically administered phenytoin to prevent post-traumatic seizures in children. Childs Brain 10:185.

## Discussion of Dr. Post's Presentation

DR. PETERSON: Is the anticonvulsant action of CBZ against cocaine kindling a result of the accumulation of the 10,11 epoxide active metabolite?

DR. POST: Yes, we've thought about that and there's a possibility but when we gave 15 mg/kg of carbamazepine an hour before, we ought to be getting more than enough epoxide out of that to be effective because I think Albright and Burnham and Susan Weiss and we have shown that the epoxide is just a little less effective than the parent compound. So with those kinds of doses that should not be the issue. It's a possibility but it doesn't look like it's the case.

DR. PETERSON: OK. Are you going to start playing the alpha beta agonist antagonist games with cocaine to try to figure out if it is catecholamine reuptake inhibition that's causing the extra death?

DR. POST: Yes, that's been on our agenda for a while but we haven't the personnel to do that. It would be interesting to find out precisely what mechanism and neurotransmitter effects are associated with that increased lethality. Because obviously that's happening not only to our rats but it's happening to a lot of famous people in the sports.

DR. PETERSON: Right, but they're checking out further things, and you might try to keep the peripheral effects separate from the central and make sure that it's not a heart attack they're dying from.

DR. POST: It doesn't seem to be. Again with our famous assay technique, the rat's hearts seem to be beating after they're dead. So that's why I thought it was a respiratory arrest but we really haven't done the physiology. That's actually one of the things we'd like to do to find out precisely what the mechanism is and then what might be the process, because obviously it has interest for cocaine related lethality and also for the mechanism of perhaps sudden death. The cocaine story looks like there are all sorts of ways that you can have a lethal event including arrythmias or MIs or strokes or hyperthermic crises but this particular one seems to be a postictal lethality.

# THE ROLE OF THE HIPPOCAMPAL SYSTEM IN THE EPILEPTIC TRANSFERENCE PHENOMENON OF KINDLING

Motoi Okamoto*, Ritsuo Nakachi*, Kiyoshi Morimoto* and Mitsumoto Sato**

*:Department of Neuropsychiatry, Okayama University Medical School, 2-5-1 Shikatacho, Okayama, 700 Japan
**:Department of Psychiatry, Faculty of Medicine, Tohoku University

## INTRODUCTION

Transference phenomenon (transfer) refers to a phenomenon where electrical or chemical kindling at one brain site is accelerated when another brain site has been kindled. Transfer is observed at brain sites which have synaptic connection with the primary kindled site. It has been considered that repeated synaptic bombardment (such as afterdischarge) from the primary focus results in a secondary functional change demonstrated by an accelerated kindling rate. Therefore, the study on the neural mechanism of transfer is important to the understanding of the mechanism of development of secondary epileptogenic focus, as well as to add to the knowledge of the neurophysiological background of memory or neuronal plasticity.

The transfer of kindling has been demonstrated in experimental studies in rats[1,2], cats[3], rhesus monkeys[4], and epileptic baboons5. In these studies, it is most clearly seen in the bilateral amygdala of cats[3] and in the bilateral dorsal hippocampus of rats[2].

It has not been well understood which interhemispheric pathway plays an important role for the development of transfer. As a result of experimental studies in rats it was suggested that either the subcortical pathway[6,7] or the hippocampal commissure[2] is important. In an early experimental study in cats, the importance of the midbrain reticular formation has been suggested[8]. Subsequent studies showed that the corpus callosum[9], hippocampal commissure[10] or massa intermedia of the thalamus[11] may be important.

Recently, Nakatsu[12] of our laboratory demonstrated in cats that transfer to the contralateral amygdala was inhibited when the bilateral ventral hippocampus had been destroyed electrolytically prior to primary site amygdala kindling. Moreover, the transfer expected to be established during kindling disappeared when electrolytic destruction was done after primary site kindling. These findings indicate that the neuronal pathway including the ventral hippocampus, plays an important role in the development of transfer. The purpose of the present study is to investigate, in more detail, the hippocampal pathway responsible for the establishment of transfer.

*Kindling 4*
Edited by J. A. Wada
Plenum Press, New York, 1990

## THE EFFECT OF IPSILATERAL VENTRAL HIPPOCAMPAL LESION ON EPILEPTIC TRANSFER TO THE CONTRALATERAL AMYGDALA

### Subjects and Methods

Five adult cats weighing 2.5 Kg or more were used. All animals were housed in an animal room maintained at 22 to 27°C under 12 hr/12 hr. dark-bright cycle. Animals had free access to food and water except for the experimental period.

Under pentobarbital anesthesia (25mg/Kg, intravenously), the left ventral hippocampus was destroyed by 12 V of direct electric current for 5 minutes. After a one-month recovery period, the animals were implanted with chronic electrodes into the bilateral amygdala (AM), dorsal hippocampus, midbrain reticular formation (MRF), and entorhinal cortex under pentobarbital anesthesia. Following a recovery period of more than one week, all animals were kindled at the left AM. Kindling was induced once per day by delivering a one sec. train of 60 Hz square wave pulse, one msec of pulse duration, at the afterdischarge threshold (ADT). Stimulation was repeated until five stage 6 generalized seizures[13] appeared. After a one-week recovery period, the right AM was stimulated in the same manner as in the left AM, and the number of afterdischarges required for the appearance of first stage 6 generalized seizure was determined.

### Results

During primary site AM kindling, stage 6 seizure appeared after 10 to 31 (mean 20.2) stimulations. The development of kindled seizure was unstable, and frequently regressed to partial seizures in 2 cats, but the electrographic and behavioral pattern of generalized convulsion was almost symmetrical in all cats. In secondary site AM kindling, stage 6 seizure appeared at the third stimulation in one cat, while 20 or more stimulations were required in the remaining 4 cats (Table 1). In the cat with positive transfer, primary site kindling developed rapidly, although the cause is unknown.

Histological examination revealed extensive destruction of the dentate gyrus and the CA field in the 4 cats without transfer (Fig. 1). In particular, it is noteworthy that destruction of the mossy fibers of dentate granule cells to the the CA field was extensive and extended for more than 5mm in antero-posterior direction in all 4 cats. In contrast, only a part of the dentate gyrus was destroyed in the animal with positive transfer.

Table 1. The effect of ventral hippocampal lesion on epileptic transfer to the contralateral amygdala

| | cat No. | H-5 | H-11 | H-12 | H-16 | H-17 |
|---|---|---|---|---|---|---|
| primary site AM | ADT(uA) | 550 | 450 | 250 | 550 | 250 |
| | kindling rate | 17 | 18 | 25 | 31 | 10 |
| secondary site AM | ADT(uA) | 200 | 300 | 250 | 350 | 200 |
| | kindling rate | 40< | 21 | 45 | 34 | 3 |
| transfer | | - | - | - | - | + |

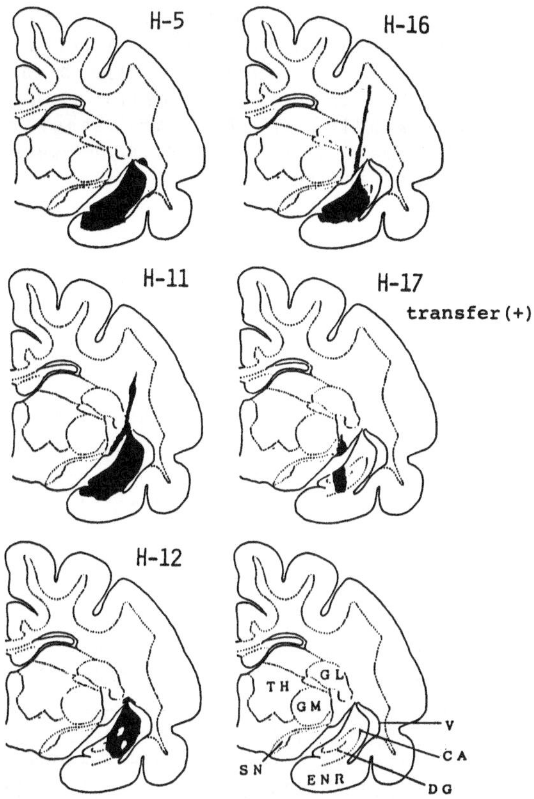


Fig.1 Schematic representation of destructive lesion in the ventral hippocampus
The shaded area shows the area with complete neuronal loss and gliosis. CA: pyramidal cell layer of the hippocampus, DG: dentate granule cell layer, V: lateral ventricle, ENR:entorhinal cortex, SN:substantia nigra, GL: lateral geniculate GM: medial geniculate, TH: thalamus

There was no difference in the ADT at the secondary site AM between the positive transfer cat and the negative transfer cats. Afterdischarge propagation to the contralateral AM during primary site kindling tended to be weak (low amplitude) in the negative transfer group, but there was no difference between the two groups in the total afterdischarge duration in the contralateral AM during primary site kindling. From these findings it is suggested that the pathway from the entorhinal cortex through the dentate granule cells and the mossy fibers to the CA field of the hippocampus plays an important role for the development of transfer.

## THE EFFECT OF SURGICAL BISECTION OF THE CORPUS CALLOSUM AND HIPPOCAMPAL COMMISSURE ON EPILEPTIC TRANSFER TO THE CONTALATERAL AMYGDALA

### Subjects and Methods

Six adult cats weighing 2.5Kg or more were used. Under pentobarbital anesthesia (25mg/Kg, intravenously), the scalp was opened and the anterior half of the corpus callosum or the posterior half of the corpus callosum including hippocampal commissure was bisected. After a 3-month recovery period, chronic electrodes were implanted into the bilateral AM, dorsal hippocampus, MRF and frontal cortex. Following a one-week recovery period, the left AM was kindled in the same manner as described in experiment I. Following primary site kindling, a one-week recovery period elapsed and the right AM was kindled.

## Results

Transfer was observed in 3 out of 6 cats, while in the remaining 3 cats development of transfer was prevented (Table 2). The development of behavioral seizure pattern during primary site kindling was unstable in 5 cats, but this was not related to the presence or absence of transfer. The electrographic and behavioral seizure patterns were largely asymmetrical and asynchronous in both primary and secondary site kindling, although stage 6 generalized seizure appeared in all 6 cats (Fig. 3). Histological examination unexpectedly demonstrated that the prevention of transfer was not related to the extent of surgical interruption of the

Table 2. The effect of commissural bisection on epileptic transfer to the contralateral amygdala

| | cat No. | R-8 | R-9 | R-20 | R-4 | R-7 | R-11 |
|---|---|---|---|---|---|---|---|
| primary site AM | ADT(uA) | 150 | 300 | 150 | 150 | 150 | 350 |
| | K.rate | 8 | 29 | 15 | 10 | 19 | 15 |
| secondary site AM | ADT(uA) | 250 | 150 | 250 | 150 | 100 | 150 |
| | K.rate | 14 | 20 | 12< | 2 | 3 | 3 |
| transfer | | - | - | - | + | + | + |

K.rate: kindling rate

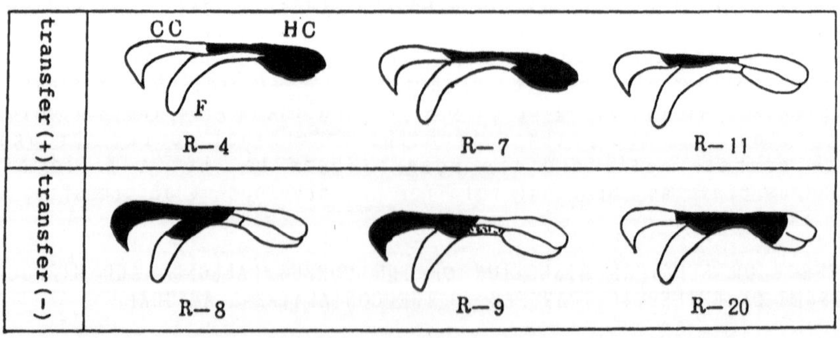


Fig.2 Schematic representation of the bisected or destructed area of the commissural pathway and fornix
Shaded area represent bisected or destructed area. The stippled area represent retrograde demyelination of fimbria fibers.
CC:corpus callosum, HC:hippocampal commissure, F:fornix

corpus callosum or the hippocampal commissure, particularly in the latter (Fig.2). The only definitive finding was that the fornix of the primary site hemisphere was destroyed in the 3 negative transfer cats, while no such destruction of the fornix was seen in the 3 positive transfer cats. Partial injury of the cingulate gyrus was observed in 3 animals, but this was not related to the presence or absence of transfer.

There were no differences in the ADT of the primary and secondary site AM, nor in the primary site kindling rate between the two groups. The propagated afterdischarge in the secondary site AM during primary site kindling was of low amplitude in those cats in which the anterior corpus callosum had been bisected. But there was no difference in total afterdischarge duration of secondary site AM during primary site kindling between the positive and negative transfer groups. These results show that hippocampal efferent via the fornix is more important than the corpus callosum and hippocampal commissure in the establishment of epileptic transfer to the contralateral AM. In order to confirm this, we examined the effect of destruction of the primary site fornix on transfer.

## THE EFFECT OF ELECTROLYTIC LESION OF THE IPSILATERAL FORNIX ON EPILEPTIC TRANSFER TO THE CONTRALATERAL AMYGDALA

### Subjects and Methods

Five adult cats weighing 2.5 Kg or more were used. Under pentobarbital anesthesia(25mg/Kg, intravenously), the left fornix (F:13.5mm, L:1.0mm, H:-13.0mm according to Jasper & Ajmone-Marsan[14]) was destroyed electrolytically (12V DC for 2 min.). After an approximate one-month recovery period, chronic electrodes were implanted into the AM, dorsal hippocampus, MRF, frontal cortex and anterior nuclei of the thalamus bilaterally under pentobarbital anesthesia. This was followed by a ono-wook or longor rocovery poriod boforo tho loft AM was kindled in the same manner as in experiment I. The right AM was kindled following a one-week recovery period after the completion of primary site kindling.

### Results

In primary site kindling, stage 6 generalized seizure appeared after 11 to 44 (mean 27) stimulations. The seizure was unstable in 3 cats, but the electrographic and behavioral pattern of generalized convulsion was symmetrical in all the animals. In secondary site kindling, stage 6 generalized seizure was induced at thesecond stimulation in 1 cat, while in the other 4 cats no transfer was seen (Table 3). Histological examination revealed various degrees of destruction of the callsal fiber, fimbria, fornix, stria terminalis, and midline thalamic structures.

Table 3. The effect of electrolytic lesion of fornix on epileptic transfer to the contralateral amygdala

| | cat No. | F-2 | F-3 | F-7 | F-8 | F-9 |
|---|---|---|---|---|---|---|
| primary site AM | ADT(uA) | 400 | 300 | 400 | 500 | 100 |
| | kindling rate | 11 | 44 | 25 | 18 | 36 |
| secondary site AM | ADT(uA) | 150 | 100 | 150 | 300 | 100 |
| | kindling rate | 14 | 45< | 23 | 11 | 2 |
| transfer | | - | - | - | - | + |

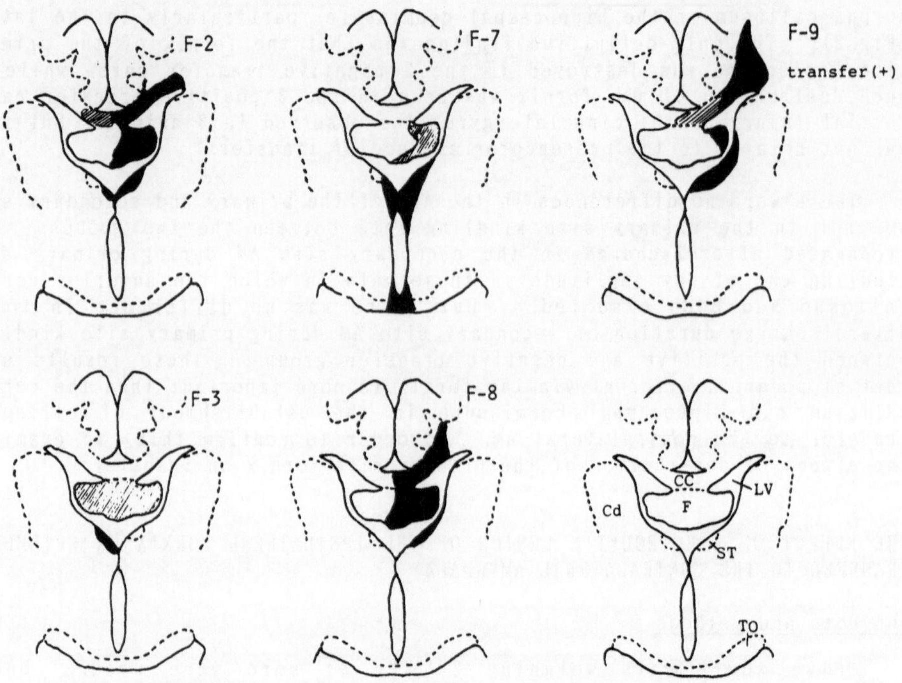


Fig. 3 Schematic representation of lesion of the fornix and adjacent brain areas
The shaded areas represent the areas with severe cell loss and gliosis. The hatching areas represent demyelination of fibers of the corpus callosum, fimbria, or fornix. CC:corpus callosum, F:fornix, Cd:caudate, ST:stria terminalis, TO:optic tract, LV:lateral ventricle

However, no consistent relation was seen between the destruction of brain areas other than the fornix and the presence or absence of transfer. However, in 4 cats with negative transfer, the left fornix was extensively destroyed and remarkable demyelination of the neuronal fibers of the left fornix and fimbria was noted. In contrast, in 1 cat with positive transfer, the left fornix showed localized injury in the dorsolateral area although the callosal fibers and stria terminalis were destroyed (Fig. 3).

The ADT in the primary site AM of 1 positive transfer cat was lower than in the 4 negative transfer cats, while there were no differences in primary site kindling rate and the ADT of the secondary site AM between the two groups. Moreover, no intergroup difference was seen in afterdischarge propagation to the secondary site AM and the total afterdischarge duration of the contralateral AM during primary site kindling. These results were considered indicating of the greater importance of the pathway via the fornix than the commissural pathway for the development of transfer among the hippocampal efferents.

DISCUSSION

The results of our study indicate that the neuronal pathway from the hippocampus through the fornix to the contralateral hemisphere plays an

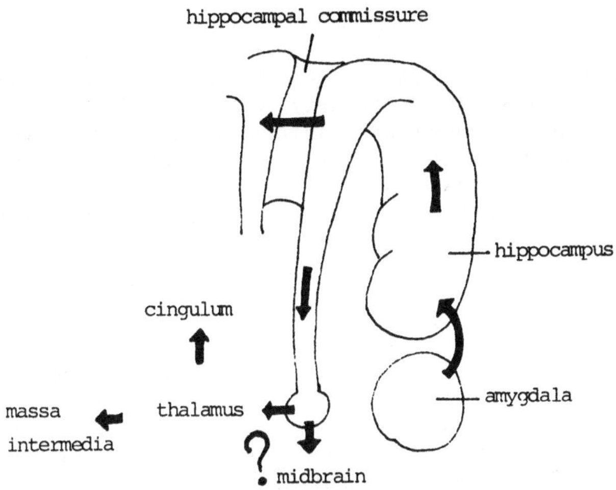


Fig. 4 Possible neuronal pathway related to development of epileptic transfer to the contralateral amygdala

important role in inducing the secondary functional change reflected as transfer in the contralateral AM. The hippocampal efferent via the fornix sends its fibers to the mammillary body, preoptic area and septum. Some of these fibers reach the anterior thalamic nucleus, and extend their synaptic connection to the cingulate gyrus. In addition, other fibers also project to the midbrain (Fig 4). In this study, we were unable to determine which of these pathways was most important for transfer. Wada and Sato[8] reported that they failed to find evidence of transfer after the electrolytic destruction of the MRF, suggesting that the pathway through the midbrain may be important. However, since their finding was obtained from only one cat, further studies are needed. The recent report of Hiyoshi and Wada[11] is more interesting. They found that transfer to the contralateral AM after unilateral AM kindling was inhibited by electrolytic destruction of the midline thalamic structures, including the massa intermedia. Considering their results, the pathway from the fornix through the thalamus may be important. However, further studies are required to elucidate in more detail the neuronal pathway responsible for the development of transfer.

Fukuda et al.[10] performed commissural bisection prior to primary site AM kindling and determined that bisection of the hippocampal commissure, but not of the anterior half of the corpus callosum, prevented the development of transfer in the contralateral AM. McIntyre and Stuckey[2] also observed that the hippocampal commissure was important for epileptic transfer to the contralateral hippocampus in dorsal hippocampal kindling in rats. Their results are not consistent with our findings. These two studies differ from ours in that closed kniff-cut commissural bisection was done without craniotomy and there was only a one-day interval between the primary site kindling and the secondary site kindling. Furthermore, the presence or absence of the fornix injury is not described in detail. However, the reason for this discrepancy remains unknown.

Heretofore it was considered that transfer is caused by propagation of the afterdischarge and that the total duration of the propagated afterdischarge during primary site kindling determines the kindling rate at the secondary brain site. However, our results showed that total afterdischarge duration in the secondary focus was not related to either the presence or the absence of transfer. This evidence indicates that the mechanism of transfer cannot be explained by the propagation of afterdischarge alone. For the propagation of afterdischarge between bilateral AM, the commissural pathway (anterior commissure and corpus callosum) appears to be important. This study shows that the hippocampal efferent via the fornix may not be important. Nonetheless, destruction of this pathway inhibited transfer. This finding indicates that the functional changes induced by transsynaptic input other than propagation of afterdischarge may play an important role in the establishment of transfer. In fact, McIntyre[7] reported that transfer can be established without afterdischarge propagation to the contralateral AM. In addition, it was demonstrated in an early experimental study[15] that the formation of a "mirror focus" can be inhibited by interrupting the subcortical pathway, even if propagation of epileptiform discharge via the commissural pathway occurs. To confirm these findings, it is necessary to determine whether or not the transference phenomenon but also other electrophysiological, microstructural, or biochemical changes in contralateral hemisphere are related to propagation of afterdischarge.

The studies in which they create surgical interruption of the interhemispheric pathway after primary site kindling reported different results depending on the animal species used. The epileptic transfer disappeared after surgical interruption in cats[9,16], and epileptic baboons[17]. In contrast, the rate of secondary site AM kindling was facilitated in rats[18]. There appears to be two reasons for this discrepancy. The first is the fact that as the animal ascend the phylogenetic scale, it takes longer to kindle[19] and to induce secondary epileptogenic change[20], indicating that transfer will take longer to occur in cats and baboons than in rats. The second reason, although speculative, is that transference phenomenon of kindling may be viewed as a developmental stage of secondary epileptogenesis[20]. The disappearance of epileptic transfer after surgical interruption of interhemispheric pathway in cats and baboon shows that the transference phenomenon to the contralateral AM is dependent on the primary site hemisphere. Transfer can be observed after the usual AM kindling to the contralateral AM in rats, cats and baboons. However, the secondary epileptogenic change in the contralateral AM is still in a dependent stage in cats and baboons, while it is in an independent stage in rats. This is the reason why transfer disappears following interruption of the interhemispheric pathway after primary site kindling in cats and baboons, but not in rats. Therefore, it seems better to consider that surgical interruption of interhemispheric pathway after primary site kindling disclose different aspects of secondary epileptogenesis from surgical interruption prior to primary site kindling.

The neuronal pathway indicated as important for transfer in our study is a part of the so called Papez's pathway[21]. This pathway is well known as playing an important role for memory. Destruction of the mammillary body and hypothalamus, or the hippocampus results in an enduring anterograde amnesia[22,23], indicating that the hippocampus and mammillary body play a critical role in the formation or consolidation of memory trace. As initially suggested by Goddard[24], it may be considered that the kindling phenomenon has a common neurophysiological mechanism with the formation of memory trace. If this is true, the transference phenomenon of kindling can be considered to have a common neurophysiological mechanism with the interhemispheric transfer of the memory trace. Therefore, the results of our study may indicate that the neuronal pathway which starts from the hippocampus and travels through the fornix to the mammillary body may play an important role for interheimispheric transfer of memory trace.

REFERENCES

1.McIntyre,D.C., and Goddard,G.V., Transfer, interference and spontaneous recovery of convulsions kindled from the rat amygdala, Electroenceph. Clin. Neurophysiol. 35:533 (1973)
2.McIntyre,D.C., and Stuckey,G.N., Dorsal hippocampal kindling and transfer in split-brain rats, Exp. Neurol. 87:86 (1985)
3.Wake,A. and Wada,J.A., Transfer and interference in amygdaloid kindling in cats, Canad. J. Neurol. Sci. 3:5 (1977)
4.Wada,J.A., Mizuguchi,T., and Osawa,T., Secondarily generalized convulsive seizures induced by daily amygdaloid stimulation in rhesus monkeys, Neurology 28:1026 (1978)
5.Wada,J.A., and Osawa,T., Spontaneous recurrent seizure state induced by daily electric amygdaloid stimulation in Senegalese baboons (papio papio), Neurology 26:273 (1976)
6.McCaughran,J.A., Corcoran,M.E. and Wada,J.A., Role of the forebrain commissures in amygdaloid kindling in rats, Epilepsia 19:19 (1978)
7.McIntyre,D.C., Split-brain rat : transfer and interference of kindled amygdala convulsions, Canad. J.Neurol. Sci. 2:429 (1975)
8.Wada,J.A. and Sato,M., The generalized comvulsive seizure state induced by daily electrical stimulation of the amygdala in split brain cats, Epilepsia 16:417 (1975)
9.Wada,J.A., Nakashima,T,, and Kaneko,Y., Forebrain bisection and feline amygdaloid kindling, Epilepsia 23:521 (1982)
10.Fukuda,H., Wada,J.A., Riche, D., and Naquet,r., Role of the corpus callosum and hippocampal commissure on transfer phenomenon in amygdala-kindled cats, Exp. Neurol., 98:189 (1987)
11.Hiyoshi,T., and Wada,J.A., Midline thalamic lesion and feline amygdaloid kindling : I.Effect of lesion placement prior to kindling, Electroenceph. Clin. Neurophysiol., 70:325 (1988)
12.Nakatsu,T., The effects of bilateral ventral hippocampal lesions on amygdaloid kindling in cats, Okayama Igakkai Zasshi 97:855 (1985)
13.Wada,J.A., and Sato,M., Generalized convulsive seizures induced by daily electrical stimulation of the amygdala in cats : correlative electrographic and behavioral features, Neurology 24:565 (1974)
14.Jasper,H.N., and Ajmone-Marsan,C., A stereotaxic atlas of the diencephalon of the cat, National Research Council of Canada, Ottawa (1954)
15.Morrell,F., Secondary epileptogenic lesions, Epilepsia 1:538 (1960)
16.Hiyoshi,T., and Wada,J.A., Midline thalamic lesion and feline amygdaloid kindling : II. Effect of lesion upon completion of primary site kindling, Electroenceph. Clin. Neurophysiol., 70:339 (1988)
17.Wada,J.A. and Komai,S., Effect of anterior two-third callosal bisection upon bisymmetrical and bisynchronous generalized convulsions kindled from amygdala in epileptic baboon, papio papio, in"Epilepsy and the Corpus Callosum" A.G.Reeves, ed., Plenum Press, New York (1985)
18.McCaughran,J.A., Corcoran,M.E., and Wada,J.A., Facilitation of secondary site amygdaloid kindling following bisection of the corpus callosum and hippocampal commissure in rats, Exp. Neurol. 57:132 (1977)
19.Wada,J.A., The clinical relevance of kindling : species, brain sites and seizure susceptibility, in:"Limbic Mechanisms", K.E.Livingston and O.Hornykiewicz, eds., Plenum Press, New York (1978)
20.Morrell,F., Secondary epileptogenesis in man, Arch. Neurol. 42:318 (1985)
21.Papez,J.W., A proposed mechanism of emotion, Arch. Neurol. Psychiat., 38:725 (1937)
22.Saunders, R.C., Impairment in recognition memory after mammillary body lesions in monkeys, Soc. Neurosci. Abstr., 9:28 (1983)
23.Zola-Morgan,S., Squire,L.R. and Amaral,D.G., Human amnesia and the mesial temporal region: enduring memory impairment following a bilateral

lesion limited to field CA1 of the hippocampus, J. Neurosci., 6:2950 (1986)
24. Goddard, G.V., McIntyre, D.C., and Leech, C.K., A permanent change in brain function resulting from daily electrical stimulation, Exp. Neurol., 25:295 (1969)

Discussion of Dr. Okamoto's Presentation

DR. MCINTYRE: I would be very interested to know what happened to your primary site after you made this lesion. Did the fornical lesion have any influence on the convulsive behaviour of your original primary site?

DR. OKAMOTO: After commissural bisection electrographic and behavioural seizure became largely asymmetrical. Fornical lesion and unilateral ventral hippocampal lesion had no obvious effect on the primary site kindled convulsion.

DR. WADA: Although a varying degree of positive transfer effect can be expected depending on the extent of primary site kindling, this absence of transfer after fornix lesion is very interesting. It is consistent with the results from studies in our laboratory and at Gif-sur-Yvette that what we regard as potential relevance to secondary epileptogenesis, this positive transfer effect, is in essence an accession to the primary site hemispheric kindled mechanism rather than a change in the opposite hemisphere, under usual primary and then secondary site kindling paradigm. I hope we will be discussing this issue again tomorrow morning.

# EROSION OF KINDLED EPILEPTOGENESIS AND KINDLING-INDUCED LONG-TERM SEIZURE SUPPRESSIVE EFFECT IN PRIMATES

Juhn A. Wada

Divisions of Neurosciences and Neurology
University of British Columbia
Vancouver, B.C. Canada V6T 2A1

## INTRODUCTION

Repeated electrical stimulation of the brain can result in the development of spontaneous recurrent seizures in dogs, cats and rats (10, 49, 51) or status epilepticus in dogs and cats (1, 49). In primates, Delgado (7) found that repeated electrical brain stimulation results in a lasting change in the EEG signature of induced seizure. Subsequently, the critical conditions necessary for progressive changes induced by repeated electrical brain stimulation were identified by Goddard who baptized it "the kindling phenomenon" (15, 16). His original studies were primarily in rodents, but he was the first to extend the observation to primates, with the specific intention of verifying the validity of the kindling phenomenon across the species. Although considerable difficulty was encountered in identifying the optimal parameters of stimulation, and despite considerable variation in the results obtained, he was able to conclude that unilateral or bilateral clonic seizure can be induced after six months of stimulation. Furthermore, once established, a rest interval of eight weeks did not weaken the kindled response in the rhesus monkey. In the meantime, spontaneous seizure development was being documented in nonepileptic patients who underwent either a series of ECTs (33), or prolonged intracerebral electrical stimulation (24, 35). All these findings strongly suggest that, in response to repeated electrical brain stimulation, the progressive development and acquisition of lasting seizure suceptibility which culminate in the emission of spontaneous seizures are the likely biological principles operating in all mammalian species, including man. Subsequent kindling studies in our laboratory in four primate species, Papio papio (PP) (45), Papio cynocephalus (PC) (5), Papio hamadryas (PH) (37), and rhesus monkeys (RM) (42) confirmed the development of spontaneous seizures, thus supporting the hypothesis that epilepsy, as defined by the spontaneous recurrent seizure state, can result from the kindling process. While the question of whether human partial epilepsy develops through a kindling-like mechanism remains open to debate, it is probably of academic interest only, since it is not possible to assess the chronological pattern of human seizure development as is done with animal subjects. Recognizing that kindling is a model of partial onset epilepsy, a more realistic question concerns the possible consequence of repeated partial onset medically-refractory seizures in man. In this regard, the significance of secondary epileptogenesis has been discussed extensively (18, 22, 28, 29, 30).

*Kindling 4*
Edited by J. A. Wada
Plenum Press, New York, 1990

The concept of secondary epileptogenesis is based primarily on electrophysiological observation; i.e., the presence of an actively discharging focus projecting epileptiform transients as evoked response to a contralateral homotopic site. As it persists, epileptiform transients independent of those of primary focus develop at the homotopic site, and eventually, under certain circumstances, EEG and behavioural seizures may emanate from this "mirror focus". The basic chronological profile of this electrophysiological development has been well established in lower animals such as frogs and rats (50). However, considerable controversy persists as to its validity in higher animals (18). For example, a study which carefully avoided potential technical problems in establishing secondary epileptogenesis in a large number of monkeys failed to confirm its development (21). Furthermore, the description of ictal behavioural events presumably originating from the mirror focus is sketchy due to obvious technical difficulties in documenting episodic behavioural events. Recently, the development of a mirror focus in man capable of generating spontaneous seizure has been reported (28, 29), emphasizing the potential hazard of uncontrolled repetition of partial seizures in man.

In the chronic kindling model, the anticipated progressive changes can be brought under precise experimental control. Therefore, kindling has an obvious advantage over the conventional model of secondary epileptogenesis, particularly for assessing both electrophysiological and behavioural chronological changes.

Kindling-induced sequential change in brain function can manifest in a number of different ways. The most obvious one is the development and subsequent dissemination of interictal spike discharge (IID). Thus, as in the secondary epileptogenesis model, epileptiform transients develop at the kindling site and propagate in time and space in association with kindling development. The development of IID is not a necessary correlate of kindling in primates, however. For example, the majority of PC never emitted IID, while they were abundantly generated in all RM, PP and PH. Nevertheless, there is a very definite chronology of IID development and dissemination with a progressive increase in quantity during kindling and a rather precipitous reduction once the animals become kindled. This diminished quantity is maintained subsequently, only to be augumented transiently by drowsiness, sleep or the recurrence of seizures. This chronological trend, initially identified in cats (47), is identical to that in primates (42, 45).

The behavioural significance of kindling-induced distant IID development may also be assessed by the transfer experiment (16). It has been demonstrated repeatedly that kindling of the secondary site, following completion of kindling at the primary site, is much more rapid than expected in naive animals. This finding has given credence to the assumption that induced susceptibility to the development of epileptiform transients at the secondary site is also associated with increased susceptibility to behavioural seizure. This phenomenon of positive transfer effect (PTE) has now been confirmed in all species of animals studied so far. While the PTE is thought to be persistent, the associated negative or interference effect (NTE) is not (6, 16, 23). Thus, both proactive NTE at the secondary site and retroactive NTE at the primary site, following completion of secondary site kindling, dissipate if a rest interval of about two weeks is imposed between primary and secondary site kindling, or vice versa. No evidence has been reported refuting the transient nature of change in the inhibitory function which decays with time in subprimate species subjected to kindling (6, 17).

Kindling-induced modification of brain function is lasting, if not permanent, and there is no known method to reverse the kindled effect. Although kindling can be prevented by phenobarbital (38, 48), seizure susceptibility once acquired by amygdaloid (AM) kindling cannot be eroded by longterm administration of phenobarbital maintained at a therapeutic plasma level (19). There is some evidence that an antiserum to nerve growth factor (NGF), having no anticonvulsant property, has a prophylactic effect on amygdaloid kindling (12). Recently, limbic kindling was reported to cause sprouting and snyaptic restructuring in the hippocampus (14, 32, 36). Since structural remodelling requires critical neurotrophic support including the NGF, it is plausible that kindling utilizes the genomic mechanism involved in cell differentiation and growth necessary for normal CNS development. Recent reports of the expression of C-fos proto-oncogen (8, 9, 27) and increased production of messenger RNA for NGF (13) by seizure induced by kindling stimulation, particularly in the hippocampus, are consistent with such an assumption. If the structural changes reported can be ascertained as permanent, it would not be surprising to find that reversing the kindled effect is difficult.

Thus, kindling is believed to cause permanent change in the response property of the synaptically-related distant site with persistently increased susceptibility both to epileptiform transients and behavioural seizures induced by electrical stimulation. Since kindling is due to afterdischarge (AD) generation rather than to electrical stimulation per se, the obvious relevance to man would be the potential consequence of repeated uncontrolled seizures causing enhanced seizure susceptibility elsewhere, conforming to the prophecy of secondary epileptogenesis (29).

Although the similarity of kindling-related phenomena across the species is remarkable, some outstanding dissimilarities are also present in primate kindling. If the similarity of findings across the mammalian species is important, then the significance of the dissimilarities must also be understood. The intention of this chapter is to review selectively the kindling effect in different primate species and at different brain sites with this particular point in mind, and to highlight those kindling effects in primates which are significantly different from those in subprimate mammals; i.e., (a) the erosion of kindled epileptogenesis, and (b) the enduring nature of the kindling-induced transhemispheric seizure-suppressive effect.

## A: PERSISTENCE OF LIMBIC BUT EROSION OF CORTICALLY-KINDLED SEIZURE SUSCEPTIBILITY

### 1) Amygdaloid kindling

In cats, kindled seizure susceptibility can be recalled after a rest interval of one year (41). In primates (PP & PC), the rest interval imposed ranges from 112-868 days. Two of the PPs with anterior 2/3 callosal section prior to kindling had kindled seizures that were contralateral hemiconvulsive, while the intact PP and PC had generalized bisymmetrical convulsions as a consequence of kindling. When restimulated after a rest interval, all the animals responded with the previously established kindled seizure except for two PPs which frequently emitted spontaneous recurrent seizures at that time. Our experience has always indicated that when stimulation is continued beyond the final stage seizure and the kindled animals suddenly fail to respond to stimulus at the established threshold, the animal has developed a spontaneous recurrent seizure state. This is consistent with the usually high afterdischarge threshold (ADT) at the seizure generating focus of medically refractory partial seizures studied by depth electrodes (4). Since spontaneous seizures in our primates were the outcome of kindling, it is clear that kindled seizure susceptibility is retained in these failed animals. Thus, AM-kindled seizure susceptibility in primates is robust and remains unchanged even after a rest period of three years.

2) Neocortical kindling

One mesial frontal (MF) kindled RM responded with kindled asymmetrical generalized seizure after a rest interval of 252 days. After a rest interval of 266 days, one PP failed to respond with kindled seizure but developed an initial Stage 1 seizure only, associated with AD when stimulated at more than twice the current intensity of the previously established generalized seizure triggering threshold (GST). Similarly, after a rest interval of seven months, one PP (W-136) kindled at the frontal polar cortex (PF) failed to respond with kindled seizure despite AD generation which required a current intensity 2.5 times higher than that of the previously established GST. In this animal, rekindling was necessary to reinstate the kindled seizure state. However, the rate of rekindling was much faster than that of the original kindling. The most accelerated portion of the rekindling was the period between clear partial motor seizure onset and the kindled generalized seizure state. In contrast, the period of nonmotor seizure prior to partial motor seizure onset was the most time-consuming part. Thus, the ratio of rekindling versus original kindling for the nonmotor seizure stage was 1:4, in contrast to 1:8 for the motor seizure stages, indicating that kindling-induced change is better retained for motor events than for those underlying nonmotor events. The emergence during rekindling of each seizure stage as a completed form from the outset, confirms that a kindled substrate for each seizure stage remained consolidated during a seven-month rest interval. Thus, the primary reason for the inability to recall kindled seizure after a rest period was due to a weakened functional linkage between the substrates for the kindled seizure stages. In addition, worthy of note is the fact that this particular epileptic baboon (PP) was emitting frequent spontaneous seizures when the original kindling was terminated (46). Chronological follow-up of this animal showed that, by the time of rekindling, there was no evidence of spontaneous seizure as verified by longterm EEG monitoring. This is the first evidence that kindled epileptogenesis capable of emitting spontaneous recurrent seizure can erode over time, eventually requiring rekindling. It should be added that upon completion of rekindling, this animal once again began to emit frequent spontaneous seizures, which abated gradually over the next four months when primary site stimulation was stopped, confirming the reversibility of kindled epileptogenesis at this particular cortical site. Obviously, the fundamental difference which exists between temporal limbic (AM) and frontal cortical kindling in primates is highly relevant to the well known clinical phenomenon of spontaneous remission in epilepsy for which no satisfactory neurobiological explanation is available.

It is possibly significant that such a spontaneous reversal occurred in the genetically-epileptic baboon, PP. A similar complete failure to reactivate kindled seizure after a rest interval was also observed in an MF-kindled epileptic baboon, PP, but not in RM. The reason for this failure in PP is not clear. However, the possibility must be entertained that primates with a genetically enhanced seizure susceptibility are endowed with an enhanced capability for seizure suppression as a natural protective mechanism. It is conceivable that any process such as kindling, which further enhances seizure susceptibility, could result in the concurrent enhancement of a naturally endowed seizure suppressive capability. Whether a similar mechanism plays a role in the medically refractory partial seizure patient becoming seizure free following resection of the seizure-generating focus, despite widespread pathophysiology as frequently revealed by depth electrode studies, remains unknown (9).

B: ENDURING SEIZURE SUPPRESSIVE TRANSHEMISPHERIC EFFECT AT THE CONTRA-LATERAL HOMOTOPIC SITE

1) AM kindling

Three species of primates (RM, PP, PH) undergoing sequential primary and then secondary site AM kindling showed PTE at the contra-lateral homotopic site. However, considerable variation in the pattern of PTE was noted according to species. All the PPs in our original AM kindling study undergoing secondary site kindling showed an abrupt development of isolated generalized convulsion or convulsions but only one subsequently showed sustained convulsive response. In contrast, PH showed either abrupt or sequential but rapid development of generalized convulsion. In this species, the pattern of secondary site seizure development was the mirror image of that of primary site kindled seizure, while in one animal the pattern was identical to that of primary site kindled seizure. This finding suggests that under certain circumstances, the secondary site response is due to reactivation of the primary site kindled neurocircuits. In RM, secondary site development was sequential at a rate about 2.5 times than at the primary site. As was the case during primary site kindling, RM showed frequent seizure stage regression which was not observed in the baboon species.

Thus, all three primate species undergoing secondary site AM kindling showed a spectrum of PTE. Since all of them showed IID at the secondary site AM, enhanced susceptibility to epileptiform transients at the "mirror focus" appears to be matched by augmented clinical seizure susceptibility. The obvious exception to this, as shown in a PH, was activation of electroclinical seizure originating in the kindled primary site hemisphere, suggesting that secondary site response does not necessarily indicate enhanced seizure susceptibility intrinsic to the secondary site hemisphere.

On the other hand, examination of the intra-ictal architecture of the evoked kindled seizure pattern at the secondary site in all primate species showed evidence of NTE with a signficantly increased latency for the onset of Stage 3 partial motor seizure when compared to the same at the primary site. However, secondary site AM kindling in PP was unique in that kindled seizure response was an isolated event with susbsequent sustained regression to a nonmotor seizure state. Even more important, no sequential seizure development occurred despite 100 daily stimulations with steady AD generation (45). Since the completion of primary site kindling required a mean of 72 stimulations in PP, it is clear that the secondary site effect was characterized by both PTE and NTE, the latter predominating over the former. Obviously, it is not a simple absence of PTE, but rather a powerful, enduring and active suppression of seizure development that is represented. This is in contrast to the generally accepted transient time-decaying process of NTE associated with kindling. PTE was totally absent at the secondary site AM in both species when surgically-prepared PP and RM (completely bisected or anterior 2/3 bisection of the corpus callosum, with or without bisection of the hippocampal commissure) underwent AM kindling (39, 40). Thus, regardless of the presence or absence of genetic seizure susceptibility,

AM kindling in primates produces both seizure facilitatory and suppressive effects which can manifest at the secondary site as PTE and apparent NTE, ranging from subtle to overt. NTE is maximal and enduring in the epileptic baboon, PP, and is probably transmitted through a subcortical route which presumably exists in the brainstem. Thus, in the absence of the forebrain commissure (most likely the CC), only the seizure suppressive effect is transmitted to the secondary site. The enduring nature of this powerful suppression of seizure development suggests that it is an aspect of kindling-induced longterm changes. Indeed, in an AM-kindled PP, W-229, the failure of sequential seizure development (beyond Stage 1) at the secondary site AM lasted nearly eight months (unpublished observations). This finding confirms that this failure is not a consequence of repeated convulsive seizures but rather of kindling-induced change. NTE is known to be transient and to decay within about two weeks to one month. It is more likely to be a direct correlate of the convulsion rather than of kindling (2, 17, 19, 23). The reason why NTE becomes maximal and enduring only in primates and in epileptic baboons particularly, is not known. However, as already discussed, it is plausible that baboons with genetically dictated seizure susceptibility are likely endowed with an enhanced seizure suppressive capacity as a natural protective mechanism. That only one out of over 100 PP kindled in this laboratory ever developed fatal status epilepticus is not inconsistent with this assumption. One may then postulate that AM kindling concurrently enhances the seizure facilitatory and the counteracting mechanisms, resulting in a spectrum between seizure facilitation and suppression, both of which are maximal in genetically epileptic primates.

2) Neocortical kindling

Among the five frontal neocortical sites studied so far, PTE was identified at the PMA, the supplementary motor area (SMA), and the motor cortex (MC), while it was totally absent at the PF and MF. In the latter two brain sites, not only was the expected PTE absent, but there was also a complete failure of electroclinical seizure development at the secondary site despite repeated afterdischarge generation. During primary site kindling at these sites, AD was readily transmitted to the secondary homotopic site from the first stimulation and there was progressive AD dissemination into a number of subcortical sites. This was associated with progressive AD growth both in duration and in the increasing complexity of the pattern. In contrast, there was no electroclinical seizure development at the secondary homotopic site despite daily AD generation, with a number of stimulations far exceeding the ones required for completion of primary site kindling. Furthermore, throughout secondary site kindling, there was no growth of AD duration or subcortical propagation. Afterdischarge transmission to the contralateral homotopic site, i.e., the primary kindled site, began only after 20 daily stimulations. Thus, in response to electrical stimulus, the secondary site cortex which is capable of producing an essential building block of kindling, i.e., AD, was unable to propagate it, temporally or spatially, beyond that site. That this temporal and spatial confinement of AD involves a much larger area than the homotopic contralateral site is suggested by one PC (W-171) among the six primates subjected to MF kindling (43). At the secondary site MF, this was the only animal which showed electroclinical seizure development, an abrupt but isolated kindled convulsive response alternating with nonconvulsive seizure. This pattern was strikingly similar to that of secondary site AM kindling in the PP mentioned above. Histological examination disclosed that the electrode tip was mislocated within the primary site hemisphere grey matter, only a few mm away from the kindled cortical site. This finding suggests that the kindled MF site is tightly enveloped by powerful intra- and inter-hemispheric seizure suppressive mechanisms, an effect which can be enduring unlike many other transient seizure refractory states such as proactive and retroactive NTE. In MF kindling these powerful seizure suppressive mechanisms were shared not only by the epileptic baboon, but also by PC and RM, further indicating that this nonmotor brain site (and possibly also the frontal polar cortical site) is strategically located to mobilize the seizure suppressive mechanisms more than other frontal sites such as the PMA, SMA and MC, all of which are more directly involved in the motor mechanism. Results of our previous cortical lesion study in AM-kindled PP are consistent with the view that the mesial frontal cortex is involved in the generation of the seizure suppressive transhemispheric effect (44).

## DISCUSSION

In the past, the persistent if not permanent modification of brain function induced by kindling emphasized the acquisition of lasting seizure susceptibility which is now well documented in all the mammalian species subjected to kindling. The mechanism for kindling-induced lasting modification of brain function is not known. However, there is evidence to indicate that kindling-induced functional change is paralleled by structural changes such as reorganization of synaptic architecture (14) and evidence of sprouting of hippocampal mossy fibers (36). Although evidence of structural modification has not yet been found outside of the hippocampus, some anatomical specificity of such an alteration is suggested since AM but not entorhinal kindling induces further infrapyramidal synaptic changes (32). However, whether these changes are permanent and critical for the kindling effect remains to clarified.

The functional/structural organizational force is conceived to be maximal during the process of CNS development based on a genetic program and influenced by environmental parameters. Since kindling-induced reorganization of brain function is extensive and widespread over both time and space, it is tempting to speculate on the involvement of genomic modifications. Indeed, AM kindling is reported to induce the expression of C-Fos, a cellular oncogen presumably involved in controlling the expression of other genes participating in cell growth and differentiation. More recent reports indicate that an antiserum to NGF exerts a prophylactic effect on AM kindling (12), while limbic seizures increase the amount of messenger RNA for NGF in the hippocampal dentate gyrus (13). Whether limbic C-Fos or NGF expression is specific to limbic, but not to other seizures such as neocortically-kindled ones, is not known. Further molecular elucidation of kindling phenomena may shed some light on why the kindled effect is permanent in the limbic but transient in some neocortical systems.

Kindling has been reported to produce longlasting facilitation associated with shorterm suppression. Indeed, seizure suppressive effects associated with kindling, when measured behaviourally, electrophysiologically and neurochemically, have been largely transient, ranging from a few hours to less than one month of enhanced inhibitory effect (17). This time-decaying suppression as measured by transcallosal response in premotor cortical kindling in primates suggests that it is the consequence of repeated convulsions rather than of kindling itself (2). Therefore, the enduring seizure suppressive state observed in secondary site kindling in primates is quite different from those extensively documented transient suppressive changes. It is very longlasting, if not permanent, and is responsible for the complete failure of kindling at the contralateral homotopic AM, PF and MF sites, despite repeated daily AD generation. In AM kindling, it is best expressed in the epileptic baboon, PP. However, contralateral suppression is also present in the nonepileptic RM, but only if the anterior 2/3 CC is sectioned, indicating a probable subcortical route of its transhemispheric mediation.

Our previous studies showed that micro-injection of gabaculine (GABA-transaminase inhibitor) or 2 amino-7-phosphonoheptanoic acid (excitatory amino acid inhibitor) into the substantia innominata (SI) reversibly blocks AM-kindled convulsion without affecting the AD generation of the kindled AM (25, 26, 31). In our attempt to understand the mechanism of this enduring suppression at the secondary site AM in kindled PP, we have evaluated the effect of intra-SI micro-injection of a number of agents, including picrotoxin (GABA antagonist), haloperidol (dopamine antagonist), propranolol (β-norepinepherine antagonist) and phentolamine (α-adrenergic antagonist). Preliminary evidence suggests that both GABAergic and α-adrenergic receptors in the SI are participating in some aspects of this seizure suppressive mechanism at the secondary site of the AM-kindled PP. The question of whether the seizure suppressive mechanism induced by neocortical kindling is shared by that of AM kindling is not known.

Although the mechanisms of the kindling-induced enduring seizure suppressive effect remain to be explored, its potential importance lies in the fact that (1) it becomes maximal in primates, (2) it is more clearly represented in genetically-epileptic primates to the extent it can completely mask the PTE, and (3) the mesial frontal cortical site is specifically involved regardless of the primate species concerned. Whether or not a similar mechanism may be involved in remission and erosion of the kindling-induced spontaneous recurrent seizure state remains unknown.

Thus, kindling in primates, as documented in lower animals, induces a lasting increase in seizure susceptibility. However, in primates, this is associated with a remarkably enhanced and enduring seizure suppressive mechanism. Such facilitation and suppression must use endogenous mechanisms normally available to the brain. Therefore, kindling-induced development and a persistent state of an unusual degree of both seizure facilitation and suppression must be accomplished at the expense of physiological function.

It is recognized that the seizure suppressive effect manifests itself maximally when nonmotor structures such as the amygdala, frontal polar or mesial frontal cortices are kindled. In contrast, kindling of the motor system such as the motor cortex, supplementary motor area and the premotor area does not produce an enduring suppressive effect on seizure. Therefore, those structures involved in emotional/cognitive/conative mechanisms, which are highly relevant to the normal organization of behaviour in primates, are the ones most susceptible to such physiological derangement. If one envisages the reorganization of function in the latter systems induced by kindling, it follows that those emotional/cognitive/conative functions regulated by these systems must also undergo modification.

In our previous chronic primate study, we found that RMs with unilateral prefrontal alumina cream lesion develop evidence of bifrontal dysfunction as measured by delayed response and delayed alteration tests. Simultaneous EEG monitoring has conclusively demonstrated that their inability to perform these tests was not due to any interictal spike discharge or electroclinical seizure, and therefore, was considered to be due to the chronic epileptogenic process (34). It is plausible that this behavioural disturbance resulted from an excess seizure suppressive effect in the contralateral prefrontal region, thus compromising the physiological function of both prefrontal cortices.

## SUMMARY

In the past, the longterm consequence of kindling emphasized persistently enhanced seizure susceptibility in all the species examined, as well as a kindling-induced transient change in the inhibitory function. This across-the-species similarity is obviously very important since it implies that the same basic neurobiological principle operates in all mammalian species. However, the difference in gross brain morphology among different mammalian species and in particular, between subhuman and human primates, is well known. Since structural difference implies functional differecne, perhaps it is not surprising that there are significant differences in kindling and the kindled state between primates and subprimates; e.g., (1) spontaneous reversal of kindled epileptogenesis occurs in frontal cortical kindling, as it does in some cases of spontaneous remission in human epilepsy, and (2) the enduring nature of the powerful seizure suppressive effect is the most striking distant consequence of kindling nonmotor structures such as the amygdala and the prefrontal and mesial frontal cortices. Finally, the kindling-induced seizure suppressive effect is maximal in genetically-epileptic Papio papio, possibly due to augmentation of a pre-existing naturally enhanced protective mechanism.

Kindling is a model of epilepsy, which is the result of a sustained disorder of brain function, at the peak of which seizures will recur. Thus, it follows that the presence of a persistently disordered brain function can also result in symptoms other than seizures. Our primate findings provide strong support for the probability that the epileptogenic process itself can have behavioural consequences which become manifest as one ascends the phylogenic scale to subhuman and human primates.

Elaboration and disclosure of mechanisms underlying both the reversal of kindled epileptogenesis and the enduring seizure suppressive effect in primates should add further insight into the kindling mechanism and its relevance to the problems of human epilepsy.

## ACKNOWLEDGEMENT

This work was supported by grants from the Medical Research Council of Canada to J.A.W.

## REFERENCES

1. Alonso-De Florida, F. & Delgado, J.M.R. (1958) Lasting behavioural and EEG changes in cat induced by prolonged stimulation of the amygdala. Am. J. Physiol. 193:223-229
2. Baba, H., Ono, K. & Wada, J.A. (1987) Transcallosal response in the chronic baboon preparation, Papio papio. II. Effect of premotor cortical kindling. Electroenceph. clin. Neurophysiol. 67/6:564-569
3. Baba, H., Sakai, S. & Wada, J.A. (1986) Premotor (area 6) cortical kindling. IN: KINDLING 3, J.A. Wada (Ed). Raven Press, NY, pp 447-469
4. Cherlow, D.G., Dymond, A.M., Crandall, P.H., Walter, R.D. & Serafetinides, E.A. (1977) Evoked response and afterdischarge

. thresholds to electrical stimulation in temporal lobe epilepsy. Arch. Neurol. 34:527-531
5. Corcoran, M.E., Cain, D.P. & Wada, J.A. (1984) Amygdaloid kindling in Papio cynocephalus and subsequent recurrent spontaneous seizures. Folia Psychiat. et Neurol. Jpn. 38/2:151-158
6. deJonge, M. & Racine, R.J. (1987) The development and decay of kindling-induced increases in paired-pulse depression in the dentate gyrus. Br. Res. 412:318-328
7. Delgado, J.M.R. (1959) Prolonged stimulation of brain in awake monkeys. J. Neurophysiol. 22:458-475
8. Dragunow, M. & Robertson, H.A. (1987) Kindling stimulation induces c-fos protein(s) in granule cells of the rat dentate gyrus. Nature 329:441-442
9. Dragunow, M., Robertson, H.A. & Robertson, G.S. (1988) Amygdala kindling and c-fos portein(s). Exp. Neurol. 102:261-263
10. Engel, J.Jr., Rausch, R., Lieb, J.P., Kuhl, D.E. & Crandall, P.H. (1981) Correlation of criteria used for localizing epileptic foci in patients considered for surgical therapy of epilepsy. Ann. Neurol. 9:215-224
11. Essig, C.F., Grose, M.E. & Williamson, E.L. (1961) Reversible elevation of electroconvulsive threshold and occurrence of spontaneous convulsions upon repeated electrical stimulation of the cat brain. Exp. Neurol. 4:37-47
12. Funabashi, T., Sasaki, H. & Kimura, F. (1988) Intraventricular injection of antiserum to nerve growth factor delays the development of amygdaloid kindling Br. Res. 458:132-136
13. Gall, C.M. & Jackson, P.J. (1989) Limbic seizures increase neuronal production of messenger RNA for nerve growth factor. Science 245:758-761
14. Geinisman, Y., Morrell, F. & de Toledo-Morrell, L. (1988) Remodelling of synaptic architecture during hippocampal 'kindling'. Proc. Nat'l. Acad. Sci. USA 85:3260-3264
15. Goddard, G.V. (1967) Development of epileptic seizures through brain stimulation at low intensity. Nature 214:1020-1021
16. Goddard, G.V., McIntyre, D.C. & Leech, C.K. (1969) A permanent change in brain function resulting from daily electrical stimulation. Exp. Neurol. 25:295-330
17. Goddard, G.V., Dragunow, M., Maru, E. & MacLeod, E.K. (1986) Kindling and the forces that oppose it. IN: The Limbic System: Functional Organization and Clinical Disorders B.K. Doane & K.E. Livingston (Eds) Raven Press, NY pp 95-108
18. Goldensohn, E.S. (1984) The relevance of secondary epileptogenesis to the treatment of epilepsy: Kindling and the mirror focus. Epilepsia 25:S156-S168
19. Hiyoshi, T. & Wada, J.A. Feline amygdaloid kindling and the sleep-waking pattern: Observations on daily 22-hour polygraphic recording. Epilepsia (In press).
20. Hiyoshi, T. & Wada, J.A. Failure of nine-month phenobarbital administration to reverse amygdaloid kindled seizure susceptibility in cats. (Submitted)
21. Lowrie, M.B. & Ettlinger, G. (1980) The development of independent secondary "mirror" discharge in the monkey: failure to replicate earlier findings. Epilepsia 21:25-30
22. Mayersdorf, A. & Schmidt, R.P. (1982) Secondary Epileptogenesis Raven Press, New York
23. McIntyre, D.C. & Goddard, G.V. (1973) Transfer, interference and spontaneous recovery of convulsions kindled from the rat amygdala. Electroenceph. clin. Neurophysiol. 35:533-543
24. Monroe, R.R. (1970) Episodic behavioural disorders: A psycho-dynamic and neurophysiologic analysis. Harvard University Press, Cambridge, Mass pp. 76-77

25. Morita, K., Okamoto, M., Seki, K. & Wada, J.A. (1985) Suppression of amygdala-kindled seizure in cats by enhanced GABAergic transmission in the substantia Innominata. Exp. Neurol. 89:225-236
26. Mori, N. & Wada, J.A. (1989) Suppression of amygdaloid kindled convulsion following unilateral injection of 2-amino-7-phosphone-heptanoic acid (2-APH) into the substantia innominata. Br. Res. 486:141-146
27. Morgan, J.I., Cohen, D.R., Hempstead, J.L. & Curran, T. (1987) Mapping patterns of c-fos expression in the central nervous system after seizures. Science 237:192-197
28. Morrell, F. (1985) Secondary epileptogenesis in man. Arch. Neurol. 42:318-335
29. Morrell, F. (1989) Varieties of human secondary epileptogenesis. J. Clin. Neurophysiol. 6/3:227-275
30. Morrell, F., Wada, J.A., & Engel, J.Jr. (1987) Potential relevance of kindling and secondary epileptogenesis to the consideration of surgical treatment for epilepsy. IN: Surgical Treatment of the Epilepsies J. Engel, Jr. (Ed) Raven Press, NY pp 701-707
31. Okamoto, M. & Wada, J.A. (1984) Reversible suppression of amygdaloid kindled seizures following unilateral gabaculine injection into the substantia nigra. Br. Res. 305:389-392
32. Represa, A., Le Gal La Salle, G. & Ben-Ari, Y. (1989) Hippocampal plasticity in the kindling model of epilepsy in rats. Neurosci. Letts. 99:345-350
33. Sato, M. & Wada, J.A. (1975) Review on the kindling preparation: A new experimental model of epilepsy. Brain & Nerve 27:257-273
34. Seino, M. & Wada, J.A. (1964) Chronic focal cortical epileptogenic lesion and behaviour: Comparison of behavioural performance in monkeys with either epileptogenic or ablatic unilateral lesion. Epilepsia 5:321-333
35. Sranka, M., Sedlack, P., & Nadvornik, P. (1977) Observation of the kindling phenomenon in treatment of pain by stimulation in the thalamus. IN: Neurosurgical Treatment in Psychiatry, Pain & Epilepsy. W.H. Sweet, S. Obrador & J.G. Martin-Rodriguez (Eds), University Park Press, Baltimore pp. 651-654
36. Sutula, T., Xiao-Xian, H., Cavazos, J. & Scott, G. (1988) Synaptic reorganization in the hippocampus induced by abnormal functional activity. Science 239:1147-1150
37. Uemura, S. & Wada, J.A. (1981) Seizure predisposition and species. Epilepsia 22:232
38. Wada, J.A. (1977) Pharmacological prophylaxis in the kindling model of epilepsy. Arch. Neurol. 34:389-395
39. Wada, J.A. & Mizoguchi, T. (1984) Limbic kindling in the forebrain-bisected photosensitive baboon, Papio papio. Epilepsia 25/3:278-287
40. Wada, J.A., Mizoguchi, T., & Komai, S. (1981) Cortical motor activation in amygdaloid kindling: Observations in nonepileptic rhesus monkeys with anterior 2/3 callosal bisection. IN: KINDLING 2, J.A. Wada (Ed) Raven Press, NY, pp 235-248
41. Wada, J.A., Sato, M. & Corcoran, M.E. (1974) Persistent seizure susceptibility and recurrent spontaneous seizures in kindled cats. Epilepsia 15:465-478
42. Wada, J.A., Mizoguchi, T. & Osawa, T. (1978) Secondary generalized convulsive seizures induced by daily amygdaloid stimulation in the rhesus monkey. Neurology 28:1026-1036
43. Wada, J.A., Mizoguchi, T. & Komai, S. (1985) Epileptogenesis in the orbital and mesial frontal cortical area of the subhuman primate. Epilepsia 26/5:472-479

44. Wada, J.A. & Okamoto, M. (1986) The differential role of mesial and lateral frontal cortices in amygdaloid kindling and kindled seizures in Senegalese baboons, Papio papio. IN: KINDLING 3, J.A. Wada (Ed). Raven Press, NY, pp 409-428
45. Wada, J.A. & Osawa, T. (1976) The generalized convulsive seizure state induced by daily electrical amygdaloid stimulation in Senegalese baboons, Papio papio. Neurology 26:273-286
46. Wada, J.A., Osawa, T. & Mizoguchi, T. (1975) Recurrent spontaneous seizure state induced by prefrontal kindling in the Senegalese baboon, Papio papio. Can. J. Neurol. Sci. 2:477-495
47. Wada, J.A. & Sato, M. (1974) Generalized convulsive seizures induced by electrical stimulation of the amygdala in cats: Correlative electrographic and behavioural features. Neurology 24:565-574
48. Wada, J.A., Sato, M., Wake, A., Green, J.R. & Troupin, A.S. (1976) Prophylactic effects of phenytoin, phenobarbital and carbamazepine examined in kindling cat preparations. Arch. Neurol. 33:426-434
49. Watanabe, E. (1936) Experimental study on pathogenesis of epileptic convulsive seizures. Psychiat. et Neurol. Jpn. 40:1-36
50. Wilder, B.J. (1972) Projection phenomena and secondary epileptogenesis mirror foci. IN: Experimental Models of Epilepsy. D. Purpura, et al. (Eds) Raven Press, New York. pp. 85-112
51. Wurz, R.H. & Olds, J. (1963) Amygdaloid stimulation and operant reinforcement in the rat. J. Comp. Physiol. Psychol. 56:941-949

Discussion of Dr. Wada's Presentation

DR. ADAMEC: Dr. Wada, when you were stimulating in the one amygdala in the *Papio*, and it is primary site kindling, have you recorded from the amygdala in the opposite hemisphere?

DR. WADA: Yes, we are always recording.

DR. ADAMEC: And you see, what? Seizure discharges propagating there or just high frequency activation of the cells? The reason why I ask is that I've done some work now in cats putting in patterns of stimulation which resemble cell firing that's produced during an amygdala discharge but without producing a seizure discharge. What I get is a long-term depression of excitatory output pathways of the amygdala. I was just curious to know whether some of you might have an inroad to explaining some of your negative transfer effects if the kind of projected activity you got to the other amygdala was perhaps producing some kind of depression in that hemisphere of excitatory outputs.

DR. WADA: That's quite possible. We always record the EEG from a large number of sites from both hemispheres but we have not done single unit recording. We have not seen any change in the AD pattern, duration, etc. I would agree that interference with excitatory amygdaloid output is a possibility but I don't think the amygdala is the site of this interference, however. It is elsewhere as indicated by the fact that AD in the amygdala can be triggered quite readily. There is no difference in the ADTs between primary and secondary sites. Clinical seizure stages 1 and then 2 develop normally but it does not progress to a partial motor seizure stage 3 despite repeated AD generation. This indicates to me that the process of interference must be operative in the mechanism functionally linking the non-motor limbic system to the motor structure responsible for motor seizure development.

DR. BERMAN: The idea exists that in the young animal there is less descending cortical inhibition, and in general young animals show greater plasticity, and kindle very rapidly. The question is whether the transfer phenomenon, negative transfer is seen in young animals, whether age is a factor?

DR. WADA: Maybe Dr. Moshe can answer this question.

DR. MOSHE: Positive transfer readily occurs in 15-day old rat pups. It has been tested two ways. first, immediately after the completion of primary amygdala kindling with positive transfer to the contralateral amygdala. The second way is months after the establishment of primary site amygdala kindling in infancy when the contralateral amygdala was tested as the rat reached adulthood. We have not performed any experiments in rat pups to test the existence of negative transfer directly. However, based on the information we obtained during kindling of two sites simultaneously (on an alternating basis) I doubt whether negative transfer exists.

DR. WADA: Well, now I'd like to call upon Dr. Kudo. In the meantime please stock up your questions and comments for the general discussion.

# CLAUSTRUM AND AMYGDALOID KINDLING

T. Kudo and J.A. Wada

Divisions of Neurosciences and Neurology, University of British Columbia, Vancouver, British Columbia V6T2A1 Canada

## INTRODUCTION

The motor cortex[1,2], basal ganglia[3,4], substantia nigra[3, 5], substantia innominata[6,7] all seem to play a respective role in generation of amygdaloid (AM) kindled convulsion. However, mechanism of functional linkage from the AM which is non motor structure to a motor mechanism responsible for the AM kindled convulsion is not known.

The claustrum has reciprocal connections with the AM[8,9], motor cortex[10], basal ganglia, substantia nigra,[11,12,13] and thalamus[14, 15, 16, 17]. Electrical stimulation of the claustrum induced contralateral version of the head and trunk associated with the extension of the contralateral foreleg in cats[18], while the injection of r-vinyl-GABA into the area of the prepyriform cortex and claustrum is reported to elevate the seizure threshold for the AM kindled seizure in rats.[19] It is plausible, therefore, that the claustrum plays a role in linking the AM with the motor mechanism.

In addition, the claustrum has bilateral reciprocal connections with the cortex across the corpus callosum[20] and the limbic system,[21, 22, 23] and receives a projection from the massa intermedia.[15] Since the corpus callosum,[24] limbic system[25] and massa intermedia[26, 27] are all implicated in the mechanism of interhemispheric positive transfer effect in AM kindling, it is possible that the claustrum also is involved in this effect.

We tested our hypothesis that the claustrum has a strategic role to play in both AM - motor linkage and the transhemispheric positive transfer effect in AM kindling by examining the effect of unilateral claustral lesion placement before and after AM kindling in cats.

## GENERAL METHOD

Technical details of surgery and the method of kindling have been published previously.[28] In brief, depth electrodes made from twisted stainless steel wires 0.35 mm in diameter insulated except for the tips were implanted bilaterally into the AM , lateral geniculate body, and midbrain reticular formation under sodium pentobarbital anesthesia. Screw

*Kindling 4*
Edited by J. A. Wada
Plenum Press, New York, 1990

electrodes 1.0 mm in diameter were placed bilaterally in the anterior sigmoid gyrus (MC), posterior lateral cortex and frontal bone. Following a two week rest period, primary site AM kindling began according to the established procedure in our laboratory.[28, 29] Briefly, the afterdischarge threshold (ADT) prior to AM kindling and the generalized seizure triggering threshold (GST) following AM kindling were determined. The pattern of seizure development by AM kindling was classified into 6 stages. Secondary site AM kindling commenced after a further two week rest interval. A day after the completion of secondary site kindling, the primary site AM was retested at the previously established GST. Five successive daily elicitations of the stage 6 seizure was arbitrarily designed to indicate the completion of primary and secondary site kindling, and primary site retest.

An electrolytic lesion was placed by passing an anodal current of 4 - 7 mA D.C. through an acutely - inserted electrode for 30 - 60 sec. under sodium pentobarbital anesthesia.

The data were analyzed statistically using the Student t test. Upon completion of the study, the animals were deeply anesthetized, and their brains were perfused with saline and 10% formalin, serially sectioned and stained with cresyl violet. Histological examination showed the recording and the stimulating electrodes were localized in the intended structures.

## EXPERIMENT I: CLAUSTRAL LESION PLACEMENT PRIOR TO AM KINDLING

We investigated the effect of the claustral lesion on the development of AM kindling and the interhemispheric positive transfer effect by placing the lesion prior to AM kindling.

### MATERIALS AND METHODS

Nine male adult cats ( Nos. 924, 925, 934, 938, 941, 952, 953, 954, 955 ) were used. Lesions were made by multiple penetration through the extent from A 19.3 to A 13.7 mm according to the atlas of Bermann et al.[30] Primary site kindling in 9 cats, and secondary site kindling and primary site retest in 7 cats were examined.

### RESULTS

Results are summarized in Table 1 and details are given below.

1) Primary site kindling

Latencies to the first stage 4 seizure, first stage 6 seizure and stable stage 6 seizure were 29.6 ± 13.5 (mean ± SD), 37.1 ± 11.4, 44.6 ± 12.1 days respectively. These latencies were significantly longer than those in the nonlesioned cat.[26] The latency from the first stage 4 seizure to the first stage 6 seizure was 8.6 ± 4.9 days and is not significantly different from that in the nonlesioned cat.[26]

The stage 3 and stage 5 seizures were absent in 2 cats and in all cats respectively. The convulsive seizure stage frequently regressed to earlier stages. The stage 4 seizure regressed to stage 2 or 3 in 4 cats and the stage 6 seizure regressed to stage 3 or 4 in 6 cats. A mean of 8.44 stimulations was required to elicit 5 consecutive stage 6 seizures. When the ADT was reduced after a stable stage 6 state, the seizure regressed from stage 6 to stage 3 or 4 in 1 cat.

Table 1. Summary of primary and secondary site kindling, and primary site retest

| | Cat No. | | | | | | | | | Mean ± S.D. | |
|---|---|---|---|---|---|---|---|---|---|---|---|
| | 924 | 925 | 934 | 938 | 941 | 952 | 953 | 954 | 955 | claustral lesioned cat | Control[a)] |
| primary site kindling | | | | | | | | | | | |
| Latency (days) to | | | | | | | | | | | |
| stage 4 | 17 | 16 | 23 | 52 | 31 | 30 | 19 | 24 | 54 | 29.6 ± 13.5[1)] | 16.0 ± 6.7 |
| stage 6 | 29 | 34 | 30 | 54 | 42 | 33 | 25 | 28 | 59 | 37.1 ± 11.4[2)] | 27.5 ± 7.4 |
| stable stage 6 | 36 | 50 | 34 | 65 | 48 | 41 | 32 | 32 | 64 | 44.6 ± 12.1[3)] | 33.0 ± 7.4 |
| stage 4 - stage 6 | 12 | 18 | 7 | 2 | 11 | 3 | 6 | 4 | 5 | 8.6 ± 4.9 | 12.0 ± 1.9 |
| secondary site kindling | | | | | | | | | | | |
| Latency (days) to | | | | | | | | | | | |
| stage 4 | 5 | - | - | 17 | 14 | 14 | 12 | 5 | 20 | 12.4 ± 5.3[4)] | 4.3 ± 3.5 |
| stage 6 | 11 | - | - | 18 | 24 | 14 | 12 | 12 | 21 | 16.6 ± 4.3[5)] | 4.3 ± 3.5 |
| stable stage 6 | 19 | - | - | 30 | 41 | 22 | 16 | 16 | 31 | 25.0 ± 8.6[6)] | 10.0 ± 4.8 |
| stage 4 - stage 6 | 11 | - | - | 2 | 11 | 1 | 1 | 8 | 2 | 5.1 ± 4.3 | |
| primary site retest | | | | | | | | | | | |
| stage 6 | 1 | - | - | 1 | 1 | 1 | 1 | 1 | 4 | | |
| stable stage 6 | 5 | - | - | 5 | 5 | 5 | 5 | 5 | 8 | | |

a; control, primary site ( Hiyoshi and Wada, 1988, n=19 )
secondary site ( Wake and Wada, 1977, n=6 )

Difference between claustral lesion and control; 1) $p < 0.01$, 2) $p < 0.05$, 3) $p < 0.01$, 4) $p < 0.05$, 5) $p < 0.01$, 6) $p < 0.01$

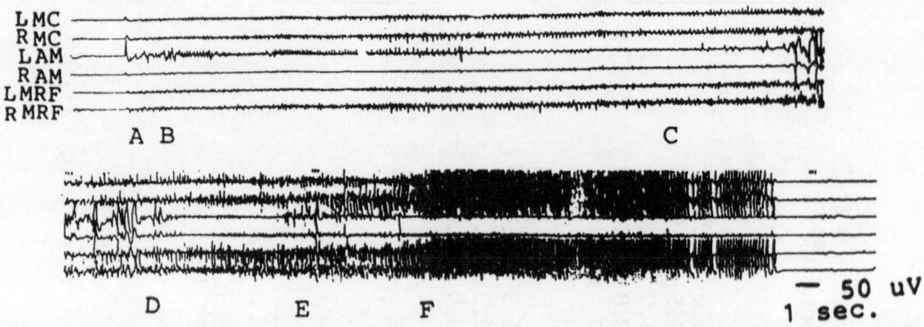


Fig. 1. Cat 954, Left AM kindling. The amplitude of selfsustained discharge is higher in the right MC than that in the left MC. Concurrent with head turning and circling to the left side, rapid discharges of increasing amplitude is recorded in the right MC ( C ). Concurrent with the axial rotation to the left side, spikes of high amplitude appeared in the right MC ( E ). Abbreviations: MC - motor cortex, AM - amygdala, MRF - midbrain reticular formation.
A: left facial twitching, B: bilateral facial twitching, C: head turning and circling to left side, D: left clonic convulsion of the left forepaw, E: axial rotation to left side, F: generalized convulsion.

There was a significant change in the stage 4 seizure pattern with mirror activity of the anticipated contralateral dominant movement characterized by the initial version of the head and trunk, circling and axial rotation, then clonic convulsion, all beginning in the limbs ipsilateral to the stimulating site before becoming generalized. Concurrent with this mirror seizure pattern, selfsustained seizure discharges appeared early in the contralateral MC with higher amplitude than in the ipsilateral MC ( Fig. 1 ).

As shown in Fig. 2, the common and critical part of the lesion involved the anterior half of the claustrum ( anterior to A 13 mm ).

2) Secondary site kindling

Latencies to the first stage 4 and 6 seizure, and the stable stage 6 state were 12.4 $\pm$ 5.3, 16.6 $\pm$ 4.3 and 25.0 $\pm$ 8.6 days, respectively. These latencies were significantly longer than those of the nonlesiond cat[26]. The latency to the first stage 4 was almost the same as the latency in the primary site kindling of the nonlesioned cat ( 16.0 $\pm$ 6.7 days).[26] On the other hand, the latency from the first stage 4 seizure to the first stage 6 seizure was 5.1 $\pm$ 4.3 days and significantly ( $p < 0.01$ ) shorter than that of primary site kindling in nonlesioned cats ( 12.0 $\pm$ 1.9 days).[27] Clinical seizure development was sequential as except for the absence of stage 5 in all cats and stage 3 in 1 cat.

3) Primary site retest

Stage 6 seizures could be reactivated at the first stimulation in 6 cats. In No. 955, 4 stimulations were required to elicit the stage 6 seizure.

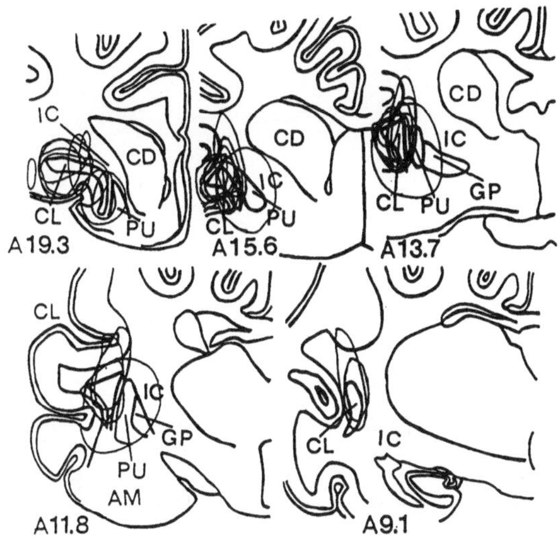


Fig. 2 The schematic representation of the lesions placed prior to AM kindling. The following abbreviations appear in Fig. 2 and Fig. 4. AM - amygdala, CD - caudate, CL - claustrum, GP - globus pallidus, IC - internal capsule, PU - putamen. The diagram of coronal sections of a stereotaxic atlas of the cat's brain are modified from Berman et al. ( 30 ).

## EXPERIMENT II: CLAUSTRAL LESION PLACEMENT AFTER COMPLETION OF AM KINDLING

The effect of the ipsilateral and contralateral claustral lesion on the previously established AM kindled seizure was studied.

### MATERIALS AND METHODS

Seven male mature cats ( Nos. 888, 889, 891, 899, 904, 905, 932 ) were used. Unilateral lesions were made by multiple penetration through A 19.3 to A 8.5 mm in 6 cats and a lesion was made in the ectosylvian gyri in NO. 932 as a control after completion of primary and secondary site kindling and primary site retest. Effects of the ipsilateral AM stimulation in 5 cats ( Nos. 889, 891, 899, 904, 905 ) and the contralateral AM stimulation in 3 cats ( Nos. 888, 904, 905 ) were examined.

### RESULT

Results are summarized in Table 2.

1) Effect of AM stimulation ipsilateral to the claustral lesion

Following lesioning, ipsilateral AM stimulation produced significant modification of the kindled seizures with regression of the previously established stage 6 seizure to the earlier stages in 4 out of 5 cats.

Table 2. The effect of the ipsilateral claustral lesion on the previously established AM kindled seizure

| Cat | ADD (sec.) before | ADD (sec.) after | latency to convulsion (sec.) before | latency to convulsion (sec.) after | stage regression |
|---|---|---|---|---|---|
| No 889 | 72.72 ± 13.04 (n=11) | 91.80* ± 11.45 (n=15) | 34.82 ± 2.28 (n=11) | 52.73* ± 4.61 (n=15) | no stage regression |
| No 891 | 60.80 ± 1.30 (n=5) | 112.50* ± 14.85 (n=2) | 20.60 ± 1.34 (n=5) | 64.00* ± 11.31 (n=2) | regression to stage 3 |
| No 899 | 67.00 ± 7.39 (n=9) | 165.57* ± 18.58 (n=8) | 36.00 ± 6.78 (n=9) | 89.71* ± 24.53 (n=8) | regression to stage 3 |
| No 904 | 72.72 ± 13.04 (n=11) | 118.79* ± 18.97 (n=12) | 34.82 ± 2.28 (n=11) | 87.29* ± 13.60 (n=12) | regression to stage 4 |
| No 905 | 88.80 ± 10.78 (n=10) | 184.00* ± 20.07 (n=4) | 68.40 ± 7.51 (n=10) | 123.00* ± 11.94 (n=4) | regression to stage 3 |

abbreviations; ADD - afterdischarge duration, before - before lesioning the claustrum, after - after lesioning the claustrum, sec. - second, n - number,
* Significantly different before and after lesioning the claustrum, $p < 0.01$,
Values are mean ± S.D., The data were analyzed using Student t test.

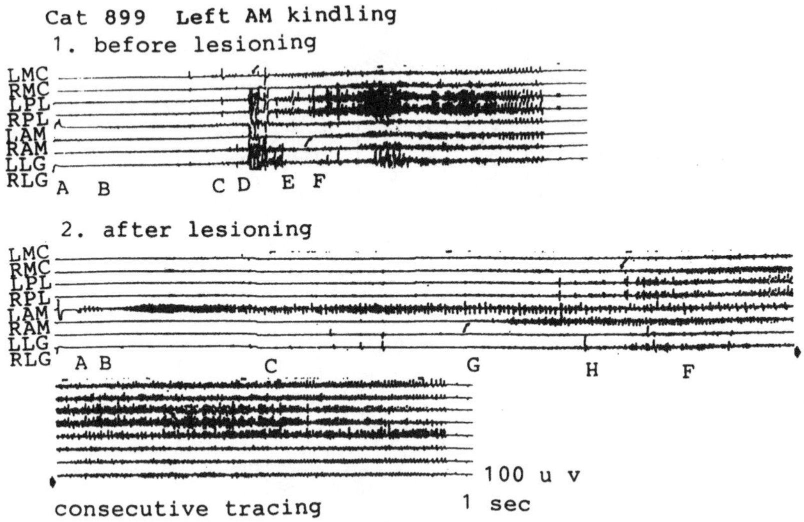


Fig. 3. Cat 899, Left AM kindling. After lesioning the left claustrum, the dominance of selfsustained afterdischarge in the left MC disappeared ( arrow ). The amplitude of AD in the left MC became smaller than that in the right MC. The latency to the onset of selfsustained discharge in the MC and contralateral AM were prolonged ( arrow ).
abbreviations; MC - motor cortex, PL - posterior lateral cortex, AM - amygdala, LG - lateral geniculate body.
A: left facial twitching, B: bilateral facial twitching, C: head nodding, D: head turning and circling to right side, and tonic extension of the right forepaw, E: clonic jumping, F: generalized convulsion, G: head turning and circling to left side, H: clonic convulsion of the left forepaw.

Both the length of the seizure ( afterdischarge duration, ADD ) the latency ( sec. ) to the onset of convulsion were significantly increased, when the seizure evolved to generalized tonic clonic convulsion ( GTC ). The seizure pattern after lesioning was strikingly different from the previously established one beginning at stage 4. Before lesioning, the animals showed contralateral headturning and circling, and subsequent tonic extension of the contralateral forepaw before GTC. The bilateral selfsustained seizure discharges ( 8 - 12 Hz, more than 100 $\mu$V ) occurred in the MC simultaneously or in the ipsilateral MC followed by contralateral MC. In contrast, post lesion animals showed mirror image the with ipsilateral headturning, circling and axial rotation evolving into an ipsilateral hemiconvulsion before GTC in 4 cats ( Nos. 889, 891, 899, 904 ). No 905 did not show the modified seizure pattern, and stage 5 absent in all cats. Higher amplitude selfsustained seizure discharge appeared earlier in the contralateral MC than ipsilateral motor cortex of 4 cats which showed the modified seizure pattern( Fig. 3 ). There was no change in the electroclinical presentation of kindled seizure in No 932.

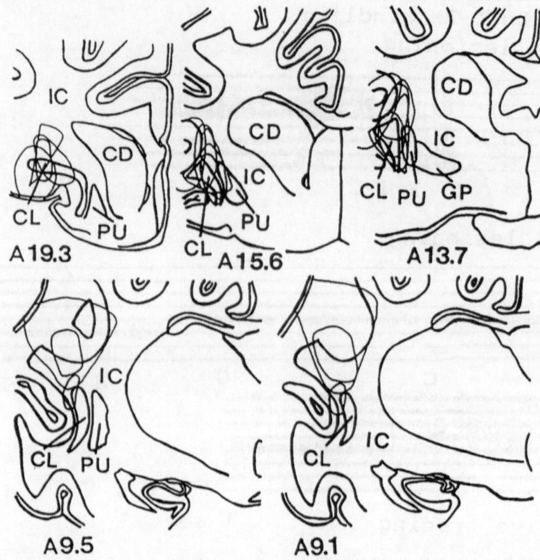


Fig. 4. The schematic representation of the effective lesions ipsilateral to the stimulating site.

As shown in Fig. 4, the common and critical part of the lesion involved the anterior part of the claustrum ( anterior to A 13.0 mm ). The anterior, middle and posterior ectosylvian gyri were lesioned partially in No 932.

2) Effect of AM stimulation contralateral to the claustral lesion

Following lesioning, contralateral AM stimulation reproduced the established kindled convulsion. Thus, there was no difference in seizure pattern, and the length of both ADD and the latency to the onset of the convulsion before and after lesioning. The lesion occupied the anterior claustrum and was larger than that of the ipsilateral group.

DISCUSSION

1) The effect of the claustral lesion on the development of the AM kindling and the established AM kindled seizure

In experiment I, the latency ( days ) to the first stage 4 seizure was significantly longer in the lesioned cat than in the nonlesioned one. Furthermore, the convulsive seizure stage regressed to the nonconvulsive seizure stages. In experiment II, after the ipsilateral lesion, the latency ( sec. ) to the convulsion was significantly increased. The established AM kindled seizure regressed to the nonconvulsive seizure. The critical lesion that induced these changes in AM kindling and the kindled seizure was in the anterior half of the claustrum. These results suggest that the claustral lesion interfered with the mechanisms of both AM seizure development and AM kindled convulsive seizure. When the convulsive seizure developed, it showed a pattern indicating initial involvement of the contralateral hemispheric motor mechanism, which was supported by the EEG findings. Therefore the claustral lesion seems to have interrupted the ictal accessing the stimulated AM to the ipsilateral hemispheric motor mechanism.

It has been reported that the ipsilateral motor cortex[1,2] and basal ganglia[3,4] participate in the AM kindled convulsion. In addition, the substantia nigra and thalamus are also involved in the development of the AM kindled seizure.[3,5] The claustrum has reciprocal connections with AM[8,9] on the one hand and with the frontal cortex[10], basal ganglia, substantia nigra[11,12,13] and thalamus[14,15,16,17] on the other. Therefore, the claustrum is strategically located for functional linkage between the AM and the cortical - subcortical motor mechanisms. The results of our experiment are consistent with this possibility.

The substantia innominata ( SI )is reported to play an important role in transforming the nonconvulsive seizure to a convulsive seizure,[6,7] since the intra - SI injections of musimol or gabaculin eliminate convulsive seizure but not limbic seizure. Whether or not kindled convulsion can occur through the contralateral motor mechanism with SI inactivation ( as in the case of claustral lesion ) is not known.

2) Transhemispheric positive transfer effect

In experiment I, the latency to the stable stage 6 seizure at the secondary site kindling was significantly increased with no evidence of a transhemispheric positive transfer effect. The latency (day) to the first stage 4 seizure was significantly prolonged and almost the same as that of primary site kindling of the nonlesioned cat. However, the latency from the first stage 4 seizure to the stage 6 seizure was significantly shorter than that of the primary site kindling in nonlesioned cats.

The corpus callosum[24], massa intermedia[26,27] and ventral hippocampus[25] are known to have a significant role in the transhemispheric positive transfer effect. However, there seem to be a different role among these three structures, because the bisection of the corpus callosum lengthened the latency ( day ) from the first stage 4 seizure to the stable stage 6 seizure state. The ventral hippocampal lesion lengthened the latency ( day ) to the first stage 4 seizure. The lesion of the massa intermedia lengthened both latencies. Therefore, the effect of the claustral lesion on the trashemispheric transfer effect was similar to the effect of the ventral hippocampal lesion. Since the claustrum has bilateral reciprocal connections with the limbic system including the entorhinal cortex, hippocampus, pyriform cortex,[21,22,23] the claustrum may participate in the interhemispheric positive transfer effect through the limbic system and its commissural extension.[31]

## CONCLUSION

1) In experiment I, the unilateral claustral lesion was place prior to ipsilateral AM kindling. Postoperative primary site AM kindling showed significantly delayed onset of stage 4 and stage 6 seizures with frequent regressions to nonconvulsive seizure stages. Most significantly, the kindled seizure pattern suggested initial involvement of the secondary site hemispheric motor mechanism. Subsequent secondary site kindling showed significantly delayed onset of stage 6, especially an increased latency to stage 4.

2) In experiment II, the unilateral claustral lesion was placed after completion of AM kindling. The ipsilateral AM stimulation resulted in regression of the kindled generalized seizure to earlier stages, prolongation of ADD , delayed onset of convulsion and the same modified seizure patterns as those observed in experiment I. The effective lesion involved in the anterior half of the claustrum in both experiments. The contralateral AM stimulation showed no change in the established seizure pattern.

3) We conclude that the anterior half of the claustrum a) participates in ictal accession of the AM onset partial seizure to the ipsilateral motor mechanism responsible for AM kindled convulsion, and b) plays an important role in the transhemispheric positive transfer effect.

ACKNOWLEDGEMENT

This work was supported by grants from the Medical Research Council of Canada to J. A. Wada.

REFERENCES

1. J. A. Wada, and A. Wake, Dorsal frontal, orbital and mesial frontal cortical lesion and amygdaloid kindling in cats, Can. J. Neurol. Sci. 4: 107 (1977)
2. J. A. Wada, T. Mizoguchi, and S. Komai, Cortical motor activation in amygdaloid kindling: observations in nonepileptic rhesus monkeys with anterior two - thirds callosal bisection,in: " Kindling 2," J. A. Wada, ed., Raven Press, New York (1981)
3. J. Jr. Engel, L. Wolfson, and L. Brown, Anatomical correlates of electrical and behavioral events related to amygdaloid kindling, Ann. Neurol. 3: 538 (1978)
4. M. Uno, and N. Ozawa, Neural mechanism of epileptic automatism in amygdaloid kindling, J. Jpn. Epil. Soc. 5: 122 (1987)(japanese)
5. J. O. McNamara, M. T. Galloway, L. C. Rigsbee, and C. Shin, Evidence implicating substantia nigra in regulation of kindled seizure threshold, J. Neurosci. 4: 2410 (1984)
6. K. Morita, M. Okamoto, K. Seki, and J. A. Wada, Suppression of amygdala - kindled seizure in cats by enhanced GABAergic transmission in the substantia innominata, Exp. Neurol. 89: 225 (1985)
7. M. Okamoto, and J. A. Wada, Reversible suppression of amygdaloid kindled convulsion following unilateral gabaculine injection into the substantia innominata, Brain Res. 305: 389 (1984)
8. I. Ishikawa, S. Kawamura, and O. Tanaka, An experimental study on the efferent connections of the amygdaloid complex in the cat, Acta Med. Okayama 23: 519 (1969)
9. J. E. Krettek, and J. L. Price, A description of the amygdaloid complex in the rat and cat with observations on intra - amygdaloid axonal connections, J. Comp. Neur. 178: 255 (1978)
10. G. Macchi, M. Bentivoglio, D. Minciacchi, and M. Molinari, The organization of the claustroneocortical projections in the cat studied by means of the HRP retrograde axonal transport, J. Comp. Neur. 195: 681 (1981)
11. D. L. Andersen, Some striatal connections to the claustrum, Exp. Neurol. 20: 261 (1968)
12. T. Arikuni, and K. Kubota, Claustral and amygdaloid afferents to the head of the caudate nucleus in macaque monkeys, Neurosci. Res. 2: 239 (1985)

13. P. Sloniewski, K. G. Usunoff, and C. Pilgrim, Diencephalic and mesencephalic afferents of the rat claustrum, Anat. Embryol. 173: 401 (1986)
14. P. Flindt - Egebak, and R. B. Olsen, Some efferent connections of the feline claustrum, Neurosci. Lett. (Suppl) 1: 159 (1978)
15. F. Hiddema, and J. Droogleever Fortuyn, The projection of the intermediary nuclear system of the thalamus and of the parataenial and paraventricular nucleus in the rat, Psychiat. Neurol. Neurochir. 63: 8 (1960)
16. J. Jr. Jimenez - Castellanos, and F. Reinoso - Suarez, Topographical organization of the afferent connections of the principal ventromedial thalamic nucleus in the cat, J. Comp. Neur. 236: 297 (1985)
17. J. L. Velayos, and F. Reinoso - Suarez, Prosencephalic afferents to the mediodorsal thalamic nucleus, J. Comp. Neurol. 242: 161(1985)
18. A. J. Gabor, and T. L. Peele, Alterations of behavior following stimulation of the claustrum, Electroenceph. Clin. Neurophysiol. 17: 513 (1964)
19. J. R. Stevens, I. Phillip, and R. Beaurepaire, r - vinyl - GABA in endopiriform area suppresses kindled amygdala seizures, Epilepsia 29: 404 (1988)
20. M. Norita, Demonstration of bilateral claustro - cortical connections in the cat with the method of retrograde axonal transport of horseradish peroxidase, Arch. histol. Jap, 40: 1 - 10 (1977)
21. H. J. Markowitsch, E. Irele, R. Bang - Olsen, and P. Flindt - Egebak, Claustral efferents to the cat's limbic cortex studied with retrograde and anterograde tracing techniques, Neurosci. 12: 409 (1984)
22. M. P. Witter, H. J. Groenewegen, and A. H. H. Lohman, Reciprocal connections of the insular and piriform claustrum with limbic cortex: an anatomical study in the cat, Neurosci. 24: 519 (1983)
23. P. Room, and H. J. Groenewegen, Connections of the parahippocampal cortex in the cat. II. subcortical afferents, J. Com. Neur. 251: 451 (1986)
24. J. A. Wada, T. Nakashima, and Y. Kaneko, Forebrain bisection and feline amygdaloid kindling, Epilepsia 23: 521 (1982)
25. T. Nakatsu, The effects of bilateral ventral hippocampal lesions on amygdaloid kindling in cats, Okayama Igakkai Zasshi 97: 855 (1985) (japanese)
26. T. Hiyoshi, and J. A. Wada, Midline thalamic lesion and feline amygdaloid kindling. I. Effect of lesion placement prior to kindling, Electroenceph. Clin. Neurophysiol. 70: 325 (1988)
27. T. Hiyoshi, and J. A. Wada, Midline thalamic lesion and feline amygdaloid kindling. II. Effect of lesion placement upon completion of primary site kindling, Electroenceph. Clin. Neurophysiol. 70: 339 (1988)
28. J. A. Wada, and M. Sato, Generalized convulsive seizures induced by daily electrical stimulation of the amygdala in cats, Neurology 24: 565 (1074)
29. A. Wake, and J. A. Wada, Transfer and interference in amygdaloid kindling in cats, Can. J. Neurol. Sci. 4: 5 (1977)
30. A. L. Berman, and E. G. Jones, "The thalamus and basal telencephalon of the cat. A cytoarchitectonic atlas with stereotaxic coordinates," The University of Wiscosin Press, Wiscosin (1982)
31. H. Fukuda, J. A. Wada, D. Riche, and R. Naquet, Role of the corpus callosum and hippocampal commissure on transfer phenomenon in amygdala-kindled cats, Exp. Neurol. 98: 189 (1987)

## Discussion of Dr. Kudo's Presentation

DR. SHOUSE: Your nice study now adds the claustrum to other extrapyramidal motor nuclei as an important structure affecting motor seizure generalization. Do you think the claustrum is more important, equally important, or less important in affecting this process?

DR. KUDO: I agree that the extrapyramidal system participates in the evolution of the seizure originating from the amygdala to generalized convulsion in amygdaloid kindled seizure. I can't say which site within the extrapyramidal system is most important in this process. However, the claustrum seems particularly involved in accession of amygdaloid onset partial seizure to the ipsilateral convulsive motor mechanism.

DR. FERNANDEZ-GUARDIOLA: I wonder if you found any EEG spiking over the temporal lobe cortex after lesioning the claustrum, particularly along the ectosylvian gyrus?

DR. KUDO: No, we have not investigated the lateral temporal cortical surface event in this experiment.

DR. MOSHE: What is the connection between the amygdala and the claustrum? Is there a topographic representation? Is the connection ipsilateral or bilateral?

DR. KUDO: There are reciprocal connections between the amygdala and the claustrum. There is a paper that reports the connection between the claustrum and the medial and posterior cortical amygdaloid nuclei. The connections are ipsilateral. I am not aware of any report on a bilateral connection between them.

# MIDLINE THALAMUS AND AMYGDALOID KINDLING

Yoshitaka Ehara and Juhn A. Wada

Divisions of Neurosciences and Neurology
University of British Columbia, Vancouver
British Columbia V6T 2A1, Canada

## INTRODUCTION

Among the midline structures, the role of the corpus callosum (CC) in the kindling model of epilepsy has been extensively investigated. Findings suggest that the CC is not important for seizure development from the amygdala (AM), hippocampus (HIPP) and temporal cortex, but that it is the major, if not the exclusive, anatomical substratum for convulsive seizure bilateralization and the development of bisynchronous bisymmetrical generalized convulsion (27,28,32, 41,52,53,55,56,57). As a first step in clarifing the possible role played by the midline structures in the elaboration of generalized seizure , we focused our attention on the non-specific thalamus which is known to participate not only in the modulation of widespread electrocortical activity (6,16,37) but also in the propagation of cortical seizure discharges (23,24,50).

In the kindling model, the results of thalamic lesioning studies suggest that the non-specific thalamic system participates in the transhemispheric positive transfer effect in cats (48,54) but not in rats (29). More recent studies in cats with massa intermedia (MI) lesion either prior to or after completion of primary site AM kindling showed : 1) a distinctively modified seizure pattern identical to that of forebrain-bisected cats; 2) the complete absence of positive transfer effects at secondary site kindling; while 3) a marked interference effect at the primary site re-test remained intact (13,14).

Midline nuclei such as the nucleus centralis medialis (NCM), rhomboideus , reuniens and parafascicularis project to the limbic system and widespread cortical areas including the motor cortex (areas 4 and 6, MC) , with a pattern of preferential projection to the prefrontal or cingulate areas (9,18,20,22,25,26,39,40,43,44). These projections are almost entirely unilateral. On the other hand, bilateral interconnections across the midline have been demonstrated among various nuclei of the intralaminar system (46). Similarly, transhemispheric motor corticosubcortical projection via the MI has been demonstrated (22,34,45).

*Kindling 4*
Edited by J. A. Wada
Plenum Press, New York, 1990

The possible role played by the MI in transhemispheric ictal transmission and positive transfer effect may involve either holizontal (interhemispheric) or vertical (intra-hemispheric) mechanisms. The present study was designed to gain insight into such mechanisms by means of intra-MI injection of ibotenic acid which is known to produce axon-sparing neuronal lesions (21). Under ideal experimental conditions, it was envisaged that ibotenic acid would destroy the MI neurons with ipsilateral projections while leaving intact the transhemispheric connections contained in the MI.

## MATERIALS AND METHODS

Twelve male cats weighing 3.8-4.6 kg were used. Technical details of surgery and electrode implantation have been published elsewhere (51). Briefly, depth electrodes made from twisted stainless steel wires, 0.35 mm in diameter and insulated except for the tips, with an interelectrode distance of 1 mm, were implanted bilaterally into the AM, HIPP and midbrain reticular formation (MRF). For the purpose of extradural cortical recording, stainless steel screws 1.0 mm in diameter were placed bilaterally in the anterior sigmoid (MC) and the anterior suprasylvian (SC) gyri. For injections into the MI, guido cannulae (23 gauge) was permanently implanted.

### Drug injections

A solution of ibotenic acid (IBO, Sigma)(5 ug/ul dissolved in phosphate buffer 50 nM, pH=7.4) was used for the lesioned group (I-group, n=6), while phosphate buffer was used for the control group (C-group, n=6). Intra-MI injections were made without anesthesia. An injection cannula (27 gauge) attached to a 5 ul Hamilton syringe by a polyethylene tube was inserted into the guido cannula and the drug was delivered in a volume of 2 ul at a rate of 0.5 ul/minute. The injection cannula was left in place for an additional 3 minutes to allow diffusion and minimize backflow of the injected drug.

### Kindling procedure

One week after the injection, primary site (left) AM kindling began. Monopolar stimulation consisting of 60 Hz sine wave lasting for 1 sec was applied. The initial AM stimulation began at 100 uA and was increased by 50 uA steps at 10 min interstimulus intervals until localized after discharge (AD) was induced. The last intensity producing AD was designated as the after discharge threshold (ADT) and was kept constant throughout kindling. The pattern of clinical seizure development by AM kindling was classified into 6 stages according to previous criteria (51). After five successive stage 6 seizures, the daily stimulus intensity was reduced by 50 uA steps until the animals ceased to respond with a kindled seizure. The last stimulus intensity which produced a stage 6 seizure was arbitrarily designated as the generalized seizure triggering threshold (GST).

After a 2-week rest interval, secondary site (right) AM kindling was started in a manner identical to primary site kindling. Twenty four hours after the completion of secondary site kindling, the primary site AM was restimulated at the previously established GST. Following five successive daily kindled convulsions, the GST was re-examined.

Upon completion of the experiment, the animals were deeply anesthetized by pentobarbital and the brains perfused with saline and 10 % formalin, serially sectioned and stained by cresyl violet or eosine and hematoxylin for histological examination. The extent of the lesions is shown in Fig. 1. The cannula tips were found in the intended structures. In 4 cats (# 921, 935, 942 and 943), the extent of cell loss and gliosis was asymmetrical with left predominance, and in 2 cats (# 930 and 931), the lesion was symmetrical.

## RESULTS

### 1) Behavior induced by inter-MI phosphate buffer or ibotenic acid injection

No behavioral or electrographic change was observed following intra-MI injection of phosphate buffer in the C-group. In contrast, a few minutes after the onset of the ibotenic acid injection all the I-group animals began to show stiffening of body associated with mydriasis, salivation, urination and defecation, as well as growling, hissing and spitting. With increased intensity, these behavioral manifestations, along with wild, aimless running and frequent bumping into objects, continued for about 30 minutes. This behavioral change was associated with a low voltage fast EEG pattern. As the running activity subsided, the animals began to lower their heads, while blinking eventually laying down after postneural adjustments, and finally curling up, accompanied by the electroclinical development of light sleep displaying miosis and full extension of the nictitating membranes. At this time, it was difficult to arouse the animals with auditory stimuli of moderate intencity. This sleeping state continued for 4 to 9 hours with EEG evidence of slow wave but not rem-sleep.

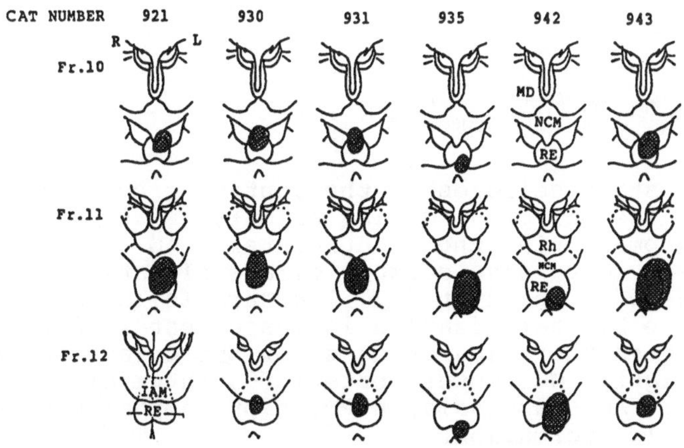


Fig.1 Schematic drawing of coronal extent with cell loss and gliosis : Numbers refer to planes of section from the atras of Jasper and Ajmone-Marsan (1954). R, right; L, left; MD, nucleus medialis dorsalis; NCM, nucleus centralis medialis; IAM, nucleus interanteromedialis; RE, nucleus reuniens; Rh, nucleus rhomboidens.

## Table 1

Primary site kindling

| ibotenic acid lesion group | | | | | | | |
|---|---|---|---|---|---|---|---|
| cat # | 921 | 930 | 931 | 935 | 942 | 942 | mean |
| A D T (uA) | 250 | 150 | 150 | 100 | 200 | 150 | 167±52 |
| G S T (uA) | 100 | 100 | 100 | 50 | 100 | 50 | 83±26 |
| latency to first stage 6 | 26 | 25 | 18 | 26 | 19 | 28 | 24±4 |
| latency to stable stage 6 | 26 | 27 | 18 | 26 | 19 | 39 | 26±8 |
| mean of AD duration | 51±4 | 110±14 | 77±28 | 64±31 | 71±42 | 71±10 | 74±30 |
| mean of axial rotation | 1±1(0-2) | 10±2(7-13) | 3±2(0-5) | 1±1(0-2) | 3±1(1-4) | 3±1(2-4) | 3±4(0-13) |

| control group | | | | | | | |
|---|---|---|---|---|---|---|---|
| cat # | 949 | 950 | 951 | 956 | 957 | 958 | mean |
| A D T (uA) | 150 | 250 | 250 | 150 | 150 | 100 | 175±61 |
| G S T (uA) | 100 | 50 | 100 | 100 | 100 | 100 | 92±20 |
| latency to first stage 6 | 31 | 28 | 17 | 25 | 36 | 20 | 26±7 |
| latency to stable stage 6 | 31 | 32 | 21 | 28 | 36 | 20 | 28±6 |
| mean of AD duration | 79±5 | 99±11 | 138±3 | 80±19 | 57±6 | 136±16 | 99±32 |
| mean of axial rotation | (0-1) | (0-1) | (0-3) | (0-1) | 0 | 0 | 0.5±0.7(0-3) |

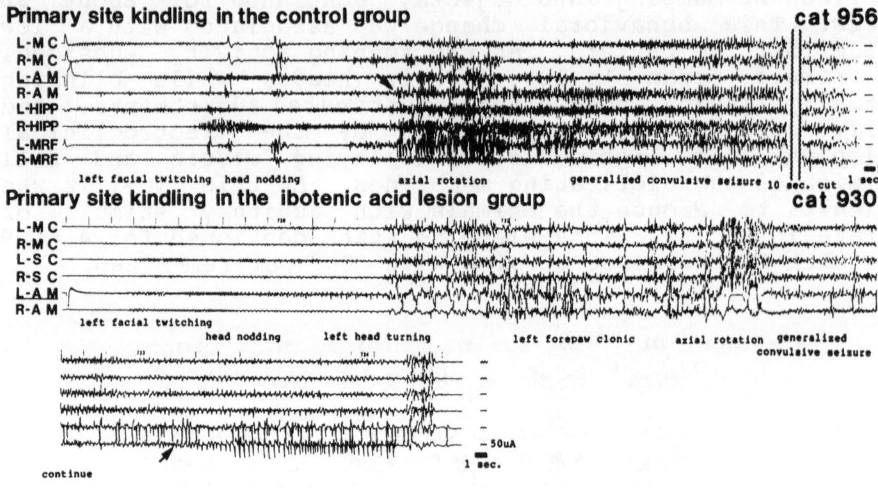


Fig.2 Final stage seizures in the control group and ibotenic acid lesion group during primary site kindling. AD development in the contralateral AM ( ↘ ) occurs before convulsive seizure bilateralization in control cat # 956, while it ( ↗ ) occurs following convulsive seizure bilateralization in lesion cat # 930. Underline indicates the stimulating site. Abbreviations: L, left; R, right; MC, moter cortex; SC, anterior suprasylvian gyrus; AM, amygdala; HIPP, hippocampus; MRF, midbrain reticular formation.

2) Primary site kindling

There was no significant difference in ADT, GST and AD duration between the I-group and the C-group. In the I-group, the mean number of AM stimulations required for the final stage 6 seizure development was 24 ( range 18-28 ) and a mean of an additional 3 stimulations was necessary to establish a stable final seizure stage for 5 consecutive days due to a marked seizure stage instability in 2/6 animals (Table 1-A). In the C-group, the mean number of AM stimulations required for the final stage 6 was 26, with an additional 2 stimulations for a stable stage in 4/6 instable animals (Table 1).

Clinically, a significant modification of the seizure patterns was noted in all the I-group animals with a complete absence of stages 2 and 5, in addition to repetitive contralateral axial rotation during stage 4. Thus, the animals elevated their contralateral forepaws far above their heads, developing clonic jerking associated with extreme contralateral head turning, resulting in axial body rotation, falling and subsequent righting. In the C-group, all the seizure stages were present and axial rotation was observed in only 4/6 animals. Although the I-group tended to show more frequent axial rotation than the C-group, there was no significant difference in the mean number of rotations during five stable stage 6 seizures between the two groups (Table 1). Electrographically, AD bilateralization involving the contralateral AM occurred during stages 3-4 in C-group, while the same was delayed till stage 6 in I-group. (Fig. 2).

3) <u>Secondary site kindling</u>

All animals eventually reached stage 6 with bisymmetrical clonic seizures. In the I-group, a mean of 23 (range 7-43) stimulations was required to reach the final stage. This was significantly longer than the mean of 6 (range 2-10) stimulations required in the C-group (P <0.05, Table. 2). Seizure stage instability with regression to

Table 2

Secondary site kindling

| ibotenic acid lesion group cat # | 921 | 930 | 931 | 935 | 942 | 942 | mean | |
|---|---|---|---|---|---|---|---|---|
| A D T (uA) | 100 | 150 | 150 | 200 | 100 | 200 | 150±45 | |
| G S T (uA) | 100 | 100 | 150 | 150 | 50 | 100 | 83±26 | |
| latency to first stage 6 | 7 | 24 | 21 | 27 | 14 | 43 | 23±12 | A |
| latency to stable stage 6 | 21 | 27 | 24 | 27 | 24 | 52 | 29±11 | B |
| mean of AD duration | 62±6 | 153±24 | 121±40 | 140±23 | 65±28 | 69±8 | 101±44 | |
| mean of axial rotation | 8±3(6-11) | 8±2(6-10) | 3±1(2-5) | 1±1(0-2) | 13±2(10-14) | 2±2(1-5) | 6±5(0-14) | |
| **control group cat #** | 949 | 950 | 951 | 956 | 957 | 958 | mean | |
| A D T (uA) | 200 | 300 | 200 | 150 | 150 | 200 | 200±55 | |
| G S T (uA) | 50 | 50 | 100 | 100 | 150 | 100 | 92±38 | |
| latency to first stage 6 | 7 | 5 | 2 | 8 | 6 | 10 | 6±3 | A |
| latency to stable stage 6 | 7 | 7 | 2 | 10 | 11 | 12 | 8±4 | B |
| mean of AD duration | 78±17 | 98±24 | 171±19 | 85±20 | 92±20 | 131±43 | 109±40 | |
| mean of axial rotation | 4±2(2-6) | 3±1(2-4) | 2±1(0-3) | 1±1(0-3) | 3±1(2-4) | 3±1(1-4) | 3±1(0-6) | |

A: P<0.05, B: P<0.01

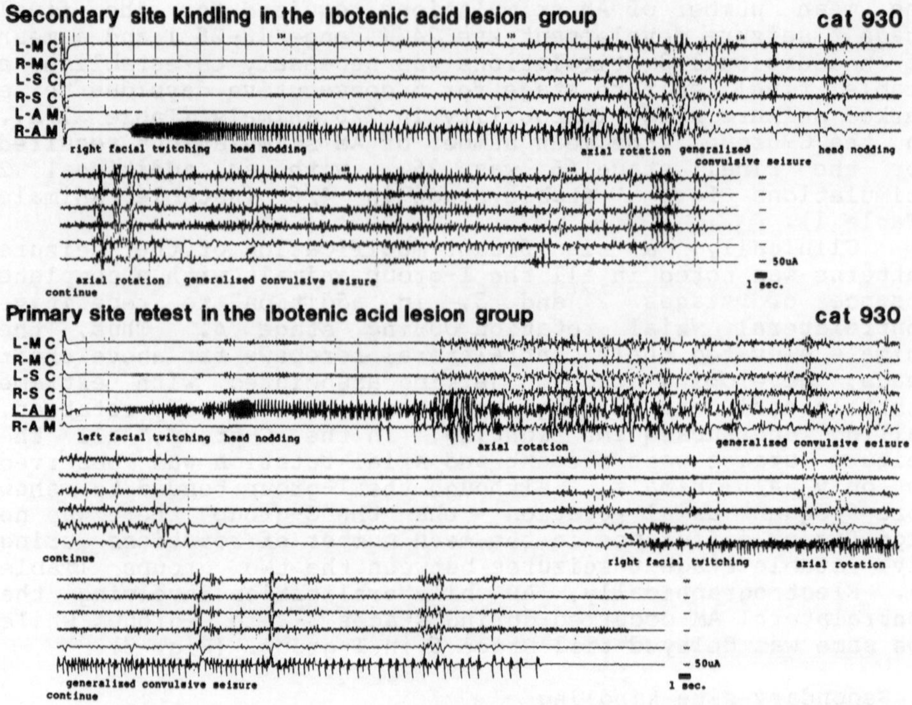


Fig. 3 Example of the predominantly hemiconvulsive seizure during secondary site kindling and primary site re-test in the ibotenic acid lesion cat. AD development in the contralateral AM (arrows) occurs following the predominantly hemiconvulsive seizure at the secondary site.

earlier stages was seen in both groups, but the mean number of stimulation required in I-group was significantly larger than that in the C-group: 29 (range 21-52) vs 8 (range 2-12)(Table 2). Thus, the positive transfer effect expected at the secondary site AM was absent in the I-group. The mean number of axial rotations at the secondary site in I-group was longer than that of the C-group. However, in both groups, the mean number of rotation was longer at the secondary site than at the primary site (Table 2).

In the I-group, 3 animals (#930, #931 and #935) showed sequential development of two independent electroclinical seizures. In these animals, AD began at the right AM (stimulating site) associated with right facial twitching and developed to predominantly hemiconvulsive seizure at the secondary site. Several seconds after the end of the first hemiconvulsive seizure, AD began in the left AM (non-stimulating site), associated with left facial twitching and progressed to a second predominantly hemiconvulsive seizure of primary site hemisphere origin (Fig.4). The seizure pattern was modified with complete absence of stages 2 and 5 in these sequential seizures. In addition both cat # 930 and # 931 showed two separate but sequentially developed of predominantly hemiconvulsive seizures, involving initially the left and then the right hemisphere in the primary site re-test. This was the mirror image of similar sequential but separate seizures observed in these animals during secondary

Table 3

Primary site re-test

| ibotenic acid lesion group cat # | 921 | 930 | 931 | 935 | 942 | 943 | mean |
|---|---|---|---|---|---|---|---|
| G S T (uA) | 200 | 150 | 150 | 100 | 100 | 100 | 133±41 |
| latency to first stage 6 | 2 | 1 | 1 | 1 | 2 | 3 | 2±1 |
| latency to stable stage 6 | 9 | 1 | 1 | 1 | 2 | 7 | 4±4 |
| mean of AD duration | 79±11 | 173±4 | 114±15 | 84±23 | 64±4 | 103±12 | 103±38 |
| mean of axial rotation | 9±2(7-14) | 12±6(8-21) | 4±2(2-6) | 3±1(2-4) | 13±2(10-16) | 5±1(3-6) | 8±5(1-21) |

| control group cat # | 949 | 950 | 951 | 956 | 957 | 958 | mean |
|---|---|---|---|---|---|---|---|
| G S T (uA) | 100 | 50 | 100 | 100 | 100 | 100 | 92±20 |
| latency to first stage 6 | 2 | 1 | 1 | 4 | 3 | 2 | 2±1 |
| latency to stable stage 6 | 2 | 1 | 1 | 4 | 3 | 2 | 2±1 |
| mean of AD duration | 107±8 | 168±60 | 181±35 | 86±26 | 95±11 | 96±31 | 121±44 |
| mean of axial rotation | 1±1(0-4) | 2±1(2-3) | 2±1(1-3) | 1±1(0-1) | 4±1(3-5) | 3±1(1-4) | 2±1(0-5) |

site kindling. In the C-group, AD propagation to the contralateral AM occurred during stages 3-4, and was identical to that observed during primary site kindling.

4) Primary site re-test

The mean number of AM stimulations required for stage 6 seizure was 2 (range 1-4) in the C-group and 2 (range 1-3) in the I-group (Table 3). Thus, the interference phenomenon expected at the primary site AM was present in both groups. There was no significant difference in AD duration and GST. However, the mean number of axial rotation in the I-group was significantly larger than that of the C-group ( $P<0.05$, Table 3).

DISCUSSION

1) Behavior after ibotenic acid injection

An motional display of fear, defense or aggression can be induced by electrical stimulation of the hypothalamus and midbrain (11,38). Although a similar rage or rage-like reaction can also be elicited by AM and HIPP stimulation (7,31,49), the quality of the emotional display appears different depending on whether it is due to hypothalamic or limbic stimulation. "Aggressive response", elicited from mid-hypothalamic stimulation sites, is said to be generally well-directed and purposeful (38), while limbic "defensive attack" is said to be characterized by aimless running (7,31). In this study, behavioral change produced by intra-MI ibotenic acid injection was characterized by agitated aimless running, threatening and autonomic disturbance. Marked sympathetic activation with "sham rage", extreme pupillary dilation, wild running and climbing were produced when stimulating the nucleus reuniens in the midline and other nuclei along the mammillo-thalamic tract on one side (15). Therefore, it is possible that the emotional display might have resulted as a consequence of direct MI stimulation. On the other hand, since it is well known that the NCM, the rhomboideus, the interanteromedialis and the reuniens project to the limbic structures (3,4,10,33,42), such behavioral display may be due to secondary activation of the limbic mechanism through primary activation of intra-MI injection of ibotenic acid, one of the excitatory amino acids. Animals in the I-group fell asleep 30 minutes after the injection, displaying only slow wave sleep (SWS) which lasted for 4 to 9 hours. Hess reported that the low frequency electrical stimulation of the lower two-thirds of the MI produced sleep in cats (12). A similar effect was also reported in dogs and rabbits (2,35). In man, stimulation of the medial thalamus more often evokes arousal than relaxation, although, sleep effects have also been reported (1,19). Possibile humoral transmission of a slow wave sleep inducing substance (DSIP) was examined by intracerebra ventricular infusion of extracorporal dialysate of blood obtained from the sinus confluens of donor rabbits whose sleeping state was maintained by continued electrical stimulation of the ventromedian intralaminar thalamus (36,47). This finding suggests that the sleep state induced by intra-MI injection of ibotenic acid may be due to a similar mechanism.

2. Electroclinical seizure development

Axial rotation, which is a reflection of a powerful adversive and asymmetrical convulsive seizure, was seen in both I- and C-groups, although the I-group a showed larger number of rotations. This feature is also seen in electrolytically MI-lesioned and forebrain-bisected cats (13,14,52). In both I- and C-groups, the repetitive axial rotation persisted throughout primary site kindling and became more pronounced during secondary site kindling and primary site re-test. Therefore, the increased number of axial rotations during secondary site kindling and primary site re-test likely represent a transhemispheric interference effect.

The pattern of seizure development in the majority of animals in I-group was significantly different with that of the C-group. The I-group cats lacked stages 2 and 5. These features are also shared by animals with forebrain-bisection undergoing AM kindling (52). In the I-group, AD development in the contralateral AM was significantly delayed and this might be related to the absence of stage 2 seizure, since ipsilateral facial twitching is known to result from AM stimulation (5,8,30,58).

In the I-group, some animals ( # 930, # 931 and # 935) showed that the seizure repetition was actually two independent predominantly hemiconvulsive seizures in secondary site kindling and primary site re-test. These findings were also seen in animals with electrolytic MI lesion. In contrast, in the control group, the independent AD in the contralateral AM occurred before seizure generalization and there was no seizure repetition. These findings suggest that two alternative routes are available for transhemispheric ictal propagation in AM kindling, because the convulsive seizure generalized with delayed AD development in the contralateral AM. The neuronal aggregate in the MI may be concerned with the ictal transmission between AM-AM via a yet-to-be identified subcortical pathway. The midline thalamic nuclei, such as the NCM, rhomboideus, reuniens and interanteromedialis are reported to project to the AM and HIPP in rats, cats and monkeys (3,4,10,34,42). and the reciprocal nature of the projection between the nucleus intraanteromedialis and the AM has also been documented (18), these projections were largely ipsilateral. Results of the present study suggest the possibility that the neuronal aggregate, rather than passing fibers in the MI, participates in AM-AM AD propagation. If this assumption is correct, the role of the MI in AM kindling is likely to be primarily intrahemispheric rather than interhemispheric.

In the animals with electrolytic MI lesion, the development of initially lateralized, but subsequently generalized, convulsive seizures were coincident with the development of sustained 8-12 Hz incremental rhythmic discharge in the MC contralaterally. This finding suggested that the MI modulates transhemispheric ictal transmission between the two MCs, presumably through the corpus callosum (13). In the present study, however, development of sustained MC discharge was not clearly demonstrated as in the previous study but the reason for this difference is not known. Based on the observation of thalamic participation in AD propagation between the two MCs following unilateral cortical placement of tungstic acid in cats, the suggestion was made that thalamocortical positive feedback plays an important role in transcallosal propagation (23,24).

3) Transhemispheric positive transfer and interference

At the secondary site AM, seizure developed very abruptly in the C-group, while the I-group required a number of AM stimulations larger than that of the primary site kindling. This finding confirmed the results of electrolytic MI-lesioned cats undergoing secondary site AM kindling and supports the theory that the MI participates significantly in the mechanism of positive transfer effect. We have indicated that the MRF plays a critical role in the kindling process, and further suggest that only through the unique participation of the brainstem (including the MRF) can animals acquire bilateral epileptogenicity (52). Therefore, it is possible that MRF may play a role in the mechanism underlying the role of the MI involved in AM kindling. The fact that upon completion of primary site AM kindling, MI lesion placement, but not CC bisection, selectively eliminates the positive transfer effects without modifying the previously established primary site AM kindled seizure pattern, suggests that the MI plays a primary role in the transhemispheric positive transfer effect (14). All the findings obtained through our previous CC bisection studies and the present series of MI lesion studies suggest that the CC is critical for the patterning of bisymmetrical convulsive seizure, while the MI is critical for making the kindled neuronal trace within the primary site hemisphere accessible to the secondary site hemisphere for a transhemispheric positive transfer effect to occur.

In this study, it was observed that both C- and I-groups showed an increased number of axial rotations during secondary site kindling and primary site re-test. Similarly, slow seizure development and delayed onset of independent AD in the contralateral AM were all present during secondary site kindling and primary site re-test. This is consistent with our previous assumption that anatomical mechanisms underlying the transhemispheric positive and negative effects are different (54).

SUMMARY

The effect on feline amygdaloid (AM) kindling of ibotenic acid injection into the massa intermedia (MI,n=6) was compared with that of phosphate buffer injection. The latter group showed the expected pattern of primary site AM seizure development and positive transfer effect at the secondary site AM. In contrast, the ibotenic acid lesioned animals showed a significantly modified pattern of seizure development with complete absence of stages 2 and 5, as well as the emergence of repetitive axial rotation. In addition, no positive transfer effect was present at the secondary site kindling. These findings are strikingly similar to those of AM kindling in forebrain bisected or electrolytic MI lesioned cats. When the primary site was re-tested, a marked interference effect was observed in all the animals. Therefore, the findings suggest that the anatomical substratum for the transhemispheric positive transfer effect and the post secondary site negative transfer (interference) effect is not the same. It is concluded that 1) the MI plays an important role for both intrahemispheric ictal propagation and for transhemispheric positive transfer effect, and 2) the neuronal aggregate, rather than passing fibers, in MI participates in these effects. It is postulated that the role of the MI in AM kindling is primarily intrahemispheric rather than interhemispheric.

ACKNOWLEDGEMENT

This study was supported by grants from the Medical Research Council of Canada to J. A. Wada.

REFERENCES

1) Akimoto, T. (1954) On the syndrome induced by electrical stimulation of the thalamus. Physiological and clinical studies on the diencephalon. Igaku Shoin, Tokyo (Japanese) (cited in ref. 4)
2) Akimoto, H., Yamaguguchi, N., Okabe, K., Nakagawa, T, Nakamura, I., Abe, K. Torii, I. and Masahashi, I. (1956) On the sleep induced through electrical stimulation on dog thalamus. Folia Psy. et. Neur. Japo., 10: 117-146
3) Aggleton, J.P., Burton, M.J. and Passingham, R.E. (1980) Cortical and subcortical afferents to the amygdala of the rhesus monkey(Macaca mulatta). Brain Res., 190: 347-368
4) Baisden, R.H., Hoover, D.B. and Cowie, R.J. (1979) Retrograde demonstration of hippocampal afferents from the interpeduncular and renuiens nuclei. Neuro. Letters, 13: 105-109
5) Baldwin, M., Frost, L.L. and Wood, C.D. (1954) Investigation of the primate amygdala: movement of the face and jaws. Neurology, 4: 586-598
6) Dempsey, E.W. and Morison, R.S. (1942) The production of rhythmically recurrent cortical potentials after localized thalamic stimulation. Am. J. Physiol., 135: 293-300
7) Fernandez de Molina, A. and Hunsperger, R.W. (1959) Central representation of affective reactions in forebrain and brain stem: electrical stimulation of

amygdala, stria teminalis and adjacent structures. J. Physiol., 145, 251-265.

8) Gastaut, G.H., Naquet, R., Meyer, A., Cavanagh, J.B. and Beck, E. (1959) Experimental psychomotor epilepsy in the cat Electroclinical and anatomopathological correlations. J. Neuropath. Exp. Neurol., 18: 270-293

9) Hendry, S.H.C., Jones, E.G. and Graham, J. (1979) Thalamic relay nuclei for cerebellar and certain related fiber system in the cat. J. Comp. Neurol., 185: 679-714

10) Herkenham, M. (1978) The connections of the nucleus reuniens thalami: evidence for a direct thalamo-hippocampal pathway in the rat. J. Comp. Neurol., 177: 589-610

11) Hess, W.R. and Brugger, M. (1943) Das subcorticale Zentrum der affektiven Abwehrreaktion. Helv. Physiol. acta, 1: 33-52

12) Hess, W.R. (1968) Hypothalamus und Thalamus. Georg Thieme Verlag, Stuttgart

13) Hiyoshi, T. and Wada, J.A. (1988) Midline thalamic lesion and feline amygdaloid kindling. 1. Effect of lesion placement prior to kindling. Electroenceph. clin. Neurophysiol., 70: 325-338

14) Hiyoshi, T. and Wada, J.A. (1988) Midline thalamic lesion and feline amygdaloid kindling. 2. Effect of lesion placement upon completion of primary site kindling. Electroenceph. clin. Neurophysiol., 70: 339-349

15) Hunter, J. and Jasper, H.H. (1949) Effect of thalamic stimulation in unanaesthetized animals. Electroenceph. clin. Neurophysiol., 1: 305-324

16) Jasper, H.H. (1949) Diffuse projection system: the integrative action of the thalamic reticular system. Electroenceph. cli. Neurophysiol., 1: 405-420

17) Jasper, H.N. and Ajmone-Marsan, C. (1954) A stereotaxic atras of the diencephalon of the cat, National Research oh Canada, Ottawa

18) Jones, E. G. and Leavitt, R.Y. (1974) Retrograde axonal transport and the demonstration of non-specific projections to the cerebral cortex and striatum from thalamic intralaminar nuclei in the rat, cat and monkey, J. Comp. Neurol., 154: 349-378

19) Jung, R. (1957) Tierexperimentelle Grundlagen und EEG-Untersuchungen bei bewusstseinsveranderungen des Menschen ohne neurologische Erkrankungen. In " Premier Congres International des Sciences Neurologiques. Rapports et Discussions," Bruxelles, Acta Medica Belgica, (ref. pp. 148-179)

20) Kaitz, S.S. and Robertson, (1981) R.T. Thalamic connections with limbic cortex. 2. Corticothalamic projections. J. Comp. Neurol., 195: 527-545

21) Kohler, C. and Schwarcz, R. (1983) Comparison of ibotenate and kainate neurotoxicity in rat brain: a histological study. Neuroscience, 8: 819-835.

22) Kunzle, H. (1976) Thalamic projections from the precentral moter cortex in Macaca fascicularis. Brain Res., 105: 253-267

23) Kusske, J.A. (1976) Interactions between thalamus and cortex in experimental epilepsy in the cat, Exp. Neurol., 50: 568- 578

24) Kusske, J.A. and Rush, J.L. (1978) Corpus callosum and propagation of afterdischarge to contralateral cortex and thalamus. Neurology, 28: 905-912

25) Macchi, G., Quattrini, A., Chinzari, P., Marchesi, G., and Capocchi, G. (1975) Quantitative data on cell loss and cellular atrophy of intralaminar nuclei following cortical and subcortical lesions. Brain Res., 89: 43-59
26) Macchi, G., Bentivoglio, M., D'Atena, C., Rossini, P and Tempesta, E. (1977) The cortical projections of the thalamic intralaminar nuclei restudied by means of the HRP retrograde axonal transport. Neurosci. Lett., 4: 121-126
27) MaCaughran, Jr., J.A., Corcoran, M.E. and Wada, J.A. (1977) Facilitation of secondary-site amygdaloid kindling following bisection of the corpus callosum and hippocampal commissure in rats. Exp. Neurol., 57: 132-141
28) MaCaugran, Jr., J.A., Corcoran, M.E. and Wada, J.A. (1978a) Role of the forebrain commissures in amygdaloid kindling in rats. Epilepsia, 19: 19-33
29) MaCaugran, Jr., J.A., Corcoran, M.E. and Wada, J.A. (1978b) Role of the nonspecific thalamus in amygdaloid kindling. Exp. Neurol., 58: 471-485
30) MacLean, P.D. and Delgado, J.M.R., (1953) Electrical and chemical stimulation of frontotemporal portion of limbic system in the waking animal. Electroenceph. clin. Neurophysiol., 5: 91-100
31) MacLean, P.D. (1957) Chemical and electrical stimulation of hippocampus in unrestrained animals. 1-11, Arch. Neurol. Phychiatr., Chicago, 78, 113-142
32) McIntyre, D.C. and Stuckey, G.N. (1985) Dorsal hippocampal kindling and transfer in split-brain rats. Exp. Neurol., 87: 86-95
33) Mehler, W.R. (1980) Subcortical afferent connections of the amygdala in the monkey, J. Comp. Neur., 190: 733-762
34) Molinari, M., Minciacchi, D., Bentivoglio, M. and Macchi, G. (1985) Efferent fibers from the moter cortex terminate bilaterally in the thalamus of rats and cats. Exp. Brain Res., 57: 305-312
35) Monnir, M.,and R. Tissot, Correlated effects in behaviour and electrical brain activity evoked by stimulation of the reticular system, thalamus and rhiencephalon in the conscious animal. In "A Ciba Foundation Symposium: On the Neurological Basis of Behaviour." London, Churchill, 1958
36) Monnir, M. and Schoenenberger, G.A. (1977) Characterization, sequence, synthesis and specificity of a delta (EEG) sleep-inducing peptide. In "Sleep 1976", eds. W.P. Koella and P. Levin, Karger, Basel, 257-263
37) Morison, R.S. and Dempsey, E.W. (1942) A study of thalamo-cortical relations. Am. J. Physiol., 135: 281-292
38) Nakao, H. (1958) Emotional behavior produced by hypothalamic stimulation. Amer. J. Physiol., 194, 411-418
39) Niimi, K., Niimi, N. and Okada, Y. (1978) Thalamic afferents to the limbic cortex in the cat studied with the method of retrograde axional transport of horseradish peroxidase. Brain Res., 145: 225-238
40) Niimi, K., Matsuoka. H., Aisaki, T. and Okada, Y. (1981) Thalamic afferents to the prefrontal cortex in the cat traced with horseradish peroxidase. J. Hirnforsch., 22: 221-241
41) Okamoto, M. (1982) An experimental study of temporal

cortical kindling in cats: the effects of midline-bisection on seizure generalization mechanism. Psychiat. Neurol. Jpn., 84: 48-67
42) Ottersen, O.P. and Ben-Ari, Y. (1979) Afferent connections to the amygdaloid complex of the rat and cat. J. Comp. Neur. 187: 401-424
43) Rinvik, E. (1968) The corticothalamic projection from the pericruciate and coronal gyri in the cat. An experimental study with silver-impregnation methods. Brain Res., 10: 79-119
44) Robertson, R.T. and Kaitz, S.S. (1981) Thalamic connections with limbic cortex. l. Thalamocortical projections. J. Comp. Neurol., 195: 501-525
45) Sakai, S.T. and Tanaka, Jr., D. (1984) Contralateral corticothalamic projections from area 6 in the racoon. Brain Res., 299: 371-375
46) Scheibel, M.E. and Scheibel, A.B. (1967) Structural organization of nonspecific thalamic nuclei and their projection toward cortex. Brain Res., 6: 60-94
47) Schoenenberger, G.A. (1984) Characterization, properties and multivariate functions of delta-sleep-inducing peptide (DSIP). Eur. Neurol., 23: 321-345
48) Sterman, M.B. and Shouse, M.N. (1981) Kindling and sleep: a new direction in the search for mechanism. J.A. Wada. Ed Kindling 2. Ravan Press, New York, 137-148
49) Ursin, H. and Kaada, B.R. (1960) Functional localization within the amygdaloid complex in the cat. EEG Clin. Neurophysiol., 12, 1-20
50) Wada, J.A. and Cornelius, L. (1960) Functional alteration of deep structures in cats with chronic focal cortical irritative lesion. Arch. Neurol., 3: 425-447
51) Wada, J.A. and Sato, M. (1974) Generalized convulsive seizures induced by daily electrical stimulation of the amygdala in cats. Neurology, 24: 565-574
52) Wada, J.A. and Sato, M. (1975) The generalized convulsive seizure state induced by daily electrical stimulation of the amygdala in split-brain cats. Epilepsia, 16: 417-430
53) Wada, J.A., Mizoguchi, T. and Komai, S. (1981) Cortical motor activation in amygdaloid kindling: observations in nonepileptic rhesus monkeys with anterior two-thirds callosal bisection. In: J.A. Wada Kindling 2 Ravan Press, New York, 235-248
54) Wada, J.A. (1982) Secondary cerebral functional alterations examined in the kindling model of epilepsy. In: A.Mayersdorf and R.P. Schmidt , Secondary Epileptogenesis. Ravan Press, New York, 45-87
55) Wada, J.A., Nakashima, T. and Kaneko, Y. (1982) Forebrain bisection and feline amygdaloid kindling. Epilepsia, 23: 521-530
56) Wada, J.A. and Mizoguchi, T. (1984) Limbic kindling in the forebrain bisected photosensitive baboon, Papio papio. Epilepsia 25: 278-287
57) Wada, J.A. and Komai, S. (1985) Effect of anterior two-thirds callosal bisection upon bisymmetrical and bisynchronous generalized convulsions kindled from amygdala in epileptic baboon, Papio papio. In: A.G. Reeves, Epilepsy and the Corpus Callosum. Plenum Press, New York, 75-97
58) Youmans, J.R. (1956) Experimental production of seizures in the Macaque by temporal lobe lesions. Neurology, 6: 179-186

Discussion of Dr. Ehara's Presentation

DR. WADA: Dr. McIntyre, you are the pioneer in this area and we are all intrigued. I so vividly remember, Dr. Burnham showing the importance of AD propagation to other sites for positive transfer to occur, and your study showed no AD propagation, yet there was a positive transfer effect; then I think Dr. Racine stood up and said, "Well, you have to tell us where the action is, brainstem, etc." Do you have any comments to make now?

DR. MCINTYRE: Well, I would be very interested in seeing some of those fornical lesion data replicated in the rat, to see if indeed similar mechanisms might be operating. Some of the lesions that Ferrendelli was describing in the mamillothalamic system might somehow interface with this as well. But really, I'm just beginning to think that a cat is not a rat!

DR. WADA: I'd like to share our experience with you about the mamillothalamic tract of Ferrendelli. We thought that maybe if we made a lesion in the mamillary body we might get more clear-cut information as to its role in positive transfer, but all the rats died with a very small electrolytic lesion for whatever reason.

Now, before the general discussion, I wish to call upon two people who are going to contribute to our symposium. First, Dr. Berman who is going to give us some information about the role of adenosine.

# ADENOSINE INVOLVEMENT IN KINDLED SEIZURES

Robert F. Berman[1], Michael F. Jarvis[2] and Carl R. Lupica[1]

[1]Department of Psychology, Wayne State Univ. Detroit, MI and [2]CIBA Geigy Corp., Summit, N.J.

Our interests have been focused on determining the role of adenosine in the development and maintenance of kindled seizures. The results from our studies, and those of others now indicate that adenosine may have a role in terminating ongoing seizure activity and may play a principal role in the development and expression of postictal events (i.e., EEG depression, spiking, refractory period). If adenosine contributes to seizure termination, it probably also serves to limit the generalization of seizures and thus the rate of kindling. If so, its role in kindling may be even more general than indicated by current available data. Of course, no single neurotransmitter or neuromodulator is likely to be solely responsible for the kindling phenomenon. In fact, several other neurotransmitters and neuromodulators are known to influence kindling and are undoubtedly involved. These include norepinephrine, acetylcholine, glutamate, GABA and probably others as well. Studies describing these systems are well represented by several papers in this volume. However, this review is focused only on adenosine, and specifically its possible role in kindling and kindling related phenomena (i.e., postictal depression, spiking). Several recent reviews have described the general physiology and pharmacology of adenosine in detail[11,31,43,51] and only a brief summary is provided below.

## ADENOSINE AS A NEUROMODULATOR

Adenosine is considered to be an important neuromodulator with anticonvulsant, hypnotic, anxiolytic and hypotensive actions. It has not achieved the status of neurotransmitter, but is widely recognized as having powerful neuromodulatory influences on synaptic transmission[11,50]. Its effects on neuronal activity are uniformly inhibitory[11]. Spontaneous neuronal firing in the cortex, thalamus, hippocampus and cerebellum is inhibited by iontophoretic application of adenosine[31]. Adenosine acts both presynaptically and postsynaptically. Presynaptically, it has the ability to inhibit release of a variety of excitatory

*Kindling 4*
Edited by J. A. Wada
Plenum Press, New York, 1990

and inhibitory neurotransmitters, including glutamate, acetylcholine, norepinephrine, and GABA[11,16]. The mechanism may be via reduction in calcium entry into presynaptic nerve terminals, thereby blocking calcium-dependent transmitter release. Postsynaptically, neurons are hyperpolarized by adenosine through an increase in potassium conductance[47]. Both mechanisms probably contribute to the inhibitory effects of adenosine, but the presynaptic effects may be primary. Adenosine is known to be localized in nerve terminals, and has been shown to be released from brain slices[21] and synaptosomes[44] in an apparently calcium-dependent fashion. A specific high affinity uptake mechanism exists[4], and the inactivation of free adenosine by adenosine deaminase is well characterized[33]. A variety of sources for released adenosine have been suggested. Most is probably released either as free adenosine, or is produced by hydrolysis of ATP, ADP or 5'-AMP by 5'-nucleotidase[29]. Other possible sources of adenosine have also been described[11,31,43,51].

The best available data indicate that basal brain levels of adenosine are approximately within the range of 0.5-2.0 micromolar[7,55]. These levels have been argued to be sufficient to provide at least partial activation of adenosine receptors under basal conditions[11]. Precise localization of adenosine pools or projections in brain has been difficult because of the ubiquitous presence of this purine, and the fact that its endogenous levels are so closely linked to ATP and AMP. However, immunoreactive adenosine is highly localized in pyramidal cells of the hippocampus, deep layers of the medial cerebral cortex, caudate nucleus and basolateral amygdala[43]. Autoradiographic mapping studies provide clear evidence for high levels of adenosine receptors in the hippocampus, striatum, cerebellum, superior colliculus, thalamus and cerebral cortex[18]. Immunoreactive neurons for adenosine deaminase, a possible marker for adenosine containing neurons, are found in the posterior basal hypothalamus. Neurons in this region have widespread projections to diverse brain regions including the cortex, striatum, and amygdala[42]. Therefore, most brain regions of interest for kindling studies show evidence of adenosine activity.

The effects of adenosine are mediated at specific cell surface adenosine receptors, A1 and A2[7]. These subtypes have been classically differentiated on the basis of their pharmacological profiles and by their differing effects on adenylate cyclase. The A1 subtype has a nanomolar affinity for adenosine and inhibits cyclase activity, while the affinity for the A2 receptor subtype for adenosine is in the micromolar range and is associated with an increase in cyclase activity. These subtypes also have a differential distribution, with the majority of the high affinity A2 sites being concentrated in the striatum[19]. The methylxanthines, including caffeine and theophylline, are potent adenosine receptor antagonists at both receptor subtypes[39]. Most of the electrophysiological effects of adenosine, at least in the *in vitro* hippocampus, have been attributed to activity at the A1 receptor.

## ADENOSINE RELEASE DURING SEIZURES AND ELECTRICAL STIMULATION

Some of the earliest evidence suggesting a role for adenosine in seizure termination came from studies demonstrating increases in brain adenosine levels following electrical brain stimulation[23,32] or electroshock seizures[40]. Seizures induced by the GABA antagonist, bicuculline, also result in increased brain levels of adenosine[52,53,30]. Within 10 sec of seizure onset brain adenosine levels are increased approximately 10-fold, and remain elevated for at least 60 sec. Direct measurements of adenosine by dialysis of interstitial fluid in frontal cortex show a 7.9-fold increase following bicuculline-induced seizures[30]. Finally, low intensity electrical stimulation of the perforant pathway in rats, a common stimulation site for hippocampal kindling and long-term potentiation, results in a release of radiolabeled adenosine from in vitro hippocampal slices[22]. Fimbria stimulation, is ineffective. Therefore, adenosine can be released following direct synaptic activation.

Episodes of cerebral hypoxia can also increase brain adenosine levels[55]. This could provide an additional mechanism for adenosine release during prolonged seizures, but is unlikely to play a major role during the relatively brief seizures typical of kindling.

## ANTICONVULSANT PROPERTIES OF ADENOSINE AND ADENOSINE ANALOGS

In an initial study, Albertson, et al.[2], reported that the adenosine analog 2-chloroadenosine produced a small reduction in seizure stage and afterdischarge (AD) duration in amygdala-kindled rats following systemic administration of the drug. These effects were antagonized by aminophylline, which had potent proconvulsive properties of its own. Later studies by Dragunow and Goddard[8] and Dragunow, et al.[9], demonstrated that adenosine (100 mg/kg, ip) reduced the AD duration in amygdala kindled animals, and that 2-chloroadenosine reduced both the AD duration and seizure stage. Again, adenosine antagonists (isobutylmethylxanthine and caffeine) greatly increased seizure severity. Papaverine, an adenosine uptake inhibitor, delayed the rate of kindling when injected systemically 20 min before daily kindling trials. These data support the argument that manipulation of endogenous adenosine can influence the development of kindling. We have observed similar phenomena using the potent adenosine analogs, L-isopropyladenosine (L-PIA) and N-ethylcarboximide adenosine (NECA). These compounds can reduce kindled seizure severity, and delay the rate of kindling[36,49]. As shown in Figure 1, the development of Stage 5 amygdala-kindled seizures was significantly slowed by systemic injections of NECA given 20 min before each daily kindling session.

The anticonvulsant effects of adenosine can also be demonstrated following central administration. Direct intracerebroventricular (ICV) injections of stable adenosine analogs (L-PIA and NECA) in amygdala-kindled rats block fully kindled seizures[3]. Chemitrode injections of L-PIA or NECA

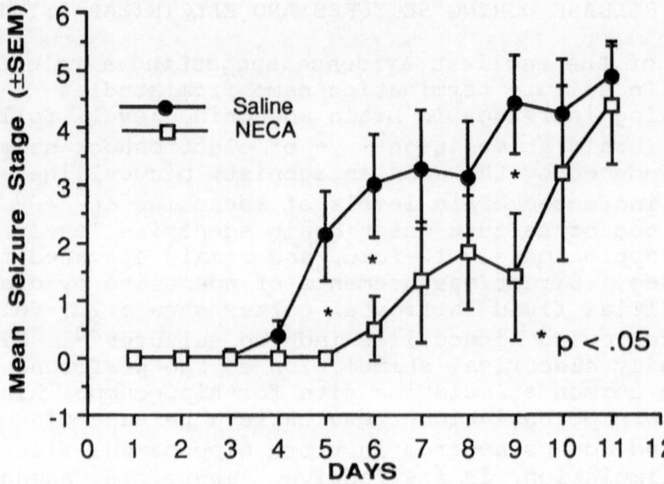


Figure 1. Daily injections of NECA (0.032 mg/kg, ip; open boxes) given 20 min before stimulation delayed the rate of amygdala-kindling in rats compared to saline injected animals (filled circles).

directly into the kindled focus are also effective, whether injected directly into a seizure focus kindled in the amygdala, hippocampus or caudate nucleus[37]. Afterdischarges can still be elicited, but only when much higher current levels are used. This indicates that one effect of activation of adenosine receptors may be to raise the kindled seizure threshold.

Adenosine has also been shown to have potent anticonvulsive effects in other seizure models, including audiogenic seizures[24], and seizures produced by pentylenetetrazol[12,28] pilocarpine[48] and electroshock[6].

METHYLXANTHINES AND KINDLING

The methylxanthines, including caffeine and theophylline, facilitate the rate of kindling and increase the severity of established seizures[2]. Both afterdischarge duration and seizure stage are dramatically increased. These findings may have clinical significance as high doses of caffeine can induce tremors and seizures in humans[35].

Originally, these effects were attributed to the ability of some methylxanthines to inhibit phosphodiesterase activity. The methylxanthines are now known to be potent adenosine receptor antagonists and the majority of their pharmacological effects are thought to occur via adenosine receptor blockade[39]. The methylxanthines not only potentiate kindled seizures, but block the anticonvulsant effects of adenosine analogs. These actions appear to be mediated by central, rather than peripheral nervous system activity, as the peripherally acting adenosine antagonist, 8-sulfophenyltheophylline (8-PST), fails to affect seizure afterdischarge duration or seizure stage when given systemically[49]. This compound also fails to block the effects of adenosine analogs on kindled seizures indicating that these effects are also centrally, and not peripherally mediated.

Data thus far indicate that a release of endogenous adenosine following seizures occurs, and that injections of adenosine or adenosine agonists reduce the severity of previously established kindled seizures. In addition, antagonists of adenosine receptors facilitate kindling and increase seizure severity. The mechanism of action at the seizure focus appears to be through an increase in the seizure threshold, but it is unclear whether the action is pre- or postsynaptic. As described above, adenosine analogs can delay the development of kindling, suggesting a role in seizure generalization. However, it may be that the delay in kindling is due to the reduction in afterdischarge activity at the focus, which only "indirectly" delays kindling.

## ADENOSINE AND POSTICTAL DEPRESSION AND SPIKING

The release of adenosine following a seizure may also contribute to, or directly mediate, the period of postictal EEG and behavioral depression typically observed after the occurrence of a fully generalized kindled seizure. During this postictal period, behavior is arrested and the EEG is dramatically reduced in amplitude and frequency; at times being nearly isoelectric[36,45]. Animals are not comatose during this postictal period, and in fact appear to be hyper-reactive to a variety of stimuli. Systemic injections of adenosine analogs (i.e., L-PIA, NECA) markedly prolong postictal EEG depression. In some cases the increases are more than 10-fold. In contrast, methylxanthines, including caffeine and theophylline, decrease its duration, often totally blocking the appearance of postictal depression while simultaneously increasing the AD duration and behavioral severity of the ictal phase. These effects also appear to be centrally mediated since 8-sulfophenyltheophylline (8-PST), with peripheral but little central activity when given systemically[41], does not affect postictal EEG depression and does not block the effects of NECA[49]. This point is important as it has been argued that the electrophysiological and behavioral effects of systemically administered adenosine analogs may not be centrally mediated[5].

The nature of postictal phenomena is poorly understood. The immediate postictal period is at least partially refractory to further seizures in kindled rats[17,26,1]. This is particularly true of the period of postictal EEG depression (Gold & Berman, unpublished observations). The typical period of postictal EEG depression in Stage 5-6 amygdala-kindled rats in our laboratory ranges from 90 to 200 sec[36]. Mucha and Pinel[26] earlier reported an almost total blockade of afterdischarges and motor seizures 90 sec into the postictal period in amygdala-kindled rats. The observation that adenosine analogs prolong postictal depression suggests that adenosine may contribute to the immediate postictal refractory period. Indirect support comes from studies demonstrating that adenosine analogs and adenosine uptake inhibitors lengthen the postseizure refractory period[6]. Moreover, high doses of aminophylline (i.e., 100 mg/kg, ip) reduce intertrial inhibition[1] in kindled rats, while theophylline shortens interictal periods in a kainate model of status epilepticus[13]. If the release of adenosine terminates seizures, as argued above, then it would be expected to contribute to the refractory period, at least immediately following a seizure.

Postictal spiking is also inhibited in amygdala-kindled rats by adenosine analogs[36]. Interictal spiking induced by penicillin in the in vitro hippocampal slice is blocked by adenosine and increased by theophylline[10]. The amygdala and pyriform cortex appear to be the generators of most interictal and postictal spiking, regardless of the kindling structure, with the pyriform cortex most reactive[20,34]. If postictal spiking represents residual epileptiform activity generated at the primary seizure focus, then these data may provide additional evidence for the anticonvulsant properties of adenosine. However, Engel and Ackermann[14] have argued that interictal spiking correlates with decreased, rather than increased epilepto-genicity in amygdala-kindled rats

## MANIPULATION OF ENDOGENOUS ADENOSINE ACTIVITY

Considered together, these results suggest that changes in the level of endogenous adenosine activity should have predictable effects on kindled seizures. Specifically, an increase in the level of endogenous adenosine or in the number or affinity of adenosine receptors should inhibit either the rate of kindling or the expression of fully kindled seizures. The available evidence supports this position. For example, Dragunow and Goddard[8] found that papaverine injections which block adenosine uptake and increase endogenous adenosine levels, delay the rate of amygdala kindling.

Our approach has been to increase adenosine receptor numbers (i.e., up-regulate) and then to determine the subsequent effects on fully kindled amygdala seizures. Chronic treatment with caffeine or theophylline can increase the number of brain adenosine A1 binding sites without affecting receptor affinity[15,27,25]. Therefore, we have chronically treated amygdala-kindled rats with theophylline for 14 days, using ip injections of 75 mg/kg for 1 week followed by 100 mg/kg for the second week. This procedure leads to a significant increase in adenosine A1 receptors. This was verified 48 hr after the end of drug treatment by receptor binding assay. The results demonstrated a significant increase in [$^{3}$H]-cyclohexyladenosine ($^{3}$H-CHA) binding to hippocampal and cerebellar membranes from theophylline treated animals compared to vehicle treated controls (Figure 2). There were no changes in receptor affinity. In a separate group of matched animals we determined the effects of receptor up-regulation on kindled seizure severity. The results are shown in Figure 3. Up-regulation of the number of A1 adenosine receptors by

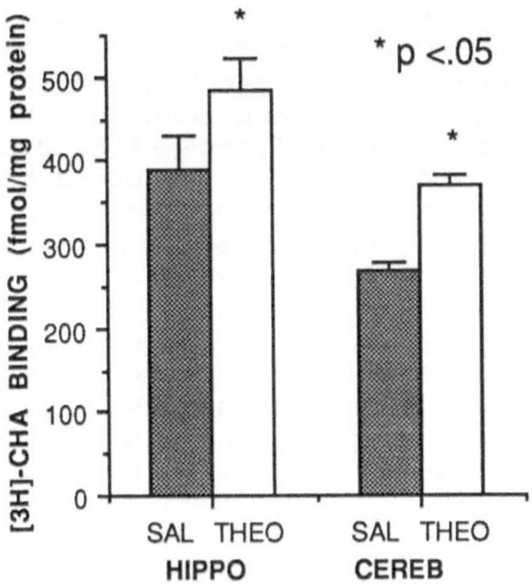


Figure 2. Up-regulation of adenosine [$^3$H]-CHA binding to adenosine A1 receptors in hippocampal and cerebellar membranes. Binding assays were carried out 48 hr following 14 days of theophylline treatment.

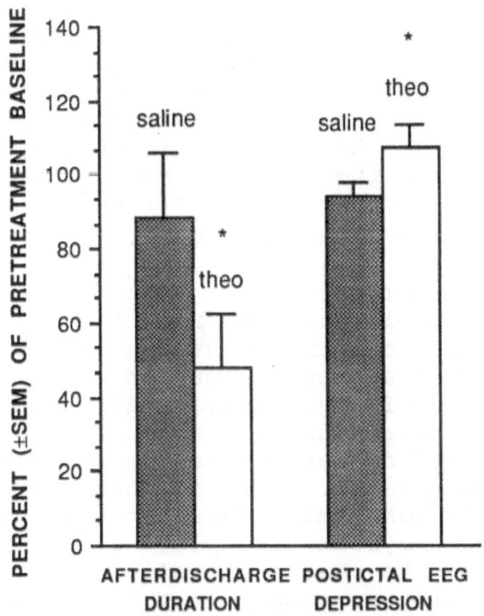


Figure 3. Effects of 14 days of theophylline (theo) or saline injections on afterdischarge duration and duration of postictal EEG depression. Animals were tested 48 hrs after last theophylline injection. (*p<.05 compared to appropriate saline injected controls).

approximately 25% resulted in a significant decrease in afterdischarge duration compared to saline-treated controls. Average seizure stage was also reduced from a mean of 5.0 to 3.1, although this effect did not reach statistical significance. The duration of postictal EEG depression was also increased as shown in Figure 3. In a similar study, Szot, et al.,[46] reported that theophylline-induced up-regulation of adenosine receptors reduced rats' sensitivity to a variety of chemical convulsants, including picrotoxin, bicuculline and pentylenetetrazol. These anticonvulsant effects closely parallel those produced by systemic injections of adenosine analogs (e.g., 2-CADO, L-PIA or NECA[36]. They indicate that endogenous adenosine activity influences seizure severity, and thus support the hypothesis that adenosine may play a role in seizure termination and expression of postictal seizure phenomena. Future experiments are needed to determine whether seizure susceptability in seizure-prone animals is related to levels of endogenous adenosine or adenosine receptors

## HYPOTHESIS

Based on data presented above, the following working hypothesis is proposed. First, that adenosine release following an initial afterdischarge serves to both terminate the ongoing seizure at the focus and to limit its spread[9,36]. Following repeated seizures, a change in neural function occurs which either engages excitatory mechanisms (e.g., glutamate system) or decreases inhibition, allowing for seizure spread to secondary sites. We suggest that a down-regulation of adenosine receptors following repeated adenosine release could contribute to a reduction in inhibition. This, in turn could facilitate seizure spread to interconnected sites. As yet, no direct support for this suggestion exists for the kindling model, but electroshock seizures have been reported to down-regulate adenosine receptors in the cerebellum in rats[54].

The phenomenon of prolonged postictal EEG depression typically appears following generalization of kindled seizures to Stages 3-5. The reduction in amplitude and frequency of the EEG suggests a period of tonic inhibition following seizure activity. Animals are also refractory to further kindled seizures during this postictal period. Because postical EEG depression can be facilitated or inhibited by adenosine agonists or antagonists, and because receptor up-regulation mimics these effects, we propose that the brain adenosine system also plays a critical role in this phenomenon. It may be that as kindling progresses, longer and more severe seizures simply lead to widespread increases in brain adenosine levels, resulting in EEG and behavioral depression. Alternatively it may be that as seizures become generalized, a critical brain region, presumedly important for purinergic neurotransmission, becomes involved mediating the period of tonic postictal inhibition. This second possibility is highly speculative at present, and no candidate sites or systems are proposed at this time. While direct evidence for this general proposal is incomplete at present, it does offer a series of testable hypotheses that we hope will lead to answers to the question; how is adenosine acting to modulate seizures and seizure susceptibility in the central nervous system?

## SUMMARY

In summary, available data support a role for endogenous adenosine activity in the expression and development of kindled seizures. Adenosine has wide-spread distribution in the central nervous system and produces both presynaptic and postsynaptic inhibitory effects on neuronal activity. Adenosine has been shown to have anticonvulsant effects in a variety of seizure models (i.e., kindling, electroshock, pentylenetetrazol, etc), and is linked to several seizure phenomena, including seizure termination, postictal EEG depression and postictal spiking. It is unlikely that adenosine is solely responsible for these phenomena, but rather contributes to these seizure manifestations along with other neurotransmitter systems. Much remains to be determined concerning adenosine's role in kindling and seizures in general. For example, the release of adenosine at the seizure focus remains to be demonstrated. Also, the physiological functions and properties of postictal EEG depression are unknown. For example, it is unknown if the period of postictal EEG depression represents a unique refractory period to further seizures, or an independent phenomenon? Our preliminary data indicate that the immediate postictal period is indeed refractory to the initiation of further kindled seizures, but a direct correlation between EEG depression and the refractory state remains to be demonstrated. Answers to these and other questions should be forthcoming and will undoubtedly clarify the role of adenosine in kindling.

## ACKNOWLEDGEMENTS

We wish to acknowledge Steven Gold and Traci Swanigan for their assistance in these experiments, as well as Cynthia Janusz, Robert J. Kleinsorge and Bonita M. Pedrosi for their helpful editorial comments during the preparation of this chapter. Supported by NIH Grant No. RR-08167 to R.F.B. Carl R. Lupica's current address is, Dept. Pharmacology, University of Colorado Health Sciences Center, Denver, Colorado.

## REFERENCES

1. Albertson, T. E., 1986, Amygdala-kindled post-ictal inhibition: Effects of intertrial intervals onrepeated days, Exp. Neurol., 92:197-206.
2. Albertson, T. E., Stark, L. G., Joy, R. M. & Bowyer, J F., 1983, Aminophylline and kindled seizures, Exp. Neurol., 81:703-713.
3. Barraco, R. A., Swanson, T. H., Phillis, J. W. & Berman,R. F., 1984, Anticonvulsant effects of adenosine analogs on amygdaloid-kindled seizures in rats, Neurosci. Lett., 46:317-322.
4. Bender, A. S., Wu, P. H. & Phillis, J. W., 1981, The rapid uptake and release of [$^3$H]-adenosine by rat cerebral cortical synaptosomes, J. Neurochem.,36:651-666.

5. Brodie, M. S., Lee, K., Fredholm, B. B., Stahle, L. & Dunwiddie, T. B., 1987, Central versus peripheral mediation of responses to adenosine receptor agonists: evidence against a central mode ofaction, Brain Res., 415:323-330.
6. Burley, C. S. & Ferrendelli, J. A., 1984, Regulatory effects of neurotransmitters on electroshock and pentylenetetrazol seizures, Fed. Proc., 43:2521-2524.
7. Daly, J. W., 1982, Adenosine receptors: Target sites for drugs, J. Med. Chem., 25:197-201.
8. Dragunow, M. & Goddard, G. V. 1984, Adenosine modulation of amygdala kindling, Exp. Neurol., 84:654-665.
9. Dragunow, M., Goddard, G. V. & Laverty, R. 1985, Is adenosine an endogenous anticonvulsant?, Epilepsia, 26:480-487.
10. Dunwiddie, T. V., 1980, Endogenously released adenosine regulates excitability in the in vitro hippocampus, Epilepsia, 21:541-548.
11. Dunwiddie, T. V., 1985, The physiological role of adenosine in the central nervous system, Internat.Rev.of Neurobiol., 27:63-139.
12. Dunwiddie, T. V. & Worth, T., 1982, Anticonvulsant effects of adenosine analogs in mouse and rat, J.Pharmacol. Exp. Ther., 220:70-76.
13. Eldredge, F. L., Paydarfar, D., Scott, S. C. & Dowell, R. T., 1989, Role of endogenous adenosine in recurrent generalized seizures, Exp. Neurol.,103:179-185.
14. Engel, J. & Ackermann, R. F., 1980, Interictal EEG spikes correlate with decreased, rather than increased, epilepto-genicity in amygdala-kindled rats, Brain Res.,190:543-548.
15. Fredholm, B. B., 1982, Adenosine actions and adenosine receptors after one week treatment with caffeine, Acta Physiol. Scand., 115:283-286.
16. Fredholm, B. B. & Hedqvist, P., 1980, Modulation of neurotransmission by purine nucleotides and nucleosides, Biochem. Pharmacol., 29:1635-1643.
17. Goddard, G. V., McIntyre, D. C. & Leech, C. K, 1960, A permanent change in brain function resulting from daily electrical stimulation, Exp. Neurol., 25:295-330.
18. Jarvis, M. F., 1988, Autoradiographic localization and characterization of brain adenosine receptor subtypes, In: Receptor Localization: Ligand Autoradiography, F. Leslie & C. A. Altar (eds.),Alan R. Liss, New York, pp. 95-113.
19. Jarvis, M. F., Jackson, R. H. & Williams, M., 1989, Autoradiographic characterization of high affinity adenosine A2 receptors in the rat brain. Brain Res., 484:111-118.
20. Kairiss, E. W., Racine, R. J. & Smith, G. K., 1984, The development of the interictal spike during kindling in the rat, Brain Res., 322:101-110.
21. Kuroda, Y. & McIllwain, H., 1977, Uptake and release of [$^{14}$C] adenine derivatives at beds of mammalian synaptosomes in a superfusion system, J. Neurochem.,22:691-699.

22. Lee, K., Schubert, P., Gribkoff, V., Sherman, B. & Lynch, G.,1982, A combined in vivo/in vitro study of the presynaptic release of adenosine derivatives in the hippocampus, J. Neurochem., 38:80-83.
23. Lewin, E. & Bleck, V., 1981, Electroshock seizures in mice:Effects on brain adenosine and its metabolites, Epilepsia,22:577-581.
24. Maitre, M., Ciesielski,, L., Lehman, A., Kempf, E. and Mandel, P. 1974, Protective effect of adenosine and nicotinamide against audiogenic seizure. Biochem. Pharmacol., 23:2807-2816.
25. Marangos, P. J., Boulenger, J. & Patel, J., 1984, Effects of chronic caffeine on brain adenosine receptors: regional and ontogenetic studies, Life Sci., 34:899-907.
26. Mucha, R. F. and Pinel, R. J., 1977, Postseizure inhibition of kindled seizures, Exp. Neurol., 54:266-282.
27. Murray, T. F., 1982, Up-regulation of rat cortical adenosine receptors following chronic administration of theophylline, Europ. J. Pharmacol., 52:113-114.
28. Murray, T. F., Sylvester, D., Schultz, C.S. and Szot, P., 1985, Purinergic modulation of seizure threshold for pentylenetetrazol in the rat. Neuropharm. 24:761-766.
29. Nagata, H., Mimori, Y., Nakamura, S. & Kameyama, M., 1984, Regional and subcellular distribution in mammalian brain of the enzymes producing adenosine, J. Neurochem., 42:1001-1007.
30. Park, T. S., Van Wyle, D. G. L., Rubio, R. & Berne, R.M., 1987, Interstitial fluid adenosine and saggital sinus blood flow during bicuculline-seizures in newborn piglets, J. Cereb. Blood Flow Metab., 7:633-639.
31. Phillis, J. W. & Wu, P. H.,1983, Roles of adenosine and adenine nucleotides in the central nervous system, in: "Physiology and Pharmacology of Adenosine Derivatives," J. W. Daly, Y. Kuroda, J. W. Phillis, H. Shimizu and M. Ui.,eds., Raven Press, New York, 219-236.
32. Pull, I. & McIlwain, H., 1972, Metabolism of [$^{14}$C]-adenine and derivatives by cerebral tissues, superfused and electrically simulated, Biochem. J., 126:965-973.
33. Pull, I. & McIlwain, H., 1974, Rat cerebral cortex adenosine deaminase activity and its subcellular distribution, Biochem. J., 144:37-41.34.
34. Racine, R. J., Mosher, M. & Kairiss, E. W., 1988, The role of the pyriform cortex in the generation of interictal spikes in the kindled preparation, Brain Res., 454:251-263.
35. Richards, W., Chuieh, J. A., Brend, D. K., 1985, Theophylline associated seizures in children, Ann. Allerg., 54:276-279.
36. Rosen, J. B. & Berman, R. F., 1985, Prolonged postictal depression in amygdala kindled rats by the adenosine analog, L-phenylisopropyladenosine, Exp. Neurol.,90:549-557.

37. Rosen, J. B. & Berman, R. F., 1987, Differential effects of adenosine analogs on amygdala, hippocampus, and caudate nucleus kindled seizures, Epilepsia, 28:658-666.
38. Rubio, R., Berne, R. M., Bockman, E. L. & Curnish, R. R., 1975, Relationship between adenosine concentration and oxygen supply in rat brain, Am. J. Physiol., 228:1896-1902.
39. Sattin, A. & Rall, T. W., 1970, The effect of adenosine and adenine nucleotides on the cyclic adenosine 3'-5'monophosphate content of guinea pig cerebral cortex slices, Mol. Pharmacol., 6:12-23.
40. Schultz, V. & Lowenstein, J. M., 1978, The purine nucleotide cycle:Studies of ammonia production and interconversion of adenine and hypoxanthine nucleotides and nucleotides by rat brain in situ, J.Biol. Chem., 253:1938-1943.
41. Seale, T. W., Abla, K. A., Shamim, M. T., Carney, J. M. & Daly, J. W., 1988, 3,7-Dimethyl-1-propargylxanthine: A potent and selective in vivo antagonist of adenosine analogs, Life Sci., 43,:671-684.
42. Senba, E., Daddona, P. E., Watanabe, T., Wu, J. Y., & Nagy, J. I., 1985, Coexistence of adenosine deaminase, histidine decarboxylase and glutamate decarboxylase in hypothalamic neurons in the rat, J.Neuroscience, 5:3393-3402.
43. Snyder, S. H., 1985, Adenosine as a neuromodulator, Ann. Rev. Neurosci., 8:103-124.
44. Stone, T.W., 1981, Physiological roles for adenosine and adenosine 5'-triphosphate in the nervous system, Neuroscience, 6:523-555.
45. Symonds, C., 1959, Excitation and inhibition in epilepsy, Brain, 82:133-146.
46. Szot, P., Sanders, R. C. & Murry, T. F., 1987, Theophylline-induced upregulation of A1-adenosine receptors associated with reduced sensitivity to convulsants, Neuropharmacology, 26:1173-1180.
47. Trussell, L. D. & Jackson, M. B., 1987, Dependence of an adenosine-activated potassium current on GTP-binding proteins in mammalian central neurons, J.Neuroscience, 10:3306-3316.
48. Turski, W. A., Cavalheiro, E. A., Ikonomidou, C., Mello, L. E., A. M., Borotolotto, Z. A. & Turski, L., 1985, Effects of aminophylline and 2-chloroadenosine on seizures produced by pilocarpine in rats: Morphological and electroencephalographic correlates, Brain Res., 361:309-323.
49. Whitcomb, K., Lupica, C. R. Rosen, J. B. & Berman, R. F., 1990, Adenosine receptors modulate postictal events in amygdala-kindled rats, Epilepsy Res., in press.
50. Williams, M., 1984, Adenosine-a selective neuromodulator in the mammalian CNS?, Trends Neurosci., 7:164-168.
51. Williams, M., 1987, Purine receptors in mammalian tissues: Pharmacology and functional significance, Ann. Rev. Pharm. Toxicol., 27:315-345.

52. Winn, H. R., Welsh, J. E., Bryner, C., Rubio, R. & Berne, R. M., 1979, Brain adenosine production during the initial 60 seconds of bicuculline seizures in rats, Acta Neurol. Scand., 72:536-537.

53. Winn, H. R., Welsh, J. E., Rubio, R. & Berne, R. M., 1980, Changes in brain adenosine during bicuculline-induced seizures in rats: Effects of hypoxia and altered systemic blood pressure, Circ. Res., 47:868-877.

54. Wybenga, M. P., Murphy, M. G. & Robertson, H. A., 1981, Rapid changes in cerebellar adenosine receptors following experimental seizures, Eur. J Pharmacol., 75:79-80.

55. Zetterstrom;, T., Vernet, L., Ungerstedt, U., Tossman, U., Jonzon, B. & Fredholm, B. B., 1982, Purine levels in the intact brain, Neurosci. Lett., 29:11-115.

## Discussion of Dr. Berman's Presentation

DR. FERNANDEZ-GUARDIOLA: I want to make clear that postictal depression is not specific to kindled seizure. Any seizure by metrazol or electroshock produces postictal depression. But I want to point out that there is another substance that blocks the postictal depression and this is naloxone. Maybe you can comment on the role of endogenous opiodes and naloxone in reverting the postical depression in relation to adenosine.

DR. BERMAN: OK. I am going to answer from my own experiments and maybe I can ask Bob Ackerman to comment from his experiments. I reported that naloxone blocks postictal EEG depression, reduces postictal spiking, but in a paper published in 1981 when I was focusing on opiodes and their role in kindling, I believe that that reduction - and I know I stated it in the paper - was due to the fact that naloxone in our hands reduced seizure severity and in many cases blocked seizures, and as a result we had very little postictal manifestations of a seizure. So that from my point of view I think that's a different mechanism and we are not able to lengthen postictal EEG depression with opioides even though we can get tremendous changes in postictal spiking. I don't know if Bob Ackerman, or anyone else, wants to comment.

DR. ACKERMAN: Well, as I am sure you know, we and others have shown, and it is very easy to demonstrate, if you give morphine ahead of a kindled seizure, the postictal depression is made more severe. The animal just lies at the bottom of cage like a rug and you get a miraculous recovery if you then inject naloxone. Frequently the depression is alleviated by the time the injection is over. We tried naloxone by itself and never could get any results that I would be willing to talk about except that the seizures seem to be, if anything, made worse, which is exactly the opposite of what you got, and since I believe me and I believe you, that's a mystery. Let's say it's a strain difference. Suffice it to say that there is some methodological difference. I have no axe to grind, by the way.

DR. BERMAN: Let me say that in all of our studies we define postictal EEG depression on the basis of electrographic data, and we don't really use behavioural depression very often. In those studies, did you measure reduction in EEG amplitude and frequency, because I think those are fundamentally different phenomena.

DR. FERNANDEZ-GUARDIOLA: I think this methodological problem is very important. We have facilitated kindling with naloxone by using repeated doses, every 15 minutes or 30 minutes, and then we have a tremendous facilitation of kindling. Another thing about these opiodes is ICV administration versus systemic administration. We get very different results. It's a mystery, I agree, but I think it's a methodological mystery.

DR. BERMAN: Shall we wrestle?

DR. CORCORAN: Rob, I want to ask you two simple questions. Could you remind us where adenosine comes from in the central nervous system, what it's contained in, and secondly, do you know if anyone has looked for changes in adenosine binding sites as a function of kindling?

<u>DR. BERMAN</u>: In answer to your first question, adenosine is in a fairly high concentration in all cells. It's been shown to be released from synaptosomes and depolarized brain slices and it's believed to be released as adenosine and also to be hydrolysed from ATP following release. We've done receptor autoradiography on kindled animals - these are preliminary studies - and we are not seeing major shifts in adenosine receptors in kindling. As far as I know we're the only people who've looked at kindling and changes in receptors and we're not seeing a great deal.

<u>DR. BURNHAM</u>: I would like to say that we have tried aminophylline and adenosine blocker in our pharmacological simulation paradigm, and the adenosine blockers and GABA blockers are so far the only drugs which more or less simulate the kindled state. Neither one of them simulates it entirely - you get a half kindled looking animal both with adenosine blockers and GABA blockers. We think it's very interesting that like the GABA system, the adenosine system, you get both reversal and at least a partial simulation, and I think the search for long-term changes would be very important in this system.

<u>DR. MODY</u>: I was wondering what the rat is doing during that depressive EEG and if that is any different from the long-term depressed EEG when you put in the adenosine agonist? And where do you think adenosine is acting, in a more generalized or a specific spot in the brain?

<u>DR. BERMAN</u>: With regard to the first question, adenosine has potent peripheral effects and in fact one of the reasons why I think that some pharmaceutical companies have lost interest in adenosine as an anticonvulsant is that when its given systemically, it has hypotensive effects, decreases locomotor activity, and a number of things go on. So that the animal is behaviourally depressed, lying down on the bottom of the cage, not doing very much at all; he's not asleep, but he's not comatose, but that's with higher levels of adenosine analogues. That's one of the reasons that we rely on the postictal EEG depression for these kinds of studies. The reason we think that it is related to the adenosine system though, is that we can give caffeine or theophylline and completely suppress the period of postictal EEG depression, even though the seizures are much more severe. The second question was, where does it act? Well, that's a tough question, too. It may act at the focus following the seizure, so that at early stages of kindling we think it may be involved in terminating the initial after-discharge; we can't prove that but that's where we're thinking. As far as the effects on postictal EEG depression in spiking, that's interesting, because you don't usually see very much postictal EEG depression until the animals are kindled to a generalized seizure, until you get to stages 3 and 4, and it's very prominent when the animals are at higher stages of kindling. That suggests that you need to be somewhere away from the initial focus. When we inject the adenosine analogues into the ventricular system, we sometimes see a slowing and decrease in amplitude of the EEG but we've never seen the induction of what we would call a period of

postictal EEG depression. Our hypothesis right now is that postictal EEG depression might be generalized by some brainstem structure or brainstem region that when the seizure generalizes to that area, may feed back to higher centres and shut down electrographic activity. That's a very speculative hypothesis right now. We don't know where it acts.

DR. CAIN: This doesn't have anything to do with adenosine but I thought I would follow up the earlier comments on opiates and kindling and route of administration, because Mike Corcoran and I have spent some years now working with opioid peptides and find specially in some more recent, as yet unpublished work, that route of administration is of course very critical. Using a given specific opioid peptide agonist in one of the routes of administration, namely, directly into various limbic brain sites you can very effectively kindle with it, but using the exact same peptide intraventricularly, for example, it's completely without any kindling effect. Other people have shown that with various routes of administration and various of these compounds you can show potent anticonvulsant effects. In 1983, as I recall, Hanan Frank wrote a very nice review pointing out these mixed effects depending on, among other things, the route of administration. It's an interesting topic, but it's a complicated one.

DR. BERMAN: My comment is that some people don't find any effect essentially of opiodes on kindling, or very little effect, and I think Drs. Stark and Albertson found very little effects of opiates on kindled seizures. So you can go anyway you want.

DR. MCINTYRE: Rob, during that behavioural postical depression, would you comment on what might be supporting the tremendous reactivity that those animals have? They are depressed spontaneously, but they certainly are not depressed when one picks them up, you suddenly experience popcorn.

DR. BERMAN: That explosive behaviour in our experience is very much restricted to that period of EEG depression and most of the time the behavioural depression recovers with the recovery of EEG activity. Sometimes it's very spontaneous, and I don't know.

DR. ACKERMAN: Why can't that reactivity just be exacerbation of sensory reflexes? Every time people get bitten it seems to be a surprise both to the experimentor and the animal.

UNIDENTIFIED: You're still being veterinary about it?

DR. ACKERMAN: No, I have Texan graduate students to handle that. I just think it's an exacerbation of reflexes. Just because there's a deaffrentation from higher centres during this depression.

DR. WADA: I just want to interject in that story of reflex. During the postictal comatose state with flat cortical EEG following amygdaloid kindled convulsion, midbrain and hypothalamic stimulation can induce appropriate emotional behaviour, for example, defense, flight or hissing, etc., identical to that elicited while animals are resting and awake. There is no change in the threshold. Therefore, at least those brainstem areas responsible for "emotional behaviour" remain functionally intact during the postictal comatose state, and presumably any appropriate afferent stimulus can activate these areas.

DR. POST: Mike Clark in our lab has found that carbamazepine's effects are A1 selective and it doesn't seem to effect any of the adenosine A2 receptors which are in that striatal picture. To the extent that carbamazepine is having A1 adenosine agonist effects selectively, it doesn't look like they're able to prevent the development of amygdala kindling. So, maybe it's the A2 effects of the adenosine compounds that you are alluding to that are more likely to inhibit the development of kindling, since carbamazepine or the A-POXIDE clearly don't inhibit the development in the rat at least.

DR. BERMAN: It could be. Our initial studies attempted to separate A1 and A2 effects. We know that A2 adenosine receptors are concentrated in the striatum, so we kindled the caudate nucleus and we injected what we thought was a selective A2 agonist, Naka, and we used LPIA as an A1 agonist. To make a long story short, both were effective in caudate nucleus in suppressing caudate kindled seizures. The more recent research by Burns et al. suggests that essentially Naka and LPIA have approximately the same nanomol affinity for the A2 receptor. And so for the A1 and A2 receptors, they may not be good agents to separate those pharmacologically right now.

DR. ADAMEC: I have a comment and perhaps a naive question. The comment is, getting back to this reactivity after a seizure, one way you could test this is, in relation to what Dr. Wada has just said, the behaviour sounds like it's very defensive, or could be, but if it is defensive in response to sensory input, then in fact you should get your largest reaction with dorsal stimulation on a rat. This is from ethological studies. Defensive responses that a rat shows are exaggerated to certain inputs. If it were reflexive, as you're suggesting, presumably any tactile input should produce a hyper reactive response. The naive question is, you are talking about using methylxanthine as adenosine blockers to relieve postictal depression. When you give a methylxanthine you're also blocking phosphodiasterase aren't you? So what you're doing is potentiating a host of cyclic and dependent processes. How do you know it's specific to adenosine.

DR. BERMAN: Well, I can see that we have other people who are willing to answer that. The affinity for the adenosine receptor, the methylxanthine affinity, is very much higher than it is for inhibiting phosphodiasterase activity. We have actually used fairly high doses, but the argument is that you're more likely to be seeing an adenosine mediated blockade and we know that the adenosine analogues are not influenced in the phosphodiasterase activity at all, yet we can reverse the adenosine mediated effects with the methylxanthine. So while we can't rule out the possibility that we may be having a phosphodiasterase effect, we are very confident that the phenomena I described are not mediated by that mechanism.

DR. MOSHE: I have a comment. In the huge literature of developmental kindling that includes all 30 papers, there is one paper by Tromer and Pasternak indicating that adenosine antagonist produced more profound, more severe effects in seizures in developing animals than in adult animals. My question for you is, have you looked at the postictal depression and psychological or behavioural changes as a function of age?

DR. BERMAN: No, we haven't examined that in anything but in adult rats.

DR. WADA: Thank you Dr. Berman. Now I would like to call upon Dr. Calvo who has something to tell us.

Dr. CALVO'S PRESENTATION AND DISCUSSION

DR. CALVO: I want to discuss a subject that presents some discrepancies and that is the relationship between sleep and epilepsy. Sleep studies of unique recordings in humans have shown that the occurrence of the seizure can delay the occurrence of sleep, and increase the latency of sleep and REM sleep. Also, in experimental epilepsy, electroshock, and with the kindling model of epilepsy there are some discrepancies. Some studies show that there is a deleterious and important effect on kindling development such as the work of Sterman and Shouse, and the one that we heard here yesterday, a deleterious effect on slow-wave sleep and REM sleep. There are other studies from Rondouin who made a longitudinal study and he did not find a real change in slow-wave sleep but he found REM sleep change. To show this we thought that the longitudinal studies are very important and to do carefully the control of these animals we developed a study massing the kindling stimulus each 2 hours and recording the animals 23 hours daily, and we found the following. May I have the first slide, please. As you can see this is the percentage of wakefulness, slow-wave sleep and paradoxical sleep through the 6 days of kindling. This is the control situation, and each day the cats received a stimulus during the light period at 2 hours intervals. You can see that through the 6 days of stimulation there are no important changes either in wakefulness, slow-wave sleep, slow-wave 2 or paradoxical sleep. And if we analyze the paradoxical sleep distribution as we can see in the next slide, this is the distribution of paradoxical sleep across 23 hours recording in the control situation, and you can see this is the light period, dark period, and the kindling stimulus is given as follows: one each hour during the light period, and it is suspended during the night period. You can see that REM sleep progressively and in a discrete manner goes to the night period and it's organized according to the stimulus in this fashion. This is day 1, day 2, day 3 of stimulation, and on day 5 all the cats are presenting generalized convulsive seizure. If you analyze the accumulation by hour of this REM sleep during these days you can see the result in the following slide. This is the accumulation in the control situation; the white squares are the accumulation of REM sleep through the 23 hours of sleep recording compared with day 1 of the kindled animals, the same kindled animals - you can see there is a shift of the colours of REM sleep during all the days of the kindling development. Even when the animals are presenting generalized convulsive seizure. One thing that I want to say is that in control situations instead of giving the electrical stimulation, we awakened the cats at the same time, that takes 1 or 2 minutes, instead of during stimulus; maybe this is important to simulate all the situation around the stimulus and these results suggest that there are the same effects from waking the cats or from kindling them and producing seizures. So I want to discuss these results with the people interested in the relationship between kindling and sleep.

DR. WADA: Dr. Shouse?

DR. SHOUSE: José, before you turn that off, can you tell me again, what does this show?

DR. CALVO: Yes, OK. This is the accumulation of REM sleep, each hour of the day, the addition of the first hour, the second hour added to the previous hours. This, we can say, is the daily quota of REM sleep of the cats.

DR. SHOUSE: OK. Are you saying that there seems to be a little bit of delay in REM just after the stimulus and then it makes itself up later?

DR. CALVO: Yes, that's it. For example, here are the stimuli during the light hours and we suspend them during the dark periods. So you see that here in the black dots the REM sleep makes a shift according to the stimulus. When the stimulus disappears, it goes up again.

DR. SHOUSE: OK. So, I'm just trying to make sense out of the various reports about REM sleep latency and what happens to REM after its stimulus. So are you agreeing with the reports from Drs. Hiyoshi, Rondouin, Tanaka, myself, Jean-Luigi Gigli, Gottman and all these folks, that seizures delay REM sleep?

DR. CALVO: Yes, but if you see here, here is the stimulus, and here are the stimuli and here is the REM sleep.

DR. SHOUSE: Yes. Are you saying that you differ from all the other reports which claim that REM rebound does not occur?

DR. CALVO: Yes. The REM rebound or the REM diminution?

DR. SHOUSE: The REM rebound.

DR. CALVO: Well, it will rebound, yes, the same day.

DR. SHOUSE: It will rebound? That's curious because no one else has seen that. That's the only unusual thing. I don't know how to explain that exactly. Suffice it to say, it's probably methodological differences, and I am quite certain that the slow-wave sleep decrements or lack of them can be explained in the same way. We know this from Gigli's work by comparison to mine, and so forth. Just to make one other comment before I let someone else talk about this, and that is I think the most important finding from yesterday's talk, and I think you commented on it, was that phenobarbital had a far worse effect on REM sleep than kindling. Would you agree with that? In other words, of all the REM suppression that we've all reported, whether it rebounds or not, in kindling, or in any other epilepsy model, if you give an animal an anticonvulsant, they have worse sleep defects than any of the normal sleep disorders that occur as a process of epilepsy by itself.

DR. CALVO: Yes, as Dr. Hiyoshi showed, for one administration I think the diminution of REM sleep will be worse, but as Dr. Hiyoshi showed, with the chronic administration it seems that REM sleep does not suffer anything. But the thing is, I think that there is no important change in REM sleep and the idea that kindling progressively diminished the REM sleep, and this is the cause of the diminution of the threshold for new seizures, that's what I disagree with.

DR. SHOUSE: Right. And of course I don't agree with that either because we've also shown that. So in a way I think we all have come to an agreement as far as the absolute percentage of REM sleep or slow-wave sleep, or even the combined findings are in and of themselves not a critical factor for seizure suppression, particularly since most anticonvulsants suppress REM sleep. Would you agree with that?

DR. CALVO: Sorry, would you repeat it?

DR. SHOUSE: I think that the data accumulated from our lab as well as everyone else's show that there is some effect on REM sleep by the creation of epilepsy. But that whatever effect that is, and however long it may last, it is not critical to the expression of seizures.

DR. CALVO: OK. I think that it is not critical for kindling.

DR. WADA: Thank you Dr. Calvo. Now, we are going into our general discussion.

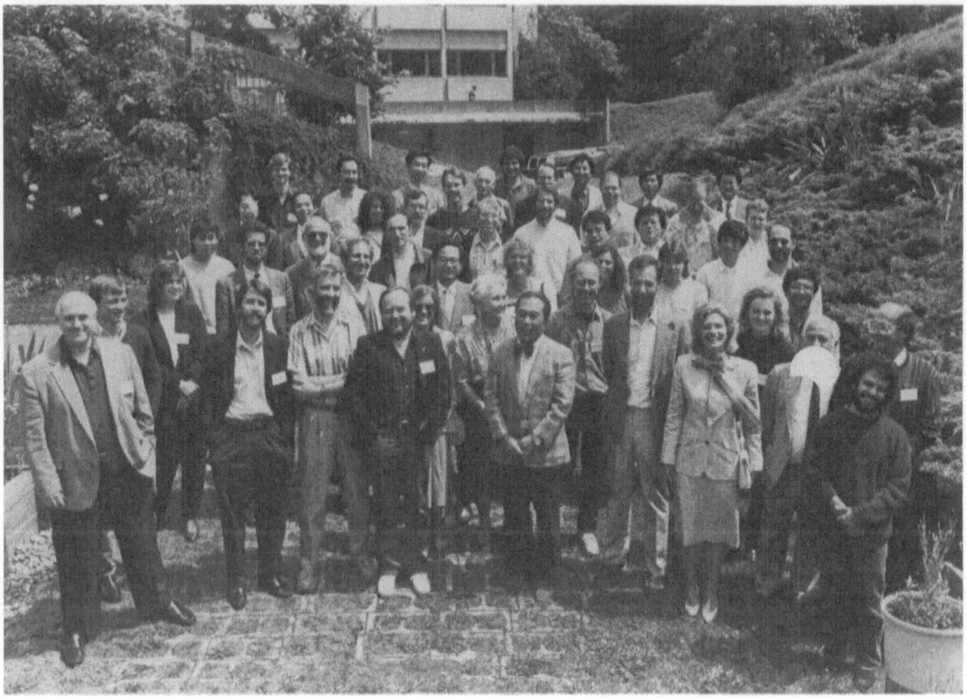

FOURTH INTERNATIONAL KINDLING SYMPOSIUM
June 15 to 17, 1989
Vancouver, British Columbia, Canada

# GENERAL DISCUSSION

DR. WADA: I would like to call on Dr. Racine to open our general discussion.

DR. RACINE: Dr. Wada asked me to start off this session with his characteristic 15 minutes notice. I was going to be quite content to ask a few mindless questions; now I realize that I have to try to be somewhat more profound. I can't tell you how deeply I appreciate this opportunity. My name is Mac Burnham. Sorry, I am coming over like a professorial Mac-attack. I must admit this has been a very impressive session. I think it's been the best yet. There is still a very clear lack of good membrane physiology, in fact this session is lacking any membrane physiology at all. I know there is some going on, and hopefully by our next meeting we'll hear about some of those studies. But other than that it has been a highly successful meeting. The historical moment, perhaps not as historical as that time a few years ago when Dr. Ruchdon from the University of Western Ontario came to U.B.C. and asked Dr. Wada if he could measure his head. Well, that's OK. You guys may have a greater cranial capacity, but you had better keep your wives away from us. Everything I just said is off the record, either that or my name is Bob Post. OK. It's just come to me. Racine.

I suppose I should make some comments on the stages. Frankly, I really don't care all that much. I was at a Neuroscience meeting a few years back with Graham Goddard, and somebody gave a talk on kindling and had modified the stages fairly substantially and was arguing that this modification should be used. I was sitting there, kind of dozing away happily, and I didn't care, but Graham got up and took the fellow to task. His point, I guess, is a reasonable one, and that is, a lot of people now have used it for a lot of years and we all understand reasonably what we mean by that, and his feeling was that it should not be changed. I don't feel quite that strongly about it, but I do feel that at this stage perhaps if it is to be changed it should be changed in some fairly formal, careful way, perhaps we should have a committee, and I nominate Dr. Wada. Seriously, I think if such a review is to take place, it should have a clinician researcher involved, and perhaps someone involved in analyzing motor control, someone familiar with notation schemes for the description of motor behaviour. Perhaps it is time to reevaluate, and maybe whoever is in charge of the next Kindling Symposium should make that attempt.

There are a couple of other observations I'd like to make about the meeting. Let's see if I can get this across coherently. I've been thinking about it for the last day or so, but I haven't thought of any very good way to say it. Something that has clearly come out of this meeting, that I think has been on our minds from the very beginning, but it's beginning to crystallize now, is that different epileptogenic agents, and the application of these agents to different sites, produce unique special temporal patterns of activation. In spite of the fact that propagation patterns can really look quite similar, and the convulsive endpoint can look quite similar, it's clear that there is some considerable specificity. Dan

McIntyre showed us something to that effect with the kindling, I mean with the transfer effect, the resistance in many cases to transfer. That would seem to be a largely transient phenomenon, but it appears that when a given seizure circuit is activated it begins to shut out alternative possible pathway seizure circuits in some way, and Burchfiel has driven that point home even more clearly with his kindling antagonism phenomenon which is really quite dramatic. We not only have these behavioural demonstrations, however, specially in this meeting we have seen that a lot of the consequences of these treatments also have unique special patterns. The metabolic patterns can be quite different, the glial development, the results of Ackerman and Buterbaugh, and so on, demonstrate that fairly clearly. Well, the question that I would like to raise, is when one of these systems is activated, one of these specific seizure systems or circuitries is activated, and appears to be shutting out other circuits, is that transient, or is that permanent? I think Burchfiel would like to say that at least in his case when you have coactive system, the shutout is largely long-lasting. Whereas we know that kindling of the transfer antagonism effect is probably transient, at least in large part. So, I guess one of the first questions that I would pose then, and maybe Dr. Burchfiel can address it, is, what convinces him that his effect is permanent and what does he think he can do to convince the rest of us.

My last point can be made a bit more briefly. And that is, I'd like to argue that we accept Mac's criteria and be fairly careful how we interpret data, pertaining to acceleration and retardation of kindling. We've seen a lot of enthusiastic work done on the NMDA system recently, and as someone who has purposely stayed away from that system until it sort of shakes itself out a little bit, I have no particular axe to grind, but looking at it from outside I can say that what strikes me is that it takes pretty large doses, doses that have long since totally blocked LTP, doses that begin to make the animal stagger around the cage, before you begin to have any very strong retardation effect on kindling.

I'll leave you with some general questions that I always kind of throw out. An area that I think we should be addressing perhaps more vigorously, again is membrane physiology. There should be more unit recording generally, extracellular recording for example in chronic animals, finding out what kind of unique events are accompanying these ictal and interictal responses. We have some very interesting data on ictal events and interictal events and now Dr. Stripling has shown us some of these, but we don't yet know very well, what the units are actually doing in some of these key areas like the dentate gyrus for example, and pyriform cortex. How does kindling induce potentiation, kindling induced potentiation relate to LTP? I don't think either one of them have all that much critically to do with kindling. Too bad McNamara is not here. I have a few other things to say about McNamara but I might as well not say them now that he's gone. I'll say them next time I see him. I know he'd enjoy them. When is someone finally going to do some proper anatomical experiment to look for degenerative mechanisms? It takes certain kinds of experiments to detect some of these events which can be quite rapid and very difficult to see where they run, like synaptic degeneration and so on. Nobody has really made any serious attempt to look for that yet. What is the substantia nigra really doing in kindling? I would like to know about that. Perhaps I can start this off, by asking Dr. Burchfiel how he is going to convince us that his effect is permanent.

DR. BURCHFIEL: Ron, I'm not quite sure what you're getting at in that, because in fact what I was trying to convince everybody of the day before yesterday, was in fact that the effects that we see in antagonism really are not permanent. I mean they are operative during the alternating paradigm, but when you stop that alternating paradigm, there is no long-lasting inhibition present.

DR. RACINE: Let me explain what I meant. If you use your alternating paradigm and you get an antagonism, particularly if you get your absolute antagonism, you led me to believe that if you then just wait, 1 week, 2 weeks, 3 weeks, 4 weeks, whatever, and then begin kindling the suppressed site, it would still take the normal time to kindle the animal, it would not show an immediate response, that's what I would call long-term.

DR. WADA: Is it not a lack of positive transfer effect?

DR. BURCHFIEL: In the absolute antagonism, there is in fact an absence of any positive transfer, this is what I am trying to say. That's the major point of all this, and I think what's happening in the antagonism paradigm is that there is a failure of some positive event to occur at the suppressed site; that the kindling just isn't reaching a certain stage and then it is stopping, it's arrested, it's halted, and it's not necessarily because some new inhibitory mechanism is holding it down, it's just that it has failed to progress beyond that particular point.

DR. WADA: I understood that when the process is halted, and you complete the dominant site kindling, and then you come back to the halted site, it takes off from there, but from that point of stage 2, or 3, to stage 5, there is no positive transfer effect. Am I right?

DR. BURCHFIEL: That's right. That suppressed site is at a stage of kindling compatible with the behaviour that it is expressing when you stimulate it, and when you then start to stimulate that site alone, on consecutive trials, it starts at that site, completes the kindling process in the same manner as if it were a single site having reached that same stage.

DR. WADA: So that there is no facilitation between that site to the final stage?

DR. BURCHFIEL: That's correct. I mean if the suppressed site is at stage 3, and the number of trials it takes to go from 3 to 5, for that suppressed site, is the same as if it were a single site also starting at stage 3 and progressing on. So it has reached a certain stage of development of kindled seizures and the subsequent development from that point on is the same as if it were a single site. So that the mechanisms that seem to operate for the suppressed site out of the antagonism paradigm are the same mechanisms that operate for a single site from a comparable intermediate stage.

DR. WADA: Thank you.

DR. ADAMEC: Actually my comment is on a different topic but it relates to something that Ron just alluded to. He said that what we need is more studies looking for neuropathology. Mac Burnham in his talk suggested that kindling may be a result of damage in the nervous system, and that wasn't a very popular hypothesis. If I look at the data that has been presented here, and by a number of people, I'd say that the bulk of the data suggest that kindling does produce damage and maybe it is damage. What we've seen is changes in numbers of synapses. If I can take liberties with Yuri Geinisman's data, change decreases in receptors, beta receptors are down, NMDA receptors may be down, muscarinic cholinergic receptors are down, the alpha 2s are down, there is a decrease in metabolism, and blood flow at the seizure focus. I guess the question is, have we reached a point where we can say that the bulk of the data suggest that kindling is due to damage, or do we have to say that seizures produce damage, and if we do our studies in fully kindled animals who have many, many seizures, we're going to see damage? But the process of kindling may be a growth process. I got into the field of kindling because I'm interested in plasticity, and it's a bit disheartening to think that what we're looking at is purely damage. I'm hoping that people will begin to look at earlier stages of kindling and anatomical studies, metabolic studies and the neurochemical studies.

QUESTION FROM THE FLOOR: Rob, why do you call it damage rather than reorganization? Reorganization of the neurocircuits and adaptive changes? I don't think it's necessarily damage.

DR. ADAMEC: Well, there is pathology that has been reported that there has been an increase, I guess there's some data suggesting kindling induces gliosis, presented at the Neuroscience Kindling Symposium last year. You could argue that C-fos induction has something to do with damage to the nervous system. The nature of receptor down-regulation isn't clear. It could be due to loss of terminals as well as a reorganization; it may not just be a loss of the number of receptors at a particular synapse during some kind of reorganization. So I agree, you could take that stand, and I would like to - I'd like to be optimistic about uncovering a growth process, but I'm not sure we're at that point.

DR. MCINTYRE: This reminds me a little bit of the developing of the nervous systems where neuroelements are dropping out of the system as the organism is growing. I see no reason why we couldn't be having a growth process that actually is a channelization of information, which in fact results in a loss of elements. I don't think it necessarily has to be pathology. It gives aging some meaning.

DR. BURNHAM: I think it's time that we begin to worry that seizures are really harming the brain. Now, we may feel that the changes are growth, there are probably growth-like changes. You might feel that the changes are destruction, there are probably destruction-like changes, but I think, and in some cases the changes are probably the brain's attempt to protect itself against further seizures, but the nature of these changes I suspect is to embarrass brain function. So we can call it harm if you don't want to call it cell death, or damage, OK? But I suspect that seizures harm the brain and that each seizure may harm the brain and I think that we see it not only in the ictal period where the focal seizure generalizes and harms the animal, but I think we're beginning to see it more and more in interictal periods. The sort of thing Dr. Wada was talking about, and Dr. Adamec has been talking about in terms of behavioural changes. I think that epilepsy may be a much more destructive disease than we have ever thought it was.

DR. STRIPLING: With regard to the distribution of seizure induced effects, I'd like to make a simple plea for the olfactory bulb. I've been talking to several people who look at the metabolic effects of seizures and anatomical distribution of status and to its damage, and certainly in a number of those cases the olfactory bulb is not included. It's a difficult structure to remove but I think that in light of the strong involvement of the olfactory cortex in limbic seizures that it would be strongly worth our while to make sure that we take a look at what's going on in the olfactory bulb as well.

DR. MOSHE: I'm quite concerned about talking about seizures producing brain damage. I've seen that as a circle around when I was training in paediatric neurology in the 60's, everything that happened to the brain produced brain damage. If you had a little movement of your finger in a little kid, that was a seizure, even if it was induced by fever. We found out that we had to give a tremendous amount of phenobarbital to control the seizures because we had been told the brain is totally destroyed by having febrile seizures and all the patients with febrile seizure are going to develop temporal lobe epilepsy. So I don't know how many kids I put on phenobarbital for four years without thinking twice. Then we found out that this is not the case. And the knee-jerk reaction of most people is if you say damage, you have to do something to protect people from this damage, that means anticonvulsant drugs, that means more damage. I'm not sure that seizures produce damage. They may reorganize the brain. They may produce different circuitries. We may see new behaviours. I'm not sure that this is damage. There are tremendous clinical implications, there are professional implications. Whether people can function appropriately if we say that this is a continuum of a disease that produces damage. We can send people back to sanatoriums where they were before, or call them insane, and I just don't like to give the feeling for that particular reason. And one more thing, in developing animals, to date there is no evidence of structural damage in the brain following seizures, which is quite the opposite of what's seen in some adult preparations.

DR. ADAMEC: I think you can have dysfunction as a result of seizures, which is not due to damage. In some reviews I have, a lot of those slides I didn't get to show you - you cut me off at a beautiful point, Juhn, and I thank you for that because it made the point more clearly - but one of the effects that I find correlates with my behavioural changes, very long-lasting changes in recurrent inhibition in the trisynaptic circuit and the ventral hippocampus. You see it in dentate, you see it in CA3, but you also see a loss of inhibition in

CA1. And these correlate with some aspects of the behavioural changes. I interpret this as a dysfunction, it's a functional dysfunction, there's no damage as far as I can tell. But what it has done, is it has attenuated one of the normal functions of the ventral hippocampal circuit in the cat which is to facilitate approach and attack behaviour. And this covaries with that element of the suppressant effects of the seizures on behaviour. I don't know whether it's causal, but it certainly is suggested it's a causal correlate. But this is a case where there was a reorganization or repatterning. I get a facilitation of outputs of the amygdala. I get an attenuation of excitability in at least part of the trisynaptic circuit. Now there's an example of the patterning, the restructuring or reorganization in one structure, although in fact it is dysfunctional if you think of the normal function of the structure. But it is not damage. And the way you'd view that in terms of perhaps therapeutically correcting it, would be quite different. For example, aggressive prophylactic anticonvulsive therapy may not be a way to go, if you could find some way to reverse. These changes may be due to excessive GABA potentiation. So the route you might go to try to reverse a change like that would be quite different if you viewed it either as damage or as a functional dysfunction. I think there is a clear empirical example of the dysfunction which is not due to damage, and something you should keep in mind when you're talking about reorganization in the brain as a consequence of kindling or partial kindling.

<u>DR. FERNANDEZ-GUARDIOLA</u>: Well, I think the problem of damage seen in epilepsy is a matter of time. I am surprised we are discussing this subject now, when from the first years of this century it was stated by Edinger in Germany that in the Sommer's sector of the hippocampus there is a strong gliosis in epileptics suffering for many years of seizures. In that case there is the damage. Of course, at the beginning of the epilepsy there is no damage, there is a cellular reorganization, but both things coexist in the same process. Kindling is not epilepsy in the real sense, kindling is another thing. We are artificially stimulating with a weak current or a strong current and I don't think we can make a comparison of kindling and epilepsy. But there is one thing that worries me, and that is, why in all the records we see in the rat mainly, we never see recordings of the temporal lobe surface. Maybe the anatomy of the rat makes it difficult to put chronic electrodes in the temporal lobe, and then we have very few data on the more important thing, on the relation between the amygdala and the temporal lobe cortex. I think we must work with other species like monkeys, cats, or try in the rat some technique, it's not impossible, to put electrodes in the deep lateral part of the scalp in order to have recordings of the temporal lobe. In that case we might get data on propagation because the propagation of the amygdala after-discharge could be cortical-cortical if the propagation from the amygdala to the temporal cortex happens at the beginning of the process. And if this is the case, as we see in our brain maps of the cat from the first day of kindling there's activation of both ectosylvian gyri, and much of the propagatioan could be cortical-cortical like in pathology. We are always looking for propagation from the amygdala but the case is that the ectosylvian and suprasylvian gyri are activated from the beginning of the process of kindling.

DR. GEINISMAN: I would like to make two points. It seems to me that one of the most direct indications of possible damage to the nervous system, to the brain, induced by kindling comes from the work of Sutula who demonstrated sprouting of mossy fibers into the dentate gyrus and the sprouting response is usually the response to the damage. However, we discussed this matter with Applegate and what is not clear from the work of Sutula is whether this particular change is typical of kindling because he sees the same sprouting response not only in kindled animals, but in animals subjected to synchronous activation. So, essentially it's not so clear, to myself at least, whether the sprouting response is specific for kindling. If it were, it would be a rather strong indication that a damage to the system occurs. OK? The second point, again, is related to the question of possible structural damage to the nervous system induced by kindling. I feel that it's necessary to make a definition of damage, what is damage, before we try to argue whether there is some damage or there is none. To me a damage is a change which could be seen directly, I guess. When you look at sections of the nervous tissue you could see that something is wrong with it. In our study, we examined electron micrographs of serial sections without a knowledge of whether they were taken from kindled or control animals. In no case, did the ultrastructure of the neuropil of the dentate molecular layer appear to be different in a given section series as compared with others. Moreover, no apparent degeneration of tissue components was observed in animals under study. It was only the quantitative analysis that allowed us to detect kindling-induced changes in synaptic ultrastructure. In the terminal synaptic field of stimulated axons, a selective loss of nonperforated axospinous synapses was found to occur following kindling. However, it was accompanied by a pronounced increase in the ratio of perforated to nonperforated synaptic contacts and by a significant enlargement of perforated postsynaptic densities. These structural modifications are reminiscent of synaptic remodelling which takes place during development and includes a selective elimination of some synapses and functional stabilization of remaining ones. Although the synaptic loss observed in kindled animals may be viewed as an indication of degeneration and damage to the nervous system due to kindling, it seems more appropriate to attribute the ultrastructural changes found so far in the terminal synaptic destruction (perhaps on an excitotoxic basis) which is not selective for synaptic type or terminal field. This would account for a loss of both synaptic subtypes in the inner molecular layer of our kindled animals.

From this point of view, therefore, I don't feel that there is an actual or a pronounced damage to the nervous system induced by kindling. If you look at the quantitative data, all the changes reported so far, are in the range of 20%. So they cannot be detected just by looking at the tissue. Quantitative analyses are needed in order to detect such changes. I would rather join the argument that we are talking about some restructuring of the nervous tissue rather than damage and with regard to that, I don't feel sorry that you have not read our paper on kindling, but we mentioned the fact that the process of synaptic reorganization reminds us very much of the reorganization during development because there is a selective elimination of nonperforated synapses which is supposed to be typical of development. And probably, there is stabilization of the remaining synaptic contacts which is believed to occur during development.

DR. WADA: Thank you, Dr. Geinisman. I'd like to just remind ourselves that yesterday and the day before yesterday, we had a number of papers for which we had no time for discussion. So it is my intention to go back paper by paper, to have anybody air their comments and discussion. Before we do that I'd like to call upon Dr. Mody once more, to make a general comment.

DR. MODY: Thank you, Dr. Wada. I would just like to answer a methodological point of view that relates to what Ron was telling us about the various stages, and maybe there is a need of reassessing them. May I just use the overhead for a second? I guess it's becoming more and more evident that there needs to be a quantification of the rate of development, I think particularly after direct studies - if we cannot use the overhead I can make the point without it. When, for instance, kindling or the development of kindling is antagonized by drugs such as MK801 or any other antagonist, and then subsequent removal of the drug, does that have an effect on the rate of development of kindling afterwards? Well, I don't know if we can agree what stage 2, 3 and so on are, but is there a reason to, when one plots for instance a straight line as a stage 1 here, and a curve that develops after the removal of the blockade of a certain receptor and this is a regular curve without the drug, how will we assess this rate versus that rate because it might be very important in finding the mechanism for it? And can we plot the points here in between the stages defined by Ron? Can we say this is a 1.5 stage, can we say there's a 2.3 stage and so on? Is there such a thing? I just want to ask what the best way of assessing this rate of development of kindling is? Can we start fitting curves to this, can we have a general agreement of sorts to measure this? That's my first point. If anybody cares to comment, then I have some others related to what I do in terms of membrane excitability.

DR. MCINTYRE: I'd like to comment on that because Jim was raising exactly this the other day, about continuous versus step functions, and of course getting a 1.5 would indicate that we almost have to be dealing with a continuous function rather than steps, but having never quantified this in any serious way, it would be my impression that much of the partial seizure activity, the stages 1 and 2, is a slow continuous building. Then when you get to the more generalizing seizures, you get into these step functions where all of a sudden you have a real massive change that steps up. So I would think that we could do 1.5 up to about 2 and then I think it all falls apart. So in my own view I think we have a problem.

DR. BURCHFIEL: I just want to reinforce what Dan is saying. Our data really indicate that there are major transitions that occur. We see essentially two major transitions, one that goes from the stage 1/2 type of behaviour which we define as more types of species-consistent automatisms, then the beginning of the stage 3 behaviour in which you begin to get some actual motor convulsive signs, the unilateral ones. And then there's another major transition from those unilateral convulsive signs to the bilateral, and our antagonism data say that those are major transitions because that is the place, the characteristic and unique place at which the process becomes halted. There is a major transition that occurs from one of those stages to the next. So I don't think that it is a continuous sort of process where there is a whole series of minor changes. I think the data are very clear in terms of what we call our absolute antagonism where you're getting a lot of these sort of minor kinds of automatisms. In terms of the kindling process, the after-discharges that are generated and are producing that kind of behavioural expression don't seem to do anything for the kindling process because once you stop that alternating paradigm where you're eliciting those behaviours very consistently and over a number of times, it really has not advanced the kindling process at all. The animal still takes the same number of subsequent stimulations to go through the subsequent stages and eventually produce generalized seizures. I would argue for a more quantal process, where you've got more discrete transitions from one state of neural organization to another. Although that's not to say that within, say, local circuits, there may be some changes occurring that will influence for instance the characteristics of the after-discharge. Even in our absolute antagonism where the behavioural seizure expression is not progressing, the after-discharge continues to grow in duration. So there are things that can affect some local circuitry, it has to do with local excitability. But it does not seem to affect the behavioural kindling process that we identify. Is that clear?

<u>DR. WADA</u>: Do you have any counter comments?

<u>DR. MODY</u>: What kind of analyses would you suggest then to compare the two rates of development?

<u>Dr. BURCHFIEL</u>: Well, if you take the kind of framework that I was talking about 2 days ago, where we've got 3 phases of kindling, with those transitions between them, then I think the analysis would be the number or the amount of time spent in each one of those phases. In other words, the quantification would be how long it takes to make those critical transitions.

<u>DR. WADA</u>: I just want to make one specific comment about the staging business. At the first Kindling Symposium in 1975 I presented one of our prefrontal kindling monkeys which had to be stimulated over 380 days to get kindled. After a 7 month rest period, the monkey would not respond with kindled seizure, and we had to rekindle. We were able to rekindle in 80 days, that is a substantial saving. The most interesting and intriguing thing was that each of the 5 different stages came on line in a consolidated form with complete behavioural manifestation, so that I tend to agree that this is not a continuous process; rather a step-wise thing.

<u>DR. STARK</u>: I would agree. I think the only way a 1.5 or 2.3 makes sense is in a relative sense. If you averaged a bunch of animals and the average sort of rank comes out to be 2.3, it's a relative way of comparing how severe the seizure was, but to try to move from one curve to another when you're starting with just a ranking scale, probably doesn't make a whole lot of sense. I understand the problem you're trying to get at, but it's difficult to make sense out of it, because I don't think it is a continuous thing. In terms of whether or

not after-discharge has anything to do with cumulative effects on kindling I think the bare bones kind of kindling phenomena that Jim is talking about with his alternating schedule may not be important, but I think we've demonstrated with chronic prophylactic drug treatment that you then take away, it appears that the after-discharge does have something that's cumulative when you take the drug away to facilitate things. That is more along the lines of what Juhn is talking about with the primates.

DR. MODY: My second comment is on things that I do in terms of membrane excitability, after kindling or even before in general. I think that when Ron was talking about the NMDA receptor channel and how come such large doses of MK801 have to be used, I think that one has to address really profoundly the mechanism of block of these drugs before one can actually make any sense of a lot of the data. These drugs, or at least the channel blockers, as we know are use-dependent. What that means probably everybody understands, that the channel needs to be opened, cannot be in a closed configuration, the drugs cannot bind there, the channel has to open first in order for the drug to go in there and bind to the channel within the channel. That means that the agonist has to bind, voltage has to change, and so on. The same is true for the unblock. The agonist has to bind at the receptor site, the channel has to mimic the open configurations, only then can the drug come out. So if we were to evoke currents successfully by applying agonists at this site, then what we have shown is the use-dependent shows up something like this which would have an inward current evoked. This is the presence of the drug, a smaller inward current, and smaller and smaller, until you have none, because now all the channels have in fact bound the drug in site. You have to apply further agonists, further excitatory amino acid or aspartate to NMDA in order to make this unblock, and then the response would recover very slowly. Actually, the recovery will be slower than the actual rate of block. This is true for MK801, it's true for ketamine, it's true for TCP and so on. Rates might be different; the phenomenon is the same. The other important process is the voltage dependence of this block. We can see that if these are inward currents, of course, caused by channel opening, and this membrane is held at -60 millivolts, then at holding the membrane at +30 millivolts would cause, of course, outward currents. Well, outward currents at +30 millivolts are not blocked by MK801, are not blocked by ketamine, are not blocked any use-dependent blocker, meaning that the mechanism of block is voltage dependent. The drug can only get into the channel when there is a driving force for the drug to get into the channel. If you plot that, on an IV type thing, where here's membrane voltage, here's the current produced by the drug application, this is where we hold the membrane potential of the cell, this is what those currents would be looking like at those various potentials. This is without magnesium, of course, otherwise you would have the voltage dependent block. If you hold the membrane at -60 you get this large current, this is an NMDA current I am talking about. Then, of course, when you hold the membrane at +30 you would get an outward going current. In the presence of ketamine and MK801 - they all use dependent blocker - this is the sort of block that you would be getting. It doesn't block when the membrane is depolarized; it only blocks at a hyperpolarized position of the membrane. This is why most of these drugs are ineffective, for instance, in certain stages of ischemic damage where there is massive release of excitatory amino transmitters. The membrane sits here, it actually depolarizes to 0 millivolts, and they don't act there; these drugs do not block the channel there, they cannot get into the channel there. It's important to realize that maybe the large amounts of drugs that we have to use in certain cases are in fact dependent upon the state of the membrane, on the state of channel activity. I could compare it to a black hole that open NMDA receptor channels, trap these drugs and they are trapped there until other activation comes about, when they can be released. So you may actually, efficiently trap these drugs - this is not true by the way for APV - you can trap these drugs at a site where NMDA channels are active and trap them there until very much or a lot more transmitter is released, dumped upon that channel, and the voltage is changed or at least try to be changed by a voltage-dependent process, until the drug can be removed from there. So these things have to be considered when one analyzes the data considering the use-dependent blockers, particularly for the NMDA channel. But back to the inhibitory mechanisms, there was only one slide I think from Nico that had baclofin on it, and I was wondering if in the GABA hypothesis of kindling, or in the GABA in general, is there any room for GABA-B receptors?

DR. BURNHAM: I don't see why not, if the GABA-B channel is associated with long lasting changes in membrane hyperexcitability, it might have an effect. Someone who knows the GABA-B system better than I do ought to speak to this.

DR. RONDOUIN: Regarding NMDA channels by ketamine, MK801 and so on, we were able to completely confirm. When you put a slice in low magnesium even if it's not free magnesium, it could be only 0.1 molar, you see a very big increase of the population spike, for example, that could be blocked for hours by the ketamine or TCP and so on. And the only way to unblock the system is, for example, to give tetanic stimulation which is able to unblock the system, and you have to put again the TCP, for example, to block it again.

DR. ADAMEC: I guess this raises a methodological consideration with the use of use-dependent blockers in kindling experiments because obviously you have to activate the NMDA channel before you're going to get the block. It's possible if you can sneak enough calcium in there to initiate those processes that are part of the cascade that is part of the permanent change in kindling - so you might expect with use-dependent blockers to get only a partial blocking effect in kindling. Whereas, I don't know if anyone has done this study, you might expect if NMDA were that critical to kindling, that if you used a very effective receptor blocker, and you blocked the receptors before you initiated the kindling process, if NMDA were critical to it, then you should block kindling. But you might not expect to block kindling with a use-dependent blocker.

DR. MODY: I agree with that and I don't think that there is any conclusive study on that using APV for instance, or CPP for that matter, but the block may be there, however. It is not going to take, with the concentration that we are using for the MK801, it's not going to take ages till the channel is blocked.

It is going to be blocked and opening. Unfortunately, there is magnesium in the brain. So we cannot make any accurate predictions or know how it's going to behave in the presence of magnesium, where so many events actually will influence it. But it is almost certain that for unblocking it, you require a lot of activity. Maybe require another seizure in fact to unblock from the drug.

DR. WADA: Dr. Moshe, please. After Dr. Moshe's comments we will return to each paper that was presented.

DR. MOSHE: Istvan I'm glad you're here because I'd like to ask you a question as a follow-up from the previous Kindling Symposium where your group indicated that there were changes in the calcium binding protein following kindling. And at that time we weren't dealing too much with NMDA receptors. Now we do more NMDA receptors than calcium influx. How does the story go?

DR. MODY: I'm so happy you asked that. We now have experimental evidence, electrophysiological evidence rather, to show that the calcium binding protein that Ken Bainbridge has shown to disappear from the dentate granule cells during the process of kindling in a progressive manner, does indeed buffer intraneuronal calcium. The calcium currents are changed in kindled granule cells. I will have to draw this again. In Toronto, with Jim Reynolds we have looked at the calcium currents of kindled granule cells; this is in a slice taken from a fully kindled animal. We put in an electrode and voltage clamped the cell and then produced steps to activate calcium currents. So, when one holds the membrane at about -80mV, puts a depolarizing step to about -10 or 0, then if all the other currents - of course in voltage clamp we are trying to block n-1 currents and look at the nth current. Well, we're blocking most of the potassium currents, if not all, and in the presence of TTX there is no sodium current, so we see a calcium current which looks something like this. And this calcium current doesn't change significantly when we put in EGTA into the electrodes, so this would be a control and this would be something like a control plus EGTA. This is in a granule cell. Why is this important? Well, EGTA binds calcium and actually removes calcium after calcium has entered the cell through the voltage dependent channel. The way that it looks is that calcium enters through the channel - a channel that opens when the membrane is depolarized. Calcium will come in, but probably calcium accumulation on this side of the membrane will stop further calcium from coming in. Well, if you have a buffer here such as EGTA, or we think it's CaBP, a calcium binding protein, removing this calcium, soaking it up like a sponge, this will allow, in fact, more calcium to enter. The inactivation of this current, therefore, is dependent upon calcium accumulation on this side. In a kindled preparation what we see is for the same current, something that looks like that. So inactivation of this calcium current is a lot faster, and why we think it's faster is because the buffer is missing. Thus more calcium can accumulate on the other side of the membrane and less calcium is let to come in into the cell during that process - such that the channel inactivates. That means that maybe even intracellular calcium levels per se are increased in the granule cell after kindling. This may then trigger all kinds of other events of course. Well, the test for this was to put in EGTA into kindled granule cells. If this was the kindled calcium current, putting in EGTA into a kindled cell made it look like a control. So by replacing the calcium buffering system in the granule cells after kindling made them behave as a control cell, so this was kindled plus EGTA. And that perhaps is an indication that this protein that disappears after kindling from these neurons, and specifically from these neurons, is indeed doing something to mop up intracellular calcium. And how this may be important could perhaps be shown here; if there are higher calcium levels inside these cells then you can think of the variety of processes that may be triggered to alter the excitability of these neurons for a long time.

? ?___________: Question from the floor.

Dr. MODY: These were done about 4 weeks after the kindling was completed. This was either amygdala or commissural kindling.

DR. WADA: Thank you, Dr. Mody. The time is just right. We have eleven o'clock. I'd like to go back to the first day to the first two papers, Dr. McIntyre and Dr. Stripling. Are there any comments?

DR. CORCORAN: I do have an axe to grind. Just for the record, it might be worth briefly reminding ourselves of Karen Gales's work which Dan very sleazily avoided talking about at any great length. As you all know, she says that there is an area in the deep prepyriform cortex that is anatomically, as far as I can tell, indistinct, but functionally very tightly organized, that may well be very sensitive for the driving of epileptiform events. That idea has driven many kindlers, most of whom are in this room I think, to try and look at the involvement of that area in actual limbic kindling and essentially the data are underwhelming. There's really not much evidence that the deep prepyriform cortex has much to do with electrically induced kindling. What I would like to ask Dan or Jeff or anybody who's looking in the olfactory cortex is whether they have explored the deep prepyriform cortex in vitro, in vivo or on the moon, whatever, however you do it, to see whether indeed there's any evidence for any kind of unusual bursting activity or any other electrophysiological event that might jibe with Gale's data and ideas.

DR. MCINTYRE: We have sliced up there occasionally but not with any particular purpose for looking for Karen's special spot. That area is undistinguished in our hands by comparison to more posterior peri amygdala pyriform, so we don't see anything that would distinguish it so far, but then we really haven't looked, we haven't got at that with that purpose in mind. It is as responsive as pyriform cortex. Jeff would like to comment.

DR. STRIPLING: We have no data directly to bear on that, but I would like to point people's attention to an article by Hoffman and Averly in the Journal of Neuroscience, January, 1989 where they found that when they put the cells in zero magnesium in a PC slice and caused bursting, they saw a late evoked EPSP. Based on a number of lines of evidence they postulate that this was coming from the deep pyriform cells or the endopyriform nucleus. They presented some papers at the

Society for Neuroscience that further shows that these deep pyriform cells have more excitable properties than the layer 2 cells. So I think that there is a lot to come there, not my data, but I think that's one place to look for what's coming.

DR. ACKERMANN: I've done some 2DG work on kainate in that area for Karen and when the stuff is affected when seizures are produced, large parts of the forebrain are involved. No, you don't get a spot, you get the entire olfactory cortex involved. So if there is a particular spot, then what you're talking about is where things are starting from, but it's not true that seizures are emanating from a spot. In that regard it looks like kindled seizures, in that when you get generalization the entire paleocortex is involved.

DR. POST: We went back and looked at our lidocain deoxyglucose data after Karen Gale's studies were published and there was a hot spot right where she said it was with the lidocain seizure with deoxyglucose.

DR. WADA: Dr. Cain, do I remember correctly, when I was in London, Dr. McLachlan of Neurology presented some data at a mini symposium for me about gabaculine injection into the substantia innominata of rats kindled at this area tempesta. You weren't there, were you? He described that the kindled seizure can be readily blocked as we previously reported in blocking of amygdaloid kindled seizure. Therefore, kindling of this particular area shares a similar mechanism with that of amygdaloid kindling.

DR. MOSHE: We've done some kindling of the area tempesta and what we found is that we put electrodes in the G spot and below the G spot and there was a difference in the kindling rate between those two areas. I don't know why but there was, minimal. Otherwise, the endopyriform cortex and the amygdala had the same kindling rates. The other interesting phenomenon we observed is that there was a very quick transfer from the pyriform cortex to the contralateral amygdala but not from the contralateral amygdala back to the pyriform cortex.

DR. WADA: Are there any further comments?

DR. SATO: As I talked yesterday, the ibotenate stimulated PI turnover was increased for 4 weeks after the last convulsion only in the amygdala plus the pyriform cortex area, not in the hippocampus or limbic forebrain. So I believe that the amygdala plus pyriform cortex area is a very important structure to maintain the long-lasting seizure susceptibility.

DR. WADA: Would anybody care to answer or comment on Dr. Sato's point? Then we will go on to Dr. Geinisman's and Dr. Post's papers since I believe that these 2 papers represent new vistas about the substrate of kindling. This morning somebody was talking about the sprouting of mossy fibers by Sutula. More recently in the Neuroscience Letter I believe, Dr. Le Gal La Salle, who unfortunately could not make it from Gif-sur-Yvette, and his colleagues have also reported sprouting due to amygdaloid but not to entorhinal kindling, and this area of sprouting is associated with markedly increased density of high affinity kainic acid binding sites which are very tantalizing.

So there is more than one report now. I remember Dr. Geinisman's remark that it is not just an isolated disclosure.

Now, as Dr. Moshe and others have shown, kindling seizure development is age-related. Kindling in immature animals can be accomplished rapidly and an acquired seizure susceptibility persists into adulthood, while the aging and aged animals are significantly less susceptible to kindling. Therefore, it is plausible that a modification of brain organization as it occurs in kindling is best accomplished during the critical period while an active organizing force involving a complex interplay of genetic and environmental parameters is at work. Even in the mature brain, Dragunow and his colleagues have described that AM kindling induces the expression of C-fos, a cellular oncogene member which may initiate later effector genes regulating the number of molecules participating in the steps necessary for the conversion of a short-term signal - kindling stimulus - into a longterm response - persistent seizure susceptibility. Kindling produces a series of very complex, widespread and extensive changes spread over weeks, months and, in the case of primates, years. Therefore, one is tempted to consider the sequential activation and modification of genetic programs involved in the natural process of plasticity in the normal maturational course of CNS development. That AM kindling induces the expression of C-fos at the site where structural evidence of plasticity has been identified, I believe, is highly significant. Dr. Post discussed C-fos, and if we realize kindling can cause restructuring of the synapse and sprouting, then we are naturally anticipating cascades of probably a very complex series of events a long way down the road.

<u>DR. POST</u>: Just in response to the methodological questions that were raised yesterday, we use two different probes for the northerns and the <u>in situ</u>. For the <u>in situ</u> there was a Hind III-BAM H 1, 4.8 kilobase fragment of mouse C-fos DNA that was used and that was labelled with 35S and for the northerns it was labelled with 32P. So for Jim and others who want more data on that, I'd be happy to give you the details and the paper as well.

<u>DR. RONDOUIN</u>: I think we have to be very cautious about sprouting in the kindling phenomenon, in the amygdaloid paradigm. Because we, with Dr. Anatoli in Montpellier, are doing similar studies as that of the team of Ben-Ari in Paris, and we don't find so clear evidence that at 15 days, that 8 days after the last class seizure, there was sprouting of the mossy fibers in C3 as they show. It seems to me that it's clear that in the kainic acid model there is sprouting. Of course, there is always damage in this model in the C3, and it is easier to understand that in these conditions sprouting occurs, but in the kindling model I think that we have to be a little bit more careful before giving a conclusion of such a specific damage. What we can think about this problem is that most of the phenomenon which occurs in kindling occurs during periods when no stimulation is given. For example, if you look at animals many months after the last seizure they will develop more serious seizure than they develop at the completion, at what we call usually full kindled seizure, and I think that could be a problem, and a subject for work and reflection.

DR. WADA: The Herculean effect you mentioned, Dr. Buterbaugh, that occurs only when you kindled the animal, that is, the animal had convulsion. Partial kindling doesn't produce the neuroprotective effect?

DR. BUTERBAUGH: That's right.

DR. WADA: If there are no other questions, we will move on to Dr. Fernandez-Guardiola's paper.

DR. RODRIGO: I'd like to make some comment about the methodology of the analysis of the kindling variables and kindling after discharge. All the methods shown in this meeting until now analyzed this evolution only in one dimension, this is only the evolution in the time. I think this is a very important lack of investigation in methods, specifically for kindling. I think the spatio-temporal analysis showed in the paper of Dr. Fernandez-Guardiola is a very important approximation of the spatio-temporal analysis method. This method can unify the traditional methods even the Fast Fourier transform, the Laplacian transformation, and all of this, and the necessary graphic representations to make the spatio-temporal unification. We propose in this paper a model of Fast Fourier and Laplacian transform to show the signals that are perpendicular to the matrix. With this method we can keep the coherence that is very complex to compute and to interpret.

DR. ONO: I liked that presentation very much indeed. My concern, however, is what does amplitude reflect in EEG signals? Certainly high amplitude EEG signal could be a reflection of neuronal involvement in seizure discharge regardless of massive excitation or inhibition. On the other hand, low amplitude EEG signal doesn't necessarily mean the intact cerebral state. For example, low amplitude EEG could be due to enhanced neuronal activity without synchronization, or simply silent. And also, I have a methodological comment to make. As far as I know, a classical coherent analysis is not applicable to such programs in investigating a non-stationary process such as after-discharge. In a more recently reported method, for example, average amount of mutual information analysis proposed by a Dutch group where time domain modelling could be applied to such a program.

DR. RODRIGO: In answer to the first part of your comment, the time domain means that you can choose or fix the time domain event. For example, this is a spike and then you can view how this spike moves in the ... space formed by a 16 electrode matrix. But with this method you can establish the relationships between the time and frequency domain, you can compute the frequency representation of this single spike and then you can establish the relationship. This is very useful to know if the propagation of this phenomena is related to this phenomena, and not to other phenomena. Dr. Fernandez will answer this.

DR. FERNANDEZ-GUARDIOLA: I want to point out that the possibility to make instantaneous averaging of many after-discharges is very useful. Of course with every stimulus, they can be randomly chosen, for example day 7, 11, 15, 20 and the average gives us a constancy of the phenomena in different animals. All this can be simultaneously viewed, and gives us another perspective, not a final solution of the problem.

DR. SHOUSE: Augusto, your presentation was fascinating. I think the problem that people are having here in looking at your pretty pictures is that there's no surface validity for the propagation from the amygdala to the contralateral frontal region. I know that you recognize that. I don't know how it is that you're going to resolve that, either by some anatomical method or some other procedure of validation. Perhaps another way of doing it would be to find perhaps a hippocampal pathway or something where the projection is straightforward and apply your analysis there and show that it is a propagation rather than some other nonspecific effect that we cannot account for.

DR. FERNANDEZ-GUARDIOLA: Thank you. The problem is, we were surprised, we had made only 4 animals and from the beginning we see this crossover of the afterdischarge, from the beginning. We had no time to make plans for more experiments because we were coming here. Of course, we are planning to make a bisection of the corpus callosum, anterior commissure, and massaintermedia. However, you remember our image, there was sort of a pathway over the cortex. I don't discard that the propagation is cortical-cortical, ipsilateral and even like, you know the phenomenon of spreading depression of leao which is a cortical-cortical phenomenon that goes slowly through the cortex when you put potassium in one point. It was well described in the forties. This is a very useful method to abolish cortical activity and see what happens in the subcortical has been used by Bures in Czechoslovakia and by Madame Fessard in Paris. It's a good method to abolish temporarily the cortical activity. We are planning to do that kind of experiment with spreading depression during amygdala propagation to see if there is cortical-cortical or subcortical-cortical propagation.

DR. ACKERMANN: The one problem you'd have with that is that Sheena Hara et al. 10 years ago did deoxyglucose studies with spreading depression and when you cut off that cortical implant you shut down those cortical cells. You have tremendous depression in sub-cortical structures.

DR. MCINTYRE: To put in a point here, Bob, not always does the 2DG represent ... we've seen that many, many times. For example, in status epilepticus we'll see a tremendous amount of deoxyglucose on the one side of the brain and if you're recording electrographically, both electrodes show very similar afterdischarge so that the glucose doesn't necessarily have to bear any relationship to what you're seeing electrographically. That was not the point I wished to make. The point I wish to make in support of your observation, did we not see a slide here showing discharge appearing first contralaterally in the cortex?

DR. WADA: Thank you. Now we move on to the following day. The first four papers related to excitatory amino acids. Does anybody have a comment or questions? Nobody? We'll move on to the next three papers, Drs. Loscher, Stark and Pinel. Are there any comments or questions? I would like to ask Dr. Stark, whose paper was very comprehensive and I learned a great deal about what people in their profession look for when one is developing new anti-epileptic compounds. At the clinical end, we say, well, the available anti-epileptic compounds are not really useful to those group of patients we already know to be refractory. And yet, those patients do respond to those compounds which are known to be effective for, say, generalized convulsion, for example, phenytoin and carbamazepine. So we can readily eliminate those secondarily generalized convulsions and many patients end up still having non-convulsive limbic seizures. And I wonder whether this problem, which we keep blaming on medication that doesn't work, maybe because we haven't really developed drugs particularly suited for this type of seizure? I refer to MES and PTZ. They largely measure convulsive seizure which we are not really addressing, and when we say kindling, we are naturally by definition using animals with kindled convulsive seizures. But our problem at the clinical end, is not those convulsions but rather limbic seizures. I agree that we need a battery of tests but couldn't we inject some more specific testing into it to see if efficacy to that non-convulsive limbic component can be specifically tested.

DR. STARK: I think you're absolutely right, and I think even the people at NIH Epilepsy Drug Development Program agree with you. They're always looking at new test procedures and things, trying to find tests that would be pre-clinically more selective for that type of agent. I think the people in the drug houses share your view and frustration that the available methods are going to continue to find the same kinds of drugs that we already have, instead of these other ones. I know, Harvey Kupperberg expresses that at every Neuroscience meeting and so on, he understands that as well. In short, I think that's one of the reasons why kindling, and I think that Dr. Loscher and I, and many others in the room would agree, that may be why kindling is still a fascination for us who are more oriented towards the drug end of it than perhaps the pathophysiology, because it at least gives us an opportunity to look not only at thresholds but expressions of things, and phase shifts between early and later stages and back shifts and so on, that this instant generalization that you get from maximal electroshock or pentylentetrazol doesn't even give you a window of opportunity to look at. So I think that's one of the reasons why we're all continuing to look at the model.

DR. LOSCHER: There's another difference between kindling and other models. We are using kindling actually as one of a model of different, models for generalized and focal seizures and what we regularly see is that the kindling model is much more resistant to both clinically established anti-epileptic drugs and new potentially anti-epileptic drugs. One of the last slides of Dr. Stark's presentation also illustrated that. If you remember he showed protective indices which is the ratio between doses of drugs causing side-effects such as sedational ataxia and the anticonvulsant effective doses, and in this slide one could see that there is practically no protective index in the kindling model, which is in contrast for instance to MES seizures or PTZ seizures in other species but also in rats. We did very similar things and up to now we compared about 30 compounds in different models including kindling and the only group of compounds which was active in the kindling model without inducing side effects such as ataxia or something like that, were selective GABA uptake inhibitors. This was interesting because these drugs were inactive in the MES model in rats, and you know, Swinyard and colleagues, Kupperberg and so on, they always proposed that drugs which are active in the MES model are also active against focal seizures in humans. So they always proposed there is no necessity to use specific models for focal seizures, but we have now several compounds which are active in the kindling model without inducing any side-effects but which are not active in the traditional models such as the MES. I guess the kindling model could be used in the search for new types of anticonvulsant drugs, which you cannot find with traditional models but what is probably important is that you differentiate when you do that in fully kindled rats, between the generalized motor seizure part of the fully kindled seizure and the focal part. What we very often see is that both clinically established drugs or new drugs are much more effective against the generalized seizures than against early stage seizures. There are some drugs which completely block all of what you can induce by the kindling stimulation, but this is not regular, only very few drugs are able to do that. And these drugs probably should be interesting for patients with limbic seizures. Thank you.

DR. MOSHE: I think the issue of the models that we use is very important. Again we'll go back to the developmental studies. PTZ and MES cannot be applied in young animals. So I guess we'll never find any drugs that control seizures in the immature brain if we continue to use those two models. I'd like also to take a minute to tell you that like Mike Corcoran I have an axe to grind. It's about the substantia nigra that Ron asked about, and suffice it to say that I think that the substantia nigra has something to do with the control of seizures, but not necessarily kindling. Seizures of any kind of form that's generalized may be controlled in the adult but it does not control those in the pup. What is the ability to control generalized convulsion is age specific and is much more limited and probably different in the young. As a result young animals may be more predisposed to generalized seizures than adults.

DR. PINEL: With respect to the development of new models for evaluation of drugs, I was reminded by a comment of Bob Post yesterday about an old study that I did that I think has a lot of relevance to this. What I did was to look at the effects of drugs on spontaneous seizures in kindled animals, and lo and behold I found an extremely large difference between the effects that you get on an elicited seizure and the effects that you get on spontaneous seizures. The one example that sticks in my mind is the drug dilantin, which in our hands at least, does not block elicited seizures until you get up to incredibly high doses. On a rather small dose, it completely obliterated spontaneous seizures during our test session. I think many people have heard me make this point many times, that if we are using kindling as a model of epilepsy I think we have to start thinking about spontaneous seizures because this of course is what clinical epilepsy is. It seems to me that in the spontaneous kindled animal we have a perfect model that duplicates clinical epilepsy in almost every respect and because it would be a new model, no people have worked with this. I think it's a model which could possibly reveal drugs that are rather effective in very subtoxic doses. I'd like to finish this point by saying that in discussing this with people over the years I've yet to find anybody that disagrees with this point, but I usually get a smirk, and yeah, but it takes 5 months to do this, and you might be insane but we're not, and we don't have the time to do it. I'd just like to point out that Dan Savage at the University of New Mexico has recently developed a technique for producing spontaneous kindled rats and it takes him about 5 weeks, and he simply has them stimulated by computer, I think about once every 3 hours, and within a few weeks, lo and behold, he has this perfect model epileptic animals. I'm really excited about his work and I think that by the time the next Kindling meeting rolls around he will have very interesting things to tell us.

Secondly, it makes sense to me that convulsions are adaptive, i.e., rewarding. The fact that many kinds of brain pathology produce convulsions suggests that the convulsion itself is an attempt of the brain to counteract pathology. I do not believe that reward reinforcement plays a role in the data that I reported on drug tolerance. We have been unable to influence kindled convulsions by using reinforcement and punishment. I prefer to interpret the results of the studies of tolerance that I have conducted with Mana and Kim in terms of the sensorimotor adaptation is reinforcing, the two interpretations become equivalent.

DR. STARK: I certainly don't mean to imply that Dr. Pinel's work in any way is excluded from what I was talking about in terms of kindled seizures. It is, until what you've just told me at least, a more labour intensive thing than what we've already dealt with in the past, but given the difference in sensitivity of drugs and other things, certainly it is an extension of the model in that sense, and I wouldn't discount that at all. In fact if Savage or somebody else has got such a reproducible method, shows that it's permanent, shows the same kind of seizure sensitivity and things that you have, NIH should be giving him the money for getting on with it.

DR. ADAMEC: I think everything that's been said here is relevant. I'd like to make a case for the other end of the process as well or to reinforce that. Certainly if you're interested in behavioural changes as a concomitant of a seizure activity in the limbic system, my work and also that of Henke in the rat. I mean a cat is not a rat but rats and cats do agree on this issue in some studies. It takes very few seizure discharges in the limbic system to alter the animal's response to stress and in the case of the cat these are very long-lasting behavioural changes. If you're interested in psychiatric concomitance of seizure disorders and perhaps in developing drugs that might short-circuit or prevent that, then certainly more information on both how to block focal discharges which are not yet clinical, let's say sub-clinical epileptiform or developing disorders, would be of great use. Also, more information in a variety of species, not just in cat, and what happens in partial kindling, what kinds of lasting changes are a consequence of that, I think that would be of great clinical relevance if your orientation were psychiatric.

DR. CORCORAN: May I change the topic, now that we've flogged that one to death. Let's flog this one to death. This is not completely a frivolous question. It's for John Pinel, and his very interesting data I think on tolerance to the anti-epileptic effects of drugs. Years ago when I became aware of his work, what struck me about it, and still strikes me, is that the literature on contingent tolerance, as I read it at least, seems to suggest that tolerance will occur to the behavioural effects of drugs in a before-treatment group as you use, when the behaviour that the drug is disrupting is an adaptive behaviour that's reinforced, when there is reinforcement involved. If one applies that rationale, and that's certainly not the only interpretation of the contingent tolerance data, but when you apply that to the kindling data you showed us yesterday, you're left with the idea that perhaps it's reinforcing to have seizures and that the reason the nervous system then develops tolerance, compensates for the anti-epileptic effect of the drug is because it wants to get back "to having seizures." So why don't you react to that?

DR. PINEL: I guess my first reaction to that is, if you think again about a clinical patient or a spontaneously seizing epileptic kindled rat, I think you can imagine a situation where really it's not the seizure that is aversive but it's the underlying pathology that is aversive. It has always impressed me, the variety of things that make an animal have a seizure. It seems to me that perhaps the seizure itself is not the pathological event but it may in some way be counteracting the underlying pathology so that in kind of a straight homeostatic kind of sense I think that it makes a lot of sense to me that perhaps seizures are reinforcing in some sort of sense. Now, my second comment is that I don't really think that reinforcement - although I think that reinforcement probably plays a role in some kinds of contingent tolerance - there's a classic example of the drunk that learns to walk around with his legs three feet apart so that he doesn't fall on his face, and I can really see that reinforcement plays a role in that kind of contingent tolerance - I don't think it really plays a role in our paradigm. The reason for that is simply because over the years I tried so many times to influence kindled seizures by using reinforcements of various kinds and I've never even had the slightest suggestion that I was having any effect with other forms of reinforcement. Doesn't rule out your interpretation, but I guess I'm not convinced of it, but I think it's a very interesting idea, nonetheless.

DR. WADA: Thank you. Now we'll move on to the next two papers by Drs. Hiyoshi and Shouse, both related to sleep and the underlying mechanisms of kindling. I might ask Dr. Shouse, you I think said in response to Dr. Moshe that your immature kittens have spontaneous seizures quite often, what kind of seizure was it? I thought that it wasn't kindled seizure, and I ask this with great concern related to the antiepileptic drug business. Dr. Moshe mentioned that all those infantile seizures are, most of them are malignant, and we have no effective drug. If we had a model for this, if the kind of spontaneously seizuring animal you produce can be a potential model for infantile spasm, or many other types of malignant seizures in immature brain? And what happens, eventually, do you sacrifice them when they grow up?

DR. SHOUSE: I talk fast but my memory is terrible. Was your first question what kinds of seizures they have? OK. We got spontaneous convulsions, that is a regular grand mal convulsion that looked just like an amygdala kindled seizure in all five of the kittens kindled or beginning kindling between 2.5 and 5 months. Anybody over 5 months did not show that effect.

DR. WADA: Does seizure start from the amygdala where you kindled?

DR. SHOUSE: These are convulsions, OK? And presumably they do start from the amygdala.

DR. WADA: You don't know?

DR. SHOUSE: As far as we can tell but I cannot guarantee that every single time that happens.

DR. WADA: You don't record from the amygdala?

DR. SHOUSE: Yes, we stimulate from it and we record from it, and in the cases where we were able to record the spontaneous convulsions, they appeared to be coming from the amygdala. Now, that is a different matter altogether from these other funny, non-convulsive seizures that we also saw. We did record from that one kitten where we obtained these complex, partial or West Syndrome-like seizures which had bilateral spiking in the amygdala. But that was not the only place we saw seizure activity. We got thalamocortical spike-wave activity that was quite odd-looking and for all we know it could have been all over the brain. We just do not know the answer to that. The seizures which we saw in this animal were nothing like an amygdala kindled seizure. As you know the kittens developed kindling just like the older ones, though the staging is a little more spread out. This seizure that we saw, they could go on for an hour and half, was absolutely nothing like any seizure I have ever seen in the cat, and not like an amygdala kindled seizure. So we do not know; maybe it is a West Syndrome-like phenomenon. Now, you asked me another question after that.

DR. WADA: What happens to them when they grow up?

DR. SHOUSE: Well, the kitten that had these seizures that everyone in the Veterans Administration Hospital observed, because it was going on all the time, we sacrificed that animal for immunohistochemistry and electron microscopy. The other kittens who did not show the catnip seizures, we sacrificed one of the other kittens; the others grew up and they are now about 7 or 8 months old and they are still having spontaneous seizures.

DR. WADA: Which kind?

DR. SHOUSE: All are convulsions. We only had one kitten with a bilateral kindling who showed these odd seizures. Is that everything? Do you have another question?

DR. WADA: No. We don't have a model for this malignant disorder of infantile spasm and the description of one of your animals is extremely intriguing. The nature of the thing is that the immature brain can respond to even partial lesion in a bilateral pattern which is age dependent. This is why I was asking with such great interest.

DR. SHOUSE: OK. Dr. Moshe has suggested that what we need to do now is, well, we're going to do some more kittens anyway, but what he thinks would be useful and I agree with him, is to use Dr. Burchfiel's model and try to really work on this West Syndrome phenomenon, because it is so interesting and because it is so devastating and the drugs we have available now I think are ... ACTH is given and it sometimes can be helpful in the treatment of it. So we will try to do that, but I don't know if it will be successful.

DR. WADA: Any other questions, comments about these two papers?

DR. CALVO: A question to Dr. Shouse. How did you detect that this was a spontaneous seizure? Did you do a longitudinal recording? How can one be sure that this is an epileptic seizure? It's very strange that epileptic patients almost never complain about sleep disorders.

DR. SHOUSE: The recordings which we conducted - this is the first question, which I understand wants to explain how we recorded the spontaneous seizures. Yes? At the VA Hospital the vivarium is called the ARF. And so the ARF people would call and say, Nick, your cat's doing it again. Then I would send somebody over to pick up the cat, we'd bring him back, and we'd hook him up. Cats would have anywhere from 2-6 more seizures in the same day. OK? And we'd hook him up and look at him throughout the period. Is that clear? So that way we could tell whether they were having recurrent seizures in the REM transition, or in slow-wave sleep or just in wakefulness, whether they went into status, which a lot of them did, and so forth. Is that what you wanted to know? What was the sleep disorders question again?

DR. CALVO: Can you explain why epileptic patients never complain about sleep disorders?

DR. SHOUSE: That's not true. I thought we kind of gave up on the sleep disorders subject. Epileptic patients complain bitterly about sleep disorders. Some clinicians report on this more commonly than do others, but in any case, since we don't know which of any of the sleep disorders is relevant to epilepsy, and it looks like a lot of them come from the antiepileptic drugs, and since we have shown that the antiepileptic drugs have a worse effect on sleep than does the epilepsy, it's not clear to me that there is any reason to pursue this subject.

DR. FERNANDEZ-GUARDIOLA: One more comment on epileptics and sleep. In our hospital, the Institute of Psychiatry, we have many sleep problems. Schizophrenia is really producing sleep disturbance, but epileptics we have a good number, but all are treated. You cannot work with non-treated epileptics because there is no ethics. And the great majority of anticonvulsants are hypnotics, then these people are sleeping very well with the hypnotics. You have data of real sleep troubles in the non-treated epileptics? Maybe in Poland, Janz? I don't know.

DR. SHOUSE: All the data I know about that have been published on sleep disorders and epileptics is contained in the sleep and epilepsy book which you are an author in. So you know about that already. And Janz, of course, has reported on the types of complaints that patients make. Two thousand patients or whatever it is, they make different complaints about their sleep and that's been duly recorded in Epilepsia now for 26 or 27 years. Is that it? OK.

DR. WADA: Thank you. Now we move on to the next two papers of Drs. Adamec and Cain. Any questions, comments, discussion? I might ask Dr. Adamec, well in general everybody, that Dr. Adamec has shown stage 2 seizures or stage 1, whatever, with a half-a-dozen to a dozen seizures, AD, which does not cause convulsion, leaving a distinct signature of behaviour disturbance in cats. Dr. Burchfiel suggested that in his paradigm so long as this antagonist paradigm is continued, the animal remains in stage 2 or 3 and as soon as the dominant side is over, then that hemisphere just goes from there as though nothing has happened before. This makes me wonder a great deal, and one of those things I keep on wondering about is that, as everybody experiences, our thinking, our future perspective perception and direction of action, are all influenced by whatever we experience every day. These events are not causing epileptic events in our brain. I hope naturally this relates to learning and memory. Dr. Augusto Fernandez-Guardiola has shown last time, that stimulation of a non-kindable area such as Raphe produced lasting change in the synaptically related distant site. And this is not an isolated event, since Dr. Stevens and colleague have shown that kindling in the conventional sense cannot be accomplished at ventral tegmental area. Yet repeated stimulation of this area causes lasting behaviour changes. And when we say kindling, with all respect to Dr. Goddard the discoverer, we define it with convulsion as the end-point. Do we really need convulsion to talk about kindling? I ask this question to provoke discussion.

DR. RACINE: Well, I think we all realize now that there are a lot of things going on in kindling, and a full definition of kindling is bound to include a lot of different variables. The way it has been defined by Goddard and most of the others that work in the field, is a definition that I am quite happy with, and that is a definition that includes the electrographic development and the end-point. Now, a definition that includes the end-point, of course, doesn't mean that everything that happens up to that end-point is not a part of kindling. Bob and others that work on partial kindling are in fact working on kindling even though there is no convulsion. If you work on an anaesthetized preparation or a paralysed preparation, well, we don't do that any more, do we, you're dealing with kindling. If you're working with an isolated slice or a slab and you get the electrographic development, you're working with kindling. The cellular changes probably are the same. We do, however, have to be aware that there are a lot of things going on and the cellular changes may not be the same in all of these various models.

DR. FERNANDEZ-GUARDIOLA: In 1981 I wrote a paper and the title was "Kindling of the Spinal Cord." I remember I was in Kyoto and Wasterlain asked me why do you call that kindling because there is no seizure at the end, but the fact is that using one second, 100 Herzt low-level stimulation every hour, you can change all the synaptic responses of the spinal cord, and have more than 1,000 increments in the polysynaptic reflex. I agree this is not kindling because there is no seizure at the end, but if we have repetitive responses that are blocked by anticonvulsants, valproate and benzodiazepines, then I think we have the right to kindle any part of the nervous system in order to gain better understanding of the process of progressive facilitation. The use of the word kindled or not is a semantic problem.

DR. BURCHFIEL: I have a question for Bob. Specifically what you're doing or what you're working with are the effects of a triggered after-discharge, of a number of them, right? And specifically from the amygdala. Do you get exactly the same behavioural changes regardless of where you trigger the after-discharges from, or is this specific to the specific circuitry involved in the after-discharge?

DR. ADAMEC: Well, if you trigger a discharge in the ventral hippocampus it doesn't stay there. I have some data now which I'm analyzing which suggests that the behavioural changes, you can trigger either from the perforant path from dentate or from amygdala, you get the same kinds of behavioural changes, although the evolution in terms of number of discharges differs a bit. I also have some data that suggests that it's the spread, the penetration of the after-discharge from the ventral hippocampal system into the amygdala that's critical for the behavioural change. I have other data which of course were slides I couldn't present, which suggest that potentiation of the output of the amygdala particularly into the ventromedial hypothalamus is at least necessary for the behavioural changes that I observe. I think that these are potentiated as a result of the spread of the discharge into that ventral amygdaloid fugal, ventromedial hypothalamic pathway.

DR. STARK: When you get the behavioural change, when you've given sufficient numbers of after-discharges, what sorts of behaviours are the after-discharges eliciting? I mean, are you getting automatisms or are you beginning to get any sort of partial convulsive effects?

DR. ADAMEC: You get absolutely nothing. All you see is absences from the beginning to the end. And the absences are only detectable, because I know when I stimulate the animal he'll turn his head, sort of look when the stimulus comes on, it'll become immobile and stare during the duration of the discharge, which can be quite long, 20-40 seconds. And then it stops, the animal will show a rebound, miaow, well it may show several miaows, and then go about it's business.

DR. STARK: Have you ever taken the animals when they begin to show the changes in their behaviour, are they really partially kindled, you get savings if you then take them on to full ...

DR. ADAMEC: I haven't tried that. I have actually argued in other papers that there is very little kindling going on, because the after-discharge doesn't lengthen. I have done power spectral analysis and often you get changes in complexity of after-discharges, at least particularly in amygdaloid kindling, and power spectral analysis indicates no change, at least in the frequency composition of the discharges from the beginning to the end of the partial kindling process. The only thing that resembles kindling in the standard sense is there is a decrease in the after-discharge threshold. But otherwise there is very little kindling going on. I think what it is, is very pre-kindling kinds of changes that are a consequence of the high frequency activation induced by the repeated discharges, that is producing some very long-lasting behavioural and physiological changes. I've measured them, there are a lot of physiological changes that last as long as behavioural changes, interictally maintained changes.

<u>DR. MODY</u>: I have a question about the FEG compound and the LTP. Did the FEG produce the same behavioural effect as the after-discharge?

<u>DR. ADAMEC</u>: It simulated it, yes.

<u>DR. MODY</u>: It simulated it, yes. Both were antagonized by RO and what I'm interested in is, could it be that there is an endogenous inverse agonist at the receptor that is being released by the AD's?

<u>DR. ADAMEC</u>: Yes. That certainly is an interpretation that there is an endogenous legand which is being potentiated. The problem I have with that is, if there were such a legand there, then why when you give RO 15-1788 to a naive animal you don't get a blockade of the animal's natural level of defensive response? It's behaviourally inactive until you change the animal's behaviour. I mean, if you wanted it to be unparsimonious you might argue somehow you've caused now the induction of a release of an endogenous legand. That's possible. But in this literature a lot of people have thought about that and in fact Poke has come up with a 3-state model which I use to explain this simply because there isn't really good evidence for an endogenous legand and RO is problematic in this because in most behavioural tests, not just in mine, it has no intrinsic activity. One would expect, I mean you can think of a yin-yang thing, you could have an endogenous anxiogen and endogenous anxiolytic compound. In that case you might expect to see movement in one way or the other, but you don't with these fairly specific receptor blockers. Which is why I did not opt for that explanation. I opt for this receptor that is probably changing state quite spontaneously and you can perhaps shift this preferred confirmation and that has a consequence on the physiological systems it's related to it.

<u>DR. WADA</u>: If there are no other questions, comments, discussion, we move on to the last four papers which are all related to transfer. Any comments, questions, expansions? If not, then we are coming to the end of this general discussion. I'm awfully glad to have practically everyone here of the participants of the Kindling 1 back in 1975 except for Dr. Goddard. I want you to know, some of the reviewers of the Kindling 3 said, well, they've kindled themselves to Kindling 3, the answers are nowhere in sight, and he hoped the next one would be the last one. I'd like to have our old-timers respond to that.

<u>DR. CORCORAN</u>: My name is Racine. I think we should go on indefinitely.

DR. MCINTYRE: I entirely agree with Mike. Thank you once again, Dr. Wada, for the invitation to come and play with the kindlers. This is always a lot of fun. It is very sad that Graham is not here because this was the meeting that he most enjoyed of all the meetings that there were. This is so concentrated and so specific that we always have quite a riotous occasion. I hope that we continue to do so without him being present, because it is lots of fun.

DR. CAIN: I don't really have a lot to add except that I find it endlessly fascinating and hope it does go on. I look forward to another Kindling conference. It focuses our interests, it brings us all together, we have a few laughs, and we learn some things. I think that's all worthwhile.

DR. FERNANDEZ-GUARDIOLA: I have been nine times in Vancouver and I like the trip, always by the same airline. My world has a very narrow pathway connecting Mexico to Vancouver, and I would like to be back here again.

DR. SATO: I think we should keep on, but as you know, there are lots of patients with intractable epilepsy, and the incidence of epilepsy is 0.8% and maybe more than 37% of them are intractable. We are not treating them well. Therefore, we have to ask ourselves and provide an answer, "Why does epilepsy become intractable?" For this purpose the kindling model is, I think, the best one. But also clinically some patients with schizophrenic disorders receiving a large number of electroconvulsive therapy have been reported to have developed spontaneous convulsion. There are more than 200 case reports of this kind in the literature. So we have to clarify the basic mechanisms of kindling. We must understand the nature of transsynaptic change, the induction mechanisms of the long-lasting transsynaptic change and the mechanisms of maintaining such changes. Furthermore, we have to know the mechanisms to produce spontaneous seizures. So we have to keep on going with our work and I hope to get together again in the future.

DR. PINEL: I think the reviewers that made that comment about Kindling 3 really were not getting at the heart of the issue. It's not whether we solved the problem or not after three meetings, I doubt if we'll have it solved after 20. The question is, are we making progress each time, and in conversations I've had with a number of people during this conference, I think that this has been a very successful conference, and probably more successful than some of the ones that we've had previously. So I would suggest that we should keep going and I think we are moving in the right direction.

DR. ADAMEC: I agree with John, and everything else that's been said. I think we are making progress. It's partly evidenced by the growing numbers of people who are attending this meeting, which brings up this plea to you, Dr. Wada. If we have it here again, let's have more days in the conference. Seven papers in the morning is a bit too much, and as our population ages, I think our attention spans won't be able to take the input.

DR. RACINE: My name is Pavlov. I'd just like to say that life is a bridge and I look forward to the next one.

DR. WADA: Well, Dr. Burnham has left because of his schedule, but he was the one who initially said, why do we keep on studying kindling? I suppose we do because of its intrinsic value as a window to our internal universe in which our future destiny lies. Because of everybody's comments, possibly, collectively we can make an arrangement next time so that Dr. Graham Goddard can be invited. He should take some holiday wherever he is at, and whatever he's doing. And I'm sure he has been watching over us and I trust that he enjoyed it. As everybody said, hopefully we can again collectively assess our progress sometime down the road. One other word about photographs. The photograph was meant for Mrs. Patricia Goddard, and I thank you for signing the card. I shall be sending your greetings and the photograph to her. And now, I'd like to thank you all for coming so far, particularly those people from across the Atlantic and Pacific, I greatly appreciate your coming here, and from the four corners of this continent. I'd like to wish you all a nice Saturday afternoon, bon voyage, and bon appetite. Lunch is right here.

DR. MOSHE: I do have another axe that I would like to grind. I would like to thank Dr. Wada for a wonderful symposium and all the people who helped to make this symposium so enjoyable. Thank you.

INDEX